Probability, Stochastic Processes, and Queueing Theory

Randolph Nelson

Probability, Stochastic Processes, and Queueing Theory

The Mathematics of Computer Performance Modeling

With 68 Figures

Springer-Verlag

New York Berlin Heidelberg London Paris
Tokyo Hong Kong Barcelona Budapest

Current address:

Randolph Nelson
Vice President of Investment Research
OTA Limited Partnership
One Manhattanville Road
Purchase, NY 10577
USA

Research done at:

Manager, Modeling Methodology
IBM T.J. Watson Research Center
P.O. Box 704
Yorktown Heights, NY 10598
USA

On the cover: Wassily Kandinsky's *Circles in a Circle*, 1923. From the Philadelphia Museum of Art: The Louise and Walter Arensberg Collection.

Library of Congress Cataloging-in-Publication Data
Nelson, Randolph.
     Probability, stochastic processes, and queueing theory: the
     mathematics of computer performance modeling/Randolph Nelson.
          p.   cm.
     Includes bibliographical references and index.
     ISBN 0-387-94452-4 (alk. paper)
       1. Probabilities.   2. Stochastic processes.   3. Queueing theory.
     I. Title.
     QA273.N354   1995
     519.2 – dc20   95-4041

Printed on acid-free paper.

Production managed by Francine McNeill; manufacturing supervised by Jacqui Ashri.
Photocomposed copy prepared using the author's LaTeX files.
Printed and bound by R.R. Donnelley and Sons, Harrisonburg, VA.
Printed in the United States of America.

9 8 7 6 5 4 3 2 1

ISBN 0-387-94452-4 Springer-Verlag New York Berlin Heidelberg

To my Mother,
and to the memory of my Father.

For Cynthia.

Preface

Notes on the Text

We will occasionally footnote a portion of text with a "**" to indicate that this portion can be initially bypassed. The reasons for bypassing a portion of the text include: the subject is a special topic that will not be referenced later, the material can be skipped on first reading, or the level of mathematics is higher than the rest of the text. In cases where a topic is self-contained, we opt to collect the material into an appendix that can be read by students at their leisure.

Notes on Problems

The material in the text cannot be fully assimilated until one makes it "their own" by applying the material to specific problems. Self-discovery is the best teacher and although they are no substitute for an inquiring mind, problems that explore the subject from different viewpoints can often help the student to think about the material in a uniquely personal way. With this in mind, we have made problems an integral part of this work and have attempted to make them interesting as well as informative.

Many difficult problems have deceptively simple statements, and to avoid the frustration that comes from not being able to solve a seemingly easy problem, we have classified problems as to their type and level of difficulty. Any classification scheme is subjective and is prone to underestimate the difficulty of a problem (all problems are easy once one knows their solution). With this in mind, ratings should only be used as a guideline to the difficulty of problems.

The rating consists of two parts: the type of problem and its difficulty.

There are three types:

E — an exercise that requires only algebraic manipulation

M — a mathematical derivation is required to do the problem

R — the problem is a research problem.

An exercise is distinguished from a derivation in that the answer proceeds directly from material in the text and does not require additional analysis. Research problems differ from derivations in that they require more independent work by the student and may require outside reading. For each type of problem, there are three levels of difficulty:

1 — easy and straightforward,

2 — moderate difficulty,

3 — hard or requires a critical insight.

Combining these parts yields nine different ratings. For example, a rating of E2 corresponds to an exercise that requires a moderate amount of calculation, and a rating of M3 is a difficult mathematical derivation that requires some clever reasoning. Research problems have the following interpretations:

R1 — A discussion problem that is posed to provoke thought,

R2 — A problem that requires a nonobvious creative solution,

R3 — A problem that could be considered to be a research contribution.

Katonah, New York   Randolph Nelson

Acknowledgments

The author gratefully thanks Cheng-Shang Chang, Rajesh Mansharamani, and Thomas Philips for the substantial time and effort that they spent reviewing the text. Not only did they read the book in its entirety, often going over a chapter several times, they also worked out examples, checked mathematical derivations, and offered extensive suggestions as to how the development might be improved. Almost all of their suggestions have made it into these pages, and I owe them a debt of gratitude.

The author also thanks Scott Baum, Richard Gail, Marvin Nakayama, Mark Squillante, Mary Vernon and David Yao for their comments and criticisms on various sections of the book that invariably led to improvements in the presentation. The anonymous referees did an outstanding job of reviewing the text in fine detail.

It is with great pleasure that I thank Armand Makowski for the many enjoyable and profitable hours that we spent discussing the philosophy of writing a book. I would like to acknowledge the help that I received from other researchers who were already authors. Harold Stone, Paul Green, and Joy Thomas helped me become familiar with the process of writing a text. Additionally, I would like to thank Joy for sharing his macro files with me. I thank Sean McNeil for writing the Perl scripts that I used to find unreferenced equations in the text.

The writing of this book was supported by Steve Lavenberg, my manager at IBM's Thomas J. Watson Research Center. I would like to thank Steve for his encouragement for this project, his guidance in my research career, and his continual efforts on behalf of my group to preserve the intellectual environment that is necessary for basic research.

It was a pleasure to work with Martin Gilchrist as the editor of the book. Not only did Martin provide critical guidance for the overall structure of the text, he also made comments on some of its mathematics. Such is the advantage of working with an editor who also has a Ph.D. in mathematics! I would also like to thank Ken Dreyhaupt at Springer-Verlag for his efforts with TEX problems.

The book would simply not have been written without the blessing of my wife, Cynthia. These pages contain more hours of time than I would like to admit that we would, otherwise, have spent together. In the spirit of this sacrifice, this is as much her book as it is mine.

Finally, I would like to acknowledge my thesis advisor, Leonard Kleinrock, for the thrill of first seeing the beauty of applied probability through his eyes.

Katonah, New York   Randolph Nelson

Contents

II Stochastic Processes   233

List of Figures

List of Tables

1

Introduction

Probability is an exquisitely beautiful field of mathematics that is rich in its depth of deductive reasoning and its diversity of applications. The theory started in the early 17th century with simple counting arguments that were used to answer questions concerning possible outcomes of games of chance. Over the ensuing years, probability has been established as a key tool in a wide range of fields that are as diverse as biology, chemistry, computer science, finance, medicine, physics, political science and sociology. This breadth of application, also found in calculus which dates from the same era, is a consequence of the fact that the theory has its roots in a branch of mathematics that is blessed with a compelling intuitive component. The full beauty of the subject can be best appreciated when its formal and intuitive aspects are melded together with applications. This motivates our goal; to derive probability in a manner that highlights the complementary nature of its formal, intuitive and applicative aspects.

We attempt to achieve this goal by making the subject as accessible as possible without unnecessarily compromising any one aspect of the theory. Our approach is to motivate probability with simple examples that appeal to intuition, and to derive results that proceed from the simple to the complex, from the specific to the general. Almost all of the results presented in the text are derived from first principles. This allows one to be able to reconstruct the deductive path starting from the axioms of probability and ending at the particular result at hand. Such a foundation removes possible doubt as to the reasoning that underpins a result. As background, we assume that the reader is familiar with elementary set theory, linear algebra, and calculus. In particular, the student should be conversant with limits.

Given this background, it might come as a surprise that we can introduce most of probability theory by investigating the mathematical properties of a simple two-person coin tossing game. We can imagine that a particular player either wins or loses a dollar depending on if heads lands up. Questions of interest include the number of dollars won (or lost!) after n tosses, the average number of tosses required to either win or lose a certain amount of money, or the probability that within a number of tosses the games reaches a break-even point. It is easy to believe that the complexity of such a simple example is not sufficient to reach very deep into the theory. This, we believe, will be the first impression of many readers and it is not without its justification. In truth, however, not only does the example have sufficient complexity to motivate combinatorics and probability theory, it also leads to renewal theory and stochastic processes. In fact, at the end of the book we find that there is still more to learn about this "simple" example and its variations. Therein lies the profound beauty of mathematics: a playful exploration of an extremely simple game that leads to unexpected results of breathtaking complexity.

The application area that we choose throughout the book is computer performance modeling. In addition to this being an area with which the author is most familiar, this choice is motivated by the fact that this area plays an important role in the field of applied probability. As the information age continues to unfold it is easy to believe that the field of computer performance modeling will only increase in importance and relevance. This choice of application domains implies that we will emphasize certain areas of probability above others and, in particular, the main focus of applications later in the book (Chapters 6 to 10) concerns queueing networks that are extensively used to model complex computer systems. The first part of the book (Chapters 2 to 5) lays the foundation of probability theory, and here the emphasis on computer performance modeling is not as specific.

We investigate those portions of probability that we feel are most vital in applications. In so doing we avoid subject areas that, although interesting in themselves, are not an integral piece of the whole puzzle. We have tried to present the theory so that its essential character remains in the student's mind long after the particularities of the subject are forgotten. Such an orientation is necessary in framing one's understanding of how probability can be applied in a wide diversity of subject areas. Throughout the first part of the book, we have alluded to interesting historical aspects of the theory. These historical notes are largely taken from fascinating books written by Bell [5], Boyer and Merzbach [8], Eves [32], and Todhunter [115].

In the next section, we outline the book and then generally discuss queueing networks at an application level. This outline may be initially skipped if one is anxious to start with the subject as quickly as possible.

1.1
Outline of
Book

It is perhaps helpful to have some orientation as to the flow and structure of the book. The book attempts to both teach and serve as a reference. With the latter in mind, we have tried to organize the text so that items can be easily found and, in particular, sometimes tables found in the main body of the text will also be found in appendices where they are more rapidly accessed. Chapters have been made to be as self-contained as possible, and a list of notation used throughout the book should aid both the student and the researcher. The book is essentially comprised of two parts. Part I (Chapters 2 to 5) covers probability, and Part II (Chapters 6 to 10) covers stochastic processes. Chapter 11 is devoted to special topics. The appendices present review material or reference material or go into greater depth in an area that lies outside the scope of the main text. We now briefly describe the motivation for each of the chapters that follows.

1.1.1
Outline of
Part I

Since throughout the book we will speak of events, such as arrivals to a computer system or the service times of customers, as being random, it is beneficial to develop the notion of randomness and probability from first principles. This is the subject of Chapter 2, and it lays the foundation for all subsequent results. We ask a very simple question in the chapter: What do we mean by the probability of a random event? We answer this question in the chapter and in the companion appendix, Appendix A, and we use the discussion to motivate two different approaches to probability. The first approach is based on the commonsense observation that, for repeatable experiments, the probability of an event can be estimated by calculating its statistical average. This **frequency-based approach** to probability, however, has several disadvantages that are bypassed by the **axiomatic approach** also considered in the second chapter. The axiomatic approach forms the basis of modern probability, and the chapter finishes with some fundamental consequences of its axioms.

The most basic form of probability arises when all events are equiprobable and we are concerned with determining the probability of a certain set of events. As an example, we could ask what is the probability of obtaining more than 5 heads in 10 tosses of a fair coin. There are 1024 different possible outcomes from flipping a fair coin 10 times. One rather laborious way to determine the required probability is to list all of these possibilities and count the number of these that have more than 5 heads. Taking the ratio of this number to 1024 yields the required probability. For such a simple experiment we do not require a sophisticated method of counting, but it does not take much effort to ask questions that are not as amenable to straightforward analysis. Developing formal counting techniques is the subject of the third chapter, in which we develop

the subject of **combinatorics**. The crux of probabilistic reasoning is
contained in this chapter, and we present an approach to combinatorics
based on recurrence relationships. This approach minimizes the number
of "tricks" used to obtain combinatoric results and motivates the use of
generating functions as formal counting mechanisms.

In Chapter 4 we develop properties of **random variables** and their
distributions. A random variable can be thought of as a variable whose
value is determined through a random experiment. For example, a random variable X could be the number of heads obtained in 10 tosses of
a fair coin. Expressions like X^2 and $3X$ make sense and explain why we
call such objects variables. Since the value of a random variable is known
only probabilistically, we must have a way to specify the values it can
take on and their corresponding probabilities. This is specified through
the use of a distribution function, and in Chapter 4 we present a unified
approach to construct a family of distribution functions that are used in
modeling. We also set up an approach that will be utilized throughout
the text to derive analogous pairs of discrete and continuous random
variables. Discrete random variables can be thought of as the outcome
of some variation of a simple coin tossing experiment and are thus easy
to understand. The continuous random variables that we consider can
be derived as a limiting form of a coin tossing experiment and, because
of this, properties that hold in the discrete case typically also hold in
the continuous case. In this way, continuous random variables are more
readily understood.

Specifying a distribution function implies that we must have a detailed
way to assign probabilities to all possible outcomes of a random experiment. Often this is not possible, and in modeling, we typically only
require summary information. For example, we might only be interested
in the average number of heads obtained in 10 tosses of a fair coin. This
motivates the definition of **expectation** given in Chapter 5, which can
be thought of as an average of a large number of experiments. Fundamental properties of probabilistic systems arise when one considers a
large number of experiments. For example, common sense tells us that if
we perform an experiment a sufficient number of times then the probability of a certain event can be obtained by forming a statistical average.
Thus to determine if a coin is fair, we can toss it a large number of times
and see if about half of the outcomes are heads. In Chapter 5 we prove
that such a commonsense notion is well founded in probability, and we
derive a large number of detailed results concerning such statistical averages. Several fundamental theorems in probability are proved in this
chapter.

Part I encapsulates our development of probability and subsequent chapters use the derived results to study **stochastic processes**. For the moment we can think of a stochastic process as being a set of outcomes of

a random experiment indexed by time. For example, $X_n, n = 1, 2, \ldots,$ could be the total number of heads obtained from the first n tosses of a fair coin in an experiment that continues for an indefinite period of time. It is clear that the value of X_{n+1} equals $X_n + 1$ if the $n + 1$st toss is heads and otherwise equals X_n. We call the set $\{X_1, X_2, \ldots, \}$ a **process** to indicate that there is a relationship or dependency between the random variables X_n. For continuous time processes we say that $X(t)$ is a stochastic process and the values of $X(t)$ and $X(t')$, for $t < t'$, have some relationship or dependency. The nature of such dependencies induces different stochastic processes, and in Part II we develop several different types of processes that are useful in modeling.

1.1.2
Outline of
Part II

Chapter 6 deals with **renewal processes** that are a particularly simple form of a process that counts the number of events, $N(t)$, that occur in a certain period of time, t. Events are assumed to occur sequentially in time, and $N(t) = n$ means that n events occurred within t units of time. The value of $N(t)$ and $N(t')$, for $t < t'$, are related to each other since events are cumulative and, in particular, $N(t') \geq N(t)$. Despite the simplicity of such a process, it is frequently used to model the arrival stream of customers to a computer system. In such models $N(t)$ counts the number of customer arrivals in t time units, and the Poisson process, which is a renewal process with a particularly simple form, is frequently used to model such arrivals. We discuss this process in detail and establish several key results of general renewal processes.

Chapter 7 studies a particular type of single server queue, the **M/G/1 Queue**, which has a wide applicability in computer performance modeling. A queue consists of an arrival stream, a waiting line, and a server that services customers. The number of customers in a queue increases with each arrival and decreases with each departure. Queues are typically used to model response times in computer systems (the duration of time to receive a response from a request) and in this chapter we provide a detailed analysis of the mathematical properties of queueing systems.

In Chapter 8 we continue our study of stochastic processes with **Markov processes**, which are the most important processes in modeling. There is a simple historical dependency relationship that is satisfied by Markov processes that is easily described for discrete time processes. For such discrete time processes, if one is given the value of X_n then the value of X_{n+1} is independent of the preceding values of X_i for $i = 1, 2, \ldots, n-1$. In other words, the process has a limited historical relationship that depends only on the value of the previous step. For example, the earlier coin toss example is a Markov process where X_n counts the number of heads in the first n tosses. Clearly, to know the value of X_{n+1} it suffices to know that of X_n and the outcome of the $n+1$st toss. One does not need to also know values X_i for $i < n$. In the chapter we derive some fundamental theorems of Markov processes that outline their probabilistic structure.

In Chapter 9 we derive the **matrix geometric** technique for solving Markov processes that have a special repetitive structure. Applications that have a matrix geometric solution form are frequently found in performance models. Such applications can typically be characterized by the fact that there is a set of resources that can have an infinite waiting line of customers.

In Chapter 10 we generalize the notion of a queue to networks of queues. In applications, such networks can model complex computer systems where customers are routed to different service centers over the course of the time they spend in the system. Queues can, for example, correspond to CPU or I/O centers and different customers can require different types of service at each of these centers. The mathematics required to determine the performance of such networks can be derived from basic principles developed in the previous chapters. In the chapter we present a unified approach to analyze a particular class of networks that possess **product form solutions**. Such networks are called **quasi-reversible queues**, and they have a solution form that is particularly simple. We show that such simplicity derives from four easily established properties.

Finally, in Chapter 11 we introduce some special topics that are natural extensions of results established in the preceding chapters of the books. These topics will be briefly discussed and references to the current literature provided.

1.1.3 Applied Probability and Performance Modeling

The application area that we consider throughout the book is computer performance modeling. We spend a few moments here discussing the relationship between applied probability and performance modeling. Computer systems are becoming increasingly complex as hardware and software systems increase in sophistication and a critical element of computer system design is to be able to analyze the performance of different system architectures to determine the best balance of performance and cost. To determine the performance of a particular design, one could *prototype* the design and then measure the performance. This is often a complex and expensive method of evaluation that does not readily lend itself to comparing and contrasting the performance of competing designs. An alternative method that does allow for such comparison is to create a detailed *simulation* of the system. A simulation can be thought of as a computer program that mimics the operations of a proposed design. The construction of such a model requires writing a computer program for each different design, and evaluation of the model requires running each simulation under identical input to gather performance statistics. Difficulties with this approach include errors in the programs that are often extremely complex and the large amount of computer time that is often required to obtain accurate estimates of performance measures. It is sometimes computationally infeasible to obtain a thorough study of the system over its complete parametric range using simulation. Additionally, it is often difficult to determine and justify simple "rules of thumb" that can act as a guide for a system designer when creating or modifying an architecture.

A third technique, which is the application area of this book, starts by creating a mathematical model of the computer system. In a model one typically characterizes system components at an abstract level where *nonessential details* of the system are not modeled. It is an art to develop good models. The objective is to capture important system features that suffice to yield insights into the performance of the system. The level of abstraction found in a model makes it easier to understand system performance without being distracted by details that are not relevant to the objective of the model. To analyze the model one ideally derives a set of equations from which performance measures can be calculated. These equations might be an exact solution to the model, or they could be an approximation based on properties of the system. In cases where the model cannot be solved mathematically, performance measures can be obtained by *simulating the model*. This approach differs from a simulation of the system because of the differing levels of abstraction between the system and the model.

The advantages of using models to analyze system performance are twofold. First, if the model can be solved analytically then one can typically easily evaluate performance measures so that a full parametric study of the system is possible. This allows a more robust comparison of performance trade-offs than possible with a prototype or a detailed system simulation. Performance measures with a simulation of the model are also more readily obtained than a detailed simulation because of the reduced computational complexity. Second, having a model provides the greatest potential for discovering rules of thumb that provide insight into qualitative aspects of the performance of the system. Since computer technology is changing rapidly, computer designs are never static and thus such rules are a vital ingredient in forming the intuition of a designer, which can be used to guide future designs or modifications.

It is important to note that the level of detail that can be modeled varies with the evaluation method selected; prototypes model fine details of the system, mathematical models are more abstract and are thus at a higher level, and system simulation (as opposed to model simulation) lies somewhere in between. It is also the case that one must distinguish between *relative* and *absolute* performance. Prototyping provides information on absolute performance measures but little on relative performance of different designs since, typically, it is difficult to change the prototype. The absolute accuracy of a simulation model depends on the level of detail that is modeled and on the validity of its input parameters. Relative performance can be obtained by comparing the outputs from different simulations under identical inputs, but this requires substantial computing time. Mathematical models provide great insight into relative performance but, often, are not accurate representations of absolute performance. This results from the fact that to create tractable mathematical models of computer systems, one frequently must abstract away certain levels of detail and make simplifying assumptions about the system. Relative performance, however, is frequently the most important measure of a set of competing systems since one is more interested in the performance advantages of one design over another than in their absolute performance. Absolute performance will depend on inputs that are often not known in advance, and one is typically concerned with selecting the "best design."

The focus of mathematical models on relative, rather than absolute, performance influences our approach to the study of performance modeling using probabilistic techniques. Relationships between different designs often follow from corresponding relationships between mathematical properties associated with their underlying models. It is partly for this reason that we emphasize the mathematical properties of probabilistic objects and explore their inter-relationships. We will always derive new mathematical structures so that they follow as consequences of properties of structures that have already been derived. An advantage of this approach is that students are never expected to believe that something is true without proof or justification. At all points it should be clear why something is true and how to derive it from first principles.

To provide a context for some of the results in the book we next present an overview of queueing theory, which is a main application area of applied probability as it relates to performance modeling. Applications using queueing theory make their appearance throughout the text but are especially found in Part II.

1.2 Queueing Systems**

A frequent application area for probability and stochastic processes is queueing theory, and here we define some standard nomenclature about queues and establish some fundamental results that will be used later in the text.

A simple single server queue, pictured in Figure 1.1, consists of a server

---

**This section can be skipped at first reading. Examples later in the book will use terminology from this section.

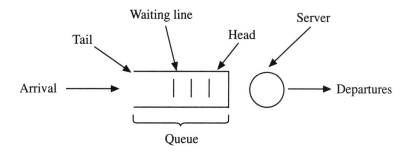

FIGURE 1.1. A Single Server Queue

that processes customer requests, a waiting line or a queue where customers wait before receiving service, and an arrival stream of customers to the system. The *head* (resp., *tail*) of the queue is the first (resp., last) position of the queue. In the simplest case, customers arrive to the tail of the queue and are served in a first-come-first-served manner (*FCFS*). In a computer model, the server could correspond to a CPU that processes customer requests. We call the portion of the queue not including the server the *waiting room*, or equivalently in computer models the *buffer*. Typically if the waiting room is finite, then any customer that arrives when all spaces are taken is assumed to be lost to the system, that is, it is as if this customer never arrived.

1.2.1 More General Queueing Systems

The simple queueing model just considered allows an astonishingly large number of variations. There could be more than one server (called *multiple server queues*); these servers could serve at the same speed (*homogeneous servers*) or different speeds (*heterogeneous servers*); customers could be served in a variety of disciplines including last-in-first-out (*LIFO*), processor sharing (*PS*), where each customer receives the same fraction of processing power, or in random order. Arrivals could correspond to more than one customer (called *batch arrival processes*) or a job[1] could consist of several tasks, each of which must be completely serviced before the job is completed.

These modifications can be combined to yield models of computer systems. For example, a simple model of a distributed parallel-processing system (called a *fork/join* queue) is shown in Figure 1.2, which consists of P homogeneous processors each with its own waiting line. Job arrivals are assumed to consist of multiple tasks that must be completed. At ar-

---

[1] To avoid repetitive nomenclature, we will interchangeably refer to arrivals to queues as either customers or jobs.

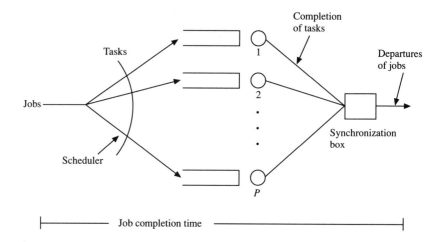

FIGURE 1.2. A Fork/Join Queue

rival these tasks are scheduled on the processors and they execute on their scheduled processors. Upon completion, tasks enter into the synchronization box as shown in the figure and wait for their sibling tasks to finish service. When all of its tasks have finished execution (this is called a *synchronization point*), the job leaves the system.

1.2.2 Queueing Networks

Queues can be joined together into networks where the output of one queue forms part of the input to another queue. Such networks model more complex interactions of a job with a computer system, and as a simple case we can consider a system comprised of a CPU and I/O servers (e.g., disk drives).

Consider the example of an open queueing network depicted in Figure 1.3. This network is a simple model of a workstation consisting of a CPU and two I/O devices, a floppy disk, and a hard disk (these are called queues or service centers). Requests are assumed to arrive from some external source (this is termed an **open model**). A request (also termed a job or a customer) consists of an alternating sequence of CPU and I/O service times (equivalently, service demands or requirements), and in the model this is depicted by the CPU- I/O loop shown in Figure 1.3. After completing service, a job leaves the system. There can be multiple jobs in the system at any time in any of the three service centers and they all compete for service. When a server finishes a request and there are other requests waiting in its queue, it uses a scheduling policy to determine the next request to serve. In our model we assume that all requests

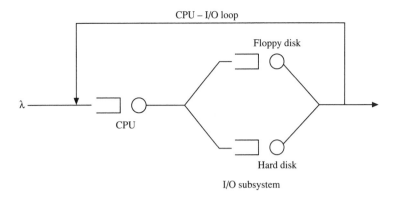

FIGURE 1.3. An Open Queueing Network

are served in first-come-first-served order. Questions of interest include the expected job response time, the expected amount of time waiting for service at the floppy and hard disk, and the maximum throughput of the system (the maximum rate at which the network can process customers).

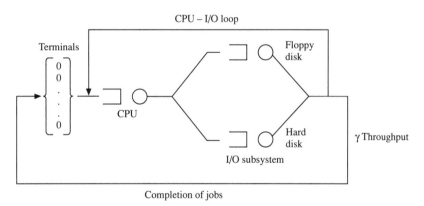

FIGURE 1.4. A Closed Queueing Network

The model shown in Figure 1.3 is an open model because arrivals are assumed to come from an external source. A closed version of this model is depicted in Figure 1.4; it consists of a fixed number of customers. This closed model can be interpreted as a set of users, often called *terminals*, that issue requests to a CPU and I/O subsystem. In a closed model jobs go through a cycle corresponding to a think time, during which a request is generated, followed by several CPU–I/O services (as shown by the loop in the figure). The number of times a job goes through the cycle is typically modeled as being random. Upon completion of a job, a new job is immediately created, which again follows the same cycle. This is an example of a closed system (i.e., there are a fixed number of total jobs in the system at all times) and variations of it are termed *central server models*. It and various modifications have been extensively analyzed. Performance measures of interest include the *throughput* of the system, γ, which is equivalent to the number of jobs returning to the thinking phase per unit time, the mean response time, $E[T]$, of jobs in the system and the utilization of the CPU, denoted by ρ. A famous result by J.D.C. Little called Little's law [80] relates these quantities, and in actuality only the throughput, γ, needs to be calculated by analysis. If γ is known and if the number of jobs in the system is given by M, then Little's law implies that $E[T] = M/\gamma$ and that $\rho = \gamma E[S]$, where $E[S]$ is the average service time at the CPU. For closed systems, like the preceding example, if there are K service centers and n_i, $i = 1, 2, \ldots, K$, for $0 \leq n_i \leq M$, is the number of customers at the ith service center, then it is clear that

$$M = \sum_{i=1}^{K} n_i.$$

If there are no restrictions on the number of customers at any service center, then the number of possible customer allocations equals the number of ways that K nonnegative integers can sum to M. This can result in a large number of different possibilities, and the complexity of analyzing closed networks is a direct consequence of this fact.

We can create more complicated examples of queueing networks that more finely model the flow of jobs through the system. For example, there can be multiple *classes* of jobs that are distinguished by the routes they take through a queueing network and by their service requirements at each server. A two-class central server type model is shown in Figure 1.5.

In such a model requests are distinguished by their service and routing characteristics. We show two queues for each service center in this figure to point out the fact that both classes of requests are processed by each server. When a server finishes serving a request at a particular service center it selects a request in its queue (if there is any) according to the scheduling discipline of that center. It might make sense, for example, to give edit jobs priority over batch jobs at the CPU since they are expected to be shorter in duration. In such a model, since resources are shared, there is a trade-off between the response time of both classes of jobs depending on the service disciplines used at the service centers and the speeds of the servers. An analysis of the system allows performance comparisons that quantify this trade-off. It may be desirable, for example, to know if certain scheduling policy assumptions lead to

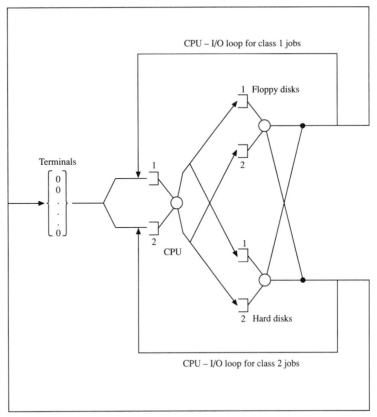

Completion of class 1 jobs

FIGURE 1.5. Two-Class Closed Queueing Model

sub-second response time for edit requests. Another question of interest is to determine the effect of increasing the transfer rate of the disk drive or to determine how the addition of another processor decreases response time. Changing the fraction of times the hard disk drive is used as compared to the floppy disk for each class of customers also changes the performance of the system, and it is desirable to know these effects before rewriting code required to make such changes. In later chapters we derive the theory that is necessary to answer questions of this type. Clearly, many more resources and different scheduling disciplines can be modeled at each queue. We postpone further discussion of such variations until Part II of the text, where our objective will be to establish the mathematics needed to determine the performance of such networks.

Bibliographic Notes

There are several books that develop the field of probability using examples from computer performance modeling. Examples of such books include Allen [2], Kobayashi [70], Molloy [82], and Trivedi [118].

Texts that assume a background in probability and cover special topics of performance modeling include Coffman and Denning [17], Gelenbe and Mitrani [37], Harrison and Patel [47], Heyman and Sobel [49], Kant [56], Kleinrock [65], Robertazzi [97], and Ross [99].

The "art" of performance modeling and the application of mathematical techniques can be found in Ferrari [36], Lazowska, Zahorjan, Graham, and Sevcik [77], Sauer and Chandy [104], and Sauer and MacNair [105].

An invaluable reference that presents a compendium of results in the field of performance modeling, queueing network theory, and simulation is found in Lavenberg [76].

Part I

Probability

2

Randomness and Probability

What do we mean by saying that an event occurs randomly with a certain probability? Answering this simple question is the objective of this chapter and our answer forms the foundation for all of the material presented in the text. The existence of randomness is often taken for granted, much like the existence of straight lines in geometry, and the assignment of probabilities to events is often assumed axiomatically. The pervasive use of the word "probability" seems to imply that randomness is the rule rather than the exception. One hears of probabilities being ascribed to such diverse things as the weather, the electability of a presidential candidate, the genetic makeup of a child, the outcome of a sporting event, the chance of winning at blackjack, the behavior of elementary particles, the future marriage prospects of two friends, quantum physics, and the wonderfully incessant gyrations of the stock market.

It would seem that we are surrounded by random events and can therefore say little that is precise and definite. Yet when people are asked for their opinion of physical reality, one finds that many believe that the world operates deterministically. Einstein said, "God does not play dice with the world," and many people similarly believe that the future is determined by the past. In other words, the states of all of the elementary particles in the world at time $t + dt$ are determined by their states at time t and the deterministic laws of physics. The world is thus a complex, deterministic, machine. Randomness arises from limits on our ability to measure the states of all elementary particles and to compute their next states. In this view randomness and unpredictability are the same phenomena, and unpredictability arises merely from lack of sufficient information and computational power.

Another worldview is that randomness is a basic building block of nature and that predictability only arises in aggregate systems where random effects cancel out. We can think of the analogy of some gas contained in a closed vessel. Each atom in the gas has a random trajectory that inflicts a random pressure every time it collides with the walls of the vessel. All of the random collisions result in an aggregate pressure that does not vary and can be measured precisely. Prediction in this view is only possible in aggregate form, and the fundamental elements are random. One doesn't question the existence of randomness, just as one doesn't question the existence of integers. This is the view, for example, of many quantum physicists.

So what is randomness: is it the essence of life or only an admission that we do not have infinite powers of calculation. How does it arise in physical and social systems? If things are indeed random, can we still make precise statements regarding eventual outcomes? These are the questions that motivate the definition of probability, and our answers explain why the field has proved to be so useful in modeling diverse types of complex systems. Often one approaches probability without addressing such questions; the reader is asked to believe that events occur randomly without further specification of what exactly is meant by a "random event." There are two good reasons for this approach. The first is that it allows one to get into the problem solving aspects of probability as quickly as possible. The second, and perhaps more important reason, is that it bypasses a thorny problem that has pestered probabilists since the very beginning of the field: circularity in the definition.

Pierre Simon Laplace (1749–1827), said to be the father of modern probability, defined the probability of an event as follows [73]:

> *The probability of the event A equals the ratio of the cases favorable for the occurrence of the event A and the total number of possible cases, provided that these cases are equally possible.*

It follows from this definition that the probability of tossing a fair coin and obtaining a head equals 1/2 since a head appears once in the two possible outcomes. The circularity in the definition arises from the fact that to calculate the probability of a set of events one must initially assume that all outcomes are random and "*equiprobable.*" This circular definition so vexed mathematicians of Laplace's time that it partly explains why the subject disappeared from mathematics shortly after Laplace published his monumental treatise, *Théorie Analytique des Probabilités*, in 1812. Luckily for us the field reappeared in the late 1800s, but still without proper definitions of randomness and probability. The need to develop a rigorous theory of probability was realized by David Hilbert, a notable mathematician who included this problem on his famous list of important and unsolved problems of mathematics in 1900.

Only recently have we been able to precisely define randomness and, in fact, it is somewhat surprising to find that in addition to physical forms of randomness such as coin tosses, there are abstract forms that arise from "deterministic" mathematical systems. Defining randomness of this last type requires a high degree of abstraction. To keep within the scope of the main body of this text we must appeal here to the reader's intuition to believe that complex systems that are governed by known, deterministic, rules can exhibit random behavior. This allows us to proceed into the practical aspects of the theory, such as how one models systems using probability, without requiring a more precise definition of what we mean by saying that something is random.

For interested readers, however, we provide in the companion appendix to this chapter, Appendix A, an introduction to recent results that define randomness and classify different types. The definition in the appendix avoids the circularity discussed here. Results presented in Appendix A include conditions under which a fair coin toss produces random outcomes and lands heads up with probability 1/2 (thus satisfying Laplace's equiprobable assumption) and the creation of a deterministic process whose outcomes are equivalent to a series of random coin tosses. Within these results are found the building blocks of the field of chaos, a subject currently receiving much attention. This appendix can be read at one's leisure.

We now turn to the question of how one determines the probability of a random event. Intuition suggests that if a large number of identical experiments are performed then the probability of an event can be obtained by calculating the average number of times the event occurs. For example, if one tossed a coin 1,000 times yielding 513 hundred heads, then it would not be unreasonable to say that the probability of heads equals 513/1,000. Intuitively, if the coin is fair then the average thus calculated should converge to 1/2 as we increase the number of tosses; a better estimate of the probability can be obtained by continuing the experiment. This approach to calculating probabilities is an extrapolation of common experience; it leads to a theory of probability called the frequency-based approach, which was developed by von Mises [120]. This approach is discussed in Section 2.1, along with a discussion of some of its limitations. These limitations suggest reasons why this approach has been largely superseded by an axiomatic theory of probability pioneered by Kolmogorov [71], which is discussed in Section 2.2. These two theories of probability are linked by the fact that the convergence of statistical averages in the frequency-based approach is expressed as a theorem (the strong law of large numbers) in the axiomatic approach. Consequences of the axioms of probability are developed in Section 2.3; they form the foundation that supports all of our results. In Section 2.4 we summarize the results of the chapter.

We close this introductory section by pointing out the hierarchy of models typically used to analyze complex systems. In the example of a computer system we first model events, such as arrivals and service times, using probability theory. That is, much like the hidden assumptions in Laplace's equiprobable assumption, we *assume, without explicit statement, that these events are random*, and then we assign probabilities to them. Once all of these random events have been specified we create an overall model of the system and analyze it using mathematical tools of probability. Most often the first part of this process, assuming that events are random, is simply taken for granted.

2.1 Probability: Frequency-Based

In this section we discuss the statistical approach to probability pioneered by von Mises [120]. We do this through the expedient of a coin tossing experiment with two players. To add a bit more interest to the development, assume that each time the coin lands heads up (resp., tails up) player 1 wins (resp., loses) a dollar. Suppose that we perform a series of tosses obtaining outcomes $X_i, i = 1, 2, \ldots$, where X_i equals 1 if on the ith toss the coin lands heads up and otherwise equals 0. The average number of times the coin lands heads up in these first n tosses is given by

$$Y_n \overset{\text{def}}{=} \frac{1}{n} \sum_{i=1}^{n} X_i. \tag{2.1}$$

Intuition suggests that if we toss the coin a sufficient number of times then we can calculate its intrinsic probability of landing heads up. In other words, if

$$\lim_{n \to \infty} Y_n = p \tag{2.2}$$

then we would say that the coin lands heads up with probability p. We are immediately confronted with problems that arise from a definition of probability based on (2.2). First, it tacitly assumes that probabilities can only be assigned to events that are repeatable. Events that depend on some value of absolute time, for example, cannot be assigned a probability, since one can never repeat the experiment. Furthermore, the number of events for which a probability can be assigned is necessarily finite. This precludes assigning probabilities to, say, events that occur on the real line. The definition, as it stands, also implies that, for any large value m, a coin that always alternates between $\lfloor mp \rfloor$ successive heads and $\lceil m(1-p) \rceil$ successive tails has a probability of p of landing heads up. (See the Glossary of Notation for the definitions of mathematical symbols like $\lfloor x \rfloor$ and $\lceil x \rceil$.) Clearly this case violates any notion of randomness.

To avoid the last type of problem, we need to also establish that the outcome of any toss is not influenced by the outcomes of other tosses. In the parlance of probability we say that outcomes are *independent* of each other. It is not clear, however, how we can show that independence holds. For any given finite sequence of outcomes it is clear that there are functions that generate exactly these outcomes. We could thus be easily fooled by outcomes that appeared to be independent but are really generated by such a function. For example, the decimal digits of π are deterministic but show no discernible pattern when carried out to more than two billion places [93]. Without knowing that the decimal digits are generated by a deterministic sequence, one is led to believe they are random. Let π_i be the ith decimal digit of π. Then if we let

$$
X_i = \begin{cases} 1, & \pi_i \text{ even,} \\ 0, & \pi_i \text{ odd,} \end{cases}
\tag{2.3}
$$

the deterministic sequence of decimal digits of π appears to generate a random sequence of coin tosses. Even if one is given a sequence of ℓ digits of π but is not told the location of the sequence, it is not possible to predict the next digit following the sequence.

Suppose we attempt to skirt the problem of establishing independence by altering the game slightly. Allow player 1 to choose which tosses are included in the sum of (2.1). Set ψ_i equal to 1 if player 1 decides the ith toss is to be included in the summation and equal to 0 otherwise. We require that player 1 decide the value of ψ_i *before* the ith toss. The player can use the values of $X_j, 1 \leq j \leq i-1$, to decide ψ_i, and we require that the player make an infinite number of choices. In other words, the player cannot, at some time, stop playing the game. Clearly player 1 wishes to maximize the number of times the coin lands heads up. Suppose that player 1 has unlimited powers of deductive reasoning. The average gain for this game is given as

$$
Y_n^\psi \stackrel{\text{def}}{=} \frac{\sum_{i=1}^n \psi_i X_i}{\sum_{j=1}^n \psi_j}.
$$

If, even under these ideal circumstances, it is the case that

$$
\lim_{n \to \infty} Y_n^\psi = p,
\tag{2.4}
$$

then we have more faith that the chance of tossing heads is given by p and that these tosses satisfy some sort of independence.

In this case, if player 1 knew that the coin tosses were determined by the decimal digits of π then on every toss the player would win one dollar. If, however, the values of X_i are chosen as in (2.3), and if player 1 is not told that the sequence is generated in this manner then, no matter what player 1 does, it appears that (2.4) holds with $p = 1/2$. In such cases the information of how tosses are generated is critical regarding the conclusion that the coin tosses are random and that each toss is independent of any other toss (see [22] for a discussion of Kolmogorov complexity, which characterizes this idea precisely). It is important to note that for (2.4) to equal (2.2) it must be the case that the sequence of values X_i shows no pattern that can help player 1 to increase the number of selected heads. For example, it cannot be the case that a 1 in the ith place is always followed by a 1 in the $10 \times i$th place, or that there are never more than 89 1s in a row, or that the pattern $1, 0, 1, 0, 1, 1$ never appears. If there were a sufficient number of such patterns then player 1 could select ψ to change the limit (2.4) to be more favorable.

We now have some sense of what we mean by probability. Of course (2.4) is a difficult operational definition since it requires a limit and only concerns events that are repeatable. Within this frequency framework one can develop algebraic properties that are satisfied by probabilities. These properties include the summation of probabilities for independent events, conditional probabilities, and combining the probability of two or more events into a joint probability for the compound event (see von Mises [120] for developments along these lines). The frequency-based approach has been largely supplanted, however, by an abstract development of probability that bypasses problems with randomness and repeatability through the means of an axiomatic framework. The point of contact between the two approaches is that the limit of (2.4) is a consequence of a theorem in the axiomatic approach. Furthermore, all of the algebraic properties satisfied by the frequency-based approach follow as consequences in the axiomatic approach.

2.2 Probability: Axiomatic-Based

The axiomatic method of probability pioneered by Kolmogorov [71] avoids the difficulties of the frequency-based approach by assuming that probabilities are given quantities. To motivate this approach we start, in Section 2.2.1, by discussing how probability theory is used in modeling. The issues raised in this investigation imply that there are certain desirable features that should be enjoyed by a mathematical theory of probability. These features deal with "mixtures of events" as described in Section 2.2.2, and they lay the foundation for the formal definition of a probability space that is given in Section 2.2.3.

**2.2.1
Probability and
Modeling**

The first step in creating a probabilistic model is to determine the random events that are important for the objective of the model. Often this requires excluding events that are "possible" but are either not of interest or are so "rare" that their effects are negligible. Even in the example of a coin tossing experiment previously discussed, we restricted the outcome to be one of two possibilities: heads or tails. For some coins, however, there is the remote chance that the coin lands upright on its side. For practical purposes, however, the occurrence of such an event is so rare, perhaps occurring 1 time out of 10^9 tosses, that it is not included in a model. For a model of the reliability of a highly reliable computer system, however, it is exactly these "rare events" that are modeled. For such systems, failures often occur at rates of 10^{-9} per second (this implies that the average time between failures is 10^9 seconds or about 31 years). Although such events might be as rare as a coin landing up heads, they are included in a model because they are important to its objective.

The key is that a practical model of a system abstracts away nonessential details. Features of a system that must be included are determined by this objective and by the performance issues that are addressed by the model. This largely depends on a detailed understanding of the system and by previous experience, which guides one's judgment in determining which features of a system are sufficiently important to be modeled. We do not, for example, need to model detailed machine instructions of a computer in a model of customer response times in a computer system, but such detail is essential if we are modeling the performance of register allocation policies.

Once the model is specified the mathematics is applied to derive results. There are certain desirable features that this mathematical structure should possess. To motivate these features, consider a simple experiment consisting of balls contained in an urn (ball and urn problems are endemic to probability models, as we will see in Chapter 3). Suppose that there are $b + w$ balls contained in an urn that are sequentially labeled using the integers. Balls $1, 2, \ldots, b$, are colored black and those with labels $b + 1, b + 2, \ldots, b + w$, are colored white. An experiment is performed where a ball is selected from the urn and then returned to the urn. We assume that all possible selections are equiprobable. We can define events that are of interest and use Laplace's definition of probability, given in the beginning of the chapter, to calculate the probability of a particular event.

For example, the probability of the event $\mathcal{A}$, that an even numbered ball is selected, according to Laplace's definition equals the number of ways that an even ball can be selected divided by the total number of ways a ball can be selected. The number of even labeled balls is $\lfloor (b + w)/2 \rfloor$

and the number of total possibilities is $b + w$. Thus we can write

$$P[\mathcal{A}] = \frac{\lfloor (b+w)/2 \rfloor}{b+w}. \tag{2.5}$$

According to the frequency-based definition of probability, we expect that if the experiment is performed a sufficiently large number of times then the actual calculated frequency of event $\mathcal{A}$ will be close to the probability specified in (2.5).

We could, of course, be interested in more than the fact that the selected ball is even. Let the set of all possible outcomes of the experiment, termed the *sample space*, be denoted by Ω. The random ball selection corresponds to a random selection of a point ω contained in Ω, and we can represent the event $\mathcal{A}$ above as

$$\mathcal{A} = \{ \omega \in \Omega : n(\omega) = 2k, \quad k = 1, 2, \ldots, \lfloor (b+w)/2 \rfloor \}.$$

In this expression $n(\omega)$ is the number of the ball selected for outcome $\omega \in \Omega$. Suppose that we are interested in the color of the ball. Define $\mathcal{B}$ to be the event that the selected ball is black. It is clear from the above that

$$\mathcal{B} = \{ \omega \in \Omega : n(\omega) \leq b \},$$

and that

$$P[\mathcal{B}] = b/(b+w).$$

A natural objective at this point is to determine the probability of "mixtures" of events $\mathcal{A}$ and $\mathcal{B}$ (and possibly of other events of interest). For example, the event corresponding to a black, even, ball being selected is given by the intersection of events $\mathcal{A}$ and $\mathcal{B}$, that is, the event $\mathcal{A} \cap \mathcal{B}$ given by

$$\mathcal{A} \cap \mathcal{B} = \{ \omega \in \Omega : n(\omega) = 2k, \quad k = 1, 2, \ldots, \lfloor b/2 \rfloor \}.$$

On the other hand, the event corresponding to selecting a black or an even ball is given by $\mathcal{A} \cup \mathcal{B}$, and that of selecting a white ball by $\mathcal{B}^c$. This motivates a requirement that we impose on events; if $\mathcal{A}$ and $\mathcal{B}$ are events then so are their intersection, union, and complements. These operations can be repeatedly applied and so doing induces a *family of sets* (this is precisely defined in Section 2.2.3).

There is a natural way to specify probabilities of events corresponding to mixtures of events; if two events have no sample points in common, that is, if their intersection is empty, then the probability of the union of these two events should be the sum of the probabilities of the events. For example, let $\mathcal{C}$ be the event of selecting ball 1. It is clear that $\mathcal{AC} = \emptyset$ and, quite naturally, we desire that

$$P[\mathcal{A} \cup \mathcal{C}] = P[\mathcal{A}] + P[\mathcal{C}] = \frac{\lfloor (b+w)/2 \rfloor + 1}{b+w}.$$

This property allows us, in Section 2.3, to derive an algebra for how one determines the probabilities of mixtures of events.

The ball and urn problem discussed above has a sample space consisting of a finite number of elements. Such sample spaces can be specified in a straightforward manner; form all possible subsets of elements of Ω. When sample spaces contain an infinite number of elements this specification is not as straightforward. Consider, for example, the problem of modeling requests to a computer system. Assume that requests can occur at any point in time and thus events for the sample space are defined on the real number system. An event of interest is the number of arrivals that occur in a certain period of time. Let $\mathcal{N}_{t_0,t_1}$ be the event that at least one arrival occurs in the time period $[t_0, t_1)$ (this corresponds to all points in the set $\{t : t_0 \leq t < t_1\}$; other variations of interval sets are listed in the Glossary of Notation). In Chapter 6 we discuss a model of an arrival process, called the Poisson process, for which the probability of event $\mathcal{N}_{t_0,t_1}$ is given by $e^{-\lambda(t_1-t_0)}$, where λ is some positive constant. Events of the type just described are defined for all possible selections of time intervals but there are other sets of time that might also be of interest. For example, one might define $\mathcal{D}$ to be the event that at least one arrival occurs at integer time epochs. A mathematical difficulty arises in infinite sample spaces, however, when we attempt to formally define what we mean by mixtures of arbitrary events and to specify probabilities for them. The definition of the probability for such events is still similar to Laplace's definition: it equals the number of favorable outcomes divided by the number of possible outcomes, but a precise definition of what is meant by "number" is not straightforward. The branch of mathematics that deals with these problems is called *measure theory* (see [7] for a reference) and lies outside the scope of this text. Although a measure theoretic specification of events and probabilities forms the foundation for probability, it does not play an important role in the practical use of the theory. In Chapter 4 we will show how random variables are used to model probabilistic systems without requiring the specification of an underlying probability space.

With these comments as motivation, we next delineate set operations that are used to define mixtures of events in Section 2.2.2, and then we give a formal definition of a probability space in Section 2.2.3.

2.2.2
Set Operations

Sets are denoted in calligraphic type, and the empty set is denoted by $\emptyset$. We denote a universal set by Ω. The complement of a set $\mathcal{A}$, denoted by $\mathcal{A}^c$, is understood to be taken with respect to Ω. It is clear that $\Omega^c = \emptyset$ and $\emptyset^c = \Omega$. It is convenient to simplify set notation by writing set operations in algebraic notation. Thus we will write

$$\mathcal{A} + \mathcal{B} \stackrel{\text{def}}{=} \mathcal{A} \cup \mathcal{B}$$

for union operations,

$$\mathcal{A}\mathcal{B} \stackrel{\text{def}}{=} \mathcal{A} \cap \mathcal{B}$$

for intersection operations, and

$$\mathcal{A} - \mathcal{B} \stackrel{\text{def}}{=} \mathcal{A} \cap \mathcal{B}^c$$

for the difference between two sets. Using this notation for sums and products we have

$$\sum_{i=1}^{n} \mathcal{A}_i \stackrel{\text{def}}{=} \bigcup_{i=1}^{n} \mathcal{A}_i$$

and

$$\prod_{i=1}^{n} \mathcal{A}_i \stackrel{\text{def}}{=} \bigcap_{i=1}^{n} \mathcal{A}_i.$$

A convenient way to picture relationships between sets is through Venn[1] diagrams.

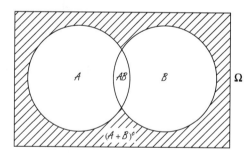

FIGURE 2.1. A Venn Diagram

The Venn diagram in Figure 2.1 shows two sets, $\mathcal{A}$ and $\mathcal{B}$, that have a non-empty intersection. The hatched region in this figure corresponds

[1]This is named after John Venn (1834–1923) who employed this device in 1876 in a paper on Boolean logic.

to $(\mathcal{A} + \mathcal{B})^c$. By shading regions of the diagram appropriately one can see (although this does not prove) the validity of De Morgan's laws:[2]

$$(\mathcal{A} + \mathcal{B})^c = \mathcal{A}^c \, \mathcal{B}^c$$

De Morgan's Laws

$$(\mathcal{AB})^c = \mathcal{A}^c + \mathcal{B}^c.$$

Intersection	**Union**
$\mathcal{A}\Omega = \mathcal{A}$	$\mathcal{A} + \Omega = \Omega$
$\mathcal{AA} = \mathcal{A}$	$\mathcal{A} + \mathcal{A} = \mathcal{A}$
$\mathcal{A}\emptyset = \emptyset$	$\mathcal{A} + \emptyset = \mathcal{A}$
$\mathcal{AA}^c = \emptyset$	$\mathcal{A} + \mathcal{A}^c = \Omega$
$\mathcal{A}(\mathcal{BC}) = (\mathcal{AB})\mathcal{C}$	$\mathcal{A} + (\mathcal{B} + \mathcal{C}) = (\mathcal{A} + \mathcal{B}) + \mathcal{C}$
$\mathcal{A}(\mathcal{B} + \mathcal{C}) = (\mathcal{AB}) + (\mathcal{AC})$	$\mathcal{A} + (\mathcal{BC}) = (\mathcal{A} + \mathcal{B})(\mathcal{A} + \mathcal{C})$
$(\mathcal{AB})^c = \mathcal{A}^c + \mathcal{B}^c$	$(\mathcal{A} + \mathcal{B})^c = \mathcal{A}^c\mathcal{B}^c$

TABLE 2.1. Algebra for Set Operations

Two sets, $\mathcal{A}$ and $\mathcal{B}$, are said to be *mutually exclusive* if they are disjoint, that is, if $\mathcal{AB} = \emptyset$, and we say that $\mathcal{A}$ *implies* $\mathcal{B}$ if $\mathcal{A} \subset \mathcal{B}$. Set operations satisfy the algebraic properties given in Table 2.1.

A *partition* of a set $\mathcal{A}$ is a collection of mutually exclusive sets whose union is $\mathcal{A}$, that is, $\mathcal{A}_i, i = 1, 2, \ldots, n$ is a partition if

$$\sum_{i=1}^{n} \mathcal{A}_i = \mathcal{A}, \quad \mathcal{A}_i\mathcal{A}_j = \emptyset, \quad 1 \le i, j \le n, \quad i \neq j. \qquad (2.6)$$

Partitions are an important set theoretic concept in probability. As an example of a partition consider the set Ω consisting of all integer pairs (x, y) where x ranges from 1 to 4 and y ranges from 1 to 6. In Figure 2.2 set $\mathcal{B}_1 = \{(1, 4), (1, 5), (1, 6)\}$ and set $\mathcal{B}_4 = \{(2, 1), (2, 2), (3, 1)\}$ and

[2]This is named after Augustus De Morgan (1806–1871) whose treatise, *Formal Logic; of the Calculus of Inference, Necessary and Probable*, in 1847, established these laws.

the collection of sets $B_1, B_2, \ldots, B_6$ form a partition of Ω. In probability theory partitions are used to account for all possible random outcomes and in the figure this corresponds to the fact that points in Ω appear in exactly one of the sets $B_i, i = 1, 2, \ldots, 6$.

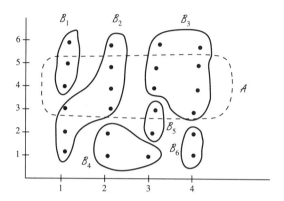

FIGURE 2.2. Partition of a Set of Pairs

2.2.3 Probability Spaces

We now will precisely define the space over which random events and probabilities are defined. Let Ω be the sample space on which are defined *random events* consisting of a family **F** of subsets of Ω. Following our discussion in the beginning of Section 2.2 we impose the following three conditions on the structure of **F**.

1. **F** contains the set Ω as one of its elements.

2. If subsets $\mathcal{A}$ and $\mathcal{B}$ are in **F** then the sets $\mathcal{A} + \mathcal{B}$, $\mathcal{A}\mathcal{B}$, $\mathcal{A}^c$, and $\mathcal{B}^c$ are also elements of **F**.

   Notice that immediate consequences of the first two conditions are that the null set $\emptyset$ is in **F** and that **F** is closed under a finite number of unions and intersections. We will often require a stronger condition than closure under a finite number of operations for infinite sample spaces.

3. If $\mathcal{A}_i, i = 1, 2, \ldots$, are elements of **F** then so are $\sum_{i=1}^{\infty} \mathcal{A}_i$ and $\prod_{i=1}^{\infty} \mathcal{A}_i$.

The family **F** defined by the first two conditions is called an *algebra of sets* and a family **F** that satisfies all three conditions is sometimes referred to as a *σ-algebra* or as a *Borel field of events*. For our purposes, the family **F** defines the events that are important to the objective of a probabilistic model of a system. Very often this set is not explicitly defined since it is understood from the context of the model. Some examples of **F** include the following:

Example 2.1 Consider three tosses of a coin. Then

$$\Omega = \{HHH, HHT, HTH, THH, HTT, THT, TTH, TTT\}.$$

The random event of "one head" corresponds to $\mathcal{A} = \{TTH, THT, HTT\}$ and at least two heads to $\mathcal{B} = \{HHT, HTH, THH, HHH\}$. One example of **F** consists of these sets and their possible unions and intersections with their complements: $\mathbf{F} = \{\Omega, \emptyset, \mathcal{A}, \mathcal{B}, \mathcal{A} + \mathcal{B}, \mathcal{A}^c, \mathcal{B}^c, (\mathcal{A} + \mathcal{B})^c\}$.

Example 2.2 (Computer Systems with Spares) To increase the reliability of a computer system one frequently uses spare components. Consider a system consisting of three processors (a main processor with two spares) and two disk drives (a main drive with one spare). The system is "down" if all processors or all disk drives fail. Failures are assumed to occur randomly. The sample space of the system consists of

$$\Omega = \{(p, d), \ p = 0, 1, 2, 3, \quad d = 0, 1, 2\}$$

where p (resp., d) corresponds to the number of failed processors (resp., disks). The random event "system down" corresponds to

$$\mathcal{A} = \{(0, 2), (1, 2), (2, 2), (3, 0), (3, 1), (3, 2)\}.$$

The event "one more processor failure causes system down" corresponds to $\mathcal{B} = \{(2, 0), (2, 1)\}$ and the event "one more disk failure causes system down" corresponds to $\mathcal{C} = \{(0, 1), (1, 1), (2, 1)\}$. The family **F** generated by $\mathcal{A}$, $\mathcal{B}$, and $\mathcal{C}$ is given by

$$\mathbf{F} = \{\Omega, \emptyset, \mathcal{A}, \mathcal{B}, \mathcal{C}, \ \mathcal{A} + \mathcal{B}, \mathcal{A} + \mathcal{C}, \mathcal{B} + \mathcal{C}, \mathcal{A} + \mathcal{B} + \mathcal{C}, \mathcal{B}\mathcal{C}, \mathcal{A}^c, \mathcal{B}^c, \mathcal{C}^c,$$
$$(\mathcal{A} + \mathcal{B})^c, (\mathcal{A} + \mathcal{C})^c, (\mathcal{B} + \mathcal{C})^c, (\mathcal{A} + \mathcal{B} + \mathcal{C})^c\}.$$

When Ω is finite the family of sets **F** can be easily generated by simply forming all possible subsets of elements of Ω. When Ω is infinite, however, such a construction is not appropriate. The reasons for this, as mentioned in Section 2.2.1, are beyond our scope. We can, however, outline a construction at an intuitive level. A typical construction for infinite sample spaces is to first define a set of "basic events" that are of interest to the model and then to enlarge this set by forming all possible countable unions and intersections. This is why we require, in the third condition listed above, that **F** be closed under countable unions and intersections.

The basic events for the arrival process discussed in Section 2.2.1, for example, consist of intervals of the form $[t_0, t_1)$ (the event $\mathcal{D}$ in that section, that is, at least one arrival at integer time epochs, is not a basic event). The family of sets **F** is enlarged by forming all possible countable unions and intersections of these basic events. Again, most probability models do not require a detailed specification of the family of sets but rather are represented in terms of random variables (discussed in Chapter 4) which are more directly related to the model.

Countable union and intersection operations, as required in the third condition on **F**, occur quite naturally in probabilistic models when the sample space is infinite. Consider, for example, an infinite sequence of coin tosses and let $\mathcal{A}_i$ be the event that a heads lands up on the ith toss, for $i = 1, 2, \ldots$. The event "no more heads after k tosses" for $k = 0, 1, \ldots$ (the case $k = 0$ means no heads is tossed), denoted by $\mathcal{B}_k$, is represented as a countable number of intersections. This event is given by

$$\mathcal{B}_k = \prod_{i=k+1}^{\infty} \mathcal{A}_i^c, \quad k = 0, 1, \ldots .$$

Similarly the event "a finite number of heads is tossed," denoted by $\mathcal{C}$, is given by

$$\mathcal{C} = \sum_{k=0}^{\infty} \mathcal{B}_k.$$

Other events of interest are addressed by Problem 2.4.

Probabilities are defined on a field of events and satisfy the following axioms:

Axiom 2.3 *With each random event $\mathcal{A}$ in **F** there is associated a non-negative number $P[\mathcal{A}]$ called the probability of $\mathcal{A}$.*

Axiom 2.4 $P[\Omega] = 1$.

Axiom 2.5 *If events $\mathcal{A}_i, i = 1, 2, \ldots, n$, are mutually exclusive, that is, if $\mathcal{A}_i \mathcal{A}_j = \emptyset$, $i \neq j$, then*

$$P\left[\sum_{i=1}^{n} \mathcal{A}_i\right] = \sum_{i=1}^{n} P[\mathcal{A}_i].$$

For infinite sample spaces we need to extend Axiom 2.5 to a countably infinite number of operations.

Axiom 2.6 *If the occurrence of event $\mathcal{A}$ is equivalent to the occurrence of at least one of the pairwise mutually exclusive events $\mathcal{A}_i, i = 1, 2, \ldots,$ then*

$$P[\mathcal{A}] = \sum_{i=1}^{\infty} P[\mathcal{A}_i].$$

We call the event Ω the *certain event* and the event $\emptyset$ the *impossible event*. The fact that $P[\Omega] = 1$ is a convenient and intuitive normalization. A *probability space* is a triple $(\Omega, \mathbf{F}, P)$ and thus requires specifying a probability to each random event in $\mathbf{F}$. An event is said to be *measurable* with respect to the probability space $(\Omega, \mathbf{F}, P)$ if it is contained in $\mathbf{F}$ and thus has an assigned probability. An event is not measurable if it is not contained in $\mathbf{F}$.

Comparing the axiomatic development of probability with that of the frequency-based approach of Section 2.1 shows that we have avoided the entire issue of randomness and the assignment of probabilities by embedding these notions in a set of axioms. The axiomatic method does not specify a method to assign probabilities to the elements of $\mathbf{F}$ but simply assumes that these probabilities are given.

The set of axioms is *consistent* in that there exist mathematical objects that satisfy the axioms. As an example, in the coin tossing experiment we can satisfy all of the axioms by setting $P[H] = p$ and $P[T] = 1 - p$ for $0 \le p \le 1$. The set of axioms, however, is not *complete* in that one can select probabilities in different ways and still satisfy the axioms. Thus in Example 2.1 we could set $P[\mathcal{A}] = 50/51$ and $P[\mathcal{B}] = 1/51$ without violating the axioms, even though it violates all intuition regarding possible outcomes of the coin tossing experiment. Indeed, it is the very fact that the axioms are not complete that makes them so useful.

In modeling, one assigns probabilities in a manner that is appropriate for the system under analysis. Thus if one tosses a fair coin, it is reasonable to assign $P[H] = P[T] = 1/2$. For finite state spaces it makes little sense to explicitly define events for which a probability of 0 is assigned. If an event has probability 0 then one eliminates it from the model. Depending on the objective of the model of the system, one also sometimes eliminates events with extremely small probabilities. As mentioned in Section 2.2.1, we do not include the event of the coin landing on its side since this has an extremely small probability. Of course, nothing precludes us from including such an event in a model, and its elimination is one of practical considerations.

For infinite sample spaces, events with 0 probability arise naturally. Recall that in the infinite sequence of coin tosses discussed above we defined the event $\mathcal{B}_k$ to be the event that no heads land up after toss k. It is easy to show that this event has a probability of 0 (for other examples of this type see Problem 2.12). A subtlety of the interpretation of probability theory arises from this fact since, conceivably, it is not "impossible" for $\mathcal{B}_k$ to occur. The fact that $\mathcal{B}_k$ has 0 probability means, in this case, that the occurrence of $\mathcal{B}_k$ is so rare that, for practical purposes, it can be assumed to *almost never occur*. We will show in (2.7) that the impossible event, $\emptyset$, also has 0 probability but here a distinction is made since the impossible event *can never occur*. We conclude that merely knowing that an event has 0 probability is not sufficient to determine the impossibility of its occurrence. A deeper discussion of this subtlety must wait until Section 4.5.2. As might be expected, contrary to finite state spaces, it is typically not possible to eliminate events having 0 probability from an infinite sample space model. For example, the event $\mathcal{B}_k$ is included in the family of sets **F** when the basic sets of the infinite coin toss experiment $\mathcal{A}_i, i = 1, 2, \ldots$, are enlarged using countable union and intersection operations. There is thus no way to eliminate the event $\mathcal{B}_k$ and still satisfy the conditions for the definition of **F**.

2.3 Probability: Immediate Consequences of the Axioms

In this section we establish some immediate consequences of the axioms and elementary set operations given in Table 2.1. It is clear that $\mathcal{A}$ and $\mathcal{A}^c$ form a partition of Ω and that

$$P[\mathcal{A}] + P[\mathcal{A}^c] = 1,$$

from which it follows, with $\mathcal{A} = \Omega$ and Axiom 2.4, that

$$P[\emptyset] = 0, \tag{2.7}$$

and thus that

$$0 \leq P[\mathcal{A}] \leq 1, \quad \text{for any } \mathcal{A} \in \Omega.$$

If the event $\mathcal{A}$ implies the event $\mathcal{B}$ then

$$P[\mathcal{A}] \leq P[\mathcal{B}], \quad \mathcal{A} \subset \mathcal{B},$$

and furthermore

$$P[\mathcal{A} + \mathcal{B}] = P[\mathcal{A}] + P[\mathcal{B}] - P[\mathcal{AB}]. \tag{2.8}$$

The last equality can be easily seen in the Venn diagram for sets $\mathcal{A}$ and $\mathcal{B}$ previously given. Its proof follows from the set equalities

$$\mathcal{A} + \mathcal{B} = \quad \mathcal{A} + (\mathcal{BA}^c), \quad \mathcal{A}(\mathcal{BA}^c) \quad = \emptyset,$$

$$\mathcal{B} = \quad (\mathcal{AB}) + (\mathcal{BA}^c), \quad (\mathcal{AB})(\mathcal{BA}^c) \quad = \emptyset. \tag{2.9}$$

Equation (2.8) is called the *inclusion–exclusion* property. A generalization of it is given in Problem 2.13.

Equation (2.8) implies, since $P[\mathcal{AB}] \geq 0$, that

$$P[\mathcal{A} + \mathcal{B}] \leq P[\mathcal{A}] + P[\mathcal{B}].$$

Furthermore, it is clear that if $\mathcal{A}_i, i = 1, 2 \ldots, n$, is a partition of Ω, then

$$P\left[\sum_{i=1}^{n} \mathcal{A}_i\right] = 1, \quad n \geq 1.$$

The next result follows immediately from the axioms of probability but is of such importance that we record it as a theorem.

Theorem 2.7 (Law of Total Probability) *Let* $\mathcal{B}_i, 1 \leq i \leq n$, *be a partition of the set* Ω. *Then for any event* $\mathcal{A}$ *we can write*

$$P\left[\mathcal{A}\right] = \sum_{i=1}^{n} P\left[\mathcal{A}\mathcal{B}_i\right], \quad n \geq 1. \tag{2.10}$$

Proof: The sets $\mathcal{A}\mathcal{B}_i, i = 1, 2, \ldots, n$, are mutually exclusive and the fact that $\mathcal{B}_i, i = 1, 2, \ldots, n$, is a partition of Ω implies that

$$\mathcal{A} = \sum_{i=1}^{n} \mathcal{A}\mathcal{B}_i, \quad n \geq 1.$$

Applying Axiom 2.5 for finite n or Axiom 2.6 for infinite n completes the proof.   ∎

Example 2.8 Consider the set shown in Figure 2.2 to be the sample space for an experiment and assume that all points have a probability of 1/24. We are interested in the probability of a set $\mathcal{A}$ defined by the set of pairs satisfying $\mathcal{A} \stackrel{\text{def}}{=} \{(x, y) : y = 3, 4, 5\}$ (within the dotted box of the figure). It is easy to see that there are 12 pairs contained in this set and thus that $P[\mathcal{A}] = 1/2$. It also follows that $P[\mathcal{A}\mathcal{B}_1] = 2/24$, $P[\mathcal{A}\mathcal{B}_2] = 4/24$, $P[\mathcal{A}\mathcal{B}_3] = 5/24$, $P[\mathcal{A}\mathcal{B}_5] = 1/24$, and $P[\mathcal{A}\mathcal{B}_4] = P[\mathcal{A}\mathcal{B}_6] = 0$. The law of total probability thus implies that

$$P[\mathcal{A}] = P[\mathcal{A}\mathcal{B}_1] + P[\mathcal{A}\mathcal{B}_2] + P[\mathcal{A}\mathcal{B}_3] + P[\mathcal{A}\mathcal{B}_5]$$

$$= 2/24 + 4/24 + 5/24 + 1/24 = 1/2,$$

as previously calculated.

In applications the law of total probability is used to simplify the calculation of the probability of event $\mathcal{A}$ by breaking down the event into simpler components. These components correspond to the intersection of $\mathcal{A}$ with sets of a partition which in Example 2.8 were the sets $\mathcal{A}\mathcal{B}_i, i = 1, 2, \ldots, 6$. A simple way to view the components $\mathcal{A}\mathcal{B}_i$ is in terms of conditional probability.

**2.3.1
Conditional
Probability**

Let Ω' be a subset of Ω that has positive probability and rewrite (2.10) as

$$1 = \sum_{i=1}^{n} \frac{P[\Omega'\mathcal{B}_i]}{P[\Omega']}, \quad n \geq 1,$$

$$= \sum_{i=1}^{n} P[\mathcal{B}_i \mid \Omega'], \tag{2.11}$$

where we have defined

$$P[\mathcal{B}_i \mid \Omega'] \stackrel{\text{def}}{=} \frac{P[\Omega'\mathcal{B}_i]}{P[\Omega']}. \tag{2.12}$$

We call (2.12) the *conditional probability* of event $\mathcal{B}_i$ given event Ω'. Since the conditional probabilities sum to 1, we can think of conditional probabilities as being defined on a new probability space with events in Ω'. The family of sets **F** of the original process can be divided into two disjoint groups; those that have no elements in common with Ω' and those that do have common elements. We denote this last set by **F'**. The new probability space Ω' has an induced family of events given by **F'** with probabilities given by the conditional probabilities

$$P'[\mathcal{A}] \stackrel{\text{def}}{=} P[\mathcal{A} \mid \Omega'], \quad \mathcal{A} \subset \mathbf{F'}.$$

Thus, from the initial probability space $(\Omega, \mathbf{F}, P)$ we have created a new probability space given by $(\Omega', \mathbf{F'}, P')$. This new probability space is called the *conditional* probability space and can be thought of as the space induced by the original process when we only view events that occur in Ω'. Some examples illustrate this concept.

Example 2.9 (Continuation of Example 2.8) Consider set $\mathcal{A}$ and the partition $\mathcal{B}_i, i = 1, 2, \ldots, 6$, given in Example 2.8. As calculated in that example $P[\mathcal{A}] = 1/2$. The figure in the example, however, clearly shows that conditioned on the event $\mathcal{B}_4$ the probability of $\mathcal{A}$ is given by $P[\mathcal{A} \mid \mathcal{B}_4] = 0$ and thus the conditional probability of $\mathcal{A}$ given event $\mathcal{B}_4$ is less than the unconditional probability of $\mathcal{A}$.

This relationship can go the other way. For example, let event $\mathcal{C}$ be the event "one or more of events $\mathcal{B}_1, \mathcal{B}_2, \mathcal{B}_3$, and $\mathcal{B}_5$ occurs." Hence $\mathcal{C} = \mathcal{B}_1 + \mathcal{B}_2 + \mathcal{B}_3 + \mathcal{B}_5$. Then, using Axiom 2.5, the conditional probability $P[\mathcal{A} \mid \mathcal{C}]$ can be calculated by

$$P[\mathcal{A} \mid \mathcal{C}] = \frac{P[\mathcal{A}\mathcal{C}]}{P[\mathcal{C}]}$$

$$= \frac{P[\mathcal{A}\mathcal{B}_1 + \mathcal{A}\mathcal{B}_2 + \mathcal{A}\mathcal{B}_3 + \mathcal{A}\mathcal{B}_5]}{P[\mathcal{B}_1 + \mathcal{B}_2 + \mathcal{B}_3 + \mathcal{B}_5]}$$

$$= \frac{P[AB_1] + P[AB_2] + P[AB_3] + P[AB_5]}{P[B_1] + P[B_2] + P[B_3] + P[B_5]}$$

$$= \frac{2/24 + 4/24 + 5/24 + 1/24}{3/24 + 7/24 + 7/24 + 2/24} = \frac{12}{19}.$$

We see that the conditional probability of A given event C is greater than the unconditional probability of A.

Example 2.10 Suppose each time a coin lands heads up, it halves its probability of landing heads up . Suppose we start with a fair coin. What is the probability that on the second toss the coin lands heads up? What is this probability given that the coin lands heads up on the first toss?

To answer the first question we note that we can partition the set of all possible outcomes according to the outcome of the first toss. If the first toss is heads then the probability that the coin will land heads up on the second toss is $1/4$, whereas if the first toss is tails then it is $1/2$. Thus using the law of total probability we have that the probability that the second toss is heads is

$$P[\text{Heads on Second Toss}] = \frac{1}{2} \cdot \frac{1}{4} + \frac{1}{2} \cdot \frac{1}{2} = \frac{3}{8}.$$

If we condition on the fact that the first outcome is heads, however, we have that

$$P[\text{Heads on Second Toss} \,|\, \text{Heads on First Toss}] = \frac{1}{4}.$$

Calculations with conditional probabilities that are difficult to interpret can proceed mechanically to yield correct results. Consider the following example.

Example 2.11 (Mechanical Calculation of Probabilities) Using the sets defined in Example 2.8 we wish to calculate $P[AB_3 \,|\, B_1 + B_2^c]$. We do this by using (2.12) to write

$$P[AB_3 \,|\, B_1 + B_2^c] = \frac{P[AB_3(B_1 + B_2^c)]}{P[B_1 + B_2^c]}.$$

We use the distributed set operation and (2.8) to calculate the numerator as

$$P[AB_3(B_1 + B_2^c)] = P[AB_3B_1 + AB_3B_2^c]$$

$$= P[AB_3B_1] + P[AB_3B_2^c] - P[AB_3B_1B_2^c]$$

$$= 0 + 5/24 - 0 = 5/24.$$

The denominator can be calculated by

$$P\left[\mathcal{B}_1 + \mathcal{B}_2^c\right] \;=\; P\left[\mathcal{B}_1\right] + P\left[\mathcal{B}_2^c\right] - P\left[\mathcal{B}_1\mathcal{B}_2^c\right]$$

$$=\; 3/24 + 17/24 - 3/24 = 17/24.$$

The resultant probability is thus given by

$$P\left[\mathcal{AB}_3 \mid \mathcal{B}_1 + \mathcal{B}_2^c\right] = 5/17.$$

The law of total probability is often stated in terms of conditional probabilities. We record this by stating it as a corollary.

Corollary 2.12 (Law of Total Probability - Conditional Form)
Let $\mathcal{B}_i, 1 \leq i \leq n, n \geq 1$, be a partition of Ω. Then for any event $\mathcal{A}$ we can write

$$P\left[\mathcal{A}\right] = \sum_{i=1}^{n} P\left[\mathcal{A} \mid \mathcal{B}_i\right] P\left[\mathcal{B}_i\right], \quad n \geq 1. \tag{2.13}$$

The usefulness of using conditional probabilities to calculate the probability of a given event cannot be overemphasized. We will use it repeatedly to demonstrate how to reduce what seems to be a complex problem into one that has a straightforward solution. As a general rule, to calculate the probability of event $\mathcal{A}$ one attempts to find a partition $\mathcal{B}_i$ such that both $P\left[\mathcal{A} \mid \mathcal{B}_i\right]$ and $P\left[\mathcal{B}_i\right]$ are easily calculated. The approach is frequently performed recursively, that is, the calculation of $P\left[\mathcal{B}_i\right]$ is also calculated by conditioning on another partition. We will consider several examples of this approach later in the book. First we establish some additional results of conditional probability.

2.3.2
Independence

Inspection of the definition given in (2.12) shows that conditional probabilities are symmetric, that is, provided that both $\mathcal{A}$ and $\mathcal{B}$ have positive probability, then

$$P\left[\mathcal{B} \mid \mathcal{A}\right] P\left[\mathcal{A}\right] = P\left[\mathcal{AB}\right] = P\left[\mathcal{A} \mid \mathcal{B}\right] P\left[\mathcal{B}\right]. \tag{2.14}$$

From this simple equation we can infer several interesting consequences. First we say that events $\mathcal{A}$ and $\mathcal{B}$ are *independent* if and only if

$$P\left[\mathcal{AB}\right] = P\left[\mathcal{A}\right] P\left[\mathcal{B}\right]. \tag{2.15}$$

It is clear from the definition that independence is a symmetric relationship. If $\mathcal{B}$ has positive probability and if $\mathcal{A}$ and $\mathcal{B}$ are independent then (2.14) shows that

$$P\left[\mathcal{A} \mid \mathcal{B}\right] = P\left[\mathcal{A}\right].$$

When two events are not independent we say that they are *dependent*.

Example 2.13 (Independence) Consider the events defined in Example 2.8 and let an experiment consist of selecting a point. Consider two experiments and let $(\mathcal{C}_1, \mathcal{C}_2)$ be the outcomes from each experiment. If the point selected in experiment i is in set $\mathcal{B}_j$ then $\mathcal{C}_i = j$. The event "both points are in set 6," denoted by $\mathcal{C}$, corresponds to $(\mathcal{C}_1, \mathcal{C}_2) = (6, 6)$. If the two experiments are not related to each other, then clearly the outcomes are independent, and thus we have

$$P[\mathcal{C}] = P[(\mathcal{C}_1, \mathcal{C}_2) = (6, 6)] = (P[\mathcal{B}_6])^2 = \left(\frac{2}{24}\right)^2.$$

If, on the other hand, the two experiments do relate then independence does not necessarily hold. For example, if the two points selected by the two experiments must be different, then

$$P[\mathcal{C}] = \frac{2}{24} \frac{1}{23} \neq (P[\mathcal{B}_6])^2$$

and thus the two experiments are dependent.

It is tempting to believe that independence of two events implies that they correspond to different experiments. This is, however, not the case, as the following example shows:

Example 2.14 (Independent but Disjoint Events) If events $\mathcal{A}$ and $\mathcal{B}$ are disjoint it implies that both events cannot simultaneously occur, and this suggests that they should be dependent. If, however, $P[\mathcal{A}] = 0$ then

$$P[\mathcal{AB}] = 0 = P[\mathcal{A}] P[\mathcal{B}]$$

and thus $\mathcal{A}$ and $\mathcal{B}$ are independent events. For example, if we modified the assignment of probabilities in Example 2.8 so that the points of set $\mathcal{B}_1$ were given 0 probability, then $\mathcal{A}$ and $\mathcal{B}_1$ would be independent but, $\mathcal{AB}_1 \neq \emptyset$.

It might appear that Example 2.14 is a contrived example since the probability of event $\mathcal{A}$ being 0 seems to imply that $\mathcal{A}$ can never occur and is therefore of no interest. This conclusion, however, is not true. In Section 4.5.2 we will provide an example of an event that has 0 probability but does occur. It is interesting to note that events can be pairwise independent without necessarily being independent. A simple example of this is given below.

Example 2.15 (Pairwise Independence $\not\Rightarrow$ Independence) Events can be pairwise independent but not independent. Consider the following example of tosses of a fair coin. Let $\mathcal{A}$ be the event that heads is obtained in the first toss, $\mathcal{B}$ that heads is obtained in the second toss, and $\mathcal{C}$ that exactly one head is obtained in both tosses. Then

$$P[\mathcal{A}] = P[\mathcal{B}] = \frac{1}{2},$$

and

$$P[\mathcal{AB}] = P[\mathcal{BC}] = P[\mathcal{AC}] = \frac{1}{4},$$

and thus all events are pairwise independent. It is clear, however, that

$$P[\mathcal{ABC}] = 0 \neq \frac{1}{8} = P[\mathcal{A}]\,P[\mathcal{B}]\,P[\mathcal{C}]$$

and thus events $\mathcal{A}$, $\mathcal{B}$, and $\mathcal{C}$ are not independent.

2.3.3
Causation

It is often tempting to express the relationship between two events in terms of cause and effect. One thinks of one event as "causing" another event. It is clear, however, that if event $\mathcal{X}$ causes event $\mathcal{Y}$ then it does not necessarily imply that event $\mathcal{Y}$ causes event $\mathcal{X}$. Thus causation is not a symmetric relationship. This points out the fallacy of thinking of probabilities in these terms. Independence or dependence between two random events is *always a symmetric relationship*, and one can never say that event $\mathcal{A}$ depends on event $\mathcal{B}$ but that $\mathcal{B}$ is independent of $\mathcal{A}$.

2.3.4
Bayes' Rule

Another consequence of the symmetry of conditional probabilities is called *Bayes' rule*, which is derivable from the law of total probability. From (2.14), we can write

$$P[\mathcal{B}\,|\,\mathcal{A}] = \frac{P[\mathcal{A}\,|\,\mathcal{B}]\,P[\mathcal{B}]}{P[\mathcal{A}]}.$$

If $\mathcal{B}_i, 1 \leq i \leq n$, is a partition then we can use the conditional form of the law of total probability (2.13) to write

$$P[\mathcal{B}_i\,|\,\mathcal{A}] = \frac{P[\mathcal{A}\,|\,\mathcal{B}_i]\,P[\mathcal{B}_i]}{\sum_{k=1}^{n} P[\mathcal{A}\,|\,\mathcal{B}_k]\,P[\mathcal{B}_k]} \quad i = 1, 2, \ldots, n, \quad \textbf{Bayes' Rule.}$$

$$(2.16)$$

The event $\mathcal{A}$ has positive probability and can occur in any of n mutually exclusive ways as indicated by events $\mathcal{B}_i, 1 \leq i \leq n$. Equation (2.16) establishes the probability that, given event $\mathcal{A}$, event $\mathcal{B}_i$ occurred. It is tempting to think, in this context, that $\mathcal{B}_i$ "caused" $\mathcal{A}$, but the above paragraphs point out the pitfalls of this viewpoint. A simple example can best explain the use of this rule.

Example 2.16 We have two coins, one that is fair and thus lands heads up with probability $1/2$ and one that is unfair and lands heads up with probability p, $p > 1/2$. One of the coins is selected randomly and tossed n times yielding n straight tails. What is the probability that the unfair coin was selected?

First, note that if one was not told the outcome of the toss, then the probability of selecting the unfair coin is $1/2$. Our question concerns how this probability changes given the outcome of the n tosses. As a simple case where these probabilities differ assume that $p = 1$. Since the coin came up tails we know immediately that the probability of the unfair coin is 0.

To answer the posed question we use Bayes' rule. We can partition the set of events into disjoint sets of events according to which coin was used in the experiment. Let $\mathcal{B}_1$ (resp., $\mathcal{B}_2$) be the event that the fair (resp., unfair) coin was selected, and let $\mathcal{A}$ be the event that n successive tails were tossed. By hypothesis we have

$$P[\mathcal{B}_1] = 1/2, \qquad P[\mathcal{B}_2] = 1/2,$$

$$P[\mathcal{A} \mid \mathcal{B}_1] = (1/2)^n, \quad P[\mathcal{A} \mid \mathcal{B}_2] = (1 - p)^n.$$

Using this in (2.16) yields

$$P[\text{unfair coin} \mid n \text{ tails tossed}] = \frac{(1 - p)^n}{(1 - p)^n + (1/2)^n}. \tag{2.17}$$

It is always good to check some cases with known solutions. For $p = 1/2$, we obtain one half which checks with intuition, and for $p = 1$ the probability equals 0 as mentioned earlier. If $p = 0$ then the unfair coin always lands tails up and substitution into (2.17) yields

$$P[\text{unfair coin} \mid n \text{ tails tossed}] = \frac{2^n}{2^n + 1}. \tag{2.18}$$

This approaches a limiting value of 1 quite rapidly. For example, for $n = 4$, (2.18) yields $16/17 = 0.94$.

2.4 Summary of Chapter 2

To model random systems one must have a concept of probability, and in this chapter we explored two main ways of defining what is meant by a "probability". Our first approach is based on statistics, where the long-term frequency of the occurrence of a set of repeated experiments was used to define probabilities. Such a definition is difficult to use in practice since it is restricted to a space consisting of a finite number of events and is defined by a limit. The axiomatic approach next discussed bypasses these problems by embedding the notions of randomness within a set of axioms. The advantage of this approach is that there is a formal method for deriving consequences of the axioms that satisfy desirable algebraic properties and link up with the frequency approach. Later this linkage is found in the strong law of large numbers (see Section 5.5).

As might be expected, defining a probability space for a performance model of a computer system is often a complicated task. Quantities such as queue lengths, server occupancies, and waiting times that form the core of a typical performance model are often difficult and tedious to specify in terms of their sample spaces and family of sets. In Chapter 4 we will introduce the expedient of a random variable that allows one to avoid the definition of a probability space by dealing directly with the quantities that are of interest. We can bypass the definition of the underlying probability space for a given set of random variables because the existence of this space is guaranteed by a basic theorem of probability (see the end of Section 4.2 and Problem 4.2). The need to specify the probability space typically only arises when one requires detailed comparisons between different probabilistic systems. For most performance models, little is gained from this specification, and one proceeds using probabilistic arguments based only on random variables. This modeling approach using random variables will become more clear as we develop Chapter 4.

Throughout the rest of the book, we consider events to be random and will study them using probability theory. Often events that we encounter that appear to "look" random are either deterministic and predictable or occur through very complicated processes that we cannot fully specify, model, or predict. The set of arrivals to a computer system is a good example of a process that cannot be precisely modeled. For example, I cannot say when I will next check my electronic mailbox. Clearly this depends on many unpredictable things. If someone calls on the phone or stops by my office, or if I decide to take an early lunch I will not read my mail. If I become involved in a tedious calculation and want a diversion, then checking my mail suddenly becomes extremely attractive. How can one create a mathematical model of when someone will walk by my office or when I will become frustrated with a calculation? Despite such difficulties, when systems are sufficiently complex or unpredictable we can often reasonably model them as if they were random.

The justification for making such assumptions and of using probability theory to model such requests lies in the fact that it works well in practice and that deterministic systems can exhibit random behavior (see Appendix A). Probability theory provides a simple tool that can be used to model complex systems that are unpredictable. Other models of greater complexity could certainly be found, but clearly at some point we reach the point of diminishing return. When choosing between two equally good models, we err on the side of the simplest. In physics this is analogous to the principle of least action and, in the philosophy of science, this is a fundamental principle referred to as *Occam's Razor* [116] after William of Occam. This philosophy can be summed up as *"The simplest explanation is best."* Occam's razor underlies everything we do in this book and explains why probability has proven to be so useful in dealing with the illusive phenomena of randomness.

2.4.1 Postscript: Calculating π to a Billion Places

In Section 2.1 we mentioned that the digits of π appear to be randomly distributed even when carried out to 2 billion decimal places. How does one generate such a sequence? (See [106] for a delightful discussion of the results that follow.) The famous Indian mathematician Ramanujan[3] is credited with several results that allow rapid approximation of π. Ramanujan had phenomenal intuition and one rather obscure equality discovered by him is

$$\frac{1}{\pi} = \frac{\sqrt{8}}{9801} \sum_{n=0}^{\infty} \frac{(4n)!\,[1103 + 26390n]}{(n!)^4 396^{4n}}. \tag{2.19}$$

This infinite series yields an accuracy of about 8 decimal digits for each term in the summation. A recursive algorithm discovered by Ramanujan, which could conceivably be programmed on a computer provided one has software that allows arbitrary precision, is given by

$$f_{n+1} = (1 + a_{n+1})^4 f_n - 2^{2n+3} a_{n+1}(1 + a_{n+1} + a_{n+1}^2), \quad n = 0, 1, \ldots, \tag{2.20}$$

where $f_0 = 6 - 4\sqrt{2}$ and

$$a_{n+1} = \frac{1 - (1 - a_n^4)^{1/4}}{1 + (1 - a_n^4)^{1/4}}, \quad n = 0, 1, \ldots$$

and $a_0 = \sqrt{2} - 1$. This *rather obvious* recurrence approaches $1/\pi$ with an error smaller than $16 \cdot 4^{n+1} \exp\{-2\pi 4^{n+1}\}$. To obtain one billion digits of accuracy one only requires 15 steps.

Bibliographic Notes

The foundations of probability are cogently delineated by von Mises [120] (frequency-based) and by Kolmogorov [71] (axiomatic-based). Feller's volume I [34] is perhaps the most cited text on probability. Non-measure-theoretic texts on probability include Chung [15], Clarke and Disney [16], Gnedenko [40], Hoel, Port, and Stone [50], and Ross [100]. Hamming [46] has an entertaining introduction to the subject geared for engineers. The treatment of probability by Rényi [95] is unique in that the sum of probabilities is not constrained to be bounded. More advanced measure-theoretic treatments of probability can be found in Chow and Teicher [14] and in volume II of Feller [35].

---

[3]Srinivasa Ramanujan (1889–1920) was a self taught mathematician who made original contributions to function theory, power series, and number theory.

2.5
Problems for
Chapter 2

Problems Related to Section 2.2

2.1 **[E2]** Prove the set equalities given in Table 2.1.

2.2 **[E2]** Simplify the following expressions:

1. $(\mathcal{A} + \mathcal{B})(\mathcal{B} + \mathcal{C})$.
2. $(\mathcal{A} + \mathcal{B})(\mathcal{A} + \mathcal{B}^c)$.
3. $(\mathcal{A} + \mathcal{B})(\mathcal{A}^c + \mathcal{B})(\mathcal{A} + \mathcal{B}^c)$.

2.3 **[E2]** Show that the following equalities hold:

1. $(\mathcal{A}_1 \mathcal{A}_2 \cdots \mathcal{A}_n)^c = \mathcal{A}_1^c + \mathcal{A}_2^c + \cdots + \mathcal{A}_n^c$.
2. $(\mathcal{A}_1 + \mathcal{A}_2 + \cdots + \mathcal{A}_n)^c = \mathcal{A}_1^c \mathcal{A}_2^c \cdots \mathcal{A}_n^c$.
3. $(\mathcal{A}^c \mathcal{B}^c)^c = \mathcal{A} + \mathcal{B}$.
4. $(\mathcal{A}^c + \mathcal{B}^c)^c = \mathcal{A}\mathcal{B}$.

2.4 **[E3]** Consider an infinite sequence of coin tosses and let $\mathcal{A}_i$ be the event that the coin lands heads up on the ith toss, for $i = 1, 2, \ldots$. Write expressions for the following events:

1. The last heads occurs on the kth step for $k = 0, 1, \ldots$. (The $k = 0$ case corresponds to no heads landing up.)

2. The coin lands heads up exactly two times.

3. The coin alternates between heads and tails.

4. The coin never lands heads up k times in a row.

Problems Related to Section 2.3

2.5 **[E1]** Numbers from the set $\{1, 2, 3\}$ are selected one at a time without replacement. Let an *occurrence* at step i, for $i = 1, 2, 3$, be the event that on the ith selection the number i is selected. What is the probability that there are no occurrences? Can you do the same problem for the set $\{1, 2, \ldots, 10\}$? (See Example 3.26 for a general solution.)

2.6 **[E2]** Suppose there are n boxes each containing $m > n$ balls. In box i, $i = 1, 2, \ldots, n$ there are i red balls and $m - i$ green balls. From one of the boxes, chosen equally likely at random, a ball is selected. It turns out that the ball is red. What is the probability that it came from the jth box?

2.7 **[E2] (Continuation)** Suppose in Problem 2.6 that $n = 4$ and $m = 5$ and 2 red balls are selected from one box. What is the probability that the third box was selected?

2.8 **[E2]** One coin has heads on both sides and another is a fair coin. A coin is selected randomly and is tossed m times yielding m heads. What is the probability that the coin is fair?

2.9 [**E3**] Suppose each time a coin is tossed it decreases its probability of coming up heads by a factor of 1/2. Assume that the coin lands heads up with probability p on the first toss.

1. What is the probability that the first m tosses are all heads?

2. Suppose that exactly one head showed up in the first m tosses. What is the probability that the head occurred at the nth toss for $1 \leq n \leq m$.

2.10 [**M3**](**Continuation**) Consider the experiment of Problem 2.9.

1. Write an expression for the probability that the coin never lands heads up. Does this event have positive probability?

2. Suppose that each time the coin is tossed and lands heads up its probability of landing heads up is reset back to p (if it lands tails up we assume that its probability of landing heads up is halved). Let $\mathcal{A}$ be the event that after some finite number of tosses the coin never lands heads up. Does $\mathcal{A}$ have positive probability?

2.11 [**E3**] Consider Example 2.8 and let event $\mathcal{C} = \{(x, y) : x = 2, 3\}$. Calculate the following probabilities.

1. $P[\mathcal{A}\mathcal{C}]$.

2. $P[\mathcal{A} + \mathcal{C}]$.

3. $P[\mathcal{A} + \mathcal{C} \,|\, \mathcal{B}_3]$.

4. $P[\mathcal{A} + \mathcal{C}^c \,|\, \mathcal{B}_2]$.

5. $P[\mathcal{A} \,|\, (\mathcal{B}_3 + \mathcal{B}_5)^c]$.

6. $P[\mathcal{C} \,|\, \mathcal{B}_6^c \mathcal{B}_4^c]$.

7. $P[\mathcal{B}_5 \,|\, \mathcal{A}\mathcal{C}]$.

8. $P[\mathcal{B}_6 \,|\, \mathcal{A}^c + \mathcal{C}]$.

2.12 [**M2**](**Continuation of Problem 2.4**) Consider the experiment in Problem 2.4 consisting of an infinite sequence of coin tosses. Suppose the coin is fair. Which of the events defined in that example have 0 probability?

2.13 [**E2**] (**Exclusion–Inclusion**) Show that equation (2.9) can be generalized to the $n = 3$ case as

$$P\left[\sum_{i=1}^{3} \mathcal{A}_i\right] = \sum_{1 \leq i \leq 3} P[\mathcal{A}_i] - \sum_{1 \leq i_1 < i_2 \leq 3} P[\mathcal{A}_{i_1} \mathcal{A}_{i_2}] + P[\mathcal{A}_1 \mathcal{A}_2 \mathcal{A}_3].$$

What is the expression for general n?

2.14 [**M3**] In the baseball world series two teams play against each other until one team wins a total of 4 games. Observe that this implies that at least 4 games and at most 7 games are played to determine the winner of the series. Suppose that team 1 has a probability of $p, 0 < p < 1$, and team 2 has a probability of $1 - p$ of winning any given game (note that teams are often evenly matched and thus $p \approx 0.5$). A few of the questions that can be asked about the probabilistic outcome of the series include (the results of future chapters allow all of these questions to be answered easily; do as many of the problems as you can at this point):

1. What is the probability that team 1 wins the series?

2. What is the probability that team 1 wins the series playing 5 games?

3. Given that team 1 won the series, what is the probability that 5 games were played?

4. What is the probability that the series lasts exactly 5 games?

5. What is the probability that no team wins two games in a row?

6. Suppose that 5 games have already been played without any team winning the series. What is the probability that team 1 will win the series?

7. What is the probability that team 1 won game 3 given that it won the series?

3

Combinatorics

Probabilistic arguments can often be reduced to techniques where one enumerates all possibilities. These techniques are based on the Law of Total Probability and are most useful when there is a set of equally probable basic events and when events of interest consist of combinations of these basic events. In this chapter we focus on such enumerative techniques and derive formal counting techniques collectively called *combinatorics*. We define four different counting paradigms in Section 3.2, which are expressed in terms of selecting distinguishable balls from an urn or, equivalently, in terms of allocating indistinguishable balls to different urns. To derive expressions for the number of different possibilities for each counting paradigm, we establish recurrence relationships between the number of possibilities in the initial problem to that of a similar, but smaller, problem. This type of argument corresponds to partitioning the counting space into disjoint sets, which are smaller and easier to count.

Counting arguments can become exceedingly complex, and the examples that we use to exemplify the different paradigms motivate the need for a systematic methodology. This leads to Section 3.3 where generating functions are defined. These functions not only allow one to systematically count combinatorial events, they are also generally useful for solving recurrence equations. In later chapters, generating functions will be used to solve problems, and their introduction here emphasizes the fact that they are simply formal counting mechanisms. This characterization is sometimes not obvious when generating functions are used in probability theory.

It is of some interest to note that most of the results in this chapter were known to mathematicians of the late 17th and early 18th centuries. Where appropriate, we will indicate some of the intriguing historical aspects of the subject. The beginning of the field of probability can be traced to a correspondence started in 1654 between Blaise Pascal (1623–1662) and Pierre Fermat (1601?–1665)[1]. The nature of their correspondence dealt with problems of gambling (to make the subject more respectable, shall we say "games of chance") posed to Pascal by his friend Chevalier de Méré, a professional gambler (the correspondence can be found in [109]). The questions posed by de Méré dealt with the **problem of points**, which required the determination of how stakes of an interrupted fair game of chance should be divided between two players (in Problem 3.1 you can try your hand on such a problem). The solution to this problem follows from commonsense reasoning but requires counting the number of ways certain outcomes can occur. The field of combinatorics was thus born from problems arising from practical considerations!

Although neither Pascal nor Fermat ever published their correspondence, the first book on probability, *Deratiociniis in Ludo Aleae* (*On Reasoning in Games of Dice*), was published by Christiaan Huygens (1629–1695) in 1657 and was based on the Pascal–Fermat correspondence. Huygen's book was included as a chapter in a book on probability authored by Jacob Bernoulli (1654–1705) called *Ars Conjectandi* that was published posthumously in 1713. This book contains the first unified treatment of combinatorics, and many of the results in this chapter are either contained in *Ars Conjectandi* or have their seeds within its pages. In 1718 Abraham de Moivre (1667–1754) published a book on probability titled *The Doctrine of Chance* (a facsimile of the third edition can be found in [27]). De Moivre made fundamental contributions to the field of probability and is now given credit for many results initially ascribed to other mathematicians. Further historical notes are found in the summary of Chapter 5.

Next we introduce the four counting paradigms considered by Bernoulli. Useful combinatorial equalities and inequalities are found in Appendix B.

---

[1]Fermat is perhaps best known for his famous *last theorem* which states that there exists no solution to the equation $x^n + y^n = z^n$ for x, y, z, and n integers and n greater than 2. Fermat claimed to have a proof that was a bit too long to fit within the margins of the page he was reading. The theorem has plagued mathematics since Fermat's time, and many great mathematicians have spent considerable efforts to find a proof. A proof for this theorem has been recently reported (June 1993); if correct, this will (sadly) set the issue finally at rest.

3.1
Different
Counting
Paradigms

To motivate the different counting paradigms, we will frequently phrase random selections in terms of how balls can be allocated to, or drawn from, urns. Balls are said to be *distinguishable* if each ball can be uniquely identified. It is convenient in this case to label the balls with sequential integers. If balls are *indistinguishable* then they cannot be uniquely identified. We make a similar distinction for urns, and when urns are distinguishable we will also label them with sequential integers. There are two related ball and urn problems that express most of the combinatorial relations that we will need.

Selection: Select k balls from an urn containing n distinguishable balls.

Allocation: Allocate k indistinguishable balls to a set of n distinguishable urns.

The relationship between these two types of problems will be made clear in what follows and, without loss of generality, we first express results in terms of the selection problem. We will consider two different replacement policies and two different methods of counting events as described below. This creates a total of four different counting paradigms.

Replacement policies are concerned with the placement of selected balls. We distinguish two such policies. If each time a ball is selected it is returned to the urn then we say that the selection is *with replacement*. With this policy it is possible to select the same ball multiple times. On the other hand, if each time a ball is selected it is removed from the urn then we say the selection is *without replacement*.

For each replacement policy, events can be counted in two different ways depending on whether we distinguish between the ordering in which balls are selected or only distinguish the set of balls selected irrespective of their ordering. With *permutations* we count the selection of ball i followed by ball j as being different from the selection of ball j followed by ball i. On the other hand, with *combinations* we do not distinguish selections by their order. Thus the selection of ball i followed by ball j is the same as the selection of ball j followed by ball i.

3.1.1
Counting
Methodology

In the next section we derive formulas for the four counting paradigms delineated above. We use a unified approach to derive these formulas and to derive solutions to combinatorial problems posed in this chapter. The unified approach is based on a very simple observation about combinatorial problems in general; one can often break a large problem into a set of similar problems that are smaller in size. In computer science this technique is typically referred to as *divide-and-conquer*. A typical example of this approach in computer science is that of sorting a list of numbers.

Example 3.1 (Recurrence for Sorting n Numbers) One way to sort

a list of n numbers is to divide the numbers into two groups of size $\lceil n/2 \rceil$ and $\lfloor n/2 \rfloor$, respectively, and then to sort each group separately. These two sorted groups can then be merged into a final sorted list. If we let $f(n)$ be the number of comparisons required to sort n numbers in this manner then it is clear that

$$f(1) \;\; = \;\; 0, \tag{3.1}$$

$$f(n) \;\; = \;\; f(\lceil n/2 \rceil) + f(\lfloor n/2 \rfloor) + n - 1, \quad n = 2, 3, \dots . \tag{3.2}$$

The boundary equation (3.1) follows from the observation that no comparisons are required to sort 1 number. The terms $f(\lceil n/2 \rceil)$ and $f(\lfloor n/2 \rfloor)$ in the general equation (3.2) count the number of comparisons required to sort the two smaller lists and the term $n - 1$ counts the number of comparisons required to merge the two lists. The solution to this recurrence is given by (see Problem 3.5)

$$f(n) = \sum_{i=1}^{n} \lceil \log_2 i \rceil . \tag{3.3}$$

In combinatorics, the technique of breaking a large problem into a set of smaller problems consists of finding a partition of the state space and then deriving recurrence relationships that express the formula in terms of these smaller problems. This method can be applied to most problems in combinatorics and often reveals a semantic interpretation of a formula. For this reason such an approach is preferable to a purely algebraic solution. This technique is based on a version of the Law of Total Probability expressed for combinatorial events, which we record as a corollary to Theorem 2.7.

Corollary 3.2 (Law of Total Probability - Combinatorics) Let

$\mathcal{B}_i, 1 \leq i \leq n$, be a partition of the set Ω. Then, for a given counting paradigm, the number of different ways event $\mathcal{A}$ can occur is given by

$$N[\mathcal{A}] = \sum_{i=1}^{n} N[\mathcal{AB}_i], \tag{3.4}$$

where $N(\mathcal{C})$ denotes the number of different ways event $\mathcal{C}$ can occur in the specified counting paradigm.

Often we calculate the probability of a certain event using counting arguments. This method is identical to that used by Laplace as mentioned in the introductory section of Chapter 2, and it is applicable if the basic events of interest are equiprobable. Thus the probability of a given event is expressed as the ratio of the number of ways the event can occur to the number of ways all events can occur. As a simple example, if one randomly selects a ball from an urn containing n_1 red balls and n_2 green balls then the total number of selections is $n_1 + n_2$. From this total, the number of possible selections of red balls is n_1, and hence the probability that the selected ball is red, is given by $n_1/(n_1 + n_2)$, provided that all balls are equally likely to be selected. We will use this method to calculate probabilities throughout this chapter.

Observe that the supposition that each basic event is selected with equal probability is a necessary condition for this method to yield correct probabilities. For example, in the above red–green ball example, if red balls were twice as likely to be selected as green balls then the probability of selecting a red ball is not given by $n_1/(n_1 + n_2)$ (see Problem 3.6).

3.2
The Four Counting Paradigms

In this section we will derive expressions for the four counting paradigms defined in the introduction. We assume that there are n balls in an urn that are numbered 1 through n from which k balls are selected. The definition of n *factorial* for nonnegative integer n is given by[2]

$$n! \stackrel{\text{def}}{=} n(n-1)(n-2)\cdots 1, \quad n = 1, 2, \dots . \tag{3.5}$$

By convention we set $0! = 1$. For large values of n, the value of $n!$ can be approximated using Stirling's formula (see (B.7)) as

$$n! \approx \sqrt{2n\pi} \left(\frac{n}{e}\right)^n . \tag{3.6}$$

3.2.1
Permutations without Replacement

We first consider the case of permutations without replacement. Denote the number of different permutations obtained in selecting k balls from an urn containing n distinguishable balls by $P(n, k)$. By convention we set $P(n, 0) = 1, n = 0, 1, \dots .$ To derive a recurrence relationship for $P(n, k)$, $k > 0$, we reduce the problem into two subproblems: the number of ways the first ball can be selected and the number of ways the remaining $k - 1$ balls can be selected. The first selection can be any of the n balls, and it occurs in n distinct ways. This leaves $n - 1$ balls in the urn from which $k - 1$ must be selected. We thus have that

$$P(n, k) = nP(n-1, k-1), \quad k = 1, 2, \dots, n, \ n = 1, 2, \dots . \tag{3.7}$$

[2]It is interesting to note that this notation was originated in 1808 by Christian Kramp (1760–1826) to avoid printing difficulties associated with a previous notation denoted by $\lfloor n$.

Iterating this equation once yields $P(n, k) = n(n-1)P(n-2, k-2)$, and continuing in this fashion yields

$$P(n, k) = (n)_k, \quad k = 0, 1, \ldots, \quad n = 0, 1, \ldots \quad (3.8)$$

where we define

$$(n)_k \stackrel{\text{def}}{=} n(n-1)\cdots(n-k+1), \quad k = 1, 2, \ldots, n, \quad n = 1, 2, \ldots . \quad (3.9)$$

By convention we set $(n)_0 = 1, \quad n = 0, 1, \ldots .$

Example 3.3 (Permutations without Replacement) Suppose that $n = 4$ and $k = 2$. Then $P(4, 2) = 12$ and the different permutations of selected balls are given by

$$12, \quad 13, \quad 14, \quad 23, \quad 24, \quad 34,$$
$$21, \quad 31, \quad 41, \quad 32, \quad 42, \quad 43.$$

We now derive a useful identity. Consider selecting k balls without replacement from an urn containing n distinguishable balls. Partition the state space into disjoint sets depending on whether a particular ball j is, or is not, selected. If ball j is not selected then the number of ways to select k balls from the remaining $n-1$ is given by $P(n-1, k)$. On the other hand, if ball j is selected then the remaining $k-1$ balls are selected from the remaining $n-1$ balls. Since ball j can be selected in any of k positions, this yields a total number of selections as $kP(n-1, k-1)$. We have thus argued that

$$P(n, k) = P(n-1, k) + kP(n-1, k-1), \quad k = 1, 2, \ldots, n, \quad (3.10)$$

which, using (3.8), implies that

$$(n)_k = (n-1)_k + k(n-1)_{k-1}, \quad k = 1, 2, \ldots, n. \quad (3.11)$$

Equation (3.10) can be established algebraically but such a derivation provides little insight into why the relationship holds. Tabulated values of $(n)_k$ are given in Table 3.1 (entries that are zero or undefined are left blank in this and in following tables).

Next we consider permutations with replacement.

3.2.2 Permutations with Replacement

Let $P^r(k, n)$ denote the number of different ways that k balls can be selected from n distinguishable balls using a replacement policy. We can picturesquely rephrase the problem as follows. Assume that balls occur in n different colors, that we have an infinite number of balls

$n \setminus k$	1	2	3	4	5	6	7	8	9
1	1								
2	2	2							
3	3	6	6						
4	4	12	24	24					
5	5	20	60	120	120				
6	6	30	120	360	720	720			
7	7	42	210	840	2,520	5,040	5,040		
8	8	56	336	1,680	6,720	20,160	40,320	40,320	
9	9	72	504	3,024	15,120	60,480	181,440	362,880	362,880

TABLE 3.1. Table of Values of $(n)_k$

of each color, and that k are selected and arranged in a sequence. The number of different possible sequences is given by $P^r(n, k)$ and obviously $P^r(n, 1) = n$. To derive the general case, we observe that the first ball can be selected in n different ways and, since the ball is replaced, this also holds for all subsequent balls. Thus

$$P^r(n, k) = nP^r(n, k - 1), \quad k = 1, 2, \ldots, \quad n = 1, 2, \ldots, \quad (3.12)$$

where we set $P^r(n, 0) = 1$ for $n = 0, 1, \ldots$. Iterating this equation once yields $P^r(n, k) = n^2 P^r(n - 2, k - 2)$ and continuing to iterate yields,

$$P^r(n, k) = \langle n \rangle_k, \quad k = 1, 2, \ldots, \quad n = 1, 2, \ldots, \quad (3.13)$$

where we define

$$\langle n \rangle_k \stackrel{\text{def}}{=} n^k, \quad k = 1, 2, \ldots, \quad n = 1, 2, \ldots \ .$$

By definition we set $\langle n \rangle_0 = 1, n = 0, 1, \ldots$. We term these expressions *factorial-R* coefficients. We write these as $\langle n \rangle_k$ instead of as n^k as a reminder that the result is obtained from a *combinatorial experiment* that uses *replacement*. This angle notation is also used when dealing with combinations with replacement in Section 3.2.10.

Example 3.4 (Permutations with Replacement) Suppose that $n = 4$ and $k = 2$. Then $P^r(4, 2) = 16$ and the different permutations of selected balls with replacement are given by

$$11, \quad 12, \quad 13, \quad 14, \quad 22, \quad 23, \quad 24, \quad 34,$$
$$21, \quad 31, \quad 41, \quad 32, \quad 33, \quad 42, \quad 43, \quad 44.$$

Tabulated values of $\langle n \rangle_k$ are given in Table 3.2.

A novel application of permutations is found in the *birthday problem*.

$n \setminus k$	1	2	3	4	5	6	7	8
1	1	1	1	1	1	1	1	1
1	2	4	8	16	32	64	128	256
3	3	9	27	81	243	729	2,187	6,561
4	4	16	64	256	1,024	4,096	16,384	65,536
5	5	25	125	625	3,125	15,625	78,125	390,625
6	6	36	216	1,296	7,776	46,656	279,936	1,679,616
7	7	49	343	2,401	16,807	117,649	823,543	5,764,801
8	8	64	512	4,096	32,768	262,144	2,097,152	16,777,216
9	9	81	729	6,561	59,049	531,441	4,782,969	43,046,721

TABLE 3.2. Table of Values of $\langle n \rangle_k$

Example 3.5 (Birthday Problem) Assume that birthdays occur uniformly over all 365 days of the year. In a group of k people what is the probability that no two people have the same birthday?

To answer the question, observe that the total number of birthdays equals the number of permutations obtained with replacement, $\langle 365 \rangle_k$, and the total number of possibilities where all birthdays are unique equals the number of permutations obtained without replacement, $(365)_k$. Thus the answer, denoted by $P(k)$, is given by

$$P(k) = \frac{(365)_k}{\langle 365 \rangle_k}$$

$$= \left(1 - \frac{1}{365}\right)\left(1 - \frac{2}{365}\right) \cdots \left(1 - \frac{k-1}{365}\right). \qquad (3.14)$$

Some typical values are illuminating (see Figure 3.1). For $k = 23$ the probability that no two people have the same birthday is less than $1/2$ and for $k = 56$ it has decreased to 0.01. Unless one has encountered this problem before these probabilities seem much too small. In Problem 3.7 we explore some consequences of relaxing the uniformity assumption on the occurrences of birthdays.

3.2.3 Coincidences

If we generalize the birthday problem so that there are n possible events then the probability that no two events are the same (such events are called *coincidences*) is given by

$$f_c(n, k) \overset{\text{def}}{=} (n)_k / \langle n \rangle_k = \left(1 - \frac{1}{n}\right)\left(1 - \frac{2}{n}\right) \cdots \left(1 - \frac{k-1}{n}\right), \qquad (3.15)$$

and it is easy to see that

$$\lim_{n \to \infty} f_c(n, k) = 1. \qquad (3.16)$$

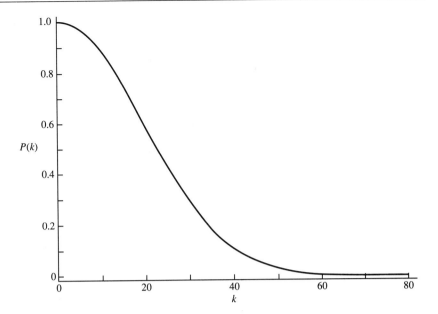

FIGURE 3.1. Probability of Unique Birthdays

Another way to interpret (3.16) is that, for any fixed k and sufficiently large n, there is little difference between permutations with and without replacement. This can be explained by observing that if n is large with respect to k then the probability of picking the same item more than once is small.

Expressions like (3.15) are often difficult to manipulate and frequently it is convenient to work with bounds that provide insight and are mathematically tractable. To establish a bound on (3.15) we use the inequality (see (B.10))

$$1 - x < e^{-x}, \quad x > 0. \tag{3.17}$$

Substituting (3.17) into (3.15) implies that

$$f_c(n, k) \; < \; e^{-1/n} e^{-2/n} \cdots e^{-(k-1)/n} \; = \; \exp\left[-k(k-1)/(2n)\right].$$

To determine an approximation for the value of k that satisfies $f_c(n, k) = q$, we can solve

$$\ln q = -\frac{k(k-1)}{2n}, \quad 0 < q < 1.$$

Simplifying this yields

$$k^2 - k + 2n \ln q = 0.$$

The quadratic formula implies that (we only take the positive root)

$$k = \frac{1 + \sqrt{1 - 8n \ln q}}{2}. \tag{3.18}$$

As an example, for $q = 1/2$ and $n = 365$ the closest integer solution to (3.18) is given by $k \approx 23$ (the actual values are $f_c(365, 23) = 0.4927$ and $f_c(365, 24) = 0.5243$).

We next consider counting arguments that only depend on the number of possibilities regardless of the ordering in which they were obtained.

3.2.4 Combinations without Replacement

Consider the case where balls are selected without replacement and let $C(n, k)$ denote the number of ways that k balls can be selected from n balls irrespective of their order. It is clear that $C(n, 1) = n$, $C(n, n) = 1$ and $C(n, k) = 0$ for $k > n$. To derive the general case we develop a recurrence relation for $C(n, k)$ by considering a partition of events. Consider a particular ball j that is either in the selected group of k or not in the group. If it is not in the group then it can be thought of as being eliminated from the urn, and thus the k balls are drawn from an urn containing $n - 1$ balls in $C(n - 1, k)$ different ways. If ball j is in the selected group then from the remaining $n - 1$ balls $k - 1$ must be selected. This situation can be obtained in $C(n - 1, k - 1)$ ways. Combining the above yields

$$C(n, k) = C(n - 1, k) + C(n - 1, k - 1), \quad k = 1, 2, \ldots, n. \tag{3.19}$$

It is interesting to see that the only difference between the recurrences for permutations (3.10) and that for combinations (3.19) is that the $(n - 1, k - 1)$ term is multiplied by k in the permutation recurrence. This arises from the fact that permutations count the k different ways that ball j can be selected.

To solve (3.19) we argue as follows. Each combination of k items can be arranged in $k!$ different permutations. Therefore there should be $k!$ times as many permutations as combinations or

$$k!C(n, k) = P(n, k).$$

Using (3.8) implies that

$$C(n, k) = \frac{(n)_k}{k!},$$

which can easily be shown to satisfy (3.19). Thus we can write

$$C(n, k) = \binom{n}{k}, \quad k = 0, 1, \ldots, n, \tag{3.20}$$

where we define

$$\binom{n}{k} \stackrel{\text{def}}{=} \frac{(n)_k}{k!} = \frac{n!}{k!\,(n-k)!}, \quad k = 0, 1, \ldots, n, \quad n = 0, 1, \ldots . \quad (3.21)$$

The expression in (3.21) is called a *binomial coefficient*.

Example 3.6 (Combinations without Replacement) Suppose that $n = 4$ and $k = 2$. Then $C(4, 2) = 6$ and the different combinations of selected balls without replacement are given by

$$12, \quad 13, \quad 14,$$
$$23, \quad 24, \quad 34.$$

Example 3.7 (A Simple Closed Queueing Network) Consider a closed queueing network consisting of n service centers and k customers for $n \geq k$. Assume that at most one customer can be at any service center at any given time. The state of the system consists of all possible allocations of customers to service centers. How many different states are there?

To answer the question, note that if we let $k_i = 0, 1$ for $i = 1, 2, \ldots, n$, be the number of customers allocated to service center i, then all possibilities satisfy $k = \sum_{i=1}^{n} k_i$. The problem is thus equivalent to the number of vectors of size n that have nonnegative components less than or equal to 1 (i.e., they are *binary vectors*) that sum to k. If we associate balls with customers and service centers with urns then it is clear that this problem is equivalent to combinations without replacement. Thus the answer is given by $C(n, k)$.

Example 3.8 (Allocation Problem) Assume that k indistinguishable bricks are allocated to n distinguishable boxes so that no box contains more than one brick. How many combinations are there? To determine the number of possible combinations we observe that if we view the experiment backwards in time then we can rephrase the question as the number of possibilities obtained from selecting k boxes from a set of n distinguishable boxes. This implies that the solution is given by $C(n, k)$.

3.2.5 Problems Equivalent to Combinations without Replacement

We can consider the brick–box problem of Example 3.8 to be the *dual problem* of the ball–urn problem where we select k distinguishable balls from an urn containing n balls. To do this, imagine that boxes correspond to balls and bricks correspond to an urn and run the brick–box experiment backwards in time. This line of reasoning and the preceding examples show that the following are equivalent problems with solution $C(n,k)$:

1. The number of ways to select, without replacement, k distinguishable balls from an urn containing n distinguishable balls.

2. The number of ways to put k indistinguishable balls into n distinguishable urns so that no more than 1 ball is in any urn.

3. The number of binary vectors $(k_1, \ldots, k_n)$ that satisfy $k = \sum_{i=1}^{n} k_i$.

Thus we see the relationship between selection and allocation problems as mentioned in the introduction.

Example 3.9 (Particular Allocations) Suppose we distribute k balls into n urns so that no urn contains more than 1 ball. For a given set of indices, $1 \le i_1 < i_2 < \cdots < i_k \le n$, what is the probability that urns $i_j, j = 1, 2, \ldots, k$ are occupied? To answer this, we first note that there are $C(n,k)$ possible ways to distribute the k balls in the n urns and each of these possibilities is equally likely. Hence the probability of any particular allocation is given by $1/C(n,k)$.

Example 3.10 (Probability of Breaking Even in Coin Tossing) We wish to determine the probability that after m tosses of a fair coin there have been the same number of heads as tails. This implies that $m = 2n$ for some n, and since each toss has a probability of $1/2$ of landing heads up we have

$$p_{2n} = \left(\begin{array}{c} 2n \\ n \end{array} \right) \frac{1}{2^{2n}}.$$

For large n we can use Stirling's approximation (3.6) to obtain

$$p_{2n} \approx \frac{\sqrt{4n\pi} \left(\frac{2n}{e} \right)^{2n}}{\left\{ \sqrt{2n\pi} \left(\frac{n}{e} \right)^{n} \right\}^{2}} \frac{1}{2^{2n}} = \frac{1}{\sqrt{n\pi}}. \tag{3.22}$$

Since each player has the same chance of winning, intuition suggests that breaking even should be a frequent event. Equation (3.22), however, shows that the probability of being even after $2n$ tosses decreases to zero at a rate of $1/\sqrt{n}$. The event is anything but frequent. After 20 tosses the probability of being even is 17.6%, and this decreases to 5.6% after 100 tosses.

Example 3.11 (Lattice Paths) Consider a path on a two-dimensional lattice where positive unit steps are taken in either the horizontal or vertical direction (possible steps from (i, j) are $(i+1, j)$ or $(i, j+1)$). How many different paths are there from the point $(0, 0)$ to the point (n_1, n_2)? To calculate this we observe that all paths have exactly $n_1 + n_2$ steps and can be distinguished by the places where horizontal steps are taken. Thus the number of paths is the number of ways we can choose n_1 horizontal steps from a total of $n_1 + n_2$ steps, that is, $C(n_1 + n_2, n_1)$.

Suppose the lattice modeled a network where cross points corresponded to switching centers that were used to route calls. Assume that an incoming call into center (i, j) can be routed to either its "up" center at $(i, j+1)$ or its "across" center $(i+1, j)$. For such a network we would be interested in the number of paths from $(0, 0)$ to (n_1, n_2) that pass through a particular point along the way. This number might, for example, provide some insight into possible congestion at that switching center.

To calculate this, observe that since we only allow unit steps, it is clear that the sum of the coordinate values along the path increases by 1 at each step. Thus all paths start out with a coordinate sum of 0 and end with a coordinate sum of $n_1 + n_2$. The number of paths that intercept the point $(i, j - i)$ for $0 \le i \le n_1$, after j steps for $i \le j \le n_2 + i$, is given by

$$\binom{j}{i} \binom{n_1 + n_2 - j}{n_1 - i}. \tag{3.23}$$

The first term $C(j, i)$ in (3.23) counts all possible paths to point $(i, j - i)$ and the second term $C(n_1 + n_2 - j, n_1 - i)$ counts the paths starting at $(i, j - i)$ and ending at (n_1, n_2). Since all paths must pass through some point $(i, j - i)$ we can count all possibilities by partitioning on j to yield the following equality:

$$\binom{n_1 + n_2}{n_1} = \sum_{i=0}^{n_1} \binom{j}{i} \binom{n_1 + n_2 - j}{n_1 - i}, \quad 0 \le j \le n_1 + n_2. \tag{3.24}$$

Suppose in the switching center one selects a path uniformly over all possible paths. Then (3.23) and (3.24) show that the probability of selecting center $(i, j - i)$ for $0 \le i \le n_1$, $i \le j \le n_2 + i$, is given by

$$\frac{\binom{j}{i} \binom{n_1 + n_2 - j}{n_1 - i}}{\binom{n_1 + n_2}{n_1}}$$

Example 3.12 (Locking in Databases) To ensure exclusive access to records in a database, requests for records must first obtain a lock for that record. Assume that a request must obtain all of the locks it requires before it can begin execution. Furthermore assume that at the end of the transaction all locks it holds are simultaneously released. Suppose that all n records of a database are accessed uniformly.

Consider a transaction, denoted by J, that arrives to a busy system. Suppose that the system has collectively assigned m locks, for $1 \leq m \leq n$, to transactions in execution and assume that J requires k locks. If $k > n - m$ then at least one of J's locks is held by a transaction already in execution and J cannot begin execution. If $k \leq n - m$ then the probability that J can obtain its locks and begin execution is given by

$$\frac{\dbinom{n-m}{k}}{\dbinom{n}{k}}, \quad 1 \leq k \leq n - m.$$

3.2.6 Basic Properties of Binomial Coefficients

Basic properties of binomial coefficients are given by (assume that $0 \leq \ell \leq k \leq n$):

Symmetry

$$\binom{n}{k} = \binom{n}{n-k}$$

Factoring

$$\binom{n}{k} = \frac{n}{k}\binom{n-1}{k-1}$$

Addition

$$\binom{n}{k} = \binom{n-1}{k} + \binom{n-1}{k-1}$$

Product

$$\binom{n}{k}\binom{k}{\ell} = \binom{n}{\ell}\binom{n-\ell}{k-\ell}$$

We can provide probabilistic interpretations of these formulas. After one selects k balls from an urn with n balls without replacement there are $n - k$ balls remaining in the urn. We can equivalently think of having selected which balls remain in the urn and thus establish the symmetry relationship above. The factoring operation shows how one should generate combinatorial values on a computer. Since the values of $n!$ increase rapidly the naive approach of calculating the value of $C(n, k)$ by forming the ratio of the factorials $n!/k!(n - k)!$ quickly overflows a computer's word length. The factoring property shows that a more efficient way to calculate a binomial coefficient is to form a product of simple ratios,

$$\binom{n}{k} = \prod_{j=1}^{k} \frac{n - j + 1}{j} \quad \textbf{Computational Method} \qquad (3.25)$$

We have already discussed the addition formula. To provide a probabilistic interpretation of the product formula consider the following experiment. We select k balls from an urn of n and put them in another urn. From this second urn we select ℓ balls from the k and place them in a third urn. This leaves a set of $k - \ell$ balls remaining in the second urn that cannot have balls in common with the ℓ balls of the third urn. Another way to count the same events is to do an experiment in which we first select ℓ balls from the set of n. We consider these to be the final selected balls but still need to make up a total of k selected balls. To do this, from the remaining $n - \ell$ balls in the urn, we select $k - \ell$. The number of possibilities for both experiments is equal since they count the same events, and thus the product formula is explained.

3.2.7 Binomial Identities

We can derive several binomial identities using (3.19). Iterating equation (3.19) shows that we can expand the recurrence in the following two ways:

$$C(n, k) = \sum_{j=k-1}^{n-1} C(j, k-1) \tag{3.26}$$

$$= \sum_{j=0}^{k} C(n-k-1+j, j). \tag{3.27}$$

Equation (3.26) classifies combinations according to the number of items selected, whereas equation (3.27) is not so easily interpreted. These equations establish the identities

$$\binom{n}{k} = \sum_{j=k-1}^{n-1} \binom{j}{k-1} \tag{3.28}$$

$$= \sum_{j=0}^{k} \binom{n+j-k-1}{j}, \tag{3.29}$$

which are less obvious when established algebraically.

Binomial coefficients have been extensively studied, and there are literally thousands of identities that they satisfy. A list of values provides some insight into the source of many of these relationships. Such a list is called "Pascal's triangle"[3] after the mathematician Blaise Pascal and is shown in Table 3.3.

Some other simple identities can be obtained by exploring Table 3.3. For example, the sum of row k, for $k = 0, 1, 2, 3, \ldots$, is seen to be $1, 2, 4, 8, \ldots$

---

[3]The triangle evidently was known to Chinese mathematicians long before its discovery by Pascal. Books published by Yang Hui in 1261 and 1275 and by Chu Shi-kié in 1299 and 1303 speak of the triangle as being ancient.

$n \setminus k$	0	1	2	3	4	5	6	7	8	9	10
0	1										
1	1	1									
2	1	2	1								
3	1	3	3	1							
4	1	4	6	4	1						
5	1	5	10	10	5	1					
6	1	6	15	20	15	6	1				
7	1	7	21	35	35	21	7	1			
8	1	8	28	56	70	56	28	8	1		
9	1	9	36	84	126	126	84	36	9	1	
10	1	10	45	120	210	252	210	120	45	10	1

TABLE 3.3. Table of Values of $C(n, k)$

thus suggesting that

$$\sum_{k=0}^{n} \binom{n}{k} = 2^n. \tag{3.30}$$

The proof of this readily follows from the binomial formula (B.4) with $x = y = 1$. Notice that the value 20 in row 6 and column 3 equals the sum of all the entries in the second column up until the fifth row, that is, $20 = 0 + 0 + 1 + 3 + 6 + 10$. This is simply (3.28) established above. Also notice that the 70 in the eighth row and fourth column equals the sum of the squares of the fourth row suggesting that

$$\binom{2n}{n} = \sum_{k=0}^{n} \binom{n}{k}^2. \tag{3.31}$$

Also observe the pattern that arises when you sum on a descending diagonal. The third diagonal, consisting of the numbers $1, 4, 10, 20, 35, 56, \ldots$, has partial sums that equal entries just below the diagonal, for example, $1 + 4 = C(5, 1)$, $1 + 4 + 10 = C(6, 2)$. Thus it appears that

$$\binom{n + m + 1}{m} = \sum_{k=0}^{m} \binom{n + k}{k}. \tag{3.32}$$

In Problem 3.18 we ask that you prove the validity of identities (3.31) and (3.32).

Binomial coefficients satisfy so many identities that we could easily tire the most persistent of readers with a list of those that are known. (See [43, 69] for an excellent treatment of the subject.) Instead of compiling such a list, we provide a general recipe by which such identities can be discovered (see Problems 3.17, 3.26, 3.27, and 3.31). This technique uses generating functions that are discussed at length in Section 3.3. Of course, the fact that we devote an entire section to generating functions suggests that they have uses in addition to creating combinatorial identities. We will extensively use them as a mathematical tool to solve problems that can be stated in terms of recurrence relationships.

3.2.8 Multinomial Coefficients

How many ways can k distinguishable balls be distributed into n different urns so that there are k_i balls in urn i? To answer the question we first note that we can pick the k_1 balls for the first urn in $C(k, k_1)$ ways. From the $k - k_1$ remaining balls, the second urn can be filled in $C(k - k_1, k_2)$ ways and the third urn can similarly be filled in $C(k - k_1 - k_2, k_3)$ ways. Continuing on yields

$$\binom{k}{k_1} \binom{k - k_1}{k_2} \cdots \binom{k_n}{k_n} = \frac{k!}{k_1! k_2! \cdots k_n!} \tag{3.33}$$

$$\stackrel{\text{def}}{=} \binom{k}{k_1, k_2, \ldots, k_n}. \tag{3.34}$$

We term the expression in (3.34) a *multinomial coefficient*.

An equivalent problem is the number of distinct sequences of k balls that can be made from $k_i, 1 \leq i \leq n$, *indistinguishable balls* of color i. To see that the solution for this problem is also given by (3.33) we argue as follows. If the k balls were distinguishable then clearly there would be $k!$ different permutations. For any given permutation, however, the k_i balls of color i could be permuted among themselves without affecting the sequence of colors in the permutation. Hence the $k!$ term that counts each of these different sequences is too large by a factor of k_i for each color i. Dividing $k!$ by all factors $k_i!$ yields (3.33).

Let us continue the above discussion to obtain yet another combinatorial identity. Suppose we do not restrict the number of balls in the sequence to be k_i but instead assume that for each color i there are an infinite number of balls of that color that we can select. Assume that balls occur in n different colors and that we have an infinite number of each color. Consider the problem of determining all possible sequences obtained from arranging k selected balls. We have already seen this problem in Section 3.2.2 when discussing permutations with replacement. Hence the number of different sequences is given by n^k. Now we can count these sequences in another way by simply enumerating all the possible ways that the k balls in the sequence can be composed of k_i balls of color i so that $k_1 + k_2 + \cdots + k_n = k$. Expressing this mathematically and equating the two counting techniques yields the following identity:

$$\sum_{k_1,k_2,\ldots,k_n} \binom{k}{k_1,k_2,\ldots,k_n} = n^k, \quad 0 \le k_i,\ 1 \le i \le n,\ \sum_{i=1}^{n} k_i = k.$$

(3.35)

Minor variable substitutions shows that (3.30) is a special case of (3.35).

3.2.9 Partial Sum of Binomial Coefficients

It is appropriate to end this section with an enigmatic problem. Equations (3.28) and (3.32) show that we can obtain closed form expressions for the partial sums of the columns and diagonals of Table 3.3. What about the partial rows of the table, that is, what is an expression for

$$\sum_{k=0}^{m} \binom{n}{k}, \quad m < n.$$

(3.36)

Try as you might you will not be able to write down a closed-form expression for this summation (see Graham et. al., [43], Section 5.7 which is based on Gosper [41] for a lucid description of why this is true). This enigmatic summation will frequently appear throughout the rest of the text.

3.2.10 Combinations with Replacement

We now consider combinations with replacement and denote the number of such combinations when k balls are selected from n balls as $C^r(n,k)$. We set $C^r(n,0) = 1$. It is clear that if $k = 1$ then there are n possibilities, and thus that $C^r(n,1) = n$. Furthermore, if $n = 1$ then, for any value of k, there is only one possibility, and thus $C^r(1,k) = 1$. To develop a general formula we again partition the possible ways in which a particular ball can be counted. Suppose among the k selections that ball j is selected at least one time. The *first time* it is selected and returned to the urn the problem is reduced to that of counting how many ways $k-1$ balls can be selected from a total of n. If, on the other hand, ball j is never selected then the problem is reduced to counting the number of ways that k balls can be selected from $n-1$ balls. As these two events are a partition of the state space, we can write

$$C^r(n,k) = C^r(n,k-1) + C^r(n-1,k), \quad k = 2,3,\ldots,n.$$

(3.37)

It is interesting to see that the only difference between the recurrence for combinations without replacement and for combinations with replacement is that the term corresponding to index values $(n-1,k-1)$ in (3.19) is replaced by a term with index values $(n,k-1)$ in (3.37). This arises from the fact that the selected ball can be chosen again in combinations with replacement.

There are two ways to solve (3.37). The first way is algebraic, where we guess the solution based on some cases that can be calculated. The structure of the answer suggests a second, more intuitive, solution. To derive the answer algebraically, consider boundary cases starting with $k = 2$. From (3.37) we have

$$C^r(n, 2) = C^r(n, 1) + C^r(n - 1, 2)$$

$$= C^r(n, 1) + C^r(n - 1, 1) + C^r(n - 2, 2). \quad (3.38)$$

Iterating (3.38) yields

$$C^r(n, 2) = C^r(n, 1) + C^r(n - 1, 1) + \cdots + C^r(2, 1) + C^r(1, 2)$$

$$= n + (n - 1) + \cdots + 2 + 1$$

$$= \binom{n + 1}{2}.$$

Similarly we can expand (3.37) for $k = 3$ using the results for $C^r(n, 2)$ above, which yields

$$C^r(n, 3) = C^r(n, 2) + C^r(n - 1, 2) + \cdots + C^r(1, 3)$$

$$= \binom{n + 1}{2} + \binom{n}{2} + \cdots + \binom{2}{2}. \quad (3.39)$$

Equation (3.39) is a summation of the form given in (3.28), which shows that

$$C^r(n, 3) = \binom{n + 2}{3}.$$

Comparing the results for these two special cases reveals a pattern that allows us to guess that

$$C^r(n, k) = \left\langle \begin{array}{c} n \\ k \end{array} \right\rangle, \quad (3.40)$$

where we define

$$\left\langle \begin{array}{c} n \\ k \end{array} \right\rangle \overset{\text{def}}{=} \binom{n + k - 1}{k}, \quad k = 0, 1, \ldots, n = 1, 2, \ldots . \quad (3.41)$$

The guess of (3.40) can be shown to be correct with an inductive proof. We term the values in (3.41) *binomial-R coefficients* (these have also been called *figurate* numbers [96]). (Recall that the angle notation is used when considering combinatorial experiments that use replacement. We only use this notation in this chapter to point out the *symmetry* between combinatorial expressions for experiments with and without replacement.) The reason we call these coefficients given in (3.40) binomial-R coefficients is clear. They correspond to binomial coefficients obtained when selection is performed *with replacement*. We summarize the formulas for the four different counting paradigms considered in Table 3.4.

	Without Replacement	**With Replacement**
Permutations	$(n)_k = n(n-1)\cdots(n-k+1)$	$\langle n\rangle_k = n^k$
Combinations	$\displaystyle\binom{n}{k} = \frac{n!}{(n-k)!k!}$	$\displaystyle\left\langle\begin{array}{c} n \\ k \end{array}\right\rangle = \frac{(n+k-1)!}{(n-1)!k!}$

TABLE 3.4. Summary of Counting Formula for Selecting k Items from n

3.2.11 Simple Derivation for Combinations with Replacement

The simple form of $C^r(n,k)$ suggests that there should be an equally simple, more intuitive, derivation. Equation (3.40) suggests that this problem is equivalent to another problem where k balls are selected without replacement from a total of $n + k - 1$ balls. To argue that this equivalence holds, number the k selected balls from 1 to k and arrange them in a nondecreasing sequence. Next modify the value of the balls by adding $i - 1$ to the number on the ith ball in the sequence. Clearly the values of balls modified in this manner cannot repeat, and the largest value of a modified ball is less than or equal to $n + k - 1$ (this is obtained if ball n is selected on the kth selection). It is clear that for each possible selection of balls in the original system there is a corresponding unique set of values in the modified version of the problem and vice-versa. A moment's reflection reveals that the modified version of the problem is equivalent to that of selecting k balls from a set of $n + k - 1$ balls labeled 1 through $n + k - 1$. The number of ways this can be done is given by $C(n + k - 1, k)$, and thus the equivalence of the original and modified system implies that this is also equal to $C^r(n,k)$.

Example 3.13 (Combinations with Replacement) Suppose that $n = 4$ and $k = 2$. Then $C^r(4, 2) = 10$ and the different combinations of selected balls with replacement are given by

$$
\begin{array}{cccc}
12, & 13, & 14, & \\
23, & 24, & 34, & \\
11, & 22, & 33, & 44.
\end{array}
$$

Example 3.14 (Closed Queueing Networks with Infinite Capacity) Consider the closed queueing network of Example 3.7 except that now the service centers have infinite capacity. How many different states are there if we disregard the ordering of customers at each service center?

To answer the question first recall that, in Example 3.7, n is the number of service centers and k is the number of customers. Note that now the number of customers allocated to service center i must satisfy $0 \le k_i \le n$ and $k = \sum_{i=1}^{n} k_i$. The problem can thus be phrased, how many possible integer vectors of size n have nonnegative components bounded by n that sum to k.

Equivalently we can phrase the problem along the lines of Example 3.8 as the number of ways that k indistinguishable bricks can be distributed into n boxes. This last problem is equivalent to the combinations with replacement problem that we have just solved. To see this, suppose we have a set of n boxes labeled 1 through n. We now select the k balls from the urn using a replacement policy. If a ball with value i is selected then we throw a brick into box i. After all k balls have been selected let k_i be the number of bricks contained in box i. It is then clear that the number of possible combinations of bricks allocated to boxes is the same as the number of possible selections of balls from the urn with a replacement policy. Thus the number of different states equals $C(n, k)$.

3.2.12 Problems Equivalent to Combinations with Replacement

These examples show that the following are equivalent problems with solution $C^r(n, k)$ (see Table 3.5 for a table of values):

1. The number of ways to select, with replacement, k distinguishable balls from an urn containing n balls.

2. The number of ways to put k indistinguishable balls into n distinguishable urns.

3. The number of vectors $(k_1, k_2, \ldots, k_n)$ with integer entries $0 \le k_i \le n$, $i = 1, 2, \ldots, n$, that satisfy $k = \sum_{i=1}^{n} k_i$.

We next consider some examples to illustrate the use of combinations in problem solving.

$n \setminus k$	0	1	2	3	4	5	6	7	8	9
0	1									
1	1	1	1	1	1	1	1	1	1	1
2	1	2	3	4	5	6	7	8	9	10
3	1	3	6	10	15	21	28	36	45	55
4	1	4	10	20	35	56	84	120	165	220
5	1	5	15	35	70	126	210	330	495	715
6	1	6	21	56	126	252	462	792	1,287	2,002
7	1	7	28	84	210	462	924	1,716	3,003	5,005
8	1	8	36	120	330	792	1,716	3,432	6,435	11,440
9	1	9	45	165	495	1,287	3,003	6,435	12,870	24,310
10	1	10	55	220	715	2,002	5,005	11,440	24,310	48,620

TABLE 3.5. Table of Values of $C^r(n,k)$

3.2.13 Basic Properties of Binomial-R Coefficients

The values of $C^r(n,k)$ for a given row k are given by the kth descending diagonal of Pascal's triangle (Table 3.3). A set of basic properties for binomial-R coefficients are given below:

Asymmetry
$$\left\langle \begin{matrix} n \\ k \end{matrix} \right\rangle = \left\langle \begin{matrix} k+1 \\ n-1 \end{matrix} \right\rangle$$

Factoring
$$\left\langle \begin{matrix} n \\ k \end{matrix} \right\rangle = \frac{n}{k} \left\langle \begin{matrix} n+1 \\ k-1 \end{matrix} \right\rangle$$

Addition
$$\left\langle \begin{matrix} n \\ k \end{matrix} \right\rangle = \left\langle \begin{matrix} n \\ k-1 \end{matrix} \right\rangle + \left\langle \begin{matrix} n-1 \\ k \end{matrix} \right\rangle$$

Binomial
$$\left(\begin{matrix} n \\ k \end{matrix} \right) = \left\langle \begin{matrix} n+1-k \\ k \end{matrix} \right\rangle$$

It is interesting to note that there does not appear to be a simple formula for the product $C^r(n,k)\, C^r(k,\ell)$ and that, using (3.32), we can write an expression for the sum (corresponding to (3.36)) as

$$\left\langle \begin{matrix} n+1 \\ m \end{matrix} \right\rangle = \sum_{k=0}^{m} \left\langle \begin{matrix} n \\ k \end{matrix} \right\rangle. \tag{3.42}$$

In contrast to binomial coefficients, (3.29) shows that there is a closed formula for partial sums of the rows of binomial-R coefficients.

Example 3.15 (Occupancy Problem)   Suppose that n distinguishable balls are allocated in n urns. What is the probability that exactly one urn has no balls?

To answer this question, we observe that there are a total of n^n possible allocations and the event of interest occurs for those allocations where $n-2$ urns contain exactly 1 ball, one urn contains 2 balls, and one urn contains no balls. We decompose the problem into two steps: select the urn occupancies and then select the ball allocations. To determine the number of ways that we can select urns so that one has 0 balls, one has 2 balls, and $n-2$ have 1 ball, we observe that it is sufficient to determine which urns have 0 and 2 balls. This, however, is equivalent to the number of different permutations obtained when 2 items are selected from n, that is, $(n)_2$. We next consider the allocation of balls to the urns. For every selection of urns there are $C(n, 2)$ ways to select the 2 balls that go in one urn and $(n-2)!$ ways to allocate the remaining balls. Combining both steps implies that the probability that exactly one urn is empty is given by

$$\frac{(n)_2 \begin{pmatrix} n \\ 2 \end{pmatrix} (n-2)!}{n^n} = \frac{n! \begin{pmatrix} n \\ 2 \end{pmatrix}}{n^n}.$$

Using Stirling's formula (3.6) for large n shows that this probability is given by

$$n(n-1)\sqrt{\frac{\pi n}{2}}e^{-n},$$

which decreases to 0 as $n \to \infty$.

Example 3.16 (Comparison of Selections with and without Replacement) Intuitively it should be the case that for large populations the number of combinations with replacement should yield similar results to the number of combinations without replacement. We showed that this is true for permutations in the discussion following equation (3.15). To show that this also holds for combinations, consider the following expression

$$\frac{\left\langle \begin{array}{c} n \\ k \end{array} \right\rangle}{\begin{pmatrix} n \\ k \end{pmatrix}} = \frac{(n+k-1)!}{n!} \frac{(n-k)!}{(n-1)!}$$

$$= \left(1 + \frac{k}{n-1}\right)\left(1 + \frac{k}{n-2}\right)\cdots\left(1 + \frac{k}{n-(k-1)}\right). \text{(3.43)}$$

As $n \to \infty$ it is clear that (3.43) has limit 1 for any fixed value k.

3.3 Generating Functions

In the previous section we saw how linear recurrences arise in combinatorial problems. To determine a solution for such recurrences, we first solved several small cases and then inferred the form of the general term. It is clear that such a solution technique has limited applicability, and in this section we present an algorithmic method for solving recurrence equations. This technique uses generating functions that are, essentially, power series expansions. The use of generating functions in combinatorics thus establishes a unified counting methodology.

There are many types of generating functions, but two types are most frequently found in combinatorics: the *exponential generating functions*, which are applicable for counting permutations, and *ordinary generating functions*, which are applicable for counting combinations. Our main focus in this section will be to develop the theory of ordinary generating functions, which we henceforth simply call *generating functions*.

3.3.1 Generating Functions for Combinations

The results in this section can be motivated by an extension to Example 3.7. We will provide the mathematical tools necessary to solve this problem in this section.

Example 3.17 (Closed Queueing Network with Finite Capacity) Consider the closed queueing network of Example 3.7 but with the modification that queues have finite capacity. Specifically, assume that service center i has a finite capacity of u_i. Hence the total number of customers at service center i satisfies $0 \le k_i \le u_i$ for $i = 1, 2, \ldots, n$, and states have the constraint that $k = \sum_{i=1}^{n} k_i$. What is the number of different states if the ordering of customers at each service center is not counted?

As a simple example of how the upper constraint complicates things, rephrase the problem equivalently to that of distributing six indistinguishable balls into three distinguishable urns where the maximal capacity of urn 1 is two balls and that of each of the remaining two urns is three balls. A way to count the number of ways the balls can be allocated to the urns is to partition all possibilities according to the number of balls in urn 1. If urn 1 has no balls, then six balls must be allocated to urns 2 and 3 so that no urn has more than three balls. There is only one way to do this, the allocation $(0, 3, 3)$. If urn 1 has one ball then the above argument implies the possible allocations are given by $(1, 2, 3)$ and $(1, 3, 2)$. Finally, for the case where urn 1 has two balls we clearly have the following possible allocations: $(2, 1, 3)$, $(2, 2, 2)$, and $(2, 3, 1)$. Thus there are six possibilities, and enumerating them was easy.

To generalize, try doing the same problem for 310 balls, where the urns have maximal capacities of 71, 137, and 193, respectively. Of course, there is no reason to stop here, and we can hardly refrain from generalizing. Suppose, for example, we put lower bounds on the values of k_i or arbitrarily restrict its value. We might say, for example, that urn i cannot have fewer than 11 or more than 213 balls and cannot have a number of balls that is evenly divisible by 7. While we are at it, we might restrict the number of balls to be equal to the product of an odd and even number and certainly not be an "unlucky" number such as 13, 51, or 71. After imagining constraints of a similar complexity on the other urns, it becomes clear that we need a new method in order to count all possible combinations. In fact, there is a simple mechanism that not only answers all of the above questions but can also be used to solve many types of probabilistic problems posed in later chapters. This technique is derived through the use of *generating functions*, which we now discuss.

Associate the power series

$$x^0 + x^1 + x^2 + \cdots = \frac{1}{1-x}, \quad 0 < x < 1,$$

with each of the n urns and form a *generating function* by multiplying these together

$$(x^0 + x^1 + x^2 + \cdots)^n = (1-x)^{-n}. \tag{3.44}$$

The variable x in this equation is called the *generating variable*. We wish to expand this as a power series, and to do this we need to determine coefficients $a_{n,k}, k = 0, 1, \ldots$, that satisfy

$$(1 + x + x^2 + \cdots)^n = \sum_{k=0}^{\infty} a_{n,k} x^k. \tag{3.45}$$

To determine these values, observe that $a_{n,k}$ counts the possible combinations of factors of x that, when multiplied together, yield x^k. For example, with $k = 4$ and $n = 2$ the possible factors are

$$(x^0 x^4, \ x^1 x^3, \ x^2 x^2, \ x^3 x^1, \ x^4 x^0),$$

and thus $a_{4,2} = 5$. Since $x^i x^j = x^{i+j}$, an equivalent way to calculate $a_{4,2}$ is to enumerate all possible ways two nonnegative integers can sum to 4. More generally, the value of $a_{n,k}$ equals the number of ways n nonnegative integers can sum to k. This was calculated in Example 3.14, and it equals the number of combinations obtained when k indistinguishable balls are placed in n distinguishable urns, that is, $a_{n,k} = C^r(n,k)$. We can thus write (3.45) as

$$(1 + x + x^2 + \cdots)^n = (1-x)^{-n} = \sum_{k=0}^{\infty} \left\langle \begin{array}{c} n \\ k \end{array} \right\rangle x^k. \tag{3.46}$$

This result shows, with (3.44), that the coefficient of the x^k term in the power series expansion of $(1-x)^{-n}$ equals the number of ways to place k indistinguishable objects into n distinguishable urns.

3.3.2 General Solution for Combinations

Let us now interpret what we have done so that we can generalize the result. Equation (3.45) shows that we associate a power series $(x^0 + x^1 + \cdots)$ with each of the n urns. This power series is used as a mechanism to facilitate counting and we can imagine that in each possible allocation only one factor of each power series is "tagged." Thus if, in a particular allocation, urn i has k_i balls then its x^{k_i} factor is tagged. Any particular allocation of k balls to n urns results in a set of n tagged factors, one for each urn, where the exponents of x of the tagged factors sum to k. Since this is true for any particular allocation, the number of all possibilities is given by the coefficient of x^k in the multiplication of all of the power series for the n urns.

To remove possible assignments of balls to urns, we eliminate factors from the urn power series. For example, suppose we remove the factors x^0 and x^1 so that all of the urns have the power series $x^2 + x^3 + \cdots$ associated with them. Consider the power series

$$(x^2 + x^3 + \cdots)^n = \sum_{k=0}^{\infty} b_{n,k} x^k. \tag{3.47}$$

It is then easy to see that the value of $b_{n,k}$ is the number of ways that n nonnegative integers that are greater than or equal to 2 can sum to k. Interpreting the results in terms of allocating balls to urns, we see that dropping the factors x^0 and x^1 from the power series associated with each urn eliminates the possibility of having 0 or 1 ball allocated to them.

The generalization is now clear. For each urn we associate a factor x^i if i balls can be allocated to the urn. Thus, if the urn must have more than 2 but less than 11 balls and not 7 or 9 balls, then the power series associated with this urn is $x^3 + x^4 + x^5 + x^6 + x^8 + x^{10}$. For each urn we form a power series (this is simply a polynomial if there are only a finite number of terms) associated with it from the constraints on that urn and then we form the generating function for the problem by multiplying the power series of all the urns. The answer for how many ways k indistinguishable balls can be distributed into n distinguishable urns subject to the specific constraints on the urns is then given by the coefficient of x^k in the generating function.

Example 3.18 How many ways are there to distribute k balls into two urns so that urn 1 cannot have two balls or more than four balls and urn 2 cannot have more than three balls? To answer the question, we form the two urn power series. For urn 1 the power series is given by $x^0 + x^1 + x^3 + x^4$, and for urn 2 by $x^0 + x^1 + x^2 + x^3$. The generating function for the system is thus given by

$$\begin{aligned}
G(x) &= (x^0 + x^1 + x^3 + x^4)(x^0 + x^1 + x^2 + x^3) \\
&= x^0 + 2x^1 + 2x^2 + 3x^3 + 3x^4 + 2x^5 + 2x^6 + x^7.
\end{aligned}$$

The answer is thus given by the coefficient of x^k. In particular, if $k = 4$ then there are three possibilities corresponding to the allocations $(1, 3), (3, 1)$, and $(4, 0)$, where the first (resp., second) element of the pair is the number of balls allocated to urn 1 (resp., urn 2).

Suppose we consider the problem posed in Example 3.17 concerning the closed queueing network with finite capacity. It is fairly clear here that, analogous to (3.45), we wish to find the power series expansion for

$$\prod_{j=1}^{n}(x^0 + x^1 + x^2 + \cdots + x^{u_j}) = \sum_{n=0}^{\infty} c_{n,k}x^k,$$

where we need to determine the values of the coefficients $c_{n,k}$. The value of $c_{n,k}$ is the total number of ways that k balls can be distributed into n urns, and thus it is also the answer to how many different possible states there are in the closed queueing network.

Consider the following examples of generating functions.

Example 3.19 Suppose we distribute k indistinguishable balls into n distinguishable urns so that each urn can contain at most one ball. The generating function for the problem is $(1 + x)^n$, which, when written as a power series using the binomial theorem (B.4), is

$$(1 + x)^n = \sum_{k=0}^{n} \binom{n}{k} x^k. \tag{3.48}$$

This shows that the number of possible combinations is given by $C(n, k)$.

Example 3.20 Consider the case where the number of balls in each urn is unlimited but must be at least m. The generating function is given by

$$(x^m + x^{m+1} + \cdots)^n = x^{mn}(1 + x + x^2 + \cdots)^n$$

$$= x^{mn}(1 - x)^{-n}.$$

Using the power series expansion of (3.46) implies that

$$x^{mn}(1 - x)^{-n} = x^{mn}\sum_{k=0}^{\infty}\left\langle \begin{array}{c} n \\ k \end{array} \right\rangle x^k$$

$$= \sum_{k=0}^{\infty}\left\langle \begin{array}{c} n \\ k \end{array} \right\rangle x^{k+mn}$$

$$= \sum_{k=nm}^{\infty}\left\langle \begin{array}{c} n \\ k - mn \end{array} \right\rangle x^k. \tag{3.49}$$

Thus the number of combinations of having k indistinguishable balls put into n distinguishable urns so that each urn has at least m balls is

$$\left\langle \begin{matrix} n \\ k - mn \end{matrix} \right\rangle, \quad k \geq mn. \tag{3.50}$$

It is easy to generalize this to the case where all urns have different minimal numbers of balls (see Problems 3.19, 3.20, 3.21, and 3.22 for generalizations).

Example 3.21 Consider an experiment where k balls are distributed into n urns so that no urn has exactly one ball. The generating function for this is given by

$$(1 + x^2 + x^3 + \cdots)^n = ((1-x)^{-1} - x)^n$$

$$= \sum_{i=0}^{n} \binom{n}{i} (-1)^i x^i (1-x)^{-(n-i)}, \tag{3.51}$$

where (3.51) follows from using the binomial theorem (B.4). Using the power series expansion of (3.46) we can write the right-hand side of (3.51) as

$$\sum_{i=0}^{n} \sum_{j=0}^{\infty} \binom{n}{i} \left\langle \begin{matrix} n-i \\ j \end{matrix} \right\rangle (-1)^i x^{i+j}.$$

The number of possible allocations is the coefficient of x^k given by

$$\sum_{i=0}^{\min\{n,k\}} \binom{n}{i} \left\langle \begin{matrix} n-i \\ k-i \end{matrix} \right\rangle (-1)^i.$$

For example, for $n = 3$ and $k = 2$ we have

$$\binom{3}{0} \left\langle \begin{matrix} 3 \\ 2 \end{matrix} \right\rangle - \binom{3}{1} \left\langle \begin{matrix} 2 \\ 1 \end{matrix} \right\rangle + \binom{3}{2} \left\langle \begin{matrix} 1 \\ 0 \end{matrix} \right\rangle = 3,$$

which corresponds to all three possible ways to allocate both balls to an urn.

Property		Sequence	Generating Function
	Definition	$f_i, i = 0, 1, \ldots$	$F(x) \overset{\text{def}}{=} \sum_{i=0}^{\infty} f_i x^i$
Addition	Addition	$af_i + bg_i$	$aF(x) + bG(x)$
	Convolution	$\sum_{j=0}^{i} f_j g_{i-j}$	$F(x)G(x)$
Shifting	Positive	$f_{i-n}, i \geq n$	$x^n F(x)$
	Negative	f_{i+n}	$\frac{F(x) - \sum_{j=0}^{n-1} f_j x^j}{x^n}$
Scaling	Geometric	$f_i a^i$	$F(ax)$
	Linear	$i f_i$	$x \frac{d}{dx} F(x)$
	Factorial	$(i)_k f_i$	$x^k \frac{d^k}{dx} F(x)$
	Harmonic	f_{i-1}/i	$\int_0^x F(t) dt$
Summations	Cumulative	$\sum_{j=0}^{i} f_j$	$\frac{F(x)}{1-x}$
	Complete	$\sum_{i=0}^{\infty} f_i$	$F(1)$
	Alternating	$\sum_{i=0}^{\infty} (-1)^i f_i$	$F(-1)$
Sequence Values	Initial	f_0	$F(0)$
	Intermediate	f_n	$\frac{1}{n!} \left. \frac{d^n}{dx} F(x) \right|_{x=0}$
	Final	$\lim_{i \to \infty} f_i$	$\lim_{x \to 1} (1-x) F(x)$

TABLE 3.6. Algebraic Properties of Generating Functions

3.3.3 Algebraic Properties of Generating Functions

We can often write down the answer to combinatorial problems by inspection using algebraic properties of generating functions. These properties are the same as properties of power series expansions; a list is given in Table 3.6. Lists of common generating functions are found in Tables C.4 and C.5. (These tables are found in Appendix C.)

3.3.4 Addition and Convolution

Let F and G be two generating functions given by

$$F(x) = \sum_{i=0}^{\infty} f_i x^i, \qquad G(x) = \sum_{i=0}^{\infty} g_i x^i. \qquad (3.52)$$

We can suppose that the values f_i and g_i have some combinatorial or probabilistic interpretation: however, they could be an arbitrary sequence of numbers. We can combine generating functions by taking their sum or product and in so doing create another generating function. Simple algebra shows that the addition of two generating functions is the generating function for the series $f_i + g_i$:

$$F(x) + G(x) = \sum_{i=0}^{\infty} (f_i + g_i) x^i, \quad \textbf{Addition Property.} \quad (3.53)$$

Similarly the product of two generating functions can be written as

$$F(x)G(x) = \sum_{i=0}^{\infty} h_i x^i, \quad \textbf{Convolution Property,} \quad (3.54)$$

where

$$h_i = f_0 g_i + f_1 g_{i-1} + \cdots + f_{i-1} g_1 + f_i g_0, \quad i = 0, 1, \ldots, \quad (3.55)$$

represents the *convolution* of the sequences f_i and g_i. We denote the convolution operator by $\circledast$ and thus can equivalently write (3.55) as

$$h = f \circledast g. \qquad (3.56)$$

The convolution equation (3.54) is extremely powerful and will be used repeatedly. In fact, we have already seen its use in Example 3.20.

Looking at (3.49) we see that two generating functions have been multiplied together: the first $F(x) = (x^m)^n$ corresponds to the allocation of balls to n urns so that each urn has exactly m balls and the second $G(x) = (1-x)^{-n}$ corresponds to the allocation of balls to n urns without any restriction. The coefficients of the power series expansion of F are given by $f_k = \delta_{mn}(k)$ where

$$\delta_{mn}(k) = \begin{cases} 1, & k = mn, \\ 0, & k \neq mn, \end{cases}$$

and from (3.46) the coefficients g_k are given by

$$g_k = \left\langle \begin{array}{c} n \\ k \end{array} \right\rangle .$$

Letting $H(x) = F(x)G(x)$ and using the convolution equation (3.54) implies that

$$h_k = \sum_{i=0}^{k} \delta_{mn}(i) \left\langle \begin{array}{c} n \\ k-i \end{array} \right\rangle = \begin{cases} \left\langle \begin{array}{c} n \\ k-mn \end{array} \right\rangle , & k \geq mn, \\ \\ 0, & k < mn, \end{cases}$$

as derived in (3.50).

The shifting and scaling properties of Table 3.6 are easily established algebraically. As a simple application of the shift operation consider the constant sequence of values $1, 1, \ldots$, and let $F(x)$ be the corresponding generating function, $F(x) = \sum_{i=0}^{\infty} x^i$. Then $xF(x)$ corresponds to the sequence $0, 1, 1, \ldots$, and thus implies that $F(x) - xF(x) = 1$. This implies that

$$F(x) = \frac{1}{1-x} = 1 + x + x^2 + \dots .$$

Other identities of this type can also be generated (see Tables C.4 and C.5.)

We can use the above identity to derive the cumulative summation property of Table 3.6. We first write out

$$\frac{F(x)}{1-x} = (f_0 + f_1 x + f_2 x^2 + f_3 x^3 + \cdots)(1 + x + x^2 + x^3 + \cdots)$$

and collect terms of like powers of x. This results in

$$\frac{F(x)}{1-x} = f_0 + (f_0 + f_1)x + (f_0 + f_1 + f_2)x^2 + \cdots$$

and thus yields the property.

The complete summation property is used later when we deal with sequences that are probabilities, since it allows one to check that probabilities sum to 1 since $F(1) = 1$ for these cases. Similarly the initial sequence property can be used to check results, since it implies that $F(0) = f_0$.

Example 3.22 Let $F(x)$ be the generating function for the sequence $f_i, i = 0, 1, \ldots$. Determine the sum of the even indexed values $f_0 + f_2 + f_4 + \cdots$.

To do this we note that

$$\frac{1}{2}\left(F(x) + F(-x)\right) = f_0 + f_2 x^2 + f_4 x^4 + \cdots.$$

Thus the sum of the even indexed values is given by

$$\frac{1}{2}\left(F(1) + F(-1)\right). \tag{3.57}$$

We can use this to write an equality for binomial coefficients. The generating function for the binomial coefficients $f_i = C(n,i)$, as given by (3.48), is $F(x) = (1+x)^n$. Substituting this into (3.57) yields

$$\sum_{i=0}^{\lfloor n/2 \rfloor} \binom{n}{2i} = 2^{n-1}. \tag{3.58}$$

3.3.5
First Passage
Times

An example that illustrates the power of generating functions and exploits several of their properties concerns the *first passage* time of a simple fair coin tossing experiment. Consider a coin tossing experiment with a fair coin and suppose that player A wins (resp., loses) a dollar if heads (resp., tails) lands facing up. Let S_n be the total sum of player A's winnings at the end of the nth toss for $n = 1, 2, \ldots$. We are interested in the first time the player's winnings equal one dollar, an event that occurs on the nth toss if

$$\{S_1 \le 0, S_2 \le 0, \ldots, S_{n-1} \le 0, S_n = 1\}. \tag{3.59}$$

This is commonly called a first passage time since it indicates the *first time* A's winnings reach one dollar.

Let $f_n, n = 1, 2, \ldots$, be the probability of event (3.59) (we set $f_0 = 0$). It is clear that the first passage occurs on the first toss if this toss lands heads up and thus $f_1 = 1/2$. The general expression for $f_n, n = 2, 3, \ldots$, follows from a simple, but surprisingly powerful, observation. If the first toss lands tails up then $S_1 = -1$ and for player A's winnings to total 1 at the nth step it must be the case that there is a first time that the player's winnings again equal 0. In other words, there must be a value i that satisfies

$$\{S_2 \le -1, S_3 \le -1, \ldots, S_{i-1} \le -1, S_i = 0\}, \quad i = 2, 3, \ldots, n-1. \tag{3.60}$$

After this return to 0 we essentially start a new version of the original first passage problem, and it is easy to see that from step $i + 1$ onward we must have

$$\{S_{i+1} \le 0, S_{i+2} \le 0, \ldots, S_{n-1} \le 0, S_n = 1\}. \tag{3.61}$$

The observation we make is that events (3.60) and (3.61) represent versions of the original first passage problem as given by (3.59). Hence, using the law of total probability by partitioning over all possible values of i yields

$$f_n = \frac{1}{2}(f_1 f_{n-2} + f_2 f_{n-3} + \cdots + f_{n-3} f_2 + f_{n-2} f_1), \quad n = 2, 3, \ldots . \quad (3.62)$$

Thus $f_n, n \geq 2$, is represented in terms of a convolution equation similar to (3.55). In words, the $1/2$ in (3.62) corresponds to the probability that the experiment does not end on the first step (i.e., the coin lands tails up the first time). Conditioned on this, the term $f_i f_{n-i-1}$, for $i = 1, 2, \ldots, n-2$, corresponds to the event that i steps are required for the first return to 0 and $n - i - 1$ steps are required from this point to first reach 1.

Whenever a convolution recurrence like (3.62) arises in a problem it is natural to approach the solution using generating functions. Let $F(x)$ be the generating function for the sequence $f_n, n = 0, 1, \ldots$, defined by

$$F(x) \stackrel{\text{def}}{=} \sum_{n=0}^{\infty} f_n x^n, \quad (3.63)$$

where we set $f_0 = 0$. It is clear that the values of f_n represent probabilities, and thus they must sum to 1 yielding

$$1 = \sum_{n=0}^{\infty} f_n = F(1). \quad (3.64)$$

Additionally, the initial property requires that

$$F(0) = f_0 = 0. \quad (3.65)$$

To obtain an expression for $F(x)$ requires some algebra (the following manipulations are representative of many generating function problems). Using (3.63) and the fact that $f_1 = 1/2$ we have

$$F(x) = \frac{x}{2} + \sum_{n=2}^{\infty} f_n x^n.$$

Substituting (3.62) into this yields

$$
\begin{aligned}
F(x) &= \frac{x}{2} + \frac{1}{2} \sum_{n=2}^{\infty} x^n \sum_{i=1}^{n-2} f_i f_{n-(i+1)} \\
&= \frac{x}{2} + \frac{1}{2} \sum_{i=1}^{\infty} f_i x^{i+1} \sum_{n=i+2}^{\infty} f_{n-(i+1)} x^{n-(i+1)} \\
&= \frac{x}{2} + \frac{1}{2} \sum_{i=1}^{\infty} f_i x^{i+1} \sum_{j=1}^{\infty} f_j x^j \\
&= \frac{x}{2} \left[1 + F^2(x) \right].
\end{aligned}
\tag{3.66}
$$

Equation (3.66) shows that to determine the generating function we must solve a quadratic equation that, using the quadratic formula, is given by

$$
F(x) = \frac{1 \pm \sqrt{1-x^2}}{x}.
$$

The initial value condition (3.65) eliminates the solution with the positive square root since it is unbounded as x approaches 0. To show that the solution with the negative square root satisfies the initial value condition, we apply l'Hospital's rule to evaluate $\lim_{x \to 0} F(x)$:

$$
F(0) = \frac{d}{dx}(1 - \sqrt{1-x^2}) = \left. \frac{x}{\sqrt{1-x^2}} \right|_{x=0} = 0.
$$

We thus have that

$$
F(x) = \frac{1 - \sqrt{1-x^2}}{x}.
\tag{3.67}
$$

It is easy to see that this equation also satisfies the normalization condition (3.64).

To determine the values of f_n for $n = 2, 3, \ldots$, we expand (3.67) in a power series and determine the coefficient of x^n. To do this we must have a way to represent a binomial coefficient where the entries are fractions, for example, $C(1/2, k)$ (this is discussed more fully in Appendix B). Substituting into the definition of binomial coefficients given by (3.21) and using the fact that $n! = n(n-1) \cdots (n-k+1)(n-k)!$, implies that

$$
\begin{aligned}
\binom{1/2}{k} &= \frac{(1/2) \cdot (1/2 - 1) \cdots (1/2 - k)!}{k!(1/2 - k)!} \\
&= (1/2)^k (-1)^{k-1} \frac{1 \cdot 3 \cdots 2k - 3}{k!}, \quad k = 1, 2, \ldots,
\end{aligned}
\tag{3.68}
$$

where $C(1/2, 0) = 1$ by convention. Equation (3.68) can be simplified as

$$\begin{pmatrix} 1/2 \\ k \end{pmatrix} = (1/2)^k (-1)^{k-1} \frac{1 \cdot 3 \cdots 2k-3}{k!} \frac{2 \cdots 2(k-1)}{2 \cdots 2(k-1)}$$

$$= \frac{1}{k} \frac{(-1)^{k-1}}{2^{2k-1}} \begin{pmatrix} 2(k-1) \\ k-1 \end{pmatrix}, \quad k \geq 1. \qquad (3.69)$$

Expanding (3.67) using the binomial formula implies that (see also (B.6))

$$\sqrt{1-x^2} = 1 - \begin{pmatrix} 1/2 \\ 1 \end{pmatrix} x^2 + \begin{pmatrix} 1/2 \\ 2 \end{pmatrix} x^4 - \begin{pmatrix} 1/2 \\ 3 \end{pmatrix} x^6 + \cdots,$$

which implies that

$$F(x) = \begin{pmatrix} 1/2 \\ 1 \end{pmatrix} x - \begin{pmatrix} 1/2 \\ 2 \end{pmatrix} x^3 + \begin{pmatrix} 1/2 \\ 3 \end{pmatrix} x^5 - \cdots.$$

Thus the general form of the values of f_n for $n = 2, 3, \ldots,$ is given by

$$f_n = \begin{cases} \begin{pmatrix} 1/2 \\ k \end{pmatrix} (-1)^{k-1}, & n = 2k-1, \\ \\ 0, & n = 2k, \quad k = 1, 2, \ldots. \end{cases}$$

This is more conveniently expressed, using (3.69), as

$$f_{2k-1} = \frac{1}{k} \frac{1}{2^{2k-1}} \begin{pmatrix} 2(k-1) \\ k-1 \end{pmatrix}, \quad k = 1, 2, \ldots. \qquad (3.70)$$

Even indexed terms equal 0 since event (3.59) can only occur for an odd number of tosses. The combinatorial expressions

$$C_k \stackrel{\text{def}}{=} \frac{1}{k} \begin{pmatrix} 2(k-1) \\ k-1 \end{pmatrix}, \quad k = 1, 2, \ldots, \qquad (3.71)$$

are called Catalan[4] numbers (see Table 3.7 for a list of some of these numbers), and they frequently arise in first passage problems (for example, they are found in an equation for the number of customers served in a busy period of an M/M/1 queue given in (7.84), which is also a first passage problem). Problem 3.43 generalizes the first passage problem just considered to the case where the coin is not necessarily fair.

---

[4]Eugene Catalan (1814–1894) wrote a paper establishing the properties of these numbers in 1838 [11]. See Problem 3.29 for an interpretation and derivation of these numbers.

i	1	2	3	4	5	6	7	8	9	10	11
C_i	1	1	2	5	14	42	132	429	1,430	4,862	16,796

TABLE 3.7. Catalan Numbers

The above steps describe the procedures for how one uses generating functions to solve probabilistic problems. We will revisit these steps in Section 3.4. Notice that (3.70), along with the normalization condition (3.64) imply that we have also derived the combinatorial identity

$$\sum_{k=1}^{\infty} \frac{1}{k} \frac{1}{2^{2k-1}} \left(\begin{array}{c} 2(k-1) \\ k-1 \end{array} \right) = 1,$$

which, again, is not obvious without knowing its probabilistic interpretation.

3.4 Solving Recurrence Equations with Generating Functions**

In the previous section we demonstrated the power of generating functions for solving combinatorial problems and we discussed several properties of generating functions. In this section we show how generating functions are used to solve recurrence equations. To describe this in a general setting and motivate the need for deriving an algebra for generating functions, we consider a simple constant coefficient, linear recurrence, and solve it using the mathematical tools previously developed.

Let f_0 and f_1 be given constants, called *boundary values*, and consider the linear recurrence

$$f_{i+1} = bf_i + af_{i-1}, \quad i = 1, 2, \ldots, \tag{3.72}$$

where non-boundary values f_i for $i \geq 2$, are called *terms* of the recurrence while a and b are two constants called the *coefficients* of the recurrence. As an example, for values $f_0 = 0$, $f_1 = 1$, and $a = b = 1$ the recurrence generates the famous Fibonacci numbers given in Table 3.8.[5]

Our objective is to derive a formula for the general recurrence defined

---

**Can be referred to as necessary.

[5]These numbers have a long and interesting history, starting when Leonardo Fibonacci (1175–1250) introduced the sequence in 1202. The sequence is frequently found in many natural systems including family trees of rabbits and bees. Many of their interesting mathematical relationships were discovered by François Eduard Anatole Lucas (1842–1891), who also has a sequence of numbers named after him (Lucas numbers are defined by $\ell_i = f_{i+1} + f_{i-1}$, $i = 1, 2, \ldots$, where f_i is the ith Fibonacci number).

i	1	2	3	4	5	6	7	8	9	10	11	12	13	14
f_i	1	1	2	3	5	8	13	21	34	55	89	144	233	377

TABLE 3.8. Fibonacci Numbers

by (3.72). To solve (3.72) we use generating functions and define

$$F(x) \stackrel{\text{def}}{=} \sum_{i=0}^{\infty} f_i x^i. \tag{3.73}$$

Recall that generating functions are used in combinatoric problems as a formal "tagging" mechanism that is used to facilitate counting. Operations that are performed on generating functions are for manipulating tags. In recurrence (3.72), for example, we associate the value of f_i with the tag x^i and use this to keep track of the number of times each boundary value is contained in each term of the recurrence. More precisely, for every term of (3.73) we can write $f_i = c_{i,0} f_0 + c_{i,1} f_1$ where the values $c_{i,0}$ and $c_{i,1}$ are the coefficients that count the number of times f_0 and f_1 are contained in the term. A list of some of these values is given in Table 3.9. (A formal method to determine the number of boundary value factors contained in f_n is to use matrix techniques. See Problems 3-39 and 3-45 for problems elucidating these techniques.)

i	f_0	f_1
0	1	0
1	0	1
2	a	b
3	ab	$b^2 + a$
4	$a(b^2 + a)$	$b(b^2 + 2a)$
5	$ab(b^2 + 2a)$	$(b^2 + a)^2 + b^2 a$

TABLE 3.9. Counting the Boundary Value Terms in a Recurrence

Tagging f_i with the value x^i allows one to account for all of these factors in a convenient manner. It is interesting to observe that as long as we only use *symbolic techniques* to manipulate the power series expansions associated with generating functions the issue of the convergence properties of these series does not arise. Convergence considerations do arise, however, when we deal with these power series expansions as real or complex valued functions. For example, convergence is clearly required if we evaluate a generating function to determine its terms using numerical techniques.

**3.4.1
An Example
Using
Generating
Functions**

We now describe a method using generating functions to solve linear recurrence relations. Such techniques will also be used when solving recurrences arising in Markov process theory. To solve (3.72) we first find an equation for $F(x)$ (3.73) that is expressible as the ratio of two polynomials (this is generally the case for linear recurrences). Next we determine the power series expansion of $F(x)$; this typically requires substantial algebraic manipulation. Some of this manipulation will seem to be ad hoc at first, but it is based on a divide-and-conquer approach where we break up a complex ratio of polynomials into a summation of simpler polynomial ratios. It is often easier to determine the power series expansions for these simpler polynomial ratios than for that of the original ratio. In this simpler form, it is thus easier to determine the equation for $f_i, i = 0, 1, 2, \dots$.

We begin by determining $F(x)$ as the ratio of two polynomials. To do this we multiply (3.72) by x^{i+1} and sum both sides starting from $i = 1$,

$$\sum_{i=1}^{\infty} f_{i+1} x^{i+1} = b \sum_{i=1}^{\infty} f_i x^{i+1} + a \sum_{i=1}^{\infty} f_{i-1} x^{i+1}. \qquad (3.74)$$

Equation (3.74) with (3.73) yields

$$F(x) - f_1 x - f_0 = bx(F(x) - f_0) + ax^2 F(x),$$

which can be rewritten as

$$F(x) = \frac{f_0 + x(f_1 - f_0 b)}{1 - bx - ax^2}. \qquad (3.75)$$

Notice that in (3.75) the numerator is a polynomial function that contains the boundary values and the coefficients of the recurrence. The denominator of (3.75) does not contain the boundary values but does contain the coefficients of the recurrence (i.e., a and b). This is a special case of the general form obtained for the generating function of linear recurrences. For example, if

$$f_{i+2} = c f_{i+1} + b f_i + a f_{i-1}$$

then the denominator of the generating function can be written down by inspection as $1 - cx - bx^2 - ax^3$ and the numerator can also be written down by inspection (after a bit of practice; see Problem 3.47), yielding

$$F(x) = \frac{(f_2 - c f_1 - b f_0)x^2 + (f_1 - c f_0)x + f_0}{1 - cx - bx^2 - ax^3}. \qquad (3.76)$$

To determine an expression for the coefficients $f_i, i = 2, 3, \dots$, we expand (3.75) in a power series. This is typically accomplished by writing it as a *partial fraction* expansion so that we can use the summation property of two power series as given in (3.53). We will later provide a formal way to create a partial fraction expansion and use the current problem as motivation for this technique. We wish to write (3.75) in the form $A/h(x) + B/g(x)$ where A and B are constants and $h(x)$ and $g(x)$ are polynomials that correspond to factors of the denominator of (3.75). To factor the denominator of (3.75) we solve the quadratic equation to obtain the two roots

$$r_0 = \frac{-b - \sqrt{b^2 + 4a}}{2a}, \quad r_1 = \frac{-b + \sqrt{b^2 + 4a}}{2a}. \tag{3.77}$$

A special case arises if both roots are the same, so for our purposes here let us assume that $r_0 \neq r_1$. With these roots we can write

$$1 - bx - ax^2 = (1 - x/r_0)(1 - x/r_1).$$

Using this we can write

$$F(x) = \frac{f_0 + x(f_1 - f_0 b)}{(1 - \alpha_0 x)(1 - \alpha_1 x)}, \tag{3.78}$$

where, for notational convenience, we define $\alpha_i \overset{\text{def}}{=} r_i^{-1}, i = 0, 1$. This allows us to write $F(x)$ as a partial fraction expansion

$$F(x) = \frac{A}{1 - \alpha_0 x} + \frac{B}{1 - \alpha_1 x}, \tag{3.79}$$

where A and B are constants. Equation (3.79) must hold for all x, and to determine the constants A and B it suffices to evaluate it for two values. Setting $x = 0$ and $x = 1$ and using (3.78) and (3.79) yields a set of linear equations

$$A + B \qquad\qquad = \quad f_0,$$

$$(1 - \alpha_1)A + (1 - \alpha_0)B \quad = \quad f_0(1 - b) + f_1,$$

which is solved by

$$A \quad = \quad \frac{f_0(\alpha_0 - b) + f_1}{\alpha_0 - \alpha_1} = \frac{-(f_0(\alpha_0 - b) + f_1)}{\sqrt{b^2 + 4a}},$$

$$B \quad = \quad \frac{-(f_0(\alpha_1 - b) + f_1)}{\alpha_0 - \alpha_1} = \frac{f_0(\alpha_1 - b) + f_1}{\sqrt{b^2 + 4a}}. \tag{3.80}$$

In these equations we have used the fact that

$$\alpha_0 - \alpha_1 = -\sqrt{b^2 + 4a}.$$

To obtain the power series expansion for (3.79) we use the identity

$$\frac{1}{1 - \alpha x} = \sum_{i=0}^{\infty} \alpha^i x^i,$$

which, using the summation property (3.53), implies that

$$F(x) = \sum_{i=0}^{\infty} \left(A\alpha_0^i + B\alpha_1^i \right) x^i.$$

Thus we finally conclude that

$$f_n = A\alpha_0^n + B\alpha_1^n. \tag{3.81}$$

As an example, the Fibonacci sequence has $f_0 = 0, f_1 = 1$ and $a = b = 1$, and thus (3.80) and (3.77) yield

$$A = -\frac{1}{\sqrt{5}}, \qquad\qquad B = \frac{1}{\sqrt{5}},$$

$$r_0 = -\phi, \qquad\qquad r_1 = \phi^{-1},$$

where $\phi = \frac{1+\sqrt{5}}{2}$ is the celebrated *golden ratio*[6] (note that $(1 - \sqrt{5})/2 = -\phi^{-1}$). Substituting these equations into (3.81) and simplifying implies that

$$f_n = \frac{1}{\sqrt{5}} \left(\phi^n - (-\phi)^{-n} \right).$$

3.4.2 General Generating Function Solution Method

Let us review the approach used to solve the linear recurrence (3.72). This technique is a standard method for solving all linear recurrence problems and can be extensively generalized. Let $f_0, f_1, \ldots, f_{d-1}$, $d \geq 1$, be the given boundary values of the recurrence and call d the *order* of the recurrence. Without loss of generality we write the recurrence in a standard form

$$f_{i+d} = a_{d-1}f_{i+d-1} + a_{d-2}f_{i+d-2} + \cdots + a_1 f_{i+1} + a_0 f_i, \quad i = 0, 1, \ldots \tag{3.82}$$

and will term the *variable* of the generating function $F(x)$ to be the x variable.

---

[6]The golden ratio, universally denoted by ϕ after the first letter of Phidias, a Greek sculptor who made artistic use of it, has also been called the *divine proportion* by Kepler. The proportion is found frequently in nature (e.g., the beautiful spirals of the chambered nautilus) and in the artistic creations of man (e.g., the front facade of the Parthenon fits within a golden rectangle).

Method for Solving Linear Recurrence Equations

1. Derive the generating function as the ratio of two polynomials $F(x) = P(x)/Q(x)$ by multiplying (3.82) by x^{i+d-1} and summing over i. The polynomial $P(x)$ has a degree of at most $d - 1$ and is a function of all the boundary values. The denominator $Q(x)$ is of degree d, has a constant coefficient of 1, and is a function of the coefficients of the recurrence.

2. Determine the d roots of $Q(x)$ denoted by $r_0, r_1, \ldots, r_{d-1}$. Suppose that there are k distinct roots and furthermore that root $r_i, 1 \le i \le k$, has a multiplicity of m_i (and thus $\sum_{i=1}^{k} m_i = d$). Then we can write $Q(x)$ in a factored form, similar to (3.78), as

$$Q(x) = \frac{\prod_{i=1}^{k}(r_i - x)^{m_i}}{\prod_{i=1}^{k} r_i^{m_i}},$$

where we have used the fact that $Q(x)$ has a constant coefficient of 1. We can rewrite this in a more convenient form by defining $\alpha_i = 1/r_i$, which yields

$$Q(x) = \prod_{i=1}^{k}(1 - \alpha_i x)^{m_i}.$$

A difficulty here is that one cannot generally obtain closed-form expressions for roots of polynomials of degree greater than four.[7]

3. Expand the ratio $P(x)/Q(x)$ by partial fractions. This can be accomplished by use of a standard technique to obtain the coefficients of the partial fraction expansion (see [92], page 15, for a derivation of this). Then

$$A_{i,j} \overset{\text{def}}{=} \frac{1}{(j-1)!} \left(\frac{-1}{\alpha_i}\right)^{j-1} \frac{d^{j-1}}{dx^{j-1}} \left[(1 - \alpha_i x)^{m_i} \frac{P(x)}{Q(x)}\right]\Bigg|_{x=1/\alpha_i}, \tag{3.83}$$

which implies that we can write

$$F(x) = \frac{P(x)}{Q(x)} = \sum_{i=1}^{k} \sum_{j=1}^{m_i} \frac{A_{i,j}}{(1 - \alpha_i x)^{m_i-j+1}}. \tag{3.84}$$

---

[7] This is a theorem due to Neils Henrik Abel (1802–1829) but is expressed more generally using results of Evariste Galois (1811–1832) that were frantically written the night before his death in a duel (see Herstein [48], page 214, for a proof of the theorem and Bell [5] for an account of the duel)!

4. Expand (3.84) into a power series using the geometric scaling property of generating functions in Table 3.6 and the fact that (see Table C.4)

$$\sum_{i=0}^{\infty} \left\langle \begin{array}{c} c \\ i \end{array} \right\rangle x^i = \frac{1}{(1-x)^c}.$$

5. The term associated with x^n in this power series expansion, f_n, is thus given by

$$f_n = \sum_{i=1}^{k} \sum_{j=1}^{m_i} A_{i,j} \left\langle \begin{array}{c} m_i - j + 1 \\ n \end{array} \right\rangle \alpha_i^n. \tag{3.85}$$

Observe that, provided one can determine the roots of $Q(x)$, the above procedure is purely mechanical and could be implemented by a symbolic algebra package. It is easy to see that (3.85) yields the solution (3.81) we previously calculated for the linear recurrence (3.72). Another example of this procedure is given in Example 3.23.

Example 3.23 (Partial Fraction Expansion of Generating Functions)
Consider the generating function

$$F(x) = \frac{x(1-3x)}{(1-2x)(1-4x)^2} \tag{3.86}$$

for which we wish to determine the coefficients $f_n, 0 \leq n$. Recognizing that the factor x corresponds to a positive shift operation that we can account for later, we concentrate on the remaining factors of (3.86), which we write in terms of a partial fraction expansion as follows:

$$\begin{aligned} G(x) &= \frac{(1-3x)}{(1-2x)(1-4x)^2} \\[2mm] &= \frac{A_{1,1}}{1-2x} + \frac{A_{2,1}}{(1-4x)^2} + \frac{A_{2,2}}{1-4x}. \end{aligned} \tag{3.87}$$

We now use (3.83) to determine these constants. The constants corresponding to the simple roots can be calculated in a straightforward manner yielding

$$A_{1,1} = (1-2x)G(x)|_{x=1/2} = \left.\frac{1-3x}{(1-4x)^2}\right|_{x=1/2} = -\frac{1}{2}$$

and

$$A_{2,1} = (1-4x)^2 G(x)\big|_{x=1/4} = \left.\frac{1-3x}{1-2x}\right|_{x=1/4} = \frac{1}{2}.$$

To determine $A_{2,2}$ we evaluate (3.83) for $j = 2$ to obtain

$$A_{2,2} = \left.\frac{-1}{4}\frac{d}{dx}(1-4x)^2 G(x)\right|_{x=1/4} = \left.\frac{1}{4(1-2x)^2}\right|_{x=1/4} = 1.$$

Substituting these constants into (3.85) and simplifying yields

$$g_n = \frac{(n+3)}{2} 4^n - 2^{n-1}. \tag{3.88}$$

We can partially check this result by using the initial sequence value property of Table 3.6, which shows that $g_0 = G(0)$. Evaluating (3.88) yields $g_0 = 1$, and this checks with (3.87), which shows that $G(0) = 1$. To determine the values of f_n we use the positive shift property of Table 3.6, which implies that $f_n = g_{n-1}, n = 1, 2, \ldots$, and thus we have

$$f_n = \begin{cases} 0, & n = 0, \\ \frac{n+3}{2} 4^{n-1} - 2^{n-2}, & n = 1, 2, \ldots. \end{cases}$$

As another example of the use of generating functions consider combinations with replacement.

Example 3.24 (Recurrence for Combinations with Replacement)
Consider the recurrence of selecting k balls from an urn with replacement as given in (3.37),

$$C^r(n, k) = C^r(n, k-1) + C^r(n-1, k), \tag{3.89}$$

and define, for a fixed value of n, the generating function

$$F_n(x) \stackrel{\text{def}}{=} \sum_{i=0}^{\infty} C^r(n, i) x^i.$$

The positive shift property (see Table 3.6) implies that the generating function of (3.89) satisfies

$$F_n(x) = x F_n(x) + F_{n-1}(x)$$

and thus

$$F_n(x) = \frac{F_{n-1}(x)}{1-x},$$

$$= \frac{1}{(1-x)^n}.$$

Table C.5 then implies that

$$C^r(n, k) = \left\langle \begin{array}{c} n \\ k \end{array} \right\rangle,$$

as we have already shown by pattern matching and induction in (3.40).

Example 3.25 (Combinatorial Identities from Generating Functions)
A summation that frequently occurs in probability (e.g., (4.12)) is

$$\sum_{n=i}^{\infty} \binom{n}{i} x^n = \frac{1}{i!} \sum_{n=i}^{\infty} (n)_i x^n.$$

In looking at Table 3.6, we see that this equals $x^i \frac{d^i}{dx^i} F(x)$ where $F(x)$ is the generating function for the sequence $f_i = 1, i \geq 0$. In Table C.4, however, we see that $F(x) = 1/(1-x)$, and thus we can write

$$\sum_{n=i}^{\infty} \binom{n}{i} x^i = \frac{1}{i!} x^i \frac{d^i}{dx^i} \left(\frac{1}{1-x} \right). \tag{3.90}$$

Some algebra shows that

$$\frac{d^i}{dx^i} \left(\frac{1}{1-x} \right) = \frac{i!}{(1-x)^{i+1}}$$

and thus substituting this into (3.90) yields

$$\sum_{n=i}^{\infty} \binom{n}{i} x^n = \frac{x^i}{(1-x)^{i+1}}. \tag{3.91}$$

**3.4.3
Solutions
without
Explicitly
Determining the
Coefficients**

Often, having the generating function is sufficient to determine the sequence. For example, in Table 3.6 properties such as linear scaling, complete summation, and initial value require knowledge of $F(x)$ rather than the set $\{f_n\}$. We will frequently use generating functions, for example, to solve queueing theoretic problems. The value of f_n can correspond to the probability that there are n customers in the queue and measures of interest include the average number of customers in the queue and the probability that the queue is empty. It follows from the linear scaling and initial value property that these measures can be determined directly from $F(x)$. The complete summation property provides a good check that $F(x)$ does actually represent queueing probabilities; $F(1) = 1$ since the sum of the probabilities must equal 1. It is also often the case that one can solve a recurrence equation without explicitly calculating an expression for $F(x)$. We see an instance of this in the following example.

Example 3.26 (Dearrangements)   As an example of the use of generating functions, consider a permutation of the values $1, 2, \ldots, n$ yielding $p_1, p_2, \ldots, p_n$. We wish to find the number, denoted by f_n, of such permutations so that $p_i \neq i$, $1 \leq i \leq n$, in other words, where no item remains in its original position. Such a permutation is called a *dearrangement*, and a variety of problems can be formulated in this manner. We define $f_0 = 1$ and $f_1 = 0$

and derive a recurrence for f_i. To do this we partition all possibilities into two disjoint sets. Consider an element j that starts out in the jth position and suppose that it gets permuted into the kth place, that is, $p_k = j$, $k \neq j$. There are $n - 1$ different ways that k can be chosen, and for each possibility there are two different dearrangements depending on where item k is permuted. In the first case, item k simply exchanges places with j and thus $p_j = k$. For this case the remaining $n - 2$ items also form a dearrangement and the total number of possibilities is given by $(n - 1)f_{n-2}$. The second possibility is that there is no exchange and thus $p_j \neq k$. The remaining $n - 1$ items also form a dearrangment yielding a total number of possibilities of $(n - 1)f_{n-1}$. We thus have argued that

$$f_n = (n - 1)f_{n-1} + (n - 1)f_{n-2}, \quad n \geq 2. \tag{3.92}$$

The difficulty in solving (3.92) lies in the multiplicative factor of $(n-1)$ in each term of the recurrence (otherwise it would look just like the Fibonacci series). A way to solve such a problem is to define $g_n = f_n/n!$ and to substitute this in (3.92), yielding

$$g_n = \left(1 - \frac{1}{n}\right) g_{n-1} + \frac{1}{n} g_{n-2}.$$

This is a convenient form, since we can arrange the equation to yield the differences in g_n as follows

$$g_n - g_{n-1} = -\frac{1}{n}(g_{n-1} - g_{n-2}).$$

Iterating this equation toward the boundary values yields

$$g_n - g_{n-1} = \frac{(-1)^n}{n!},$$

which implies that

$$g_n = (-1)^n \left[\frac{1}{n!} - \frac{1}{(n-1)!} + \frac{1}{(n-2)!} - \cdots\right]$$

$$= 1 - \frac{1}{1!} + \frac{1}{2!} - \frac{1}{3!} + \cdots + (-1)^n \frac{1}{n!}.$$

Observe that this is the initial portion of the power series expansion of e^{-1}. Restated in terms of the original sequence implies that

$$f_n = n! \left(1 - \frac{1}{1!} + \frac{1}{2!} - \frac{1}{3!} + \cdots + (-1)^n \frac{1}{n!}\right). \tag{3.93}$$

For large n we can approximate (3.93) as

$$f_n \approx n! e^{-1}. \tag{3.94}$$

Note that in Example 3.26 we did not need to actually calculate an expression for the generating function to derive an expression for g_n. It is often the case that useful properties of the sequence can be obtained even if one does not have a specification of the generating function. It is also interesting, from an applications point of view, to calculate the probability that a permutation of n items corresponds to a dearrangement. This probability is given by $f_n/n!$, which, by (3.94), approximately equals $1/e$ for large n.

Term (n,k)	Coefficients of recurrence			Combinatorial Interpretation
	$(n-1,k)$	$(n,k-1)$	$(n-1,k-1)$	
$(n)_k$	1	0	k	*Factorial Coefficients* - Number of ways to select k items from a total of n distinguishable items without replacement.
$\langle n \rangle_k$	0	n	0	*Factorial-R Coefficients* - Number of ways to select k items from a total of n distinguishable items with replacement.
$\binom{n}{k}$	1	0	1	*Binomial Coefficients* - Number of ways to select k items from a total of n indistinguishable items without replacement.
$\left\langle \begin{array}{c} n \\ k \end{array} \right\rangle$	1	1	0	*Binomial-R Coefficients* - Number of ways to select k items from a total of n indistinguishable items with replacement.

TABLE 3.10. Coefficients for Combinatorial Recurrences

3.5 Summary of Chapter 3

In this chapter we have derived expressions for the number of different permutations or combinations that can arise when experiments are performed with or without replacement. These expressions form the kernel of all counting arguments and can be extensively generalized. One such generalization deals with generating functions that, in this chapter, are used as formal counting mechanisms. We derived procedures to solve recurrence equations and derived some properties of generating functions. Various examples showed that generating functions can yield solutions even when they cannot be expressed explicitly. We will revisit the results of this chapter frequently in the rest of this book since they are fundamental in solving problems that arise in probabilistic systems. Special numbers that frequently occur in applications, such as the Fibonacci and Catalan numbers, have been examined in this chapter and will be referred to later. A table summarizing the recurrences that have been developed in this chapter is given in Table 3.10.

Bibliographic Notes

Riordan's classic text [96] is probably the most frequently referenced book on combinatorics. Excellent introductory texts include Cohen [18], Constantine [20], and Vilenkin [119]. This last text contains numerous interesting examples and problems. The entertaining text by Graham, Knuth, and Patashnik [43] has a wealth of insights and derivations of properties of special numbers such as Fibonacci and Catalan numbers.

3.6
Problems for
Chapter 3

Problems Related to Section 3.1

3.1 **[R1] The Problem of Points.** This is essentially the problem posed by Chevalier de Méré to Pascal that initiated the Pascal/Fermat correspondence that started the field of probability. We are interested in determining the division of the stakes of an interrupted game of chance between two equally skilled players knowing the scores of the players at the time of interruption and the number of points needed to win the game. We can think of the game as that of tossing a fair coin with the first player winning if heads lands up. In one of the letters, Fermat discussed the case where player 1 needs two points to win and player 2 needs three. The solution here follows from a simple enumeration. It is clear that at most four more games are needed to decide the outcome of the game. Considering all possibilities for four games shows that 11 of the outcomes are favorable to player 1 and 5 to player 2. Hence the stakes should be divided into the ratio 11:5. What is the solution if player 1 needs n_1 points to win and player 2 needs n_2 points to win?

3.2 **[M3]A Problem from Ars Conjectandi in 1713.** In the first book on probability, Huygen's 1657 book *Deratiociniis in Ludo Aleae*, the student is left with five unsolved problems in probability. These problems are solved by Jacob Bernoulli (1654–1705) in his book on probability, *Ars Conjectandi*, published in 1713. You can try your hand here on the fifth problem. Players 1 and 2 each take 12 counters and play with three dice. If 11 is thrown, then player 1 gives a counter to player 2 and if 14 is thrown then player 2 gives a counter to player 1. The game ends when one player has won all the counters. Show that player 1's chance of winning as compared to player 2's chance is as 244,140,625 is to 282,429,536,481.

Actually Bernoulli solves a more general problem. Let players 1 and 2 have m and n counters, respectively, and let their respective chances of winning be given by a to b (i.e., with probability $a/(a+b)$ player 1 wins). As above, the game ends when one player has won all the counters. Bernoulli showed that the probability that player 1 wins is given by

$$\frac{a^n(a^m - b^m)}{a^{m+n} - b^{m+n}}.$$

Derive Bernoulli's solution.

3.3 **[M2]Another Problem from Ars Conjectandi in 1713.** Players 1 and 2 play with two dice, and player 1 wins if a 6 is thrown and player 2 wins if a 7 is thrown. Player 1 first has one throw. This is followed by two throws from player 2. Player 1 then has two throws, followed by three throws from player 2. This pattern continues until one player wins. Show that player 1's chance to win when compared to player 2's chance is 10,355 to 12,276.

3.4 **[E2]** Use Stirling's approximation (3.6) to derive an equation for an approximation for the probability that there are twice as many heads as tails in $m = 3n$ tosses of a fair coin.

3.5 [**M3**] Show that the solution to the recurrence relationship given in (3.1) and (3.2) is given by (3.3).

3.6 [**E1**] There are n_1 red balls and n_2 green balls in an urn from which one is randomly selected. What is the probability that a red ball is selected given that red balls are twice as likely to be selected as green balls.

Problems Related to Section 3.2

3.7 [**R1**] The birthday problem of Example 3.5 assumed that birthdays were equally likely to occur on any of the 365 days of the year. In actuality births are not uniformly distributed. Let $p'(k)$ denote the probability that, of k people, no two have the same birthday when the frequency of birthdays is not uniform. What do you think is the relationship between $p(k)$ given in (3.14) and $p'(k)$? Can you provide a simple example that supports your supposition?

3.8 [**E3**] In Example 3.5 we showed that $(n)_k/\langle n \rangle_k$ has limit 1 as $n \to \infty$ (see (3.16)). Show that

$$\left(1 - \frac{k-1}{n}\right)^{k-1} \leq \frac{(n)_k}{\langle n \rangle_k} \leq \left(1 - \frac{1}{n}\right)^{k-1}.$$

3.9 [**E2**] Prove the set of basic properties for binomial and binomial-R coefficients given in Sections 3.2.6 and 3.2.13, respectively.

3.10 [**E1**] An urn contains n distinguishable balls from which k balls are selected without replacement. These balls are returned to the urn, and we again make a selection of k balls without replacement. What is the probability that the two samples have exactly ℓ balls in common?

3.11 [**E2**] Suppose that an urn contains N balls, M of which are defective where $M \leq N$. A total of $n, n \leq N$, balls are selected at random without replacement from the urn. What is the probability that $m, 1 \leq m \leq M$, of the selected balls are defective?

3.12 [**E1**] There are two decks of 52 cards that have each been randomly shuffled. Cards from each deck are turned over one at a time and compared. A match occurs at a given step if the cards from both decks are the same. What is the probability of having no matches?

3.13 [**M2**] Consider a two-dimensional path on the lattice that starts at point $(0,0)$ and ends at point (n,n) by taking unit steps either horizontally or vertically. How many such paths are there between these two points that do not touch point (i,i) for $i = 1, 2, \ldots, n-1$?

3.14 [**E2**]**Liebnitz's Formula.** Let $D^n f$ denote the nth derivative of a function f. Prove that

$$D^n fg = \sum_{k=0}^{n} \binom{n}{k} D^k f \, D^{n-k} g.$$

Generalize the result to $D^n fgh$.

3.15 [**M2**] A chessboard of size n is a board consisting of n^2 squares arranged in n rows and columns. A rook is a chess piece that when put on square (i, j) of the chess board can attack pieces on any square in the ith row or the jth column. How many different ways can n rooks be placed on a chessboard of size n so that no two rooks can attack each other?

3.16 [**E2**] Show that

$$(n + m)_k = \sum_{j=0}^{k} \binom{j}{k} (n + j)_{k-j} (m - j + 1)_j.$$

3.17 [**E2**] Establish the following combinatorial identities.

$$n \binom{n}{k} = (k + 1) \binom{n}{k + 1} + k \binom{n}{k},$$

$$n \binom{n}{k} = k \binom{n + 1}{k + 1} + \binom{n}{k + 1},$$

$$\binom{n}{2} \binom{n}{k} = \binom{k}{2} \binom{n + 2}{k + 2} + 2k \binom{n + 1}{k + 2} + \binom{n}{k + 2}.$$

3.18 [**E3**] Prove the identities given in (3.31), (3.32), and (3.42).

3.19 [**E3**] How many ways can k indistinguishable balls be put into $2m$, $m \geq 1$, distinguishable urns so that even (resp., odd) urns only have an even (resp., odd) number of balls.

3.20 [**M1**] How many ways can k indistinguishable balls be put into n urns so that no urn has m balls.

3.21 [**M2**] Suppose k balls are randomly thrown into n urns. What is the probability that m, for $1 \leq m \leq n - 1$, urns are empty?

Problems Related to Section 3.3

3.22 [**M1**] How many possible ways are there to distribute k indistinguishable balls into n distinguishable urns so that urn i has at least k_i balls?

3.23 [**M1**] Let $F(x)$ be the generating function for the sequence $f_0, f_1, \ldots$. Write an expression in terms of F that yields $f_1 + f_3 + f_5 + \cdots$.

3.24 [**M2**] Write an expression for the coefficients of (3.47). Hint: Observe that

$$(x^2 + x^3 + \cdots)^n = \frac{x^{2n}}{1 - x},$$

and use (3.46).

3.25 [**E2**] What is the coefficient of x^k in $(1 + x + x^2)^m (1 + x)^m$?

3.26 [**E2**] Use the identity $(1+x)^n(1+x^{-1})^n = x^{-n}(1+x)^{2n}$ to show that

$$\sum_{k=0}^{n} \binom{n}{k}^2 = \binom{2n}{n}.$$

3.27 [**E2**] Use the identity $(1-x^2)^{-n-1} = (1+x)^{-n-1}(1-x)^{-n-1}$ to show that

$$\sum_{k=0}^{2\ell} \binom{n+2\ell-k}{n} \binom{n+k}{n} (-1)^k = \binom{n+\ell}{\ell}.$$

3.28 [**E3**] Use the equality $(1+x)^n(1+x)^{-k-1} = (1+x)^{n-k-1}$ to derive a combinatorial identity.

3.29 [**M3**] An arithmetic unit can sum two integers at a time. Suppose that n integers are to be summed sequentially. As an example, four numbers can be summed sequentially in five different ways $(x_1 + x_2) + (x_3 + x_4)$, $((x_1 + x_2) + x_3) + x_4$, $(x_1 + (x+2+x_3)) + x + 4$, $(x_1 + ((x_2+x_3) + x_4))$, and $x_1 + (x_2 + (x_3 + x_4))$. We wish to calculate the number of ways in which this can be performed in general.

1. Let f_n be the number of ways n integers can be summed. Show that

$$f_n = \sum_{k=1}^{n-1} f_k f_{n-k}.$$

2. Let $F(x)$ be the generating function of f_n. Show that the generating function satisfies $F(x) - x = F^2(x)$ and thus that

$$F(x) = \frac{1 - \sqrt{1-4x}}{2}.$$

3. Using $F(x)$ show that

$$f_n = \frac{1}{n} \binom{2(n-1)}{n-1}$$

and thus f_n is the nth Catalan number (see (3.71)).

3.30 [**E2**] Consider sums of values $a_i, i = 1, 2, \ldots$ so that each a_i is equal to 1 or -1 and let $S_n = a_1 + a_2 + \ldots + a_n$. How many different possible values of S_n are there?

3.31 [**M2**] Use $\langle n \rangle_2 = 2\binom{n}{2} + \binom{n}{1}$ and (3.28) to show that

$$\sum_{k=1}^{n} k^2 = \frac{n(n+1)(2n+1)}{6}.$$

Problems Related to Section 3.4

3.32 [**M3**] Consider a linear recurrence defined by

$$f_{i+1} = \begin{cases} bf_i + af_{i-1}, & 1 \le i \le n-1, \\ b'f_i + a'f_{i-1}, & i \ge n, \end{cases}$$

where $n > 2$ and f_0 and f_1 are given boundary values. Write an expression for $f_n, n = 0, 1, \ldots$.

3.33 [**M3**] Consider a linear recurrence defined by

$$f_{i+1} = \begin{cases} bf_i + af_{i-1}, & i \text{ odd}, \\ b'f_i + a'f_{i-1}, & i \text{ even}, \end{cases}$$

where f_0 and f_1 are given boundary values. Write an expression for $f_n, n = 0, 1, \ldots$.

3.34 [**M1**] Consider a linear recurrence defined by

$$f_{i+1} = c + bf_i + af_{i-1}, \quad i \ge 1,$$

where f_0 and f_1 are given boundary values. Write an expression for $f_n, n = 0, 1, \ldots$.

3.35 [**E3**] The sequence of numbers $1, 5, 19, 65, 211, \ldots$ comes from the recurrence

$$f_{i+1} = bf_i + af_{i-1}, \quad i = 1, 2, \ldots,$$

where a and b are constants. Determine the coefficients of the recurrence and write an expression for f_n.

3.36 [**E2**] Let the generating function for the sequence $f_0, f_1, \ldots$, be given by

$$F(x) = (1-x)\left((x^3 - 2x^2 + 5x)^{1237}(x^3 - 8x^2 + 7)^2 - (x^2 + 18x)^{913}\right).$$

What is f_0, f_1, $\sum_{i=0}^{\infty} f_i$, and $\sum_{i=0}^{\infty} \sum_{j=0}^{i} f_j$?

3.37 [**M2**] How many different ways can you make \$1.23 from an unlimited supply of pennies, nickels, dimes, and quarters? Hint: Use generating functions.

3.38 [**M3**] The Fibonacci numbers satisfy an amazing number of equalities. In this problem let f_i be the ith Fibonacci number. Establish as many of the following equalities as you can.

1. $f_2 + f_4 + \cdots + f_{2n} = f_{2n+1} - 1$.
2. $f_1^2 + f_2^2 + \cdots + f_n^2 = f_n f_{n+1}$.
3. $f_1 f_2 + f_2 f_3 + \cdots + f_{2n-1} f_{2n} = f_{2n}^2$.
4. $f_1 f_2 + f_2 f_3 + \cdots + f_{2n} f_{2n+1} = f_{2n+1}^2 - 1$.
5. $f_3 + f_6 + \cdots + f_{3n} = (f_{3n+2} - 1)/2$.

3.39 [M3]**Matrix Representation of a Linear Recurrence.** Consider the linear recurrence

$$f_{n+1} = 2f_n - f_{n-1}, \quad n = 1, 2, \ldots . \tag{3.95}$$

At this point we do not specify the initial values f_0 and f_1. Observe that we can rewrite (3.95) in matrix form as

$$\begin{bmatrix} f_{n+1} & f_n \end{bmatrix} = \begin{bmatrix} f_n & f_{n-1} \end{bmatrix} M \quad n = 1, 2, \ldots, \tag{3.96}$$

where

$$M \stackrel{\text{def}}{=} \begin{bmatrix} 2 & 1 \\ -1 & 0 \end{bmatrix} .$$

Successive substitution into (3.96) implies that

$$v_{n+1} = v_1 M^n, \quad n = 1, 2, \ldots,$$

where

$$v_{n+1} \stackrel{\text{def}}{=} \begin{bmatrix} f_{n+1} & f_n \end{bmatrix} .$$

Observe that the determinant of M is nonzero, and thus M has an inverse.

1. Write an expression for the boundary values of v_1 in terms of the values v_{n+1}.

2. Prove or disprove: For any value f_n there exist boundary values f_1 and f_0 so that the recurrence (3.95) starting with these values yields f_n.

3. Write a general expression for M^n.

4. Use this to show that the sum of the first and second rows of M^n are equal to $2n + 1$ and $-(2n - 1)$, respectively.

5. Show that

$$f_{n+1} + f_n = (2n + 1)f_1 - (2n - 1)f_0, \quad n = 1, 2, \ldots .$$

6. Show that $f_{n+1} = (n + 2)f_1 - (n - 1)f_0$.

3.40 [M3] Let f_i^d be the number of ways the integer i can be written as the sum of *distinct* integers and let f_i^o be the number of ways i can be written as the sum of *odd* integers. For example,

$f_5^d = 3$	$f_5^o = 3$
5	5
$4 + 1$	$3 + 1 + 1$
$3 + 2$	$1 + 1 + 1 + 1 + 1$

We wish to show that $f_i^d = f_i^o$, a result initially discovered by Euler.

1. Show that the generating function for f_i^d is given by $F^d(x) = (1+x^1)(1+x^2)(1+x^3)\cdots$.

2. Show that the generating function for f_i^o is given by

$$F^o(x) = \frac{1}{(1-x)(1-x^3)(1-x^5)\cdots}.$$

3. Using the fact that $(1+x^i)(1-x^i) = (1-x^{2i})$ show that $F^d(x) = F^o(x)$.

3.41 [E3] Let

$$F(x) = \frac{4x^2(1-8x)}{1-8x+20x^2-16x^3}.$$

Find an expression for f_n.

3.42 [M2] In terms of the generating function for f_i denoted by $F(x)$ write expressions for the generating functions of

1. $(i+1)f_{i+1}$.
2. $f_{ki}, k = 0, 1, \ldots$.
3. $f_{ki+j}, k = 0, 1, \ldots, 0 \le j < k$.
4. $f_{i+1} - f_i$.
5. f_i/i^2.

3.43 [M2] In the first passage problem considered in Section 3.3.5 we assumed that the coin used in the experiment was fair. In this problem we ask you to reproduce the derivation for an unfair coin. Let the probability of heads be given by p. Show that the generating function for the first passage time is given by

$$G(x) = \frac{1 - \sqrt{1 - 4p(1-p)x^2}}{2px},$$

and thus that

$$f_n = \begin{cases} \frac{(-1)^{k-1}}{2(1-p)} \dbinom{1/2}{k} (4p(1-p))^k, & n = 2k-1, \ k = 1, 2, \ldots, \\ \\ 0, & n = 2k, k = 1, 2, \ldots . \end{cases}$$

Show that this can be written as

$$f_n = \begin{cases} \frac{1}{2^{2k-1}} \dbinom{2(k-1)}{k-1} \frac{1}{2k(1-p)} (4p(1-p))^k, & n = 2k-1, \\ \\ 0, & n = 2k. \end{cases}$$

3.44 [M3] Use an inclusion–exclusion argument to determine the number of dearrangements as posed in Example 3.26.

3.45 **[M3]** In this problem we use properties of matrix multiplication to obtain inequalities for Fibonacci numbers. Let $|M|$ be the determinant of a matrix M. Recall that

$$\left|M^{m+n}\right| = |M^m||M^n|.$$

Set

$$M = \begin{bmatrix} 1 & 1 \\ 1 & 0 \end{bmatrix}.$$

Many equalities can be derived from properties of matrix multiplication and the above relationship. Show that the following equalities are among them.

1.

$$M^n = \begin{bmatrix} f_{n+1} & f_n \\ f_n & f_{n-1} \end{bmatrix}.$$

2. $f_{n+1}f_{n-1} - f_n^2 = (-1)^n$.

3. $f_{n+1}f_m + f_n f_{m-1} = f_n f_{m+1} + f_{n-1}f_m$.

4. $f_{n+m+1} - f_{n+m-1} = f_{n+1}f_{m+1} - f_{n-1}f_{m-1}$.

5. $f_{n-1}(f_{n+1} - f_{n-1}) = f_n(f_n - f_{n-2})$.

3.46 **[E2]** Let $f_{n+1} = f_n + f_{n-1}$ be a recurrence relationship and assume that the starting values, f_1 and f_0, are arbitrarily selected positive integers. What is $\lim_{n\to\infty} f_{n+1}/f_n$?

3.47 **[M1]** Write the generating function as in (3.76) for the following recurrence

$$f_{i+k} = a_1 f_{i+k-1} + a_2 f_{i+k-2} + \cdots + a_k f_i, \quad i = 0, 1, \ldots, k \geq 1.$$

4

Random Variables and Distributions

This chapter continues developing the theory of probability by defining random variables that are functions defined on the sample space of a probability space. Most stochastic models are expressed in terms of random variables: the number of customers in a queue, the fraction of time a processor is busy, and the amount of time there are fewer than k customers in the network are all examples of random variables. It is typically more convenient to work with random variables than with the underlying basic events of a sample space. Unlike a variable in algebra, which represents a deterministic value, a random variable represents the outcome of a random experiment and can thus only be characterized by its probabilistic outcome. To characterize these possibilities, we specify a distribution function that expresses the probability that a random variable has a value within a certain range. For example, if we let X_k be the number of heads obtained in k tosses of a fair coin, then the value of X_k is random and can range from 0 to k. The probability that there are less than or equal to ℓ heads, denoted by $P[X_k \leq \ell]$, can be calculated by enumerating all possible outcomes and is given by

$$P[X_k \leq \ell] = 2^{-k} \sum_{i=0}^{\ell} \binom{k}{i} \tag{4.1}$$

(notice that this is the enigmatic summation of (3.36)). The set of values of $P[X_k \leq \ell]$ for all ℓ is called the *distribution function* of the random variable X_k. In this chapter we will show that a distribution function can be used in place of a probability space when specifying the outcome of a random experiment. This is the basis for most probabilistic models that specify the distributions of random variables of interest but *do not specify* the underlying probability space. In fact, it is very rare in models to find a specification of a probability space since this is usually understood from the context of the model.

In Section 4.1 we discuss properties of random variables, and this is followed in Section 4.2 by a discussion of distribution functions. Section 4.3 presents a unified framework to derive distributions that are commonly found in modeling. Properties of random variables and distributions are more accessible when there is a discrete set of possible outcomes, as in the example of the coin tossing experiment just mentioned. This is the reason we first derive discrete forms of distributions in Section 4.4. Properties of these distributions can be obtained through the expedient of coin tossing experiments. For each discrete distribution, there is an associated limiting form obtained by an appropriate scaling operation. This leads us in Section 4.5 to derive a continuous version of each discrete distribution discussed in Section 4.4. Examples of continuous random variables include interarrival times of requests to a computer system and service times at a processor. Useful continuous random variables and their distributions are elucidated in this section. We summarize properties of random variables and particular distributions in Section 4.6.

4.1
Random
Variables

Recall that a function $y = f(x)$ is a mapping that, for each value x, assigns a single value to y. Random variables are functions that are defined on the sample space Ω of a probability space $(\Omega, \mathbf{F}, P)$. More precisely, a random variable $X(\omega)$ is an association of a real valued number to every random outcome $\omega \in \Omega$. (It is usual to suppress the functional dependency on ω; we typically write X for $X(\omega)$.) One can think of a random variable as a real number that results from a random experiment. It is perhaps more appropriate to call such an object a *function* rather than a *variable* since the independent "variable" is a point in the sample space of an experiment rather than the value of a function. The use of the term "variable" is best explained by the fact that we use random variables in mathematical expressions much as variables are used in algebra. For example, we will write expressions such as

$$2X - YZ \qquad (4.2)$$

where X, Y, and Z are random variables.

As an example of a random variable, recall that in Example 2.8 the sample space consisted of all integer pairs (x, y) for which $1 \le x \le 4$ and $1 \le y \le 6$. The random experiment discussed in the example consists of selecting a pair randomly where all pairs are equally likely. We can define a random variable X for this experiment to be the sum of the x and y

coordinates of the randomly selected point. It is clear that $X = 4$ if any one of the points $(3,1), (2,2)$, or $(1,3)$ is selected and thus $P[X = 4] = 1/8$. We can express this relationship more precisely by letting $\mathcal{A}$ be a set of real valued numbers and defining $X^{-1}(\mathcal{A}) \stackrel{\text{def}}{=} \{\omega \mid X(\omega) \in \mathcal{A}\}$ to be the set of outcomes that result in assigning X a value contained in the set $\mathcal{A}$. This implies that

$$P[X \in \mathcal{A}] = P\left[X^{-1}(\mathcal{A})\right]. \tag{4.3}$$

For us to write an expression like (4.3) it must be the case that the event $X^{-1}(\mathcal{A})$ is *measurable* in the probability space $(\Omega, \mathbf{F}, P)$. This means the event must be contained in the family of events $\mathbf{F}$ defined for the probability space,

$$X^{-1}(\mathcal{A}) \in \mathbf{F}.$$

As an example, if $\mathcal{A} = \{2, 4\}$ then $X^{-1}(\mathcal{A}) = \{(1,1), (3,1), (2,2)(1,3)\}$ and thus $P[X \in \mathcal{A}] = P[\{(1,1), (3,1), (2,2), (1,3)\}] = 1/6$. Observe that if we are told the values of $P[X \in \mathcal{A}]$ for all sets $\mathcal{A}$ then we would not need to specify the underlying probability space. An expedient we typically use along these lines is to specify the *distribution* of a random variable given by

$$P[X \le x] = P\left[X^{-1}(-\infty, x]\right], \quad -\infty < x < \infty.$$

This implies that the event $X^{-1}(-\infty, x]$ is measurable for all x. A formal definition of a random variable can now be given.

Definition 4.1 *A random variable X defined on a probability space $(\Omega, \mathbf{F}, P)$ is a real valued mapping that assigns the real number $X(\omega)$ to every random outcome $\omega \in \Omega$ so that for every x, $-\infty < x < \infty$, $\{\omega \mid X(\omega) \le x\}$ is an event contained in $\mathbf{F}$.*

We call a random variable *discrete* if it can only take on a countable number of values, otherwise we say that it is a *continuous* random variable. Random variables that have discrete and continuous components are termed *mixed* random variables. If the sample space is countable then random variables defined on it are necessarily discrete, but discrete random variables can also be defined on uncountable sample spaces. A simple example of this is a random variable fixed at 1 for all random outcomes, that is, $X(\omega) = 1$ for all ω. Such a random variable is clearly discrete even if Ω is uncountable. To make these definitions more concrete consider the following examples.

Example 4.2 (Degenerate Random Variables) The simplest example of a random variable is a constant, that is, $X(\omega) = c$, $\omega \in \Omega$. We call such a random variable a *degenerate random variable* and note that it corresponds to a deterministic value for every random experiment. This is equivalent to a variable that would be defined in algebra, and thus such variables are a special case of random variables. Note that if Z_1 and Z_2 are degenerate random variables that are both equal to c, then the following equality holds trivially:

$$P[Z_1 = Z_2] = 1. \tag{4.4}$$

Also observe that degenerate random variables satisfy the following relationship:

$$P[Z_1 = x_1, Z_2 = x_2] = P[Z_1 = x_1] P[Z_2 = x_2], \qquad \text{for all} \quad x_1, x_2. \tag{4.5}$$

Example 4.3 (Continuation of the Birthday Problem) In Example 3.5 we derived the probability that a group of k people have unique birthdays. This probability was given by $p(k) = (365)_k / 365^k$. Let X be a random variable that equals the number of different birthdays in a set of 23 people. Then $P[X = 23]$ is the probability that all 23 people have different birthdays, that is, $P[X = 23] = p(23)$.

The next two simple examples will be used to delineate all of the possible equality relationships that can exist between two random variables. There are eight such relationships, which are summarized in Table 4.1.

Example 4.4 (A Simple Urn Model) Consider a random experiment consisting of an urn containing one ball, B_a, labeled 0; two balls, B_b and B_c, labeled 1; and one ball, B_d, labeled 2. In the experiment we randomly select a ball and let Y_1 be a random variable that equals the label of the ball. Clearly, we have

$i =$	0	1	2
$Y_1^{-1}(i)$	$\{B_a\}$	$\{B_b, B_c\}$	$\{B_d\}$
$P[Y_1 = i]$	1/4	1/2	1/4

Another random variable is Y_2, which equals 0 if the label on the selected ball is even and 1 if it is odd. For this case, we have

$i =$	0	1
$Y_2^{-1}(i)$	$\{B_a, B_d\}$	$\{B_b, B_c\}$
$P[Y_2 = i]$	1/2	1/2

If the ball selected is B_a, that is, if $\omega = B_a$, then we have $Y_1 = Y_2$. This equality does not hold generally, however. To see this, define random variable Y_3 to be the difference between Y_1 and Y_2. (The value of Y_3 is determined only after performing experiments to determine the values of Y_1 and Y_2.) Then we have

$i =$	0	2
$Y_3^{-1}(i)$	$\{B_a, B_b, B_c\}$	$\{B_d\}$
$P[Y_3 = i]$	$3/4$	$1/4$

Example 4.5 (Continuation of Example 2.1) In Example 2.1 we considered the sample space consisting of all possible outcomes of three tosses of a fair coin. Recall that the sample space for this experiment is given by

$$\Omega = \{HHH, HHT, HTH, THH, HTT, THT, TTH, TTT\}.$$

Many random variables can be defined for this system where, in all cases, we assume the coin is fair:

1. Let X_1 be a random variable that equals the total number of heads obtained in the first and third tosses. This random variable satisfies

$i =$	0	1	2
$X_1^{-1}(i)$	$\{THT, TTT\}$	$\{HTT, TTH, HHT, THH\}$	$\{HTH, HHH\}$
$P[X_1 = i]$	$1/4$	$1/2$	$1/4$

Comparing X_1 with random variable Y_1 of example Example 4.4 shows that

$$P[X_1 = i] = P[Y_1 = i], \; i = 0, 1, 2.$$

We say that X_1 and Y_1 are *stochastically equal* or, equivalently, *equal in distribution*. Notice that two random variables can be stochastically equal even though they are defined on different probability spaces.

2. Let X_2 equal the number of heads obtained in the second toss. Then we have

$i =$	0	1
$X_2^{-1}(i)$	$\{HTH, HTT, TTH, TTT\}$	$\{HHH, HHT, THH, THT\}$
$P[X_2 = i]$	$1/2$	$1/2$

3. Finally we let X_3 be 1 minus the number of tails obtained on the second toss. It satisfies

$i =$	0	1
$X_3^{-1}(i)$	$\{HTH, HTT, TTH, TTT\}$	$\{HHH, HHT, THH, THT\}$
$P[X_3 = i]$	$1/2$	$1/2$

Note that for all possible outcomes ω, $X_2 = X_3$. When this is satisfied we say that the random variables are *equal for all sample points*. In this case it is also clear that random variables X_2 and X_3 are stochastically equal, that is, that $P[X_2 = i] = P[X_3 = i]$ for i.

4.1.1 Forms of Equality

Comparing the preceding examples shows that there are two different forms of equality for random variables. Random variables that are defined on the same sample space can be equal for all possible random outcomes as with X_2 and X_3 of Example 4.5. In this case the random variables are also stochastically equal. On the other hand, random variables can be defined on the same or different probability spaces and yet still be stochastically equal. One such case is that of X_1 and Y_1 of Examples 4.5 and 4.4, respectively. A precise definition of stochastic equality is given in the following definition:

Definition 4.6 *Two random variables X and Y are said to be* stochastically equal, *denoted by $X =_{st} Y$, if*

$$P[X \le a] = P[Y \le a], \quad \text{for all} \quad a. \tag{4.6}$$

4.1.2 Independence

Two random variables X and Y are independent if

$$P[X \le x, Y \le y] = P[X \le x] P[Y \le y], \quad -\infty < x, y < \infty. \tag{4.7}$$

This implies that

$$P[X \le x \mid Y \le y] = P[X \le x].$$

In Example 4.5 it is clear that X_1, which counts the total number of heads in the first and third tosses, and X_2, which counts the number of heads in the second coin toss, are independent random variables. Random variables X_2 and X_3, however, are not independent. As with random events, if two random variables are not independent we say that they are *dependent*.

A special case of independence occurs between degenerate random variables that are equal to each other. In Example 4.2 random variables Z_1 and Z_2 are clearly equal to each other for every random experiment as shown by (4.4) but they are also independent of each other as shown by (4.5). This is the only case where random variables are equal for all sample points but also independent. We summarize all possible relationships between two random variables X and Y in Table 4.1.

It is important to realize that mathematical expressions involving random variables are interpreted differently than in classical algebra, which deals with deterministic quantities. For example, suppose that X_1 and X_2 are independent random variables that satisfy $X_1 =_{st} X$ and $X_2 =_{st} X$. It is tempting, then, to infer that

$$X_1 + X_2 =_{st} X + X = 2X. \tag{4.8}$$

Such a conclusion is, of course, complete nonsense. A simple counterexample is to assume that X can take on the values 0 or 1. Then $2X$ is necessarily equal to 0 or 2 but $X_1 + X_2$ can equal these values as well as 1 ($X_1 = 0$ and $X_2 = 1$, or vice-versa). Hence stochastic equality is not possible. If we change the assumptions so that X_1 and X_2 are not independent but, in fact are equal for all sample points, then (4.8) is valid. These simple examples point out some subtleties in dealing with mathematical expressions involving random variables.

Sample Spaces	Equality	Stochastic Equality	Independence
Same	$X(\omega) = Y(\omega)$	$X =_{st} Y$	Independent
Same	$X(\omega) = Y(\omega)$	$X =_{st} Y$	Dependent
Same	$X(\omega) \neq Y(\omega)$	$X =_{st} Y$	Independent
Same	$X(\omega) \neq Y(\omega)$	$X =_{st} Y$	Dependent
Same	$X(\omega) \neq Y(\omega)$	$X \neq_{st} Y$	Independent
Same	$X(\omega) \neq Y(\omega)$	$X \neq_{st} Y$	Dependent
Different	Not Comparable	$X =_{st} Y$	Independent
Different	Not Comparable	$X \neq_{st} Y$	Independent

TABLE 4.1. Enumeration of all Possible Relationships Between Two Random Variables

4.1.3 Examples of Random Variables

Example 4.7 (Random Variables Associated with Queueing Systems) As mentioned in Chapter 1, queueing systems are often used to model computer systems. For a single server first-come-first-served queue with infinite waiting room, random variables that are of interest include: S_n the service time of the nth customer; A_n, the interarrival time between the nth and $n-1$st customer; W_n, the time spent waiting to receive service by the nth customer; and T_n, the response time of the nth customer ($n = 1, 2, \ldots$, and by convention we set $W_0 = T_0 = 0$). Waiting times and response times (recall from Chapter 1 that the waiting time of a customer is the time it spends in the system prior to receiving service; the response time is the waiting time plus the service time) are calculated values and a random experiment ω for such a system corresponds to an infinite set of randomly selected values $\{S_n, A_n\}$ for $n = 1, 2, \ldots$. The response time of the nth customer is composed of the time the customer spends waiting for and receiving service, and thus

$$T_n = W_n + S_n, \quad n = 1, 2, \ldots . \tag{4.9}$$

To determine a recursion for W_n we consider two cases. The nth customer immediately goes into service if it arrives after the $n-1$st customer has left the system, that is , if $A_n \geq T_{n-1}$. Thus $W_n = 0$ if $A_n \geq T_{n-1}$. On the other hand, if $A_n < T_{n-1}$ then customer n must wait before beginning service. At the time of customer n's arrival the $n-1$st customer has already been in the system A_n time units, and thus there are $T_{n-1} - A_n$ time units left in its response time. This corresponds to the waiting time of customer n, and thus $W_n = T_{n-1} - A_n$ if $A_n < T_{n-1}$. These two cases are succinctly written in one equation, as follows:

$$W_n = (T_{n-1} - A_n)^+$$

$$= (W_{n-1} + S_{n-1} - A_n)^+, \qquad (4.10)$$

where $(x)^+ \overset{\text{def}}{=} \max\{0, x\}$. This recurrence differs from the recurrences considered in Chapter 3 since it is expressed in terms of random variables rather than in terms of deterministic quantities. Solutions to such equations are considered in Chapter 7.

Example 4.8 (The Geometric Random Variable)   Consider an experiment consisting of an urn containing one red ball and $k-1$ green balls, for $k > 1$, and a coin that lands heads up with probability p. The experiment consists of simultaneously selecting a ball from the urn (using replacement) and tossing the coin. Let X_i be a random variable that equals 1 if the ith coin toss lands heads up and equals 0 otherwise. Similarly let Y_i equal 1 if a red ball is chosen from the urn on the ith selection and equal 0 if a green ball is selected. The experiment proceeds in the following fashion. At step 1 we toss the coin and select a ball from the urn (which we then replace). If the ball is red then we stop the experiment. If the ball is green we continue the experiment by again tossing the coin and selecting a ball. We continue in this manner until a red ball is selected and denote this step by $n, 1 \leq n$. The value of n is called a *stopping time* (in Definition 5.13 we provide a precise definition of a stopping time), and it is clear that we can define a random variable N on the probability space associated with the urn so that the event $N^{-1}(n)$ corresponds to selecting a red ball for the first time on step n. Obviously if $N = n$ then $Y_1 = 0, Y_2 = 0, \ldots, Y_{n-1} = 0$ and $Y_n = 1$. Define X to be the cumulative values of X_i up to and including the stopping point of the experiment,

$$X = X_1 + X_2 + \cdots + X_N.$$

We wish to calculate the probability that $X = i$.

Consider first the specific case where $p = 1$ and hence the coin lands heads up on every toss. Then, from above, we can write

$$P[X = i] \;\; = \;\; P[N = i]$$

$$= \;\; P[Y_1 = 0, Y_2 = 0, \ldots Y_{i-1} = 0, Y_i = 1].$$

Since the random variables Y_i are independent of each other, this implies that

$$P[N = i] \;\; = \;\; P[Y_1 = 0] \, P[Y_2 = 0] \cdots P[Y_{i-1} = 0] \, P[Y_i = 1]$$

$$= \;\; (1 - \alpha)^{i-1} \alpha, \;\; i = 1, 2, \ldots \, , \tag{4.11}$$

where $\alpha = 1/k$. Hence, for the case where $p = 1$ the random variable X has a distribution given by

$$P[X = i] = (1 - \alpha)^{i-1} \alpha.$$

This distribution is termed the *geometric distribution with parameter α*. We also say that X is a *geometric random variable*.

The case for general p can be determined by using the conditional form of the law of total probability. If we condition on the event $N = n$ then $P[X = i \mid N = n]$ can be calculated by counting all possible ways i heads can occur from n coin tosses. This is given by

$$P[X = i \mid N = n] = \binom{n}{i} p^i (1 - p)^{n-i}, \;\; 0 \le i \le n,$$

and thus we have

$$P[X = i] \;\; = \;\; \sum_{n=i}^{\infty} P[X = i \mid N = n] \, P[N = n]$$

$$= \;\; \sum_{n=i}^{\infty} \binom{n}{i} p^i (1 - p)^{n-i} (1 - \alpha)^{n-1} \alpha, \;\; i \ge 1. \tag{4.12}$$

Using the identity in (3.91) and a bit of algebra yields

$$P[X = i] = \frac{\alpha (1 - \alpha)^{i-1} p^i}{(1 - (1 - \alpha)(1 - p))^{i+1}}.$$

Notice that if $p = 1$ then X is geometrically distributed and if $p = 0$ then X is a degenerate random variable.

4.1.4 Joint Random Variables

We next turn our attention to the notion of joint random variables. Consider an example of a random experiment that assigns values to two random variables X and Y and a two-dimensional set of values $\mathcal{B}$ that corresponds to some of the possible outcomes of a random experiment. The probability that the *joint values* of X and Y lie within $\mathcal{B}$ is given by

$$P\left[(X,Y) \in \mathcal{B}\right] = P\left[\{\omega : (X(\omega), Y(\omega)) \in \mathcal{B}\}\right].$$

The following example clarifies this concept:

Example 4.9 (Joint Random Variables) Consider a random experiment defined on the sample space consisting of the outcome of a coin toss and the selection of a ball from an urn as in Example 4.4. This sample space is given by

$$\Omega = \{HB_a, TB_a, HB_b, TB_b, HB_c, TB_c, HB_d, TB_d\}.$$

Define X to be a random variable that counts the number of heads and random variable Y to be the value of the selected ball obtained in a random experiment. A random experiment on this sample space simultaneously assigns a value to X and one to Y. Let $Z \stackrel{\text{def}}{=} (X, Y)$ be a random vector. The random experiment assigns Z a value according to the following rules:

$(x, y) =$	(0,0)	(0,1)	(0,2)	(1,0)	(1,1)	(1,2)
$Z^{-1}(x,y)$	$\{TB_a\}$	$\{TB_b, TB_c\}$	$\{TB_d\}$	$\{HB_a\}$	$\{HB_b, HB_c\}$	$\{HB_d\}$
$P\left[Z = (x,y)\right]$	1/8	1/4	1/8	1/8	1/4	1/8

It is clear that, over the same sample space as in the preceding example, we can define many different random variables, and we can define mathematical expressions involving these random variables. Consider the following example:

Example 4.10 (Continuation of Example 4.9) For Example 4.9 let $Z = XY$. This sample space for Z is the same as that in the example and we have the following values:

$i =$	0	1	2
$Z^{-1}(i)$	$\{HB_a, TB_a, TB_b, TB_c, TB_d\}$	$\{HB_b, HB_c\}$	$\{HB_d\}$
$P\left[Z = i\right]$	5/8	2/8	1/8

4.2
Distribution
Functions

As previously mentioned, the value of a random variable can only be specified probabilistically using distribution functions. We say that a function $H(x)$ is a *distribution function* if $H(x)$ is nondecreasing and assumes the values $H(-\infty) = 0$ and $H(\infty) = 1$. We term the function $\overline{H}(x) \stackrel{\text{def}}{=} 1 - H(x)$ the *survivor function* of distribution H and say that a function $h(x)$ is a *density function* if $h(x)$ is nonnegative and $\int_{-\infty}^{\infty} h(z)dz = 1$. For every density function h there is a corresponding distribution function H given by

$$H(x) = \int_{-\infty}^{x} h(z)dz. \tag{4.13}$$

If $H(x)$ is a differentiable distribution function,[1] then

$$h(x) = \left. \frac{d}{dz}H(z) \right|_{z=x}. \tag{4.14}$$

is a density function. Note that there are cases where H satisfies the definition of a distribution function but is not differentiable anywhere. In such cases there is no corresponding density function. We will typically assume that a random variable has a density function.

The distribution function of random variable X is denoted by F_X and has the interpretation

$$F_X(x) = P[X \leq x] = \int_{-\infty}^{x} f_X(z)dz,$$

where $f_X(z)$ is the density function of X. This implies that

$$P[a < X \leq b] = F_X(b) - F_X(a).$$

We can interpret the density function of X as

$$f_X(x)dx \approx P[x < X \leq x + dx] \tag{4.15}$$

for small dx. For a given value x and distribution function we call the interval $(-\infty, x)$ the *initial portion of the distribution* and (x, ∞) the *tail of the distribution*. The statistics of the tail of the distribution are typically used to determine asymptotic properties of X since, for large x, they represent the statistics of extreme values.

The above definitions can be generalized, and we say that a function H is a *joint distribution* if it is nondecreasing, satisfies $H(-\infty, -\infty) = 0$ and $H(\infty, \infty) = 1$, and if $H(x, \infty)$ and $H(\infty, y)$ are distribution functions. For simplicity we discuss the special case of $\mathbb{R}^2$ (the extension to $\mathbb{R}^d$ is

---

[1]More precisely, H must be right-differentiable.

straightforward). In this case we require that H must be *supermodular*, that is, for $-\infty < x_1 \le x_2 < \infty$ and $-\infty < y_1 \le y_2 < \infty$ it must satisfy

$$H(x_2, y_2) + H(x_1, y_1) - H(x_1, y_2) - H(x_2, y_1) \ge 0. \qquad (4.16)$$

A *joint density function* h is a nonnegative function that satisfies

$$\int_{-\infty}^{\infty} \int_{-\infty}^{\infty} h(x, y) dx dy = 1.$$

If h is a density function then H is a distribution function where

$$H(x, y) = \int_{-\infty}^{x} \int_{-\infty}^{y} h(w, z) dw dz,$$

and if H is differentiable then

$$h(x, y) = \left. \frac{\partial^2}{\partial w \partial z} H(w, z) \right|_{(w,z)=(x,y)}$$

is a density function.

The joint distribution function for the *joint random variable* (X, Y) denoted by $F_{(X,Y)}$ has the interpretation

$$F_{(X,Y)}(x, y) = P[X \le x, \ Y \le y] = \int_{-\infty}^{x} \int_{-\infty}^{y} f_{(X,Y)}(w, z) dw dz,$$

where $f_{(X,Y)}$ is the density function of (X, Y). This implies that

$$P[x_1 < X \le x_2, \ y_1 < Y \le y_2] = \int_{x_1}^{x_2} \int_{y_1}^{y_2} f_{(X,Y)}(w, z) dw dz.$$

The requirement of supermodularity in (4.16) follows probabilistically from the fact that

$$P[x_1 < X \le x_2, \ y_1 < Y \le y_2] \ge 0. \qquad (4.17)$$

To see this, observe that we can write (4.17) as

$$F_{(X,Y)}(x_2, y_2) + F_{(X,Y)}(x_1, y_1) - F_{(X,Y)}(x_1, y_2) - F_{(X,Y)}(x_2, y_1) \ge 0.$$

The joint density function can be interpreted as

$$f_{(X,Y)}(x, y) \, dx \, dy \approx P[x < X \le x + dx, \ y < Y \le y + dy].$$

The *marginal density* of Y (the marginal density of X is defined analogously) is determined by

$$f_Y(y) = \int_{-\infty}^{\infty} f_{(X,Y)}(x, y) dx. \qquad (4.18)$$

Note that if X and Y are independent random variables then

$$F_{(X,Y)}(x,y) = F_X(x)F_Y(y) \quad \text{and} \quad f_{(X,Y)}(x,y) = f_X(x)f_Y(y).$$

The *conditional distribution* of X given Y is defined as

$$F_{X \mid Y}(x \mid y) = \frac{F_{(X,Y)}(x,y)}{F_Y(y)} = P[X \leq x \mid Y \leq y] \qquad (4.19)$$

provided that $F_Y(y) > 0$, and the conditional density function is defined by

$$f_{X \mid Y}(x \mid y) = \frac{f_{(X,Y)}(x,y)}{f_Y(y)} \qquad (4.20)$$

provided that $f_Y(y) > 0$. We defer an interpretation of (4.20) to Example 4.15. These definitions of the distribution and density functions are graphically shown in Figure 4.1.

The definitions given above are most suited for continuous random variables since quantities are expressed in terms of integrals and derivatives.

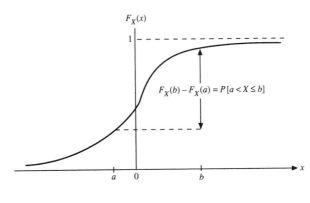

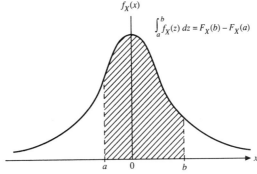

FIGURE 4.1. Distribution and Density Functions

For discrete random variables (without loss of generality we will typi-cally assume that discrete random variables are defined on the integers) the density function of X is interpreted as

$$f_X(x) \stackrel{\text{def}}{=} \begin{cases} P[X=x], & \text{for } x \text{ integer}, \\ \\ 0, & \text{for } x \text{ noninteger}, \end{cases}$$

and the distribution function is given by

$$F_X(x) \stackrel{\text{def}}{=} P[X \leq x] = \sum_{i \leq x} f_X(i).$$

This is a special case of the above definitions for continuous random variables if the value of $f_X(x)$ is 0 for x noninteger and for integer x satisfies

$$\int_x^{x+dx} f_X(z)dz = P[X=x].$$

Hence $f_X(x)$ is a function that integrates to $P[X=x]$ over the range $[x, x+dx]$ for $x = 0, 1, \ldots$. Such a function is defined using an *impulse function*, which we will briefly discuss.

4.2.1
Impulse
Functions

An impulse function, denoted by $\delta_0(t)$, can be thought of as the density function of a degenerate random variable. This function satisfies

$$\delta_0(t) = \begin{cases} \infty, & t = 0, \\ \\ 0, & t \neq 0, \end{cases} \tag{4.21}$$

with the constraints that

$$\int_{-\infty}^{\infty} \delta_0(t)dt = 1, \quad \int_{-\infty}^{x} \delta_0(t)dt = \begin{cases} 1, & x \geq 0, \\ \\ 0, & x < 0. \end{cases} \tag{4.22}$$

We will not dwell on the mathematical delicacies associated with im-pulse functions but will think of such functions as a limiting form of a symmetric function around 0 that has unit area. The particular shape of the symmetric function is not important and one example is that of a triangle with a base $[-b, b]$ on the horizontal axis and a height of $1/2b$ on the vertical axis. The equation for the triangle is given by

$$h_b(x) = \begin{cases} \frac{1}{b}\left(1 - \frac{|x|}{b}\right), & |x| \leq b, \\ \\ 0, & |x| > b. \end{cases}$$

It is clear that this triangle has unit area for all b and that the limiting form as $b \to 0$ satisfies equations (4.21) and (4.22).

We graphically represent the function $a\delta_0(x - b)$ as an arrow at point b that has a height of a units. As an example of how this is used to represent discrete probabilities, consider Example 4.5 where X_1 is the number of heads in three fair coin tosses. It is clear, since each coin toss has a probability of $1/2$ of landing heads or tails, that each of the eight possible outcomes has the same probability of $2^{-3} = 1/8$. To determine the probability that i heads are contained in the three tosses is a matter of counting all possibilities as exemplified in Chapter 3. We thus have that the distribution function is given by (see (4.1))

$$F_{X_1}(x) = \int_{-\infty}^{x} f_{X_1}(x) = \frac{1}{8} \sum_{i \leq x} \binom{3}{i}.$$

We plot these density and distribution functions in Figures 4.2 and 4.3, respectively.

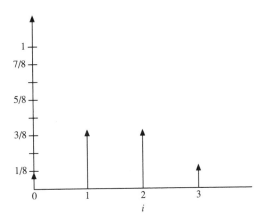

FIGURE 4.2. Density of the Number of Heads in Three Tosses of a Fair Coin

When it is clear that a random variable is discrete we typically simplify notation by dropping the use of the impulse function. For the coin toss example, we would typically write

$$f_{X_1}(i) = \frac{1}{8} \binom{3}{i}, \quad i = 0, 1, 2, 3$$

instead of

$$f_{X_1}(x) = \frac{1}{8} \binom{3}{x} \delta_0(x - i), \quad i = 0, 1, 2, 3.$$

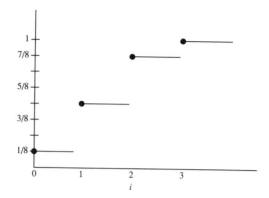

FIGURE 4.3. Distribution of the Number of Heads in Three Tosses of a Fair Coin

Mixed distributions, consisting of continuous and discrete components, frequently occur in practice as shown in the following example.

Example 4.11 (Single Server Queueing System) Consider a single server queueing system that serves customers in first-come-first-served order. Let W be a random variable representing the waiting time of an arriving customer (the time the customer spends in the system prior to receiving service). If there are no customers in the system at the time of arrival, then $W = 0$. If there are N, for $N = 1, 2, \ldots$, customers in the system at time of arrival, then W equals the time needed to completely service these customers, a random variable denoted by Z_N. If the probability that there are no customers in the system is $1 - \rho$, for $0 \le \rho < 1$, then the density of W is given by

$$f_W(x) = (1 - \rho)\delta_0(x) + \rho f_{Z_N}(x), \quad x \ge 0. \tag{4.23}$$

Thus W is a random variable that has an impulse of size $(1 - \rho)$ at the origin and is continuous over $(0, \infty)$. The distribution function of W is given by

$$F_W(x) = 1 - \rho + \rho F_{Z_N}(x), \quad x \ge 0.$$

4.2.2 Probability Generating Functions

The *probability generating function* of a nonnegative integer valued discrete random variable X, denoted by $\mathcal{G}_X$, is the ordinary generating function associated with the sequence of values $P[X = i], i = 0, 1, \ldots,$ and is defined by

$$\mathcal{G}_X(z) \stackrel{\text{def}}{=} \sum_{i=0}^{\infty} P[X = i] z^i, \tag{4.24}$$

provided that the sum converges in some interval about 0. The quantity $\mathcal{G}_X(z)$ is also called the *z-transform* and is frequently used in signal analysis. If $h_i = P[X = i]$, $i = 0, 1, \ldots$, is a sequence of values corresponding to a discrete density function then it is clear that the ordinary generating function of the sequence, $H(x) = \sum_{i=0}^{\infty} h_i x^i$, equals the probability generating function of X evaluated at x, that is, $H(x) = \mathcal{G}_X(x)$. To preserve the semantics associated with random variables, we make it customary to use the "$\mathcal{G}_X(z)$ notation" when dealing with a sequence that corresponds to the discrete density function of a random variable and use the "$H(x)$ notation" when the sequence is not associated with a density. As in Chapter 3, probability generating functions greatly simplify solutions of problems in probability theory. The mathematical foundation for using probability generating functions lies in the fact that discrete distributions are uniquely defined by their probability generating function. The proof of this follows from results in complex analysis, which show that if

$$\sum_{i=0}^{\infty} a_i z^i = \sum_{i=0}^{\infty} b_i z^i, \quad -z_0 < z < z_0,$$

then we can equate coefficients of z^i to conclude that $a_i = b_i$, $i = 0, 1, \ldots$. We record this, without proof, in the following theorem:

Theorem 4.12 *Let X and Y be discrete random variables with probability generating functions $\mathcal{G}_X(z)$ and $\mathcal{G}_Y(z)$, respectively. If $\mathcal{G}_X(z) = \mathcal{G}_Y(z)$ for some z_0, $-z_0 < z < z_0$, then $X =_{st} Y$.*

4.2.3 Existence of Probability Spaces for a Given Distribution

It is clear that, by definition, a random variable has an associated density and distribution function. What is perhaps not so clear is that for any given distribution (or density) function there exists a random variable with that distribution. This means that for any function that is nondecreasing and satisfies $H(-\infty) = 0$ and $H(\infty) = 1$ there exists a probability space and a random variable X defined on that space so that the distribution function of X is given by H. This is an important result because it justifies the common practice of defining a random variable by specifying its distribution function *without ever making reference to an underlying probability space*. For discrete distributions it is easy to construct a probability space and define a random variable on this space with this distribution (see Problem 4.2). The construction is not straightforward, however, for continuous distribution functions, and discussion of this point lies outside the scope of this text.

A distribution completely characterizes a random variable and we can equivalently talk about problems in terms of either random variables or distributions. To avoid repetitious language we will use this equivalence to switch from one nomenclature to the other. For example, we will say that X is a geometric random variable or, equivalently, that X has a geometric distribution.

4.3 Unified Derivation of Common Distributions

We can now specify distribution functions that are commonly found in applications. In doing this we present a unified method for deriving the distributions frequently used in modeling. This approach highlights the relationships between random variables and facilitates a description of their properties. We first discuss deterministic and indicator random variables from which other discrete distributions can be easily derived. These distributions include the discrete uniform, binomial, geometric, and negative binomial distributions. Associated with each of these discrete distributions is an analogous limiting form of the distribution: the continuous uniform, Poisson, exponential, and Erlang distributions, respectively. These four pairs of distributions

$$\text{discrete uniform} \longleftrightarrow \text{continuous uniform}$$

$$\text{binomial} \longleftrightarrow \text{Poisson}$$

$$\text{geometric} \longleftrightarrow \text{exponential}$$

$$\text{negative binomial} \longleftrightarrow \text{Erlang}$$

are extensively used in applications including computer performance modeling, and each pair shares similar mathematical properties. For the discrete versions, properties are readily derived through the expedient of simple coin tossing experiments. Analogous properties for the corresponding limiting forms then follow through limiting arguments. An important pictorial summary of the relationships of all of the above distributions is found in Figure 4.4. One of our key objectives in this chapter is to derive the relationships depicted in this figure.

4.3.1 Indicator Random Variables

In this section we lay the groundwork that allows us to derive the discrete distributions that will be used throughout the text. We start with the simplest random variable: a *degenerate random variable* that takes on only one value for all $\omega \in \Omega$. We let D_c be a degenerate random variable that has value c. The degenerate random variable D_c has the following distribution and density functions:

$$F_{D_c}(x) = \begin{cases} 1, & x \geq c, \\ 0, & x < c, \end{cases} \tag{4.25}$$

and

$$f_{D_c}(x) = \delta_0(x - c). \tag{4.26}$$

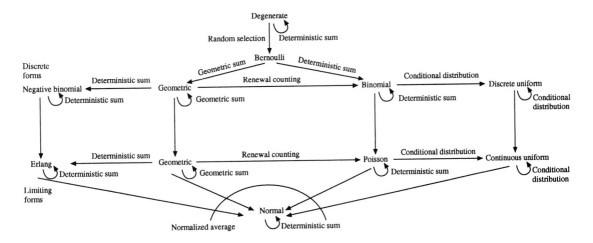

FIGURE 4.4. Distributions and Their Relationships

Naturally a richer set of properties is found in *nondegenerate* random variables and the first generalization we consider is that of an *indicator random variable*. This random variable "indicates" the occurrence of a random event and is defined by

$$I_A(\omega) = \begin{cases} 1, & \omega \in \mathcal{A}, \\ 0, & \omega \notin \mathcal{A}, \end{cases} \tag{4.27}$$

where ω is the outcome of a random experiment. It is clear that (4.27) is a function defined on a sample space Ω and is thus a random variable. Furthermore, the probability that the event $\{I_{\mathcal{A}} = 1\}$ equals the probability of event $\{\mathcal{A}\}$, that is,

$$P[I_{\mathcal{A}} = k] = \begin{cases} 1 - P[\mathcal{A}], & k = 0, \\ P[\mathcal{A}], & k = 1. \end{cases}$$

4.3.2 Bernoulli Random Variables

Indicator random variables are often used as selection mechanisms and it is here that we see that they are generalizations of deterministic random variables. For example, suppose that a random experiment has two possible outcomes, 0 and 1, and let outcome 1 correspond to event $\mathcal{A}$. The outcome from a *random selection*, denoted by random variable X, is given by

$$X = 0 \cdot I_{\mathcal{A}^c} + 1 \cdot I_{\mathcal{A}}$$

$$= I_{\mathcal{A}}$$

and equals an indicator random variable. We term such a random variable a *Bernoulli random variable*[2] *with parameter p*, where $p = P[I_{\mathcal{A}}]$. It is clear that if one sets $p = 0$ or $p = 1$ then one recaptures a degenerate random variable. We let $B_{1,p}$ denote a Bernoulli random variable with parameter p and index a set of independent Bernoulli random variables with parameters $p_i, i = 1, 2, \ldots$, by $B_{1,p_i}^i, i = 1, 2, \ldots$ (this generalized notation is explained by the fact that a Bernoulli random variable is a special case of a binomial random variable that is denoted by $B_{n,p}$). We pictorially represent a Bernoulli random variable in Figure 4.5.

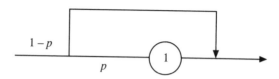

FIGURE 4.5. Pictorial Representation of a Bernoulli Random Variable

---

[2]Named after Jacob Bernoulli (1654–1705).

The distribution function of $B_{1,p}$ is given by

$$F_{B_{1,p}}(x) = \begin{cases} 0, & x < 0, \\ 1 - p, & 0 \le x < 1, \\ 1, & x \ge 1, \end{cases} \tag{4.28}$$

and the density function is given by

$$f_{B_{1,p}}(x) = \begin{cases} 1 - p, & x = 0, \\ p, & x = 1. \end{cases} \tag{4.29}$$

The probability generating function of a Bernoulli random variable is given by

$$\mathcal{G}_{B_{1,p}}(z) = P[X = 0] z^0 + P[X = 1] z^1 = 1 - p(1 - z). \tag{4.30}$$

It is clear that we can use indicator random variables more generally. To do this let $\mathcal{A}_i, 1 \le i \le n$, be a partition of the sample space and set

$$X = 1 \cdot I_{\mathcal{A}_1} + 2 \cdot I_{\mathcal{A}_2} + \cdots + n \cdot I_{\mathcal{A}_n}.$$

We call X a *Bernoulli-type random variable* and note that its distribution is given by

$$F_X(x) = \sum_{j \le x} P[\mathcal{A}_j]$$

and its density by

$$f_X(x) = \begin{cases} P[\mathcal{A}_i], & x = i, \quad i = 1, 2, \ldots, n, \\ 0, & \text{otherwise.} \end{cases}$$

The random variable thus defined corresponds to the outcome of a multi-sided coin toss where the value i is obtained with probability $P[\mathcal{A}_i]$.

4.3.3 Discrete Uniform Distribution

We distinguish the special case of a Bernoulli-type random variable where $P[\mathcal{A}_i] = 1/n, i = 1, 2, \ldots, n$. In such a case each set of the partition occurs with equal probability, and we have

$$P[X = i] = \frac{1}{n}, \quad i = 1, 2, \ldots, n. \tag{4.31}$$

We call this the *discrete uniform distribution* over the set $\{1, 2, \ldots, n\}$. Such a random variable corresponds to a random selection of an item from a set of n items. For future reference let U_n denote a discrete uniform random variable defined over the set $\{1, 2, \ldots, n\}$ and index a set of independent uniform random variables over sets $\{1, 2, \ldots, n_i\}, i = 1, 2, \ldots$, by $U_{n_i}^i, i = 1, 2, \ldots$. The probability generating function of a discrete uniform is given by

$$\mathcal{G}_{U_n}(z) = \frac{z(1 - z^n)}{n(1 - z)}. \tag{4.32}$$

$I_\Omega = 1$	$I_\emptyset = 0$
$I_{A^c} = 1 - I_A$	$I_A I_\Omega = I_A$
$I_A I_A = I_A$	$I_A I_B = I_{AB}$
$I_A I_{A^c} = 0$	$I_A + I_{A^c} = 1$
$I_{A+B} = I_A + I_B - I_{AB}$	$I_A(I_B I_C) = (I_A I_B) I_C$
$I_A(I_B + I_C) = I_A I_B + I_A I_C$	$I_{(AB)^c} = 1 - I_A I_B$

TABLE 4.2. Algebra for Indicator Random Variables

4.3.4 Properties of Indicator Random Variables

Indicator random variables are the fundamental building blocks of distributions and some equalities satisfied by them are given in Table 4.2. Because Bernoulli random variables are indicator random variables they are easily manipulated using the results of Table 4.2. For example, in a coin toss where X equals 1 if heads lands up, an event denoted by H, we can write $X = I_H$. If X_1 and X_2 correspond to independent coin tosses then the random variable X, defined by $X = X_1 X_2$, can be written as

$$X = I_{\{H_1\}} I_{\{H_2\}} = I_{\{H_1, H_2\}},$$

where $\{H_1, H_2\}$ denotes the event of obtaining heads on both tosses. The joint distribution of this experiment is given by

$$F_X(x_1, x_2) = \begin{cases} 0, & x_1 < 0, x_2 < 0, \\ (1-p)^2, & 0 \le x_1 < 1, \ 0 \le x_2 < 1, \\ (1-p), & x_1 \ge 1, \ 0 \le x_2 < 1, \ \text{or} \ 0 \le x_1 < 1, x_2 \ge 1, \\ 1, & x_1 \ge 1, \ x_2 \ge 1. \end{cases}$$

The corresponding *joint density* is given by

$$f_X(x_1, x_2) = \begin{cases} (1-p)^2, & x_1 = 0, x_2 = 0, \\ p(1-p), & x_1 = 0, x_2 = 1, \quad \text{or } x_1 = 1, x_2 = 0, \\ p^2, & x_1 = 1, x_2 = 1. \end{cases}$$

4.4 Constructing Discrete Distributions from Indicator Random Variables

From indicator random variables we can construct many of the distributions that we need in the text by the application of a few basic operations. These operations provide insight into relationships between different random variables and establishes a framework within which to consider stochastic models based on these distributions.

As an example, the arrivals of requests to a computer system often arise from a large number of users, each of which issues requests infrequently. To model such an arrival process, one divides time into equal length slots and assumes that each user generates requests in each slot with a small probability. Under these assumptions, the total number of requests has a binomial distribution (all distributions mentioned in these motivating comments will be derived later). If the probability of generating a request is sufficiently small then the distribution of the number of requests received from a large number of users is approximated by a limiting form of the binomial distribution, namely, the Poisson distribution. The fact that many computer systems satisfy these assumptions explains why it is typical to model arrivals as if they were Poisson. Our derivation of the Poisson distribution presented later in Section 4.5.3 uses such a limiting process. This highlights how one links up the mathematical properties of a model with the features of an application.

A distribution closely related to the Poisson distribution, the exponential distribution, also plays a key role in mathematical modeling. We show that this distribution is obtained as a limiting form of coin tossing, where an experiment ends the first time the coin lands heads up. Example 4.8 showed that the duration of the experiment is given by the geometric distribution. It is clear that the probability that the coin lands heads up does not depend on the outcomes of previous tosses. Each time we toss the coin we, in a sense, start a new game, and the duration of the remaining portion of the game after tossing k tails does not depend on k. In other words, the probability that the game lasts five trials is the same as the probability that the game lasts five trials given that it has already lasted k trials. We say that the geometric distribution is memoryless. Since the exponential distribution is a limiting form of a geometric distribution, it is also memoryless.

4.4.1 Operations Used to Form Random Variables

We now describe some basic operations that are used to derive distributions for random variables or to establish relationships between them. These operations will become more readily understood when they are later used in this section and we record them here for future reference.

1. **Random Summation:** Let $X_i, i \geq 1$, be a sequence of independent and identically distributed random variables and let N be an integer random variable. Define

$$X \stackrel{\text{def}}{=} X_1 + X_2 + \cdots + X_N.$$

We wish to calculate the distribution of X under two different cases: *deterministic sum* when N is a degenerate random variable and thus $N = n$ for some constant integer n, and *geometric sum* if N has a geometric distribution.

2. **Renewal:** Consider a sequence of independent and identically distributed discrete random variables X_i where $X_i \geq 1$ for all $i = 1, 2, \ldots$. Let $N(m)$ be a random variable that counts the number of completed events in m time units. We are interested in the distribution of $N(m)$, which is given by

$$P\left[N(m) \leq n\right] = P\left[X_1 + X_2 + \cdots + X_{n+1} > m\right]. \qquad (4.33)$$

To explain (4.33), note that the event $N(m) \leq n$ implies that the $n+1$st event cannot occur within m time units. If the $n+1$st event occurs after time m, however, it must be the case, since $X_n \geq 1$, that no more than n events occurred prior to time m. The explanation of the nomenclature *renewal* will be deferred until Section 4.4.15.

3. **Conditional Distribution:** For given random variables Y and Z we define X to be the value of Y given the value of Z, which is denoted by $X = Y \mid Z$. We wish to calculate the distribution of X.

We now use these operations to derive a family of distributions that will be used throughout the text. The inter-relationships between distributions and their intrinsic properties can be found in Figure 4.4.

4.4.2 Discrete Uniform Random Variables

We use the conditional distribution operation to show that discrete uniform distributions are invariant under conditioning. Consider a discrete uniform random variable defined over the integers $\{1, 2, \ldots, n\}$, denoted by U_n. We claim that the distribution of U_n conditioned on the event

$\{U_n \leq k\}$ is uniformly distributed over $\{1, 2, \ldots, k\}$. To show this we use the conditional form of the law of total probability to write

$$P[U_n = i \mid U_n \leq k] = \frac{P[U_n = i, U_n \leq k]}{P[U_n \leq k]}$$

$$= \frac{1/n}{k/n} = \frac{1}{k}, \quad i = 1, 2, \ldots, k.$$

To interpret this result, consider an experiment where we have a set of n urns into which we randomly place one ball. The probability that the ball is found in any given urn equals $1/n$. For a smaller version of the experiment, say one with $n - 1$ urns, the probability of selecting the urn containing the ball equals $1/(n-1)$. Suppose that in the original experiment we select urn i and find that the ball is not in the urn. The ball must be in one of the urns $1, 2, \ldots, i - 1, i + 1, \ldots, n - 1$, and the distribution of which urn it is contained in is exactly the same as that of an experiment consisting of only $n - 1$ urns. Hence the conditional probability of selecting the urn containing the ball is given by $1/(n-1)$.

4.4.3 Binomial Random Variables

Our first use of the deterministic sum operator is to create the *binomial distribution* (see Figure 4.4). Let X be the deterministic sum of n independent Bernoulli random variables with parameter p given by

$$X = B_{1,p}^1 + B_{1,p}^2 + \cdots + B_{1,p}^n.$$

We wish to determine the distribution of X. Clearly the value of X equals k if exactly k of the Bernoulli random variables equal 1. For such a case there are $\binom{n}{k}$ different ways to select which Bernoulli random variables are equal to 1, and each such possibility has the same probability of $p^k(1-p)^{n-k}$. Hence we have

$$F_X(x) = \sum_{k \leq x} \binom{n}{k} p^k (1-p)^{n-k}, \qquad 0 \leq x. \qquad (4.34)$$

This distribution is called the *binomial distribution with parameters* (n, p). We let $B_{n,p}$ denote a binomial random variable with parameters (n, p) and index a set of independent binomial random variables with parameters $(n_i, p_i), i = 1, 2, \ldots,$ by B_{n_i,p_i}^i. The density function of a binomial random variable is given by

$$f_{B_{n,p}}(x) = \binom{n}{x} p^x (1-p)^{n-x}, \qquad x = 0, 1, \ldots, n, \qquad (4.35)$$

and is plotted in Figure 4.6.

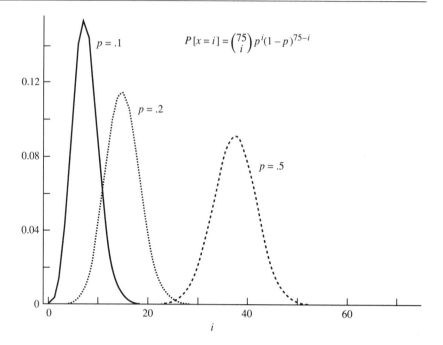

FIGURE 4.6. Binomial Density Function

We note that the probability generating function of $B_{n,p}$ is given by

$$
\mathcal{G}_{B_{n,p}}(z) = \sum_{k=0}^{n} \binom{n}{k} p^k (1-p)^{n-k} z^k
$$

$$
= (1 - p(1-z))^n. \tag{4.36}
$$

The convolution property of generating functions discussed in Section 3.3.4 (see (3.54) and (3.55)) implies that the probability generating function of the sum of independent random variables equals the product of their individual probability generating functions. Thus (4.36) shows that

$$
B_{n,p} =_{st} B^1_{1,p} + B^2_{1,p} + \cdots + B^n_{1,p}. \tag{4.37}
$$

We represent the equality (4.37) pictorially in Figure 4.7.

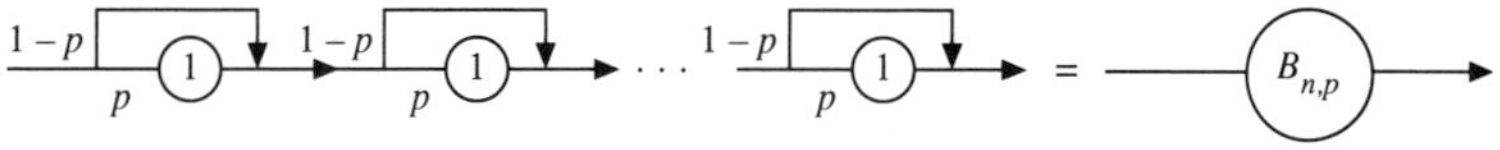

FIGURE 4.7. The Sum of Bernoulli Random Variables Is Binomial

4.4.4 Sum of Binomial Random Variables with the Same Parameter

It is interesting to note that the deterministic sum of a set of independent and identically distributed binomial random variables is also binomially distributed. To see this, let

$$X = B_{n_1,p} + B_{n_2,p} + \cdots + B_{n_k,p}.$$

To calculate the density of X it is convenient to use probability generating functions. The convolution property of generating functions implies that the probability generating function of X is given by

$$\mathcal{G}_X(z) = (1 - p + pz)^{n_1+n_2+\cdots+n_k}.$$

Comparing this with (4.36), shows that this is the probability generating function for the distribution

$$P\left[X = i\right] = \binom{n}{i} p^i (1 - p)^{n-i}, \quad i = 0, 1, \ldots, n,$$

where $n \stackrel{\text{def}}{=} n_1 + n_2 + \cdots + n_k$. Thus the sum of k binomial variables with parameters $(n_i, p), i = 1, 2, \ldots, k$, is binomial with parameters $(n_1 + \cdots + n_k, p)$. We record this as

$$B_{n_1+\cdots+n_k,p} =_{st} B_{n_1,p} + \cdots + B_{n_k,p}. \tag{4.38}$$

4.4.5 Intuitive Argument for Sum of Binomial Random Variables

The above derivation implies that binomial random variables can be viewed as counting the number of successful outcomes obtained in a series of independent Bernoulli experiments. This interpretation allows us to derive certain results by appealing to intuition. For example, suppose we take n balls and sequentially put them into an urn after coloring each ball red independently with probability p. The total number of red balls in the urn is binomially distributed with parameters (n, p). It is clear, however, that we could consider this experiment as proceeding in two stages: the placement of the first n_1 balls in the urn, followed by the placement of the remaining $n_2 = n - n_1$ balls. The final distribution obviously is not altered by viewing the experiment in this manner, and thus the final distribution of the number of red balls for the two-stage experiment is also binomial with parameters (n, p). Since each step of the experiment is simply a smaller version of the total experiment, it follows that the number of red balls placed in the urn in each step of the experiment is binomially distributed with parameters (n_1, p) and (n_2, p), respectively. This simple reasoning establishes that the sum of two binomial random variables with parameters (n_1, p) and (n_2, p), respectively, is binomially distributed with parameters $(n_1 + n_2, p)$. Taking the above line of reasoning to the extreme, we can consider the experiment with n balls as n experiments each with one ball. Each of these experiments, then, is binomially distributed with parameters $(1, p)$ and, substituting into (4.35), we obtain the Bernoulli distribution as expected. Hence the Bernoulli distribution is simply a special case of the binomial distribution.

**4.4.6
Sum of Binomial
Random
Variables with
Different
Parameters**

An experiment similar to that just discussed lends intuition into why the sum of binomial random variables with different probability parameter values is not binomially distributed. Indeed, suppose that this were the case. Then if we view each experiment as consisting of one ball, this implies that the probability we use to color balls could be varied with each step of the experiment. For example, ball i could be colored red with probability $p(i), i = 1, 2, \ldots, n$. If the number of red balls in the urn under these conditions is binomially distributed, then all possible ways of coloring balls and putting them into an urn yield a binomial distribution. Hence we would be forced to conclude that all discrete distributions are binomial!

This conclusion is clearly ridiculous but it is interesting to contrast this with a result we establish in Section 4.5.5, that shows that there is a limiting form of such a statement which is true. To describe this result, suppose that we sum n Bernoulli random variables, each with a possibly different parameter value as above. For a given value of n, let $p_n(i)$, for $i = 1, 2, \ldots, n$, be the probability that the ith Bernoulli random variable equals 1. Suppose that for each value i, the value of $p_n(i)$ satisfies $p_n(i) \to 0$ as $n \to \infty$. This implies that for large n the probability that any one of the n Bernoulli random variables is nonzero is small. With this assumption suppose that

$$p_n(1) + p_n(2) + \cdots + p_n(n) \longrightarrow \lambda, \quad \text{as } n \longrightarrow \infty, \quad (4.39)$$

where λ is some positive constant. In Section 4.5.5 we show that the sum of the n Bernoulli random variables has a limiting distribution that depends only on λ and is independent of the particularities of the functions $p_n(i)$. This limiting form is the celebrated Poisson distribution, which is widely used in computer performance modeling. We are forced to conclude that under the assumption of (4.39), the limiting sums of Bernoulli random variables with different probability parameters do have the same distribution!

Example 4.13 (Requests from Terminals) One typically models the arrival of requests from terminals in discrete time systems as the summation of Bernoulli random variables. Time is slotted into unit intervals, and one assumes that each terminal issues a request (at most one) in each slot with a certain probability. The above argument shows that in such systems one must make uniform assumptions about the terminals (specifically that the probability of generating a request is the same in each terminal) in order to have a simple model of the total arrival of requests in each slot. For example, if there were two classes of terminals, say n_1 terminals that generated requests with probability p_1 and a second class of n_2 terminals that generated requests with probability $p_2 \neq p_1$, then the total arrival of requests in each slot is given by the convolution of two binomial distributions. If X is the number of requests in a given slot, then for $i = 0, 1, \ldots, n_1 + n_2$ we have

$$P[X = i] = \sum_{k=0}^{i} \binom{n_1}{k} p_1^k (1 - p_1)^{n_1-k} \binom{n_2}{i-k} p_2^{i-k} (1 - p_2)^{n_2-i+k}.$$

This distribution is decidedly more difficult to deal with than a simple binomial distribution. As the number of terminals becomes large and their corresponding probability of generating a request in each slot becomes small, a tractable model is created, namely, that of a Poisson distribution.

Example 4.13 is our first glimpse into a phenomena intrinsic to many probabilistic models - that they often become more tractable as the number of "objects" increases. Such simplifications arise from limiting theorems in probability that will be discussed in Section 5.5. The above simplification of an arrival process when the number of terminals increases is an example of this phenomena.

4.4.7 Geometric Random Variables

Let $G_{1,p}$ be a geometric random variable with parameter p and index a set of independent geometric random variables with parameters $p_i, i = 1, 2, \ldots,$ by $G^i_{1,p_i}, i = 1, 2, \ldots$ (this notation is explained by the fact that a geometric random variable is a special case of a negative binomial random variable denoted by $G_{1,p}$). Recall that the density of a geometric random variable is given by (this is shown graphically in Figure 4.8)

$$f_{G_{1,p}}(i) = (1 - p)^{i-1} p, \quad i = 1, 2, \ldots, \tag{4.40}$$

and the probability generating function is given by

$$\mathcal{G}_{G_{1,p}}(z) = \frac{pz}{1 - (1 - p)z}. \tag{4.41}$$

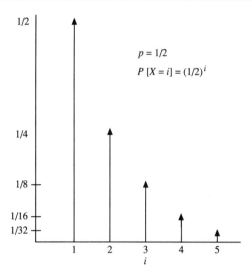

FIGURE 4.8. Geometric Density Function

FIGURE 4.9. Pictorial Representation of a Geometric Random Variable

We represent a geometric random variable pictorially in Figure 4.9.

4.4.8 Geometric Sum of Geometric Random Variables

Let X be the geometric sum of geometric random variables defined by

$$X = G_{1,p}^1 + G_{1,p}^2 + \cdots + G_{1,p}^N,$$

where N is a geometric random variable with parameter α. We show here that X is geometric with parameter αp. The proof of this is simplified by the use of generating functions. To calculate the probability generating function of X we use the convolution property of generating functions. If we condition on $N = n$ (which occurs with probability $(1 - \alpha)^{n-1}\alpha$) then we can write

$$
\begin{aligned}
\mathcal{G}_X(z) &= \sum_{n=1}^{\infty} \left(\frac{pz}{1 - (1-p)z} \right)^n (1 - \alpha)^{n-1}\alpha \\
&= \frac{\alpha p z}{1 - (1 - \alpha p)z}.
\end{aligned}
$$

This is the same form as in (4.41), which shows that X is geometric with parameter αp. Thus we have shown that if $N =_{st} G_{1,\alpha}$ then

$$G_{1,\alpha p} =_{st} G_{1,p}^1 + G_{1,p}^2 + \cdots + G_{1,p}^N. \tag{4.42}$$

We represent this equality pictorially in Figure 4.10.

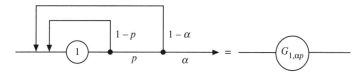

FIGURE 4.10. The Geometric Sum of Geometric Random Variables Is Geometric

4.4.9 Starting Geometric Distributions from 0

Sometimes the domain of geometric random variables starts from 0 rather than from 1. We can define $Y = G_{1,p} - 1$ to be a geometric random variable starting from 0 with a density given by

$$P[Y = i] = (1 - p)^i p, \quad i = 0, 1, \ldots . \tag{4.43}$$

This random variable corresponds to an experiment that can stop before the first step. When we say a distribution is geometric without specifying if it starts from 1 or 0, we shall always assume that it starts from 1. Note that the survivor function of Y is given by

$$P[Y > n] = \sum_{i=n+1}^{\infty} p(1 - p)^i = (1 - p)^{n+1}, \quad n = 0, 1, \ldots . \tag{4.44}$$

4.4.10 Memoryless Property of the Geometric Distribution

A property of geometric random variables that is crucial in modeling systems is the *memoryless property*. This property can best be described in terms of a coin tossing experiment. Consider a sequence of coin tosses where each toss lands heads up with probability p. We stop the experiment the first time heads lands up. The duration of the game is geometric with probability p, and hence the probability that the game lasts more than n tosses can be written as

$$P\left[G_{1,p} > n\right] = \sum_{i=n+1}^{\infty} (1 - p)^{i-1} p$$

$$= (1 - p)^n, \quad n = 1, 2, \ldots . \tag{4.45}$$

Suppose that the coin is tossed m times and lands tails up each time. This implies that the duration of the game is greater than m tosses. We wish to determine the probability that the game continues for more than n additional tosses. Intuitively, it is clear that the fact that tails have landed up in the previous m tosses has no effect on the probability that heads will land up on any succeeding toss. In essence, at each toss we can consider that we start the game over, and thus the probability that it continues for n more tosses is the same as the probability that a new game lasts more than n tosses. We thus have argued that

$$P\left[G_{1,p} > m + n \mid G_{1,p} > m\right] = P\left[G_{1,p} > n\right].$$

We can formally prove this using (4.45) as follows:

$$
\begin{aligned}
P\left[G_{1,p} > m + n \mid G_{1,p} > m\right] &= \frac{P\left[G_{1,p} > n + m, G_{1,p} > m\right]}{P\left[G_{1,p} > m\right]} \\
&= \frac{\sum_{i=n+m+1}^{\infty}(1-p)^{i-1}p}{(1-p)^m} \\
&= (1-p)^n.
\end{aligned}
\tag{4.46}
$$

This property is called the *memoryless property* since past outcomes do not influence future probabilities. This property plays a key role in modeling since it allows one to analyze a system without having to keep track of events that have occurred in the past.

4.4.11 Discrete + Memoryless $\Longrightarrow$ Geometric

It is natural to question if there are other discrete distributions that are memoryless. To show that this property is unique to geometric random variables, let Y be a random variable that equals the number of steps of an experiment that stops on step i with probability $\beta_i, i \geq 1$. The probability the experiment stops at step 1 is thus given by β_1. If Y is memoryless, the conditional probability that the experiment stops at step i, given that it has not stopped by step $i - 1$, is also equal to β_1. Using the above definitions we can write

$$\alpha_{i-1} \stackrel{\text{def}}{=} P\left[Y > i - 1\right] = \beta_i + \beta_{i+1} + \cdots, \quad i \geq 1. \tag{4.47}$$

The assumption that Y is memoryless implies that

$$\beta_1 = P\left[Y = i \mid Y > i - 1\right] = \frac{\beta_i}{\alpha_{i-1}}, \quad i \geq 1. \tag{4.48}$$

Equation (4.47) implies that

$$\beta_i = \alpha_{i-1} - \alpha_i, \quad i \geq 1, \tag{4.49}$$

which, in turn, implies that (4.48) can be rewritten as

$$\frac{\alpha_i}{\alpha_{i-1}} = 1 - \beta_1, \quad i \geq 1.$$

Iterating this equation results in

$$\alpha_i = (1 - \beta_1)^i, \quad i \geq 1,$$

and, using (4.49), yields

$$\beta_i = (1 - \beta_1)^{i-1}\beta_1,$$

which establishes that Y is geometric as claimed. A random variable that is discrete and memoryless is thus necessarily geometrically distributed.

4.4.12 Negative Binomial Distribution

We next derive the negative binomial distribution from the geometric distribution using the deterministic sum operation. Let X be the deterministic sum of n independent geometric random variables with parameter p. To determine the distribution of X we use the convolution property of generating functions and (4.41) to write

$$\mathcal{G}_X(z) = \left(\frac{pz}{(1 - (1-p)z)}\right)^n. \tag{4.50}$$

To invert this we use (3.46) to write

$$\left(\frac{pz}{(1 - (1-p)z)}\right)^n = (pz)^n \sum_{j=0}^{\infty} \binom{n+j-1}{j} (1-p)^j z^j.$$

A simple change of variables allows us to determine the coefficient of z^i, which yields

$$P[X = i] = \begin{cases} 0, & i = 0, 1, \ldots, n-1, \\[2ex] p^n \binom{i-1}{i-n} (1-p)^{i-n}, & i \geq n, \end{cases} \tag{4.51}$$

which is termed the *negative binomial distribution with parameters* (n, p) (this distribution is pictured in Figure 4.11).

Using our previous notation, we let $G_{n,p}$ be a negative binomial random variable with parameters (n, p) and index a set of independent negative binomial random variables with parameters $(n_i, p_i), i = 1, 2, \ldots$, by

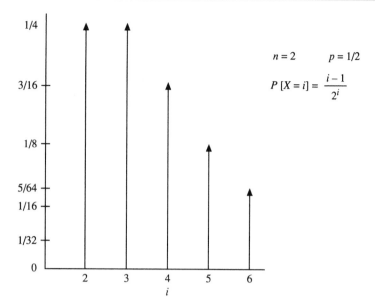

FIGURE 4.11. Negative Binomial Density Function

$G_{n_i,p_i}^i, i = 1, 2, \dots$. The name "negative binomial" for the distribution arises from the equality

$$\binom{-n}{i} = \frac{(-n)!}{i!(-n-i)!}$$

$$= (-1)^i \frac{(n)(n+1)\cdots(n+i-1)}{i!}$$

$$= (-1)^i \binom{n+i-1}{i},$$

which shows that we can write (4.51) as

$$P[G_{n,p} = i] = p^n \binom{i-1}{i-n}(1-p)^{i-n}$$

$$= p^n \binom{-n}{i-n}(-1)^i(1-p)^{i-n}, \quad i = n, n+1, \dots.$$

We record the probability generating function of a negative binomial random variable with parameters (n, p) as

$$\mathcal{G}_{G_{n,p}}(z) = \left(\frac{pz}{1-(1-p)z}\right)^n \tag{4.52}$$

and note that the sum of n geometric random variables with parameter p has a negative binomial distribution with parameters (n, p):

$$G_{n,p} =_{st} G^1_{1,p} + G^2_{1,p} + \cdots + G^n_{1,p}. \tag{4.53}$$

This equality (4.53) is pictorially represented in Figure 4.12.

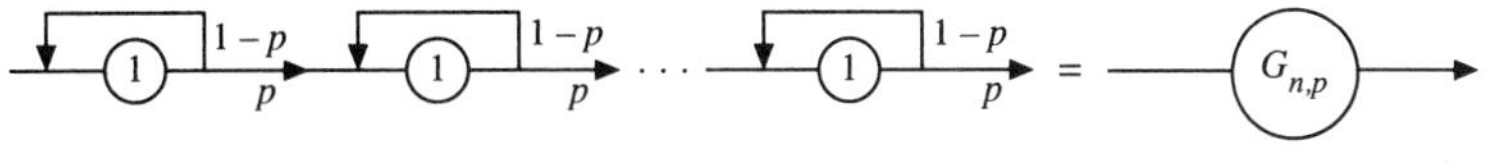

FIGURE 4.12. The Sum of Geometric Random Variables Is Negative Binomial

4.4.13 Sum of Negative Binomial Random Variables

It is easy to see that the deterministic sum of random variables that have a negative binomial distribution also has a negative binomial distribution. Define X by

$$X = G_{n_1,p} + G_{n_2,p} + \cdots + G_{n_k,p}.$$

The probability generating function for X is the product of terms of the form (4.50) and thus equals

$$\mathcal{G}_X(z) = \left(\frac{pz}{1 - (1-p)z} \right)^{n_1+n_2+\cdots+n_k},$$

which clearly has the same form as a negative binomial. Hence the sum has a negative binomial distribution with parameters $(n_1 + n_2 + \cdots + n_k, p)$, which we record as

$$G_{n_1+n_2+\cdots+n_k,p} =_{st} G_{n_1,p} + G_{n_2,p} + \cdots + G_{n_k,p}. \tag{4.54}$$

4.4.14 Starting Negative Binomial Distributions from 0

As with geometric random variables, sometimes negative binomial random variables are assumed to start from 0 rather than from n. Random variable $Y = G_{n,p} - n$ is called a negative binomial random variable that starts from 0. Clearly Y has a distribution given by

$$P[Y = i] = p^n \binom{n+i-1}{i} (1-p)^i$$

$$= p^n \binom{-n}{i} (-1)^i (1-p)^i, \quad i = 0, 1, 2 \ldots . \tag{4.55}$$

Unless specified otherwise, when we say that a distribution is negative binomial with parameters (n, p) we implicitly mean that it starts from n. We use the negative binomial distribution in the next section to establish relationships between geometric and binomial random variables. Although negative binomial distributions are not frequently found in applications, their limiting form, Erlang random variables, are widely used and will be used in Markov process theory to model computer systems (see Chapter 8).

4.4.15 Relationship Between Geometric and Binomial Distributions

In this section we use the renewal operation to derive an important relationship between the geometric and binomial distributions. Let $X_i, i \geq 1$, be a set of independent and identically distributed integer valued random variables. It is convenient to think of the values of X_i as the lifetime of some component that needs replacement or *renewal* at random times. We form the partial sums

$$S_n = X_1 + X_2 + \cdots + X_n, \quad n \geq 1,$$

and note that the values of $S_1, S_2, \ldots,$ form an increasing integer valued sequence. The value of S_n is called the time of the nth *renewal epoch*. We define the integer valued random variable $N(m)$ to be the number of *renewal epochs* found in m time units. Random variable $N(m)$ is termed the *renewal counting process* for random variables X_i. These definitions are shown graphically in Figure 4.13.

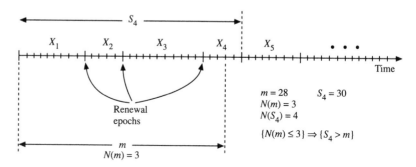

FIGURE 4.13. Renewal Counting Process

If we select m to be equal to the time of the nth renewal then it is clear that the number of renewals up to time m is given by n, that is, $N(m) = n$. We are interested in determining the distribution of the number of renewals for an arbitrarily selected time. It is clear that the event $\{N(m) \leq n\}$ occurs if the $n + 1$st renewal takes place after time m. This probability is given by (4.33), repeated here as

$$P[N(m) \leq n] = P[S_{n+1} > m]. \tag{4.56}$$

This simple but important relationship and its continuous time counterpart (see (4.96)) have profound consequences in applied probability. Our first use of the relationship is to show that the renewal counting process for geometric random variables is binomially distributed.

4.4.16 Renewal Counting Process of Geometric Random Variables

Let $X_i, i \geq 1$, be a set of independent geometric random variables with parameter p. For a particular value of n it follows from Section 4.4.12 that $S_{n+1} \stackrel{\text{def}}{=} X_1 + X_2 + \cdots + X_{n+1}$ has a negative binomial distribution with parameters $(n+1, p)$. Using (4.56) allows us to write the distribution of $N(m)$ as

$$P\left[N(m) \leq n\right] = \sum_{j=m+1}^{\infty} p^{n+1} \left(\begin{array}{c} j-1 \\ j-n-1 \end{array} \right) (1-p)^{j-n-1}, \quad n = 0, 1, 2, \ldots .$$

We claim that this implies that $N(m)$ is binomially distributed with parameters (m, p). To establish this we must show that

$$P\left[N(m) \leq n\right] = \sum_{i=0}^{n} \left(\begin{array}{c} m \\ i \end{array} \right) p^i (1-p)^{m-i},$$

which implies we that need to establish the identity

$$\sum_{i=0}^{n} \left(\begin{array}{c} m \\ i \end{array} \right) p^i (1-p)^{m-i} = p^{n+1} \sum_{j=m-n}^{\infty} \left(\begin{array}{c} n+j \\ j \end{array} \right) (1-p)^j, \quad m \geq n.$$
$$(4.57)$$

In terms of random variables (4.57) can be written as

$$P\left[B_{m,p} \leq n\right] = P\left[G_{n+1,p} > m\right]. \tag{4.58}$$

Observe that the summation on the left-hand side of (4.57) has been previously encountered in equations (3.36) and (4.1) (in both cases set $p = 1/2$ and eliminate constant terms). Our comments at the end of Section 3.2.9 stated that it is impossible to find a closed-form expression for the sum and thus we must content ourselves with the above identity. Validation of the identity in (4.57) is left to the reader in Problem 4.8.

4.4.17 Intuitive Argument for the Renewal Counting Process

The above analysis showing that $N(m)$ is binomially distributed requires that we establish a rather obscure identity (4.57). It is not surprising, however, for this special case of a renewal counting process, that there is a simple intuitive argument that establishes the relationship. This argument is summarized pictorially in Figure 4.14, which summarizes the

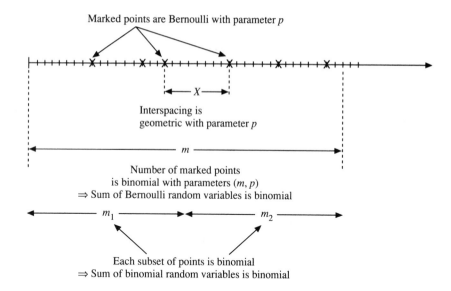

FIGURE 4.14. Relationship among Bernoulli, Geometric, and Binomial Random Variables

relationships among Bernoulli, geometric, and binomial random variables.

We consider the points $1, 2, \ldots$ and mark each point independently with probability p (thus they are Bernoulli random variables with parameter p). It is clear that the number of intervening points between two marked points is geometrically distributed with parameter p and that the number of marked points contained in the first m points is binomially distributed with parameters (m, p). These two observations establish that the renewal counting process for geometrically distributed random variables is binomially distributed. In addition, as shown in Figure 4.14, this implies that the sum of Bernoulli random variables is binomially distributed. If we partition m points into m_1 and m_2 points so that $m_1 + m_2 = m$, then it is also clear that each subset is binomially distributed.

The simplicity of this argument follows from the fact that both the geometric and binomial distributions arise from probabilistic experiments where points are independently marked with probability p and we cannot expect such a simple argument to hold in other cases. For example, the renewal counting process for random variables that have a negative binomial distribution is not as easily determined or interpreted (see Problem 4.9).

**4.4.18
Relationship
Between
Uniform and
Binomial
Distributions**

We next establish a relationship between the discrete uniform distribution and the binomial distribution. Consider the points $1, 2, \ldots, m$, which are independently marked with probability p, and let Y be a random variable that counts the number of marked points; thus Y is binomially distributed with parameters (m, p). We define an indicator random variable $Z_k(i_1, i_2, \ldots, i_k)$ to equal 1 if points i_j are marked for $1 \leq i_1 < i_2 \cdots < i_k \leq m$. For example, if $k = 2$ and $m \geq 7$, then $Z_2(3, 7) = 1$ if points 3 and 7 are marked. We wish to determine the density function of Z_k. To do this we observe that, although the value of Z_k depends on the particular values of $i_j, j = 1, 2, \ldots, k$, its density does not and thus we can suppress the dependency and write

$$P[Z_k = 1] = p^k.$$

Now consider the density of Z_k conditioned on having k marked points, which we write as

$$P[Z_k = 1 \mid Y = k].$$

Consider first the special case of one marked point for which we can write

$$
\begin{aligned}
P[Z_1 = 1 \mid Y = 1] &= \frac{P[Z_1 = 1, Y = 1]}{P[Y = 1]} \\[2mm]
&= \frac{p \ (1 - p)^{m-1}}{\binom{m}{1} p^1 (1 - p)^{m-1}} \\[2mm]
&= \frac{1}{m},
\end{aligned}
$$

which is the discrete uniform distribution over the integers $\{1, 2, \ldots, m\}$. This is an intuitive result since each point has the same probability of being marked.

More generally, for $1 \leq k \leq m$ we have

$$
\begin{aligned}
P[Z_k = 1 \mid Y = k] &= \frac{P[Z_k = 1, Y = k]}{P[Y = k]} \\[2mm]
&= \frac{p^k \ (1 - p)^{m-k}}{\binom{m}{k} p^k (1 - p)^{m-k}} \\[2mm]
&= \frac{1}{\binom{m}{k}}.
\end{aligned}
$$

We have seen this result before in Example 3.9, where it corresponded to the probability that a particular set of k urns are occupied when k balls are allocated to m urns so that no urn contains more than one ball. Hence, if we start with a binomial distribution and condition on the event of k marked points then the distribution of these marked points is the same as the distribution obtained if we uniformly distributed k points among the m points.

We have now established all of the relationships depicted in Figure 4.4 for discrete random variables. Next we establish corresponding relationships for the limiting distributions.

4.5 Limiting Distributions

The connection between the discrete distributions discussed in Section 4.4 and the distributions discussed in this section is made precise through the use of limiting operations. In these operations we change the time and scale of a discrete random variable (these notions will be made precise in this section) to derive their limiting forms. All of the properties previously established for discrete random variables hold analogously by their limiting counterparts.

Our derivations in this section depend on a type of convergence of random variables defined below. (For other types of convergence for random variables see Definitions 5.38 and 5.39.)

Definition 4.14 *A sequence of random variables $X_n, n = 1, 2, \ldots$, converges in distribution to a random variable X if $\lim_{n \to \infty} F_{X_n}(x) = F_X(x)$ for all points x where the distribution function F is continuous. We denote this by $X_n \overset{D}{\longrightarrow} X$.*

In this section we create sequences of random variables that converge in the sense of Definition 4.14 by scaling the domain of discrete random variables by $1/n$ and then letting $n \to \infty$. The continuous uniform case makes this clear.

4.5.1 Continuous Uniform Distribution

In this section we scale the value of the discrete uniform distribution over the integers $1, 2, \ldots, n$, by $1/n$ to create a continuous distribution over the interval $[0, 1]$. Let us first define $U_{[0,1]}$ to be a continuous uniform distribution defined over the interval $[0, 1]$ with distribution

$$
F_{U_{[0,1]}}(x) = \begin{cases} x, & 0 \le x \le 1, \\ \\ 1, & 1 \le x < \infty, \end{cases}
\tag{4.59}
$$

and density

$$f_{U_{[0,1]}}(x) = \begin{cases} 1, & 0 \leq x \leq 1, \\ \\ 0, & 1 \leq x < \infty. \end{cases} \tag{4.60}$$

We now define

$$X_n \stackrel{\text{def}}{=} \frac{1}{n}U_n, \tag{4.61}$$

where we recall that U_n is a discrete uniform random variable over $\{1, 2, \ldots, n\}$. Observe that $1/n \leq X_n \leq 1$. We are interested in the limiting distribution of X_n as $n \to \infty$. To determine this we calculate its distribution as

$$P[X_n \leq x] = P[U_n \leq nx] = \frac{\lfloor nx \rfloor}{n}, \quad 0 \leq x \leq 1.$$

From this we obtain

$$\lim_{n \to \infty} P[X_n \leq x] = x$$

and thus that the limiting distribution is the continuous uniform distribution defined in (4.59). We record this result as

$$\frac{1}{n}U_n \xrightarrow{D} U_{[0,1]} \quad \text{as } n \longrightarrow \infty.$$

Equation (4.61) is representative of the type of scaling used in this section to obtain the limiting forms of the discrete distributions derived in Section 4.4. We next discuss a paradox that arises from this scaling.

4.5.2 Probability of Zero $\not\Rightarrow$ Impossibility

Because the random variable $U_{[0,1]}$ is defined over a continuous domain we are confronted with a paradox. To explain this, let $x = i/n$ and thus

$$P[X_n = x] = P[U_n = nx] = \frac{1}{n}.$$

It is clear that this has a limit of 0 as $n \to \infty$ and it is equally clear that such a limit exists for all x. We are led to conclude that the probability of the event $\{X_n = x\}$ as $n \to \infty$ equals 0 for all x. This seems to imply that all outcomes are impossible! It is difficult to reconcile this conclusion with the fact that it is also the case that (4.60) defines a density function since it is nonnegative and integrates to 1.

These statements, which apply to all continuous distributions, appear at first to be contradictory. They are, however, logically consistent. To resolve this paradox we note that in the limiting operation used to derive the continuous uniform distribution each individual point probability decreases to 0 as n increases to infinity in a way so that X_n always has a probability distribution. Since every random experiment must result in some outcome, we conclude that an event having a probability of 0 *does not necessarily* mean that it cannot occur, that is, that it is the impossible event.

We can complimentarily argue that an event with probability 1 does not necessarily imply that it always occurs, that is, that it is the certain event. For example, if we let $I\!R$ be the set of real numbers between 0 and 1 then

$$P\left[U_{\{0,1\}} \in I\!R\right] = 1. \tag{4.62}$$

For $\mathcal{R}$ being any finite or countably infinite set of real values, however, it is clear from the above that

$$P\left[U_{\{0,1\}} \in \mathcal{R}\right] = 0 \tag{4.63}$$

and that

$$P\left[U_{\{0,1\}} \in I\!R - \mathcal{R}\right] = 1. \tag{4.64}$$

Clearly $\{U_{\{0,1\}} \in I\!R\}$ and $\{U_{\{0,1\}} \in I\!R - \mathcal{R}\}$ are different events but, as equations (4.62) and (4.64) show, they cannot be distinguished by their probability of occurrence. We distinguish between these two events by saying that the event $\{U_{\{0,1\}} \in I\!R - \mathcal{R}\}$ occurs *almost everywhere*, or equivalently that it occurs *almost surely*. This explicitly means that the value of $U_{\{0,1\}}$ might fall outside a given range but will only do so for a set of points that collectively have zero probability (i.e., satisfy (4.63)). For example, if Y is a continuous random variable defined on $I\!R$ and X satisfies

$$X = \begin{cases} c, & Y \text{ noninteger,} \\ 0, & Y \text{ integer,} \end{cases}$$

then it follows that

$$X = c \quad \text{almost surely.}$$

This means that, although the event $\{X \neq c\}$ is not impossible, it occurs on a set that collectively has a probability of 0. These notions are used later in the text to define a type of convergence for a sequence of random variables (see Definition 5.39 of "almost sure convergence") .

Example 4.15 (Conditional Density Functions) Recall that the conditional density is defined by (assume that $f_Y(y) \neq 0$)

$$f_{X \mid Y}(x \mid y) = \frac{f_{(X,Y)}(x,y)}{f_Y(y)}. \qquad (4.65)$$

Using the interpretation given in (4.15), it is tempting to interpret this as

$$f_{X \mid Y}(x \mid y)dx \approx P\left[x \leq X \leq x + dx \mid Y = y\right]. \qquad (4.66)$$

There is no problem with such an interpretation if Y is a discrete random variable, but the above discussion shows that for continuous random variables a problem arises since $P\left[Y = y\right] = 0$ for y. It is, however, clear that (4.65) defines a density function since it is nonnegative and satisfies

$$\int_{-\infty}^{\infty} f_{X \mid Y}(x \mid y)dx = \int_{-\infty}^{\infty} \frac{f_{(X,Y)}(x,y)}{f_Y(y)}dx = \frac{f_Y(y)}{f_Y(y)} = 1,$$

where the last equality comes from (4.18). In analogy with the discrete case, the interpretation provided in (4.66) makes sense if

$$P\left[x_1 \leq X \leq x_2 \mid Y = y\right] = \int_{x_1}^{x_2} f_{X \mid Y}(x \mid y)dy, \quad x_1 \leq x_2, \qquad (4.67)$$

a result we can establish by a limiting argument. We write (4.67) as

$$P\left[x_1 \leq X \leq x_2 \mid Y = y\right] = \lim_{\Delta y \to 0} P\left[x_1 \leq X \leq x_2 \mid y - \Delta y \leq Y \leq y + \Delta y\right]. \qquad (4.68)$$

We can write the right-hand side of this equation in terms of $f_{(X,Y)}$ as

$$\lim_{\Delta y \to 0} \frac{\int_{y-\Delta y}^{y+\Delta y} \left(\int_{x_1}^{x_2} f_{(X,Y)}(w,z)dw \right) dz}{\int_{y-\Delta y}^{y+\Delta y} \left(\int_{-\infty}^{\infty} f_{(X,Y)}(w,z)dw \right) dz} = \frac{\frac{1}{2\Delta y}\int_{y-\Delta y}^{y+\Delta y} \left(\int_{x_1}^{x_2} f_{(X,Y)}(w,z)dw \right) dz}{\frac{1}{2\Delta y}\int_{y-\Delta y}^{y+\Delta y} \left(\int_{-\infty}^{\infty} f_{(X,Y)}(w,z)dw \right) dz}.$$

The numerator of this converges to $\int_{x_1}^{x_2} f_{(X,Y)}(w,y)dw$ and the denominator converges to $f_Y(y)$. Hence (4.68) is given by

$$P\left[x_1 \leq X \leq x_2 \mid Y = y\right] = \frac{\int_{x_1}^{x_2} f_{(X,Y)}(w,y)dw}{f_Y(y)}$$

$$= \int_{x_1}^{x_2} f_{X \mid Y}(x \mid y)dy, \quad x_1 \leq x_2,$$

as desired.

Example 4.16 (Conditional Uniform Densities) Let Y be uniform over $[0,1]$ and, for a given value of Y, let X be uniform over $[0,Y]$. Then

$$F_{X \mid Y}(x \mid y) = \frac{x}{y}, \quad 0 \leq x \leq y,$$

and the joint distribution of (X, Y) is given by

$$f_{(X,Y)}(x, y) = \frac{1}{y}, \quad 0 \le x \le y \le 1.$$

It thus follows that the marginal distribution of X is given by

$$f_X(x) = \int_0^1 f_{(X,Y)}(x, y) dy = \int_x^1 \frac{1}{y} dy = -\ln x, \quad 0 < x \le 1.$$

4.5.3
The Poisson
Distribution

Our interest in this section is to determine the limiting form of a binomial distribution. Consider a set of points $\{0, 1, \ldots, n\}$, each of which is marked with probability $\min\{\lambda/n, 1\}$. Each point corresponds to a Bernoulli random variable with parameter $\min\{\lambda/n, 1\}$. We are concerned with the random variable obtained when $n \to \infty$. For sufficiently large n, the value of λ/n is less than 1, and thus the number of marked points is binomially distributed with parameters $(n, \lambda/n)$. From (4.35) we have that

$$P\left[B_{n,\lambda/n} = i\right] = \binom{n}{i} \left(\frac{\lambda}{n}\right)^i \left(1 - \frac{\lambda}{n}\right)^{n-i}, \quad \lambda/n < 1. \tag{4.69}$$

Our interest lies in the distribution of $B_{n,\lambda/n}$ for large n. To determine this distribution it is more convenient if we rearrange the terms of this expression as follows. First we write the combinatorial coefficient as $(n)_i/i!$ and multiply (4.69) by 1 in the form n^i/n^i. Rearranging the terms yields

$$P\left[B_{n,\lambda/n} = i\right] = \frac{(n)_i}{n^i} \frac{\lambda^i}{i!} \frac{\left(1 - \frac{\lambda}{n}\right)^n}{\left(1 - \frac{\lambda}{n}\right)^i}, \quad \lambda/n < 1.$$

We have already shown that $\lim_{n\to\infty} (n)_i/n^i = 1$ (see (3.16)), and it is clear since i is fixed that $\lim_{n\to\infty} \left(1 - \frac{\lambda}{n}\right)^i = 1$. Recall that

$$\lim_{n\to\infty} \left(1 - \frac{\lambda}{n}\right)^n = e^{-\lambda}. \tag{4.70}$$

Combining this with the above limits shows that

$$\lim_{n\to\infty} P\left[B_{n,\lambda/n} = i\right] = \frac{\lambda^i e^{-\lambda}}{i!} \tag{4.71}$$

which is called the *Poisson distribution*[3] *with parameter* λ. This distribution and the exponential distribution related to it are, by far, the

---

[3]Named after the French mathematician Siméon Poisson (1781–1840). For further historical notes on Poisson, see the conclusions of Chapter 5.

most important distributions in stochastic modeling. We let P_λ be a Poisson random variable with parameter λ and index a set of independent Poisson random variables with parameters $\lambda_i, i = 1, 2, \ldots$, by $P_{\lambda_i}^i, i = 1, 2, \ldots$. We have thus just shown that

$$B_{n,\lambda/n} \xrightarrow{D} P_\lambda \quad \text{as} \quad n \longrightarrow \infty. \tag{4.72}$$

We represent this equality in Figure 4.15.

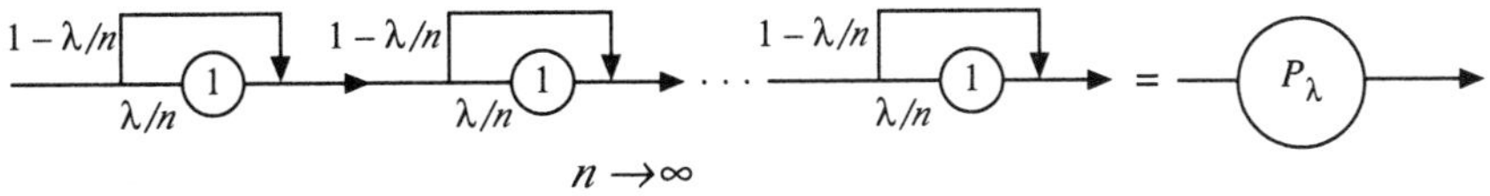

FIGURE 4.15. The Limiting Sum of Bernoulli Random Variables Is Poisson

Note that a Poisson random variable is a discrete random variable that can take on nonnegative integer values. The distribution and density functions of a Poisson random variable are given by

$$F_{P_\lambda}(i) = \begin{cases} 0, & i < 0, \\ \sum_{k \leq i} \frac{\lambda^k e^{-\lambda}}{k!}, & i = 0, 1, \ldots, \end{cases} \tag{4.73}$$

and

$$f_{P_\lambda}(i) = \begin{cases} \frac{\lambda^x e^{-\lambda}}{i!}, & i = 0, 1, \ldots, \\ 0, & \text{otherwise.} \end{cases}$$

In Figure 4.16 we plot the Poisson density function for various values of λ.

The probability generating function of the Poisson distribution is given by

$$\mathcal{G}_{P_\lambda}(z) = \sum_{k=0}^{\infty} z^k \frac{\lambda^k e^{-\lambda}}{k!} = e^{-\lambda(1-z)}. \tag{4.74}$$

**4.5.4
Poisson
Approximation
of a Binomial
Distribution**

For large n, binomial densities, as given by (4.35), are computationally intractable. As seen above, however, the Poisson distribution is obtained as a limiting form of a binomial distribution. This suggests we can approximate the binomial density with a Poisson density which is more computationally tractable. The above derivation shows that the density

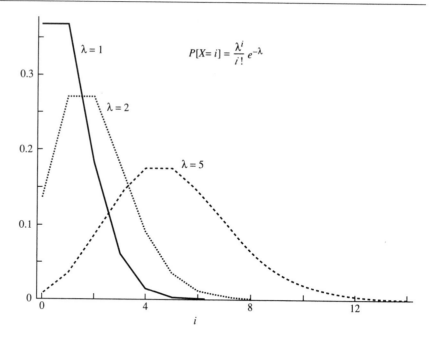

FIGURE 4.16. Poisson Density Function

of a binomial with parameters (n, p), for large n and small p, is approximately equal to the density of a Poisson with parameter np, that is, that

$$\binom{n}{i} p^i (1-p)^{n-i} \approx \frac{(np)^i e^{-np}}{i!} \tag{4.75}$$

for large values of n. This is indeed an accurate approximation for large n and small p.

4.5.5 Sum of Poisson Random Variables

Recall that the deterministic sum of two binomial random variables with a common probability parameter is also binomially distributed. Since the Poisson distribution is a limiting form of a binomial distribution, it is not surprising that the Poisson distribution is also invariant to deterministic summations. This can easily be proved through generating functions. The convolution of Poisson random variables with parameters λ_i, $i = 1, 2, \ldots, n$, has the probability generating function

$$e^{-\lambda_1(1-z)} e^{-\lambda_2(1-z)} \cdots e^{-\lambda_n(1-z)} = e^{-(\lambda_1 + \lambda_2 + \cdots + \lambda_n)(1-z)},$$

which, upon comparison to (4.74), shows that the sum is Poisson with parameter $\lambda = \lambda_1 + \lambda_2 + \cdots + \lambda_n$. Hence the sum of two Poisson random variables with *possibly different* parameter values is Poisson, and we record this by

$$P_{\lambda_1 + \lambda_2 + \cdots + \lambda_n} =_{st} P_{\lambda_1} + P_{\lambda_2} + \cdots + P_{\lambda_n}. \qquad (4.76)$$

Observe that, in comparison to binomial random variables, the invariance of the distribution with respect to deterministic sums *does not require that the parameter values be equal*. This is a truly remarkable and important property of the Poisson distribution. It implies that the distribution can be obtained in a more general limiting form.

4.5.6 Generalized Limit that Leads to the Poisson Distribution

To derive the general form of the limit that leads to the Poisson distribution let us continue the scenario at the end of Section 4.4.6. Assume that n Bernoulli random variables with parameters $p_n(i), i = 1, 2, \ldots, n$, are summed yielding a random variable X_n. We are interested in determining the distribution of $X = \lim_{n \to \infty} X_n$. Suppose that, for each i, the value of $p_n(i)$ decreases to 0 in a such a way as to satisfy

$$p_n(1) + p_n(2) + \cdots + p_n(n) \longrightarrow \lambda \quad \text{as} \quad n \longrightarrow \infty, \qquad (4.77)$$

where λ is a positive constant. The probability generating function of X_n is given by

$$\mathcal{G}_{X_n}(z) = \prod_{i=1}^{n} (1 - p_n(i)(1 - z)). \qquad (4.78)$$

We are interested in determining the limiting form of $\mathcal{G}_{X_n}(z)$. Since products are difficult to work with, we reduce the problem to one of summations by using logarithms. Note that the Taylor expansion of $\ln(1 - z)$ is given by

$$\ln(1 - z) = -z - \frac{z^2}{2} - \frac{z^3}{3} - \cdots,$$

which, for small z implies that[4]

$$\ln(1 - z) = -z - o(z^2).$$

Taking the natural logarithm of (4.78) yields

$$\ln \mathcal{G}_{X_n}(z) = \sum_{i=1}^{n} \ln(1 - p_n(i)(1 - z))$$

$$= -\sum_{i=1}^{n} (1 - z)p_n(i) + o\left([(1 - z)p_n(i)]^2\right). \qquad (4.79)$$

---

[4]Recall that a function $f(x)$ is said to be $o(x)$, written $f(x) = o(x)$, if $f(x)/x \to 0$ as $x \to 0$ (see Appendix D).

Under the conditions for $p_n(i)$, that is,

$$\lim_{n \to \infty} \sum_{i=1}^{n} p_n(i) = \lambda$$

and $p_n(i) \to 0$ it follows that

$$\lim_{n \to \infty} \sum_{i=1}^{n} p_n^2(i) \quad \leq \quad \lim_{n \to \infty} \max_i \{p_n(i)\} \sum_{i=1}^{n} p_n(i)$$

$$= \quad \lim_{n \to \infty} \max_i \{p_n(i)\} \lambda = 0.$$

Thus (4.77) and (4.79) imply that

$$\ln \mathcal{G}_X(z) = \lim_{n \to \infty} \ln \mathcal{G}_{X_n}(z) = -\lambda(1 - z),$$

which implies that

$$\mathcal{G}_X(z) = e^{-\lambda(1-z)}. \tag{4.80}$$

We thus conclude that the sum of a large number of Bernoulli random variables, each with a small probability, has a Poisson distribution *regardless of the exact values of their parameter values*.

4.5.7 Difference Between Binomial and Poisson Random Variables

The conclusion that the sum of two Poisson random variables is Poisson even when their parameter values differ seems to run counter to the conclusion that the sum of two binomial random variables with different parameter values is not binomially distributed. To resolve this paradox we argue that as n gets large all of the Bernoulli probabilities are small and thus can be thought of as being *approximately equal*. In particular, $p_n(i) - p_n(j) \to 0$ as $n \to \infty$ for all $1 \leq i, j \leq n$ and, in this sense, all of the binomial random variables have the "same" parameter values.

The above derivation implies that, in practice, the Poisson distribution arises as the limit of a large number of independent trials each of which has a small probability of success. As mentioned in the introductory comments to this chapter, arrivals of requests to computer and communication systems are often modeled by random variables with a Poisson distribution. Other phenomena that are modeled by a Poisson distribution include radioactive decay, chromosome interchanges in cells, the flow of traffic, the distribution of plants and animals in space or time, the occurrence of accidents and the spread of epidemics. Places where a large number of people gather are typically fertile grounds for observing Poisson random variables. It would not be surprising, for example, to find that in a small time interval the total number of coughs in a large audience (after subtracting out particularly persistent coughers!) is approximately Poisson. Mathematical systems and models where the Poisson distribution plays an important role include reliability modeling, inventory theory, and queueing theory.

4.5.8
The Exponential
Distribution

We next determine the limiting form of the geometric distribution. To derive this consider a scaled time version of the geometric process. For integer n consider a countably infinite set of points $1/n, 2/n, \ldots$, each of which is marked with probability $\min\{\lambda/n, 1\}$, where λ is a fixed constant. Each point corresponds to a Bernoulli random variable with parameter $\min\{\lambda/n, 1\}$ and we can think of the marking as a coin tossing experiment in which every $1/n$ time units we toss a coin that lands heads up with probability $\min\{\lambda/n, 1\}$. We are interested in the time of the occurrence of the first head in this experiment for large values of n. If n is sufficiently large, then $\lambda/n < 1$ and the time of the first head, a random variable denoted by X_n, is given by

$$X_n = \frac{1}{n} G_{1,\lambda/n}.$$

We are interested in the statistics of X_n for large n. Note that the event $\{X_n > x\}$ occurs if and only if $\{G_{1,\lambda/n} > \lfloor nx \rfloor\}$ where $\lfloor x \rfloor$ is the largest integer smaller than or equal to x. Hence the survivor function of X_n is given by

$$P\left[X_n > x\right] = P\left[G_{1,\lambda/n} > \lfloor nx \rfloor\right] = \left(1 - \frac{\lambda}{n}\right)^{\lfloor nx \rfloor},$$

and thus

$$\lim_{n \to \infty} P\left[X_n > x\right] = e^{-\lambda x}. \tag{4.81}$$

Equation (4.81) is the survivor distribution for a continuous random variable, termed an *exponential distribution with parameter* λ. We let $E_{1,\lambda}$ denote an exponential random variable with parameter λ (this generalized notation is explained by the fact that an exponential random variable is a special case of an Erlang random variable denoted by $E_{n,\lambda}$) and we index a set of independent exponential random variables with parameters $\lambda_i, i = 1, 2, \ldots$, by $E_{1,\lambda_i}^i, i = 1, 2, \ldots$. Expressed in these terms, (4.81) implies that

$$\frac{1}{n} G_{1,\lambda/n} \xrightarrow{D} E_{1,\lambda} \quad \text{as} \quad n \longrightarrow \infty.$$

The distribution and density functions for this random variable are given by

$$F_{E_{1,\lambda}}(x) = \begin{cases} 0, & x < 0, \\ 1 - e^{-\lambda x}, & 0 \le x, \end{cases} \tag{4.82}$$

and

$$f_{E_{1,\lambda}}(x) = \begin{cases} \lambda e^{-\lambda x}, & 0 \le x, \\ 0, & \text{otherwise}, \end{cases} \tag{4.83}$$

and we plot the density in Figure 4.17.

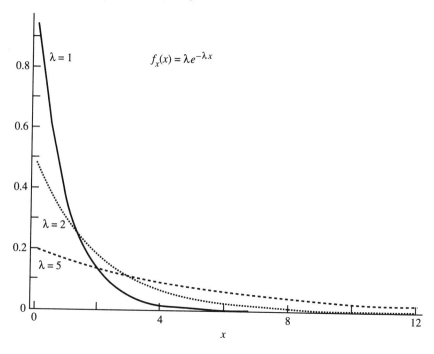

FIGURE 4.17. Exponential Density Function

Example 4.17 (Scaling of Exponential Random Variables) The parameter λ in $E_{1,\lambda}$ can be viewed as a scale parameter. To see this we will show that if one alters the parameter by a factor of α then the original distribution can be recouped by simply scaling the value of the resultant random variable by the same factor α. More precisely, let Y_α be a random variable defined by

$$Y_\alpha \overset{\text{def}}{=} \alpha E_{1,\alpha\lambda}$$

for any value $\alpha > 0$. We then claim that $Y_\alpha =_{st} E_{1,\lambda}$. This follows from a simple calculation

$$P[Y_\alpha > x] = P[\alpha E_{1,\alpha\lambda} > x]$$

$$= P[E_{1,\alpha\lambda} > x/\alpha]$$

$$= e^{-(\alpha\lambda)\,(x/\alpha)} = e^{-\lambda x}.$$

We record this result by

$$\alpha E_{1,\alpha\lambda} =_{st} E_{1,\lambda}. \tag{4.84}$$

Not all random variables are scalable in this fashion.

4.5.9 Continuous + Memoryless $\Longrightarrow$ Exponential

The above derivation of the exponential distribution corresponds to a continuous coin tossing experiment in which an experiment ends the first time heads lands up. In the limiting form we toss the coin n times faster per unit time while simultaneously decreasing the probability that it lands heads up inversely with n. The experiment ends on the first head and, in the limit, is measured in continuous time. Because it is essentially coin tossing, the exponential distribution shares many of the properties of the geometric distribution. Its most salient property is the memoryless property that was discussed in Section 4.4.10. We can easily provide another argument that the exponential distribution is memoryless by simply calculating, as in (4.46), the conditional probability

$$P\left[E_{1,\lambda} > x + z \,|\, E_{1,\lambda} > z\right] = \frac{P\left[E_{1,\lambda} > x + z, E_{1,\lambda} > z\right]}{P\left[E_{1,\lambda} > z\right]}$$

$$= \frac{e^{-\lambda(x+z)}}{e^{-\lambda z}}$$

$$= e^{-\lambda x}. \tag{4.85}$$

Analogous to the geometric case, the only continuous distribution that is memoryless is the exponential distribution, and this can be used as a defining property of the distribution. To establish this, assume that X is a nonnegative continuous random variable that is memoryless. Then it must be the case that X satisfies

$$P[X > x + z \,|\, X > z] = \frac{P[X > x + z, X > z]}{P[X > z]}$$

$$= P[X > x].$$

Rewriting this in terms of the survivor function of X yields

$$\overline{F_X}(x+z) = \overline{F_X}(x)\overline{F_X}(z), \qquad (4.86)$$

where $\overline{F_X}$ is nonincreasing and continuous with boundaries $\overline{F_X}(0) = 1$ and $\overline{F_X}(\infty) = 0$. We let $\overline{F'}(0) = -\lambda$ where λ is a positive constant. If $\overline{F_X}(x) = 0$ or if $\overline{F_X}(x) = 1$ for all x then equation (4.86) is satisfied, but such solutions are eliminated because they do not satisfy the boundary conditions. To find a solution that does satisfy the boundary conditions we use (4.86) to write

$$\frac{\overline{F_X}(x+\epsilon) - \overline{F_X}(x)}{\epsilon} = \frac{\overline{F_X}(x)\overline{F_X}(\epsilon) - \overline{F_X}(x)}{\epsilon} = \overline{F_X}(x)\frac{\overline{F_X}(\epsilon) - 1}{\epsilon}.$$

Letting $\epsilon \to 0$ implies that

$$\frac{d\overline{F_X}(x)}{dx} = -\lambda\overline{F_X}(x)$$

where we have used l'Hospital's rule [5] to evaluate the limit. This linear differential equation has a unique solution that satisfies the boundary conditions given by

$$\overline{F_X}(x) = e^{-\lambda x}.$$

This establishes that the exponential is the only continuous memoryless distribution.

4.5.10 Geometric Sum of Exponential Random Variables

Another property inherited by exponential random variables from their relationship to geometric distributions is that the distribution of an exponential random variable is invariant under the geometric sum operation. Recall that in Section 4.4.8 we proved this for geometric random variables using the convolution property of generating functions. Generating functions, as developed in Chapter 3, however, are essentially a counting mechanisms for *discrete systems*. This confronts us with an immediate problem in attempting this solution approach because the exponential distribution is defined over a continuous, not discrete, domain. The ordinary generating functions that we derived for combinatoric arguments cannot be used, and for now we must content ourselves with a direct calculation of the convolution of two exponential densities. (In

---

[5] Named after Marquis de l'Hospital (1661–1704), who wrote the first textbook on calculus in 1696. The actual discovery of how to determine the indeterminate form 0/0 was made by Johann Bernoulli (1667–1748) who, under a curious financial arrangement, made a pact to send all of his mathematical discoveries to l'Hospital to be used as the Marquis wished!

Section 5.4 we develop the continuous counterpart of a discrete generating function, the Laplace transform, which, like generating functions, can be used to derive the density function of the convolution of two density functions.)

Let $f_i(t)$ be the density function of the sum of i exponential random variables with parameter λ. We have already expressed f_1 in (4.83). If we know f_{i-1} then we can calculate f_i through convolution

$$f_i(t) = \int_0^t \lambda e^{-\lambda x} f_{i-1}(t-x) dx.$$

Starting with $i = 2$ yields

$$f_2(t) = \int_0^t \lambda e^{-\lambda x} \lambda e^{-\lambda(t-x)} dx = \lambda^2 t e^{-\lambda t}$$

and using this for $i = 3$ yields

$$f_3(t) = \int_0^t \lambda e^{-\lambda x} \lambda^2(t-x)e^{-\lambda(t-x)} dx = \frac{\lambda^3 t^2 e^{-\lambda t}}{2!}.$$

The pattern that emerges suggests that

$$f_i(t) = \frac{\lambda^i t^{i-1} e^{-\lambda t}}{(i-1)!}, \tag{4.87}$$

which is easily proved by induction. This density is called the *Erlang* density with parameters (i, λ). (A generalization of this density, the *gamma density*, is explored in Problem 4.23.) We let $E_{i,\lambda}$ be an Erlang random variable with parameters (i, λ) and index a set of independent Erlang random variables with parameters $(n_i, \lambda_i), i = 1, 2, \ldots,$ by $E_{n_i,\lambda_i}^i, i = 1, 2, \ldots$. We record the fact that the sum of exponential random variables is Erlang by

$$E_{n,\lambda} =_{st} E_{1,\lambda}^1 + E_{1,\lambda}^2 + \cdots + E_{1,\lambda}^n. \tag{4.88}$$

It can be shown that

$$\frac{1}{n} G_{m,\lambda/n} \xrightarrow{D} E_{m,\lambda} \quad \text{as} \quad n \longrightarrow \infty.$$

In Figure 4.18, we plot the Erlang density function for various parameter values.

Having established an equation for the density function of the sum of exponential random variables we can now show that a geometric sum of exponential random variables is exponential. This follows from a simple conditioning argument. Let N be a geometric random variable with parameter α and define X by

$$X = E_{1,\lambda}^1 + E_{1,\lambda}^2 + \cdots + E_{1,\lambda}^N.$$

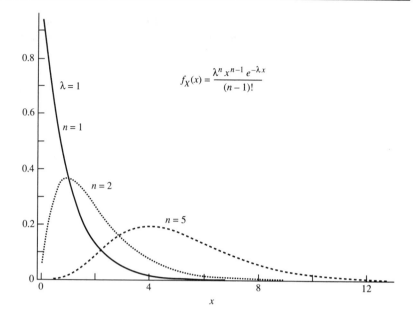

FIGURE 4.18. Erlang Density Function

Using (4.87) we calculate

$$
\begin{aligned}
f_X(t) &= \sum_{i=1}^{\infty} (1-\alpha)^{i-1} \alpha \frac{\lambda^i t^{(i-1)} e^{-\lambda t}}{(i-1)!} \\
&= \lambda \alpha e^{-\lambda t} \sum_{i=1}^{\infty} \frac{(\lambda t(1-\alpha))^{i-1}}{(i-1)!} \\
&= \lambda \alpha e^{-\lambda \alpha t},
\end{aligned}
$$

and thus X is exponentially distributed with parameter $\lambda\alpha$. We conclude that the geometric sum of exponential random variables is exponentially distributed, and we record this result by

$$
E_{1,\alpha\lambda} =_{st} E_{1,\lambda}^1 + E_{1,\lambda}^2 + \cdots + E_{1,\lambda}^N, \qquad N =_{st} G_{1,\alpha}. \tag{4.89}
$$

Pictorially we represent this equality in Figure 4.19.

Example 4.18 (Number of Poisson Arrivals During an Exponential Service) Let random variable X be exponential with parameter μ and let $Y_i, i = 1, 2, \ldots$, be independent exponential random variables with parameter λ that is also independent of X. We think of X as a service time in a single server queue and $Y_i, i = 1, 2, \ldots$, as the interarrival times of customers to the

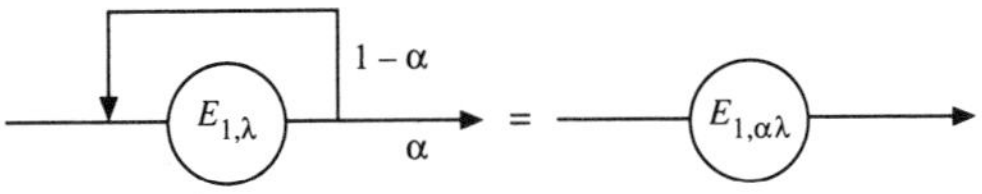

FIGURE 4.19. The Geometric Sum of Exponential Random Variables Is Exponential

queue after the initiation of service. We are interested in the number of arrivals that occur during the service interval, a random variable we denote by N. The event $\{N > n\}$ occurs if

$$\{X > Y_1 + Y_2 + \cdots + Y_{n+1}\}, \quad n = 0, 1, \ldots,$$

and thus

$$P[N > n] = P[X > Y_1 + Y_2 + \cdots + Y_{n+1}], \quad n = 0, 1, \ldots. \quad (4.90)$$

The sum $Y_1 + Y_2 + \cdots + Y_{n+1}$ is an Erlang random variable and thus from (4.88) and (4.87) we can calculate (4.90) as

$$\begin{aligned} P[N > n] &= \int_0^\infty e^{-\mu t} \frac{\lambda^{n+1} t^n e^{-\lambda t}}{n!} dt \\ &= \frac{\lambda^{n+1}}{n!} \int_0^\infty t^n e^{-(\mu+\lambda)t} dt. \end{aligned}$$

To evaluate the integral, note that (see (B.12))

$$n! = \int_0^\infty x^n e^{-x} dx. \quad (4.91)$$

With some algebra this yields

$$P[N > n] = \left(\frac{\lambda}{\lambda + \mu}\right)^{n+1}$$

which, after comparison to (4.44), shows that N is geometric with parameter $\mu/(\lambda + \mu)$ and starts from 0,

$$P[N = n] = (1 - \beta)^n \beta, \quad n = 0, 1, \ldots$$

where $\beta \stackrel{\text{def}}{=} \mu/(\lambda + \mu)$.

The simple geometric form of N suggests an equally simple derivation. The probability that a service time X does not complete before the first interarrival time is given by

$$P[X > Y_1] = 1 - \beta.$$

If the service time does not complete before the first interarrival time then, because X is memoryless, the probability that it does not complete before the second interarrival time is still $1 - \beta$. Thus, from (4.85), we have

$$P[X > Y_1 + Y_2 \mid X > Y_1] = 1 - \beta = P[X > Y_2].$$

The problem can thus be rephrased as a coin tossing experiment, where heads lands up with probability $1 - \beta$, and we play the game until the first head. Random variable N is then equal to one less than the number of tosses in the game and is thus geometric with parameter β and starts from 0.

Example 4.19 (Hyperexponential Distribution)   A random variable that frequently occurs in modeling arises from a random selection of exponential random variables. We let $X = E_{1,\lambda_i}$ with probability α_i, where with $\alpha_1 + \alpha_2 + \cdots + \alpha_n = 1$. The distribution of X is said to be *hyperexponential* and, for the case of $n = 2$, is pictured in Figure 4.20.

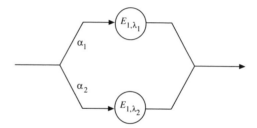

FIGURE 4.20.   Pictorial Representation of a Hyperexponential Random Variable

Later we will use hyperexponential distributions to model arrival and service processes for queueing systems.

Example 4.20 (Minimum of Exponentials is Exponential)   A property of exponential random variables that is frequently required in applications (see Example 5.3) is that the minimum of independent exponential random variables is exponentially distributed. This follows from the memoryless property of the exponential distribution and can be easily derived. Suppose that $E_{1,\lambda}$ and $E_{1,\mu}$ are independent. Then it follows that

$$P\left[\min\{E_{1,\lambda}, E_{1,\mu}\} > t\right] = P\left[E_{1,\lambda} > t\right] P\left[E_{1,\mu} > t\right]$$

$$= e^{-\lambda t} e^{-\mu t} = e^{-(\lambda+\mu)t},$$

which shows that the minimum of $E_{1,\lambda}$ and $E_{1,\mu}$ is exponentially distributed. More generally we record this fact as

$$E_{1,\lambda_1+\lambda_2+\cdots+\lambda_n} =_{st} \min\left\{E_{1,\lambda_1}, E_{1,\lambda_2}, \ldots, E_{1,\lambda_n}\right\}. \tag{4.92}$$

Example 4.21 Consider the set of independent exponentially distributed random variables $E_{1,\lambda_i}, i = 1, 2, \ldots, n, n+1$ and assume that E_{1,λ_i} is the service time of the ith server in a queueing system. We are interested in the probability that the ith server is the first to finish processing, that is, that E_{1,λ_i} has the minimal value among all of the random variables. We let M be a random variable that equals the index of the minimal exponential, and thus the event of interest is given by $\{M = i\}$. Without loss of generality set $i = n + 1$. We can thus write

$$P[M = n+1] = P\left[E_{1,\lambda_{n+1}} < E_{1,\lambda_j}, \; j = 1, 2, \ldots, n\right]$$

$$= P\left[E_{1,\lambda_{n+1}} < \min_{j=1,2,\ldots,n}\{E_{1,\lambda_j}\}\right]. \tag{4.93}$$

Using the results of Example 4.20 (see (4.92)) we have that the minimum of exponentials is exponential and thus (4.93) can be written as

$$P[M = n+1] = P\left[E_{1,\lambda_{n+1}} < E_{1,\lambda_1+\lambda_2+\cdots+\lambda_n}\right].$$

Conditioning on the value of $E_{1,\lambda_{n+1}}$ and calculating implies that

$$P[M = n+1] = \int_0^\infty \lambda_{n+1} e^{-\lambda_{n+1}t} \, e^{-(\lambda_1+\lambda_2+\cdots+\lambda_n)t} dt$$

$$= \frac{\lambda_{n+1}}{\lambda_1 + \lambda_2 + \cdots + \lambda_n + \lambda_{n+1}}. \tag{4.94}$$

Thus the probability in question equals the ratio of the rate λ_{n+1} to the sum of the rates of all the exponential random variables (including λ_{n+1}). In general, the probability that the ith exponential random variable has the minimal value is simply equal to the ratio of its rate to that of the sum of the rates of all of the exponential random variables. This property plays a key role when we discuss Markov processes in Section 8.4.

Example 4.22 (Expression for the Maximum of Exponential Random Variables) The memoryless property of exponential random variables allows us to write the following expression for the maximum of n independent exponential random variables with parameter λ:

$$\max\left\{E_{1,\lambda}^1, E_{1,\lambda}^2, \ldots E_{1,\lambda}^n\right\} =_{st} E_{1,\lambda} + E_{1,2\lambda} + \cdots E_{1,n\lambda}. \tag{4.95}$$

To establish this, consider the simple case $n = 2$. We decompose the maximum into the sum of the intervals of time required for the smaller of the two random variables to finish, followed by the time required for the larger of the two to finish conditioned on the time of the smaller. From Example 4.20 the time for the smaller to finish is exponentially distributed with parameter 2λ. From the memoryless property of exponential random variables, however, the remaining time for the larger of the two to finish conditioned on the time of the smaller is identical to its initial distribution, that is, it is exponential with parameter λ. Hence we obtain (4.95) for the special case of $n = 2$. The argument for the general case proceeds in a similar manner.

4.5.11 Renewal Counting Process of Exponential Random Variables

Because geometric random variables are defined for continuous time we first adapt our previous discrete time definitions of a renewal counting process given in (4.56). Let $N(t)$ count the number of renewals within t time units for $t \geq 0$ and let S_n be the sum of n independent and identically distributed continuous random variables. For continuous random variables we can write

$$P[N(t) \leq n] = P[S_{n+1} > t], \quad t \geq 0. \qquad (4.96)$$

We let S_n be the sum of independent exponential random variables with parameter λ. Using the fact that the deterministic sum of exponential random variables is distributed as an Erlang random variable we write

$$P[N(t) \leq n] = \int_t^\infty \frac{\lambda^{n+1} t^n e^{-\lambda t}}{n!} dt.$$

We claim that the right-hand side of this equation equals the distribution function of a Poisson random variable with parameter λ; hence

$$P[N(t) \leq n] = \sum_{k=0}^n \frac{(\lambda t)^k e^{-\lambda t}}{k!}.$$

This implies that we need to establish the following identity:

$$\sum_{k=0}^n \frac{(\lambda t)^k e^{-\lambda t}}{k!} = \int_t^\infty \frac{\lambda^{n+1} t^n e^{-\lambda t}}{n!} dt. \qquad (4.97)$$

Expressed in terms of random variables, (4.97) implies that

$$P[P_{\lambda t} \leq n] = P\left[E_{n,\lambda} \geq t\right]. \qquad (4.98)$$

Notice that this development is analogous to that of Section 4.4.16 that describes the relationship between binomial and negative binomial random variables. In particular, we see that (4.97) and (4.98) are analogous to (4.57) and (4.58). Analogous to the discrete case, we ask the reader to prove the identity given by (4.97) (see Problem 4.16)!

4.6
Summary of
Chapter 4

This chapter defined random variables and established their mathematical properties. Essentially one can think of random variables as being generalizations of variables used in algebra. Instead of having fixed values as in algebra, random variables have values that are determined by a random experiment and can be described by their distribution functions. Because of this and the fact that random variables are defined on a probability space, the notions corresponding to equality between random variables are also generalized. In particular, two random variables can be equal if they are defined on the same probability space and are equal for all random outcomes, or they can satisfy a weaker form of equality if they have the same distribution function. Analogies between variables in classical algebra and random variables are found in Table 4.3 (convergence in probability and convergence with probability 1 listed in the table will be discussed in Chapter 5).

Property	Classical Algebra	Random Variables	Interpretation
Equality	$x = y$	$X(\omega) = Y(\omega)$ $X =_{st} Y$	Equal for all $\omega \in \Omega$ Equal in distribution
Value	x	X $E[X]$, σ_X^2, C_X^2 F_X, f_X	Sample from a random experiment Statistical values Distribution and density
Convergence	$x_n \to x$	$X_n \xrightarrow{D} X$ $X_n \xrightarrow{P} X$ $X_n \xrightarrow{a.s.} X$	Convergence in distribution Convergence in probability Convergence with probability 1
Summation	$x_1 + x_2 + \cdots + x_n$	$X_1 + X_2 + \cdots + X_N$	Deterministic sum: $N =_{st} D_n$ Geometric sum: $N =_{st} G_{1,p}$

TABLE 4.3. Analogies Between Algebra and Random Variables

This chapter started with the simplest distribution, that of a degenerate random variable that can only take on one value with probability 1. This corresponds exactly to the notion of a variable in algebra, and using this as a starting point we derived the indicator random variable (essentially a Bernoulli random variable), which was used to generate a family of distributions that are commonly used in modeling applications. We first derived four discrete distributions - the uniform, geometric, binomial, and negative binomial distributions - and then derived their corresponding limiting forms - the continuous uniform, exponential, Poisson, and Erlang distributions. All of these distributions are mathematically interrelated and their properties that prove to be invaluable in modeling applications were established in this chapter. The summary of these relationships is given in Figure 4.4, and in Table 4.4 we summarize the derivations and properties linking discrete distributions and their continuous counterparts.

Bibliographic Notes

A thorough compendium of discrete and continuous distributions can be found in the two texts by Johnson and Kotz [54, 55]. Haight [45] provides a wealth of information about the Poisson distribution. Since all books on probability have sections on random variables, the bibliographic notes of Chapter 2 should also be consulted for further reading.

Discrete Form	Limit	Continuous Form
Discrete uniform	$\frac{1}{n} U_n \xrightarrow{D} U_{[0,1]}$	Continuous uniform
Binomial $B_{n,p} =_{st} B^1_{1,p} + \cdots + B^n_{1,p}$ $B_{n_1 + \cdots + n_k, p} =_{st} B_{n_1,p} + \cdots + B_{n_k,p}$	$B_{n, \lambda/n} \xrightarrow{D} P_\lambda$	Poisson $P_{\lambda_1 + \cdots + \lambda_n} =_{st} P_{\lambda_1} + \cdots + P_{\lambda_n}$
Geometric $G_{1,\alpha p} =_{st} G^1_{1,p} + \cdots + G^N_{1,p}, \qquad N =_{st} G_{1,\alpha}$	$\frac{1}{n} G_{1,\lambda/n} \xrightarrow{D} E_{1,\lambda}$	Exponential $\alpha E_{1,\alpha\lambda} =_{st} E_{1,\lambda}$ $E_{n,\lambda} =_{st} E^1_{1,\lambda} + \cdots + E^n_{1,\lambda}$ $E_{1,\alpha\lambda} =_{st} E^1_{1,\lambda} + \cdots + E^N_{1,\lambda}, \qquad N =_{st} G_{1,\alpha}$ $E_{1,\lambda_1 + \cdots + \lambda_n} =_{st} \min\left\{ E_{1,\lambda_1}, \ldots, E_{1,\lambda_n} \right\}$ $\max\left\{ E^1_{1,\lambda}, \ldots, E^n_{1,\lambda} \right\} =_{st} E_{1,\lambda} + \cdots + E_{1,n\lambda}$
Negative binomial $G_{n,p} =_{st} G^1_{1,p} + \cdots + G^n_{1,p}$ $G_{n_1 + \cdots + n_k, p} =_{st} G_{n_1,p} + \cdots + G_{n_k,p}$	$\frac{1}{n} G_{m,\lambda/n} \xrightarrow{D} E_{m,\lambda}$	Erlang $E_{n,\lambda} =_{st} E^1_{1,\lambda} + \cdots + E^n_{1,\lambda}$

TABLE 4.4. Summary of Analogous Discrete and Continuous Random Variables

4.7
Problems for
Chapter 4

Problems Related to Section 4.1

4.1 **[M3] Buffon's Needle Problem.** This problem was proposed and solved by Comte de Buffon in 1777. Let a needle of length ℓ be thrown at random onto a horizontal plane that is ruled with parallel lines a distance $a > \ell$ apart. We wish to determine the probability, p, that the needle will intersect a line. Let x denote the distance of the center of the needle from the nearest of the parallel lines.

1. Show that the needle will intersect a line if and only if $x < (1/2)\ell \sin \phi$ where ϕ is the angle that the needle makes with the parallel lines.

2. Show that the desired probability is given by

$$p = \frac{\frac{l}{2} \int_0^\pi \sin \phi d\phi}{\frac{\pi a}{2}} = \frac{2\ell}{\pi a}.$$

Hence estimation of p provides an estimation of π. The best estimation using this technique appears to be one given by the Italian mathematician Lazzerini in 1901 who, from 3,408 tosses of the needle, determined π to an accuracy of six decimal places. It is clear that better techniques exist for the calculation of the decimal digits of π, and for an entertaining discussion of the current state of the art see Preston's article "The Mountains of π"[93].

4.2 **[M2]** In this problem we show that given a discrete density function $f(x)$ there exists a probability space $(\Omega, \mathbf{F}, P)$ and a random variable X defined on that space having density f. This justifies the common use of defining a random variable by specifying its distribution function without specifying an underlying probability space. Let $f(x)$ be a discrete density function so that on values $x_1, x_2, \ldots, f$ takes on non-zero values. Let $\Omega = \{x_1, x_2, \ldots\}$ and let $\mathbf{F}$ be the set of all subsets of Ω. Define a probability measure over Ω, $P[\omega], \omega \in \Omega$ and a random variable X so that $f(x_i) = P[X = x_i]$.

4.3 **[M2]** Consider the experiment given in Example 2.8 and let X be a random variable that equals the sum of the x and y coordinates of a randomly selected point. Show that the distribution of X is given by

$$P[X \le x] = \frac{1}{24} \begin{cases} 0, & x < 2, \\[2mm] \frac{i(i-1)}{2}, & i \le x < i+1, \quad i = 2, 3, 4, 5, \\[2mm] 14, & 6 \le x < 7, \\[2mm] 14 + \frac{(i-6)(15-i)}{2}, & i \le x < i+1, \quad i = 7, 8, 9 \\[2mm] 24, & 10 \le x. \end{cases}$$

Problem Related to Section 4.2

4.4 **[E1]** It might seem reasonable to suppose that knowing the marginal densities f_X and f_Y allows one to construct the joint density function $f_{(X,Y)}$. In this problem we show that there are infinitely many joint densities with the same marginals. To do this consider a joint density defined over integers $x = 0, 1$ and $y = 0, 1$:

$$f_{(X,Y)}(x,y) = \begin{cases} 1/4 + \epsilon, & x = y, \\ 1/4 - \epsilon, & x \neq y. \end{cases}$$

Show that the result follows from the fact that the marginal densities f_X and f_Y are independent of ϵ.

Problem Related to Section 4.3

4.5 **[E1]** Prove the equalities for indicator random variables given in Table 4.2.

Problems Related to Section 4.4

4.6 **[R1]** A fair coin has been tossed 10 times yielding 10 straight heads. A person naive in probability argues that they are betting that the next time the coin is tossed it will lands tails up because they have never seen a coin land heads up 11 times in a row. They make the bet and after winning say, "There that proves it." How do you show that they are not justified in thinking that they are using correct reasoning?

4.7 **[M2]** Let $X = B_{1,1-p} + G_{1,p}$. Calculate the density of X.

4.8 **[M3]** Prove the identity in (4.57) that was used to establish that the process that counts the number of renewals of a geometric random variable is binomially distributed:

$$\sum_{i=0}^{n} \binom{m}{i} p^i (1-p)^{m-i} = p^{n+1} \sum_{j=m-n}^{\infty} \binom{n+j}{j} (1-p)^j, \quad m \geq n.$$

4.9 **[M2]** Let $X_i, i \geq 1$, be a set of independent and identically distributed random variables that have a negative binomial distribution with parameters (n, p). Determine the distribution of the renewal counting process $N(m)$.

4.10 **[E3]** Let $X_i, i = 1, 2, \ldots, m, m > 1$, be a set of independent Bernoulli random variables with parameter p. Let $i_j, j = 1, 2, \ldots, k, 1 \leq k \leq m$, be a random selection of k unique index values from the set $\{1, 2, \ldots, m\}$ where we assume that all such possible selections are equally likely. Define a random variable $N = X_{i_1} + X_{i_2} + \cdots + X_{i_k}$. In a similar fashion, for a different random selection of index values, define the random variable N'. Determine the densities of the following random variables:

1. N and N'.

2. $N \mid N'$. (Hint: First establish if they are mutually independent).

3. $N + N'$. (Hint: Condition on the number of common index values).

4. $N \mid N + N'$.

5. $N - N'$.

4.11 **[M2]** Consider two independent binomial random variables X_1 and X_2 with parameters (n_1, p_1) and (n_2, p_2), respectively. Show that the conditional random variable $X_1 \mid (X_1 + X_2)$ has a density function given by

$$P[X_1 = i \mid X_1 + X_2 = k] = \frac{f(i)\alpha^i}{\sum_{j=a}^{b} f(j)\alpha^j}, \quad a \le i \le b, \tag{4.99}$$

where $a = \max\{0, k - n_2\}$, $b = \min\{n_1, k\}$,

$$\alpha = \frac{p_1(1 - p_2)}{p_2(1 - p_1)},$$

and $f(i)$ is the hypergeometric distribution

$$f(i) = \frac{\binom{n_1}{i}\binom{n_2}{k - i}}{\binom{n_1 + n_2}{k}}.$$

Provide a probabilistic interpretation of (4.99).

4.12 **[M2] (Continuation)** Assume, as in Problem 4.11, that X_1 and X_2 are independent Poisson random variables with parameters p_1 and p_2, respectively. Show that the conditional random variable $X_1 \mid (X_1 + X_2)$ has a binomial distribution. Provide a probabilistic interpretation.

4.13 **[M3]** Consider a coin tossing experiment where the coin lands heads up with probability p. Let $X_i = 1$ if heads lands up on the ith toss and otherwise let it be equal to 0. Define the events

$$I_k = \{\min i \ : \ X_i = 1, X_{i+1} = 0, X_{i+2} = 0, \ldots, X_{i+k-1} = 0, X_{i+k} = 1\},$$

for $k \ge 1$.

1. Determine the distribution of $I_k, k \ge 1$.

2. Are the random variables $I_k, I_{k'}$ independent?

Problems Related to Section 4.5

4.14 **[M1]** Let X and Y be continuous random variables that are independent and identically distributed. What is the probability that $P[X > Y]$?

4.15 **[M2]** Let two points be randomly selected on the interval $[0, 1]$. Write an expression for the distribution of the distance between them.

4.16 [**E2**] Prove the identity

$$\sum_{k=0}^{n} \frac{(\lambda t)^k e^{-\lambda t}}{k!} = \int_t^{\infty} \frac{\lambda^{n+1} t^n e^{-\lambda t}}{n!},$$

which was used to show that the renewal counting process for exponential random variables is the Poisson distribution. (Hint: Integrate the right-hand side by parts n times and use the fact that $\int_0^x \lambda \exp[-\lambda y] dy = 1 - \exp[-\lambda x]$).

4.17 [**M2**] Let X and Y be independent Poisson random variables with parameters λ and μ, respectively. Let $Z = X - Y$. Show that

$$P[Z = z] = \begin{cases} \lambda^z e^{-\lambda + \mu} \sum_{i=0}^{\infty} (\lambda \mu)^i / (i!(i+z)!), & z = 0, 1, 2, \dots \\ \mu^{-z} e^{-\lambda + \mu} \sum_{i=0}^{\infty} (\lambda \mu)^i / (i!(i-z)!), & z = 0, -1, -2, \dots . \end{cases}$$

4.18 [**E2**] Prove the following equalities for the Poisson distribution:

$$\sum_{i=0}^{\infty} \frac{\lambda^{2i}}{(2i)!} e^{-\lambda} = \frac{1 + e^{-2\lambda}}{2},$$

and

$$\sum_{i=0}^{\infty} \frac{\lambda^{2i+1}}{(2i+1)!} e^{-\lambda} = \frac{1 - e^{-2\lambda}}{2}.$$

4.19 [**M2**] Let $X_i, i \geq 1$, be independent Bernoulli random variables with parameter p and let $N(t)$ be Poisson with parameter λt. Calculate the density function for

$$X(t) = X_1 + X_2 + \cdots + X_{N(t)}.$$

Explain how you could have obtained this answer using the limiting arguments that were used to derive the Poisson distribution from the binomial distribution.

4.20 [**E2**] Let X_1 and X_2 be independent exponential random variables with parameter λ. Derive the density for $\max\{X_1, X_2\}$.

4.21 [**E2**] Let Y be an exponential random variable with parameter λ and let X, conditioned on Y, be a uniform random variable over the interval $[0, Y]$. Determine the joint distribution of (X, Y) and the conditional distribution of Y given X.

4.22 [**M2**] Let X and Y be random variables and define $Z_{min} = \min\{X, Y\}$ and $Z_{max} = \max\{X, Y\}$. In terms of f_X and f_Y write the distribution and density functions of Z_{min} and Z_{max} in the following cases.

1. Random variables X and Y are both discrete.

2. Random variables X and Y are both continuous.

3. Random variable X is discrete and random variable Y is continuous.

4.23 [**E3**] Recall that an Erlang random variable has two parameters (i, λ) where i is integer. The density function for an Erlang is given by

$$f_i(t) = \frac{\lambda^i t^{i-1} e^{-\lambda t}}{(i-1)!}.$$

The *gamma* density is a generalization of this where i can take on noninteger values. In particular, a gamma density has two positive parameters (α, λ) and a density function given by

$$f_\alpha(t) = \frac{\lambda^\alpha t^{\alpha-1} e^{-\lambda t}}{\Gamma(\alpha)},$$

where

$$\Gamma(\alpha) \overset{\text{def}}{=} \int_0^\infty x^{\alpha-1} e^{-x} dx$$

is termed the *gamma function*. (See Appendix B for the history of the gamma function.)

1. Show that

$$\int_0^\infty x^{\alpha-1} e^{-\lambda x} dx = \frac{\Gamma(\alpha)}{\lambda^\alpha}.$$

2. Show that the gamma function satisfies the same recurrence as that of factorials, that is, $\Gamma(\alpha + 1) = \alpha \Gamma(\alpha)$.

3. Show that $\Gamma(1) = 1$ and thus that $\Gamma(n) = (n-1)!$ if n is an integer.

4.24 [**M2**] Consider an infinite freeway with an infinite number of cars. Cars are assumed to be lined up in a lane and are initially at rest. Each driver independently selects a speed from an exponential distribution with parameter λ and begins driving at the minimum of this selected speed and the speed of the car in front of them. In this way, cars are bunched up into groups led by where the lead car is the slowest car in the group. What is the distribution of the group size?

Suppose speeds are selected from an Erlang random variable with parameters (n, k). Now what is the distribution of group size? Justify your answer.

5

Expectation and Fundamental Theorems

In the previous chapter we defined random variables and established some of their properties. Like variables in algebra, random variables have values but these values can only be determined from random experiments. To characterize all possible outcomes from such an experiment we use distribution functions, which provide a complete specification of a random variable without the necessity of defining an underlying probability space. Specifying distribution functions poses no inherent mathematical difficulties. However, in applications it is frequently not possible to determine the complete distribution function. Often a simpler representation is possible; instead of specifying the entire function we specify a set of *statistical values*. A statistical value is the average of a function of a random variable. This can be thought of as the value that is obtained from averaging the outcomes of a large number of random experiments as discussed in the frequency-based approach to probability of Section 2.1. In the axiomatic framework of Section 2.2 such an average corresponds to the operation of *expectation* applied to a function of a random variable.

The statistical concepts of a mean and variance (to be precisely defined) arise from expectation as do other fundamental, but perhaps less intuitive, properties of random variables. These include Markov's inequality, Chebyshev's inequality, Chernoff's bound, the law of large numbers, and the central limit theorem; all of which are discussed and derived in Section 5.5. It is important to note that the expectation operation essentially transforms a problem that deals with random phenomena characterized by a distribution function into a problem that deals with statistical values that are deterministic. It is perhaps for this reason, combined with the fact that distributions for random variables corresponding to real physical phenomena are difficult to obtain, that expectation is so widely used in mathematical modeling.

In Section 5.1 we define expectation and derive some of its basic properties. In Section 5.2 we extend this notion to the expectation of a function of a random variable. This leads to a discussion of the moment generating function in Section 5.3. Moment generating functions are important mathematical tools and are used extensively in modeling. Closely related to the moment generating function is the Laplace transform considered in Section 5.4. Laplace transforms are the continuous analog of generating functions first considered in Chapter 3 and can be used to solve the analog of discrete recurrence equations - differential equations. In Section 5.5 we establish fundamental bounds on distributions and limiting results for the statistical average of a set of random variables. The limits derived here form the link between the frequency-based approach to probability and the axiomatic approach mentioned in Chapter 2.

5.1
Expectation

To motivate the definition of expectation, consider an experiment that stops on the ith trial with probability β_i, $i = 1, 2, \ldots$. We are interested in the average number of trials needed before stopping.[1] To derive an expression for this, consider performing a large number of experiments. Intuitively, the percentage of these experiments that stop on the ith trial is given by $100\beta_i\%$. Thus, in a typical experiment, the average number of trials before stopping the experiment is given by

$$1 \cdot \beta_1 + 2 \cdot \beta_2 + 3 \cdot \beta_3 + \ldots + \ .$$

This motivates the following definition of *expectation*. (Recall that we typically assume that discrete random variables take on integer values. It is trivial to modify the following definition to relax this assumption.)

Definition 5.1 *The expectation of a discrete random variable X that can assume values $i, i = \ldots, -1, 0, 1, \ldots$, denoted by $E[X]$, is defined by*

$$E[X] \stackrel{\text{def}}{=} \sum_{i=-\infty}^{\infty} i f_X(i), \qquad (5.1)$$

provided the sum converges absolutely.

---

[1]You might sense the flavor of a game of chance in the definition of expectation. Indeed the first definition of expectation is given by Huygens in his 1657 book on probability that answered gambling questions posed by Chevalier de Méré to Pascal (see Chapter 3).

If X is continuous then its expectation is defined by

$$E[X] \stackrel{\text{def}}{=} \int_{-\infty}^{\infty} x f_X(x) dx, \tag{5.2}$$

provided the integral converges absolutely.

When the summation in (5.1) or the integral in (5.2) does not converge absolutely we say that the expectation of X is undefined.

Absolute convergence is required in Definition 5.1 to avoid the possibility of ill-defined expressions such as $E[X] = \infty - \infty$. Positive random variables do not have an expectation if the sum (or integral) diverges to infinity. In such cases, instead of saying that the expectation is not defined we will say that the expectation is infinite. The expectation of X will also sometimes be called the *average* or *mean* of X.

**5.1.1
Properties of
Expectation**

From Definition 5.1 it is clear that properties that hold for summation (equiv. integral) operations also hold for expectations. An immediate consequence of this observation is that expectation is a linear operator and satisfies the properties given in Table 5.1 (we suggest the reader prove these in Problem 5.2).

Note that the definition given in (5.1) is equivalent to defining a weighted sum with respect to weights specified by a density function. The notation $E[X]$ does not explicitly specify the weights by which the expectation of X is calculated and, by convention, these weights are assumed to be the density of X. A more precise notation makes this explicit,

$$E_X[X] \stackrel{\text{def}}{=} \sum_{i=-\infty}^{\infty} i f_X(i).$$

More generally, we can change the "thing" measured in an expectation, so that an expression like $E[X^2]$ means

$$E[X^2] = E_X[X^2] = \sum_{i=-\infty}^{\infty} i^2 f_X(i)$$

and the expression $E[h(X)]$, for a function h, means

$$E[h(X)] = E_X[h(X)] = \sum_{i=-\infty}^{\infty} h(i) f_X(i).$$

Similar expressions hold for continuous random variables.

#	Property
1	$E[X + Y] = E[X] + E[Y]$, provided the expectations of X and Y exist.
2	$E[cX] = cE[X]$ if c is a constant.
3	$E[X] = c$ if $X = c$ deterministically.
4	$E[X] \leq E[Y]$ if $P[X \leq Y] = 1$.
5	$|E[X]| \leq E[|X|]$.
6	$|E[X]| \leq c$ if $P[|X| \leq c] = 1$.
7	$E[I_A] = P[A]$.
8	$E[X_1 + X_2 + \cdots + X_N] = E[N]E[X]$ if the random variables X_i are independent and identically distributed with expectation $E[X]$ and N is a stopping time (defined in Definition 5.13) for the sequence $X_i, i = 1, 2, \ldots$. This is Wald's equality and is proved in Theorem 5.18

TABLE 5.1. Properties of Expectation

The seventh property of expectation (see Table 5.1) implies that we can view the probability of any event as the expectation of an indicator random variable of that event. This interpretation is often used to provide a different semantic meaning for the probability of an event. Consider the following example:

Example 5.2 Consider a single server queue and let $\mathcal{E}$ be the event that the server is busy and let N_s be the number of customers in the server (N_s equals 1 or 0 depending on if there is a customer in service). To write an expression for $P[\mathcal{E}]$ let I_s be an indicator random variable that equals 1 if the server is busy and equals 0 otherwise. Property 7 implies that

$$P[\mathcal{E}] = E[I_s].$$

A moment's reflection, however, shows that $E[I_s] = E[N_s]$, since the server can hold at most one customer (customers that are not being served must wait in the queue). Thus the probability that the server is busy equals the expected number of customers in the server, that is,

$$P[\mathcal{E}] = E[N_s].$$

5.1.2 Expectation of Common Distributions

The expectation of the distributions derived in Sections 4.4 and 4.5 can be easily calculated. We start with discrete distributions.

Bernoulli: Since a Bernoulli random variable is simply an indicator random variable it immediately follows from property 7 of expectation that

$$E[B_{1,p}] = p. \tag{5.3}$$

Binomial: A binomial random variable is the sum of Bernoulli random variables, and thus it follows from (4.37) and property 1 of expectation that

$$E[B_{n,p}] = np. \tag{5.4}$$

Geometric: The expectation of a geometric random variable follows from simple algebra:

$$E[G_{1,p}] = \sum_{n=1}^{\infty} n(1-p)^{n-1}p = -p\frac{d}{dp}\sum_{n=1}^{\infty}(1-p)^n$$

$$= -p\frac{d}{dp}\frac{1-p}{p} = \frac{1}{p}. \tag{5.5}$$

Observe that if X is a geometric random variable with parameter p that starts from 0 then

$$E[X] = E[G_{1,p} - 1] = \frac{1-p}{p}.$$

Negative Binomial: Since the sum of geometric random variables is negative binomial, it follows from (4.53) and (5.5) that

$$E[G_{n,p}] = \frac{n}{p}. \tag{5.6}$$

If X is a negative binomial random variable with parameters (n, p) and starts from 0, then

$$E[X] = E[G_{n,p} - n] = \frac{n(1-p)}{p}.$$

Poisson: A simple calculation shows that the expectation of a Poisson random variable is given by

$$E[\,P_\lambda\,] = \sum_{k=0}^{\infty} k \,\frac{\lambda^k e^{-\lambda}}{k!} = \lambda e^{-\lambda} \sum_{k=1}^{\infty} \frac{\lambda^{k-1}}{(k-1)!} = \lambda. \qquad (5.7)$$

Exponential: Using (4.91) implies that

$$E\left[E_{1,\lambda}\right] = \int_0^\infty x\lambda e^{-\lambda x} dx = \frac{1}{\lambda}. \qquad (5.8)$$

Erlang: Since the sum of exponential random variables is Erlang, we have from (4.88) and (5.8)

$$E\left[E_{n,\lambda}\right] = \frac{n}{\lambda}. \qquad (5.9)$$

Example 5.3 (Expected Response Time for a Parallel-Processing System) Consider a simple model of the execution of a single job in a parallel-processing system. There are $P \geq 1$ homogeneous processors that serve a single job consisting of $b \geq 1$ tasks with independent exponentially distributed service times with parameter μ. The tasks are queued in a common facility that is accessed by processors as they finish execution. The job is considered to be completed when all of its b tasks have finished execution (we can think of executed tasks as waiting to be *synchronized* with the unserviced tasks). In Figure 5.1 we represent the system modeled.

We are interested in the expected response time of the job, a quantity denoted by $T(b)$. There are two cases to consider.

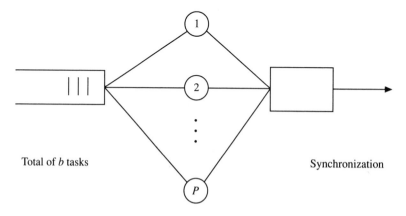

FIGURE 5.1. Execution of Multiple Tasks on Parallel Processors

Case 1: $(b \leq P)$ In this case it is clear that each task is immediately assigned to a processor, and thus

$$E\left[T(b)\right] = E\left[\max\left\{E^1_{1,\mu}, E^2_{1,\mu}, \ldots, E^b_{1,\mu}\right\}\right].$$

To calculate an expression for this we note that (4.95) implies that

$$E\left[\max\left\{E^1_{1,\mu}, E^2_{1,\mu}, \ldots, E^b_{1,\mu}\right\}\right] = E\left[E_{1,\mu} + E_{1,2\mu} + \cdots E_{1,n\mu}\right].$$

Using (5.8) and the definition of the harmonic function, given by

$$H_b \stackrel{\text{def}}{=} \sum_{i=1}^{b} 1/i,$$

implies that

$$E\left[T(b)\right] = \frac{1}{\mu}H_b.$$

Euler discovered that the harmonic function can be approximated by

$$H_b \approx \ln(b) + \mathbf{C} + \frac{1}{2b}, \tag{5.10}$$

where

$$\mathbf{C} \stackrel{\text{def}}{=} \lim_{n \to \infty} H_n - \ln(n) \approx 0.57721$$

is *Euler's constant*. From this, we can conclude that the time needed to serve b tasks on an infinite number of processors (this way we are assured that each task is immediately assigned a processor) grows logarithmically with b.

Case 2: $(b > P)$ In this case some tasks must wait before being assigned a processor. After $b - P$ completions the remaining P tasks are assigned a processor. The expected time of each of the first $b - P$ completions is given by $1/\mu P$ since each completion is distributed as the minimum of P exponential random variables. The last P tasks are each assigned a processor, and thus case 1 shows that the expected time to complete them is given by H_P/μ. We thus have

$$E\left[T(b)\right] = \frac{b-P}{\mu P} + \frac{1}{\mu}H_P, \quad b \geq P.$$

Combining both cases shows that

$$E\left[T(b)\right] = \begin{cases} H_b/\mu, & b \leq P, \\ (b-P)/(\mu P) + H_P/\mu, & b > P. \end{cases}$$

5.1.3 Examples of Random Variables that Do Not Have Expectations

In this section we list some random variables that do not have expectations. Consider the density function of an integer random variable X given by

$$f_X(i) = \frac{1}{2\,|i|\,(|i|+1)}, \qquad i \neq 0. \tag{5.11}$$

To see that this sums to 1 and is thus a density function, we write

$$\frac{1}{2\,|i|\,(|i|+1)} = \frac{1}{2\,|i|} - \frac{1}{2(|i|+1)}$$

and thus

$$
\begin{aligned}
\sum_{i \neq 0} f_X(i) &= 2\sum_{i=1}^{\infty} \frac{1}{2i(i+1)} \\
&= \sum_{i=1}^{\infty} \left[\frac{1}{i} - \frac{1}{(i+1)} \right] \\
&= (1 - 1/2) + (1/2 - 1/3) + \cdots = 1.
\end{aligned}
$$

Random variable X does not have an expectation, however, since

$$\sum_{i \neq 0} |i|\, f_X(i) = \sum_{i=1}^{\infty} \frac{1}{i+1}$$

and it is well known that the harmonic series $\sum_{i=1}^{\infty} 1/i$ does not converge (equation (5.10) shows that it grows logarithmically).

A continuous distribution that does not have an expectation is the *Cauchy distribution.*[2] This density is given by

$$f(x) = \frac{1}{\pi(1+x^2)}, \qquad -\infty < x < \infty. \tag{5.12}$$

We can show that this integrates to 1 as

$$\int_{-\infty}^{\infty} \frac{dx}{\pi(1+x^2)} = \frac{1}{\pi} \arctan(x) \Big|_{-\infty}^{\infty} = \frac{1}{\pi} \left\{ \frac{\pi}{2} - \frac{-\pi}{2} \right\} = 1.$$

We can also show that (5.2) does not converge absolutely:

$$\frac{1}{\pi}\int_{-\infty}^{\infty} \left| \frac{x}{1+x^2} \right| dx = \frac{2}{\pi}\int_{0}^{\infty} \frac{x}{1+x^2}\, dx = \frac{2}{\pi}\ln(1+x^2)\Big|_{0}^{\infty} = \infty.$$

---

[2]Named after Augustin-Louis Cauchy (1789–1857) who was so prolific (in addition to several books, he wrote 789 papers) that to save money the Academy of Sciences where he worked put a restriction of 4 pages on all published papers. The rule stands to this day.

Examples where random variables do not have expectations (or have infinite expectations) often occur in nonobvious ways. In Problem 5.4 the reader is asked to show that the first passage problem of Section 3.3.5 does not have an expectation. In that problem, we considered a series of fair coin tosses where a player wins or loses a dollar if the coin lands heads or tails up, respectively. The event of interest is the time for the player to win exactly one dollar. On initial encounter it is not at all obvious that the expected number of tosses before this occurrence is infinite.

5.1.4
Expectation of a Function

Properties 1 and 2 in Table 5.1 show that expectation is a linear operator. More precisely, suppose that h is a linear function given by

$$h(x_1, x_2, \ldots, x_n) = a_1 x_1 + a_2 x_2 + \cdots + a_n x_n + b,$$

where $a_i, i = 1, 2, \ldots, n$, and b are constants. Then it follows that, for random variables $X_1, X_2, \ldots, X_n$,

$$E\left[h\left(X_1, X_2, \ldots, X_n\right)\right] = h\left(E\left[X_1\right], E\left[X_2\right], \ldots, E\left[X_n\right]\right). \quad (5.13)$$

There are two important things to observe about this equality:

1. Equation (5.13) holds regardless of the relationship of random variables $X_i, i = 1, 2, \ldots, n$ (e.g., X_1 could be discrete, X_2 could be continuous, and the set of random variables could be mutually dependent).

2. Equation (5.13) holds for linear functions but *does not generally hold* if h is nonlinear. A simple example of a function that illustrates the possibility of all forms of equality/inequality between $E\left[h(X)\right]$ and $h(E\left[X\right])$ is given in Example 5.4.

Example 5.4 Consider the function $h(x) = (x - 1/2)^3$ and let X be a Bernoulli random variable with parameter p. Then $h(X)$ equals $1/8$ (resp., $-1/8$) with probability p (resp., $(1 - p)$), and thus

$$E\left[h(X)\right] = \frac{-(1-p)}{8} + \frac{p}{8} = (2p - 1)/8.$$

On the other hand, $h(E\left[X\right]) = (p - 1/2)^3$. Some simple calculations imply that all possible relationships can hold between $E\left[h(X)\right]$ and $h(E\left[X\right])$:

$$E\left[h(X)\right] \begin{Bmatrix} < \\ = \\ > \end{Bmatrix} h(E\left[X\right]) \text{ if } \begin{Bmatrix} 0 < p < 1/2 \\ p = 0, \ 1/2, \text{ or } 1 \\ 1/2 < p < 1 \end{Bmatrix}.$$

We can use property 4, however, to derive an extremely useful inequality for the special case where h is a convex (or concave) function. This is called *Jensen's inequality*, and is derived later.

5.1.5 Jensen's Inequality

We define convexity as follows:

Definition 5.5 *A function $h(x)$, for $x \in \mathbb{R}^n$, is said to be* convex *if*

$$h(\alpha x_1 + (1 - \alpha)x_2) \leq \alpha h(x_1) + (1 - \alpha)h(x_2), \quad 0 \leq \alpha \leq 1. \qquad (5.14)$$

A function h is concave if $-h$ is convex.

There are two special cases of convex functions that are of interest. If $x \in \mathbb{R}$ and h has a second derivative then it can be shown that h is convex if

$$h^{(2)}(x) \geq 0, \quad \text{for all } x. \qquad (5.15)$$

(It is interesting to note that all convex functions defined on $\mathbb{R}$ are continuous, see Problem 5.5.) If h is defined on the integers, $x \in \mathbb{N}$, then it can be shown that h is convex if

$$h(x + 1) - 2h(x) + h(x - 1) \geq 0, \quad x \text{ integer.} \qquad (5.16)$$

Define the linear function g that is tangent to h at the point a as follows:

$$g(x, a) \stackrel{\text{def}}{=} h(a) + h^{(1)}(a)(x - a). \qquad (5.17)$$

We then have the following lemma, which is proved in Problem 5.6:

Lemma 5.6 *Let h be a convex function. For a fixed constant, a, let g be defined as in (5.17). Then*

$$g(x, a) \leq h(x), \quad \text{for all } x. \qquad (5.18)$$

In Figure 5.2 we graphically show the inequality (5.18). This provides a "picture proof" of Jensen's inequality, which is analytically proved in the following theorem:

Theorem 5.7 (Jensen's Inequality) *Suppose that h is a differentiable convex function defined on $\mathbb{R}$. Then*

$$E\left[h(X)\right] \geq h(E\left[X\right]).$$

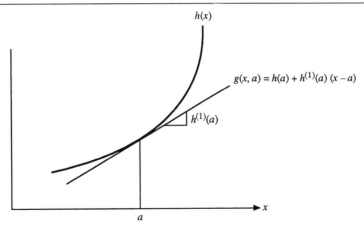

FIGURE 5.2. Pictorial Proof of Jensen's Inequality

Proof: Let g be defined by (5.17). It is clear that $h(X)$ and $g(X, E[X])$ define random variables that satisfy $P[g(X, E[X]) \leq h(X)] = 1$. Property 4 of expectation thus implies that

$$E[h(X)] \geq E[g(X, E[X])]. \tag{5.19}$$

Writing out the expectation of $g(X, E[X])$ using properties 1 and 3 of expectation yields

$$E[g(X, E[X])] = E\left[h^{(1)}(E[X])(X - E[X]) + h(E[X])\right]$$

$$= h(E[X]),$$

which, combined with (5.19), completes the proof.   ∎

Jensen's inequality is widely used to establish bounds on the performance of systems. We should note that the inequality holds for all convex functions (we only proved the theorem for the continuous case). Consider the following applications of the inequality to parallel-processing systems.

Example 5.8 (Continuation of the Parallel-Processing Model) In case 1 of Example 5.3, a job consisting of b tasks was scheduled on a set of parallel processors so that all tasks were allocated a processor. The response time of the job was given by the maximum of the execution times for the b tasks. Let $X_i, i = 1, 2, \ldots, b$, be random variables indicating the service time of the tasks. These could have different distributions, but we assume that they are independent. Since the maximum function is convex, Jensen's inequality implies that

$$E[\max\{X_1, X_2, \ldots, X_b\}] \geq \max\{E[X_1], E[X_2], \ldots, E[X_b]\}.$$

This expresses the intuitive fact that the expected response time is greater than or equal to the largest expected task service time.

Example 5.9 (Lower Bound on an Optimal Parallel Schedule) Consider scheduling n jobs, each consisting of b tasks on P processors. Task service times are independent and identically distributed and a job is considered to be completed only after all of its b tasks have finished execution. Let X denote the task service time. A *schedule*, ψ, is an assignment of all of the tasks to the processors. Let $L_{i,p}^{\psi}$, for $i = 1, 2, \ldots, n$ and $p = 1, 2, \ldots, P$, be the last position of tasks from job i that are scheduled by ψ on processor p. We set $L_{i,p}^{\psi} = 0$ if no tasks from job i are scheduled on processor p by policy ψ. The sum of the expected response times of all of the jobs in the system can be written as

$$E\left[T^{\psi}\right] = \sum_{i=1}^{n} E\left[\max_{p}\left\{S\left(L_{i,p}^{\psi}\right)\right\}\right],$$

where $S(j)$ is a random variable that equals the sum of the service times of j independent tasks. A particular schedule for four jobs consisting of two tasks on two processors is shown in Figure 5.3.

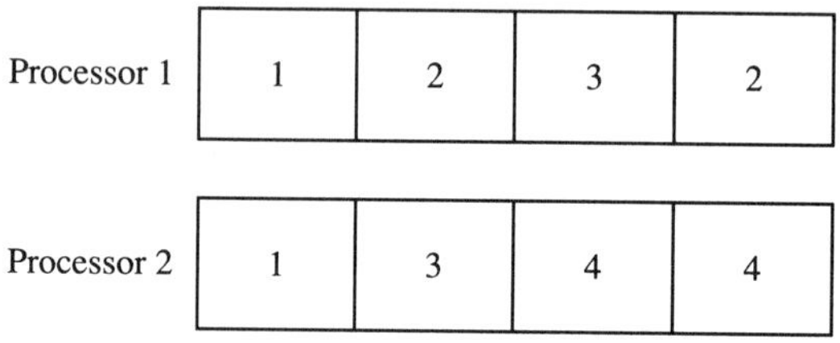

FIGURE 5.3. A Particular Parallel-Processing Schedule

For the schedule depicted in the figure, it is clear that

$$L_{1,1}^{\psi} = 1, \ L_{1,2}^{\psi} = 1 \qquad L_{2,1}^{\psi} = 4, \ L_{2,2}^{\psi} = 0$$

$$L_{3,1}^{\psi} = 3, \ L_{3,2}^{\psi} = 2 \qquad L_{4,1}^{\psi} = 0, \ L_{4,2}^{\psi} = 4,$$

and thus that

$$E\left[T^{\psi}\right] = E\left[\max\left\{S(1), S(1)\right\}\right] + E\left[\max\left\{S(3), S(2)\right\}\right] + 8E\left[X\right].$$

Let π denote an optimal scheduler. Our objective is to derive a lower bound on the sum of the expected response times for jobs scheduled by π. To derive this bound observe that the maximum function is convex and thus Jensen's inequality implies that we can lower bound the performance of any particular schedule ψ by assuming that service times are deterministic. For deterministic service times it is clear that

$$E\left[\max_p \left\{S\left(L_{i,p}^\psi\right)\right\}\right] = \max_p \left\{L_{i,p}^\psi\right\} E\left[X\right].$$

Applying this to the schedule ψ shown in Figure 5.3 yields

$$E\left[T^\psi\right] \geq 12 E\left[X\right].$$

Let $E\left[T^{\pi_d}\right]$ be the performance of an optimal scheduling policy π_d when service times are deterministic. Then Jensen's inequality implies that the performance of π is lower bounded by

$$E\left[T^\pi\right] \geq E\left[T^{\pi_d}\right]. \tag{5.20}$$

We next claim that policy π_d is achieved by scheduling tasks from jobs in a round-robin fashion on the processors. Specifically, we take the tasks from the first job and place the first task on processor 1, the second task on processor 2, and continue on in this fashion possibly wrapping around the processors (if $b > P$) until we allocate all b tasks. We then take the tasks from job 2 and continue this round-robin scheduling where we left off. The performance of this policy is given by

$$E\left[T^{\pi_d}\right] = E\left[X\right] \sum_{i=1}^n \lceil ib/P \rceil. \tag{5.21}$$

To see that π_d is the best schedule for deterministic service times observe that the time required to service k jobs with deterministic task times is lower bounded by $\lceil kb/P \rceil$ since a total of kb tasks must complete execution on P processors. Equation (5.21) shows that π_d achieves this lower bound. Combining (5.20) and (5.21) yields the lower bound

$$E\left[T^\pi\right] \geq E\left[X\right] \sum_{i=1}^n \lceil ib/P \rceil.$$

Observe that we used two steps to derive this lower bound. The first step used Jensen's inequality to show that for any particular schedule a lower bound is achieved by keeping the same allocation of tasks but assuming that they have deterministic service times. The second step specified the optimal policy when service times are deterministic. It would be difficult to establish the first step without using Jensen's inequality.

The following example provides a useful way to remember the correct direction of Jensen's inequality:

Example 5.10 (Variance) Since the function x^2 is convex, Jensen's inequality implies that

$$E\left[X^2\right] \geq \left(E\left[X\right]\right)^2.$$

This implies that the variance (to be discussed in the following paragraphs) of random variable X, denoted by σ_X^2, is nonnegative:

$$\sigma_X^2 \overset{\text{def}}{=} E\left[X^2\right] - (E\left[X\right])^2 \geq 0. \tag{5.22}$$

Equation (5.22) provides a mnemonic device for easily recalling the direction of Jensen's inequality. It is clear that x^2 is convex and thus the fact that the variance is nonnegative implies $E\left[X^2\right] \geq (E\left[X\right])^2$. More generally, since x^{2n} is a convex function, Jensen's inequality implies that

$$E\left[X^{2n}\right] \geq (E\left[X\right])^{2n}.$$

Example 5.11 (Moment Generating Function) The moment generating function of a random variable X is defined by $\mathcal{M}_X(\theta) = E\left[\exp(\theta X)\right]$ for some value $\theta \geq 0$ (this function will be discussed in Section 5.3). Since the exponential function is convex, this implies that

$$\mathcal{M}_X(\theta) \geq e^{\theta E[X]}.$$

Example 5.12 (Jensen's Inequality for Concave Functions) If h is a convex function then $g(x) \overset{\text{def}}{=} -h(x)$ is concave and thus Jensen's inequality implies

$$E\left[h\left(X\right)\right] \geq h\left(E\left[X\right]\right) \Longrightarrow E\left[-h\left(X\right)\right] \leq -h\left(E\left[X\right]\right)$$

and thus

$$E\left[g\left(X\right)\right] \leq g\left(E\left[X\right]\right),$$

which is Jensen's inequality as stated for concave functions. As an example, the minimum function is concave and thus

$$E\left[\min\{X_1, X_2, \ldots, X_n\}\right] \leq \min\{E\left[X_1\right], E\left[X_2\right], \ldots, E\left[X_n\right]\}.$$

We now turn our attention to properties of expectation that require additional assumptions on the underlying random variables.

5.1.6 Expectation of Products of Random Variables

If X and Y are independent random variables then $E\left[XY\right] = E\left[X\right]E\left[Y\right]$. We show this for the discrete case. Define $Z \overset{\text{def}}{=} XY$ to be a random variable with density $f_Z(z)$. (To avoid ambiguities, we write the values of random variables X and Y as x and y, resp.) Independence implies that

$$f_Z(z) = \sum_{(x,y)\,:\,xy=z} f_X(x) f_Y(y).$$

We first show that Z has an expectation provided that X and Y have expectations. This follows from

$$\sum_z |z| f_Z(z) = \sum_z \sum_{(x,y)\,:\,xy=z} |x|\,|y|\,f_X(x) f_Y(y)$$

$$= \left(\sum_x |x| f_X(x) \right) \left(\sum_y |y| f_Y(y) \right) < \infty.$$

To calculate the expectation of Z we write

$$E[Z] = \sum_z \sum_{(x,y)\,:\,xy=z} (xy) f_X(x) f_Y(y)$$

$$= \sum_{(x,y)} (xy) f_X(x) f_Y(y)$$

$$= \left(\sum_x x f_X(x) \right) \left(\sum_y y f_Y(y) \right) = E[X]\,E[Y].$$

A similar argument clearly holds if X and Y are discrete, continuous, or mixed.

5.1.7 Survivor Representation of Expectation

If X is nonnegative and has an expectation, then we can write

$$E[X] = \begin{cases} \sum_{j=1}^{\infty} P[X \geq j], & X \quad \text{discrete,} \\[2ex] \int_0^{\infty} \overline{F_X}(x)\,dx, & X \quad \text{continuous,} \end{cases} \tag{5.23}$$

where we recall that $\overline{F_X}(x) \stackrel{\text{def}}{=} 1 - F_X(x)$ is the survivor function of X. We show this for the discrete case. Note that we can write $i = \sum_{j=1}^{i} 1$. Using this in the definition of expectation, (5.1) yields

$$E[X] = \sum_{i=1}^{\infty} i P[X = i] = \sum_{i=1}^{\infty} P[X = i] \sum_{j=1}^{i} 1$$

$$= \sum_{j=1}^{\infty} \sum_{i=j}^{\infty} P[X = i] = \sum_{j=1}^{\infty} P[X \geq j].$$

The proof for the continuous case is similar (see Problem 5.9).

5.1.8 Stopping Times

We have already seen an example of a stopping time in Example 4.8. In that example, N was a random variable that, at each step, independently *stopped* an experiment with probability p. Another example of a stopping time is the first time that the sum of independent random variables $X_1, X_2, \ldots$, surpasses a value of x. Event $\{N = n\}$ occurs if

$$X_1 + X_2 + \cdots + X_{n-1} \leq x$$

and

$$X_1 + X_2 + \cdots + X_n > x.$$

In this case, it is clear that the value of N can be determined by forming the partial sums

$$X_1, \quad X_1 + X_2, \quad X_1 + X_2 + \cdots + X_k,$$

and the experiment is stopped the first time one of these partial sums surpasses the value of x. Obviously if the event $\{N = n\}$ occurs then the values of $X_{n+1}, X_{n+2}, \ldots$ are not included in the calculation needed to determine N. The values of $X_i, 1 \leq i \leq n$, however, are included in the calculation that determines N since their sum determines when the experiment is stopped. This motivates the following definition:

Definition 5.13 *An integer valued random variable N is a stopping time for a sequence of random variables $X_1, X_2, \ldots$ if, for all possible values of n, $n = 0, 1, \ldots$, the event $\{N = n\}$ is "determined" by the values $X_1, X_2, \ldots, X_n$. By this we mean that if we know the values of $X_1, X_2, \ldots, X_n$ then we can decide if the event $\{N = n\}$ has occurred. It is important to note that the random variables $X_1, X_2, \ldots$, may be dependent random variables.*

Stopping times typically arise in random experiments that proceed in a sequential fashion where knowledge of X_n is obtained only after the experiment has produced values for $X_1, X_2, \ldots, X_{n-1}$. Random variables $X_1, X_2, \ldots, X_{n-1}$ and X_n may be dependent. The stopping time $\{N = n\}$, however, depends on the *values* of $X_1, X_2, \ldots, X_n$ that are obtained in the experiment and *at step n* these values are independent of the values of $X_{n+1}, X_{n+2}, \ldots$. In other words, there is an indicator function, $I_{\{N=n\}}$, that satisfies

$$P[N = n \mid X_1 = x_1, \ldots, X_n = x_n, X_{n+1} = x_{n+1}, \ldots] = I_{\{N=n\}}(x_1, \ldots, x_n).$$
(5.24)

The indicator function in (5.24) depends only on the values $x_1, \ldots, x_n$. Consider the following example:

Example 5.14 (First Passage Times) Let $X_i = 1$ if the ith toss of a fair coin lands heads up and equal to -1 otherwise. Suppose that a gambler wins one dollar on the ith toss if head lands up and loses one dollar otherwise. Then the random variable defined by

$$N \stackrel{\text{def}}{=} \min\{n : X_1 + X_2 + \cdots + X_n = 1\} \tag{5.25}$$

is a stopping time that denotes the first time the gambler wins exactly one dollar. An expression for the indicator function of (5.24) is given by

$$I_{\{N=n\}}(x_1, x_2, \ldots, x_n) = I_{\{x_1 < 1\}} \cdots I_{\{x_1 + x_2 \cdots + x_{n-1} < 1\}} I_{\{x_1 + x_2 \cdots + x_n = 1\}}. \tag{5.26}$$

This stopping time is equivalent to the first passage time considered in Section 3.3.5 (see (3.59)) and thus N has a density given by (3.70). Observe that $N' = \max\{n : X_1 + X_2 + \cdots + X_n = 1\}$ is not a stopping time.

It is tempting to believe that if N is a stopping time for the sequence $X_1, X_2, \ldots$, then the event $\{N = n\}$ is independent of the random variables $X_{n+1}, X_{n+2}, \ldots$. This, however, is not generally true and highlights a subtle, but important, property of stopping times. Consider the following example:

Example 5.15 (First Passage Times - Continued) Consider the experiment of Example 5.14, but where the outcome of the coin tosses are dependent random variables. In particular, assume that the experiment starts with a fair coin but, after tossing the first head, changes to a coin that always lands heads up. The event $\{N = n\}$ of (5.25) is still a stopping time since it is determined by the values of $X_1, X_2, \ldots, X_n$ and the (5.26) is still valid for this modified experiment. The event $\{N = n\}$, however, is *not independent* of random variables $X_{n+1}, X_{n+2}, \ldots$. To see this observe that if $X_{n+1} = -1$ then $X_i = -1, i = 1, 2, \ldots, n$ which implies that $\{N \neq n\}$.

Example 5.15 shows that if random variables $X_1, X_2, \ldots$, are dependent random variables then the stopping time $\{N = n\}$ is not generally independent of $X_{n+1}, X_{n+2}, \ldots$. A special case where independence is achieved is given in the following lemma:

Lemma 5.16 Let $X_1, X_2, \ldots$, be a sequence of independent random variables and let N be a stopping time for this sequence. Then the event $\{N = n\}$ is independent of $X_{n+1}, X_{n+2}, \ldots$.

This lemma is easily proved by noting that the event $\{N = n\}$ is determined by the values of $X_1, X_2, \ldots, X_n$ that are independent of random variables $X_{n+1}, X_{n+2}, \ldots$. We now continue another example of a stopping time:

Example 5.17 (Stopping Time) For the same random variables of Example 5.14, let

$$N \stackrel{\text{def}}{=} \min\{n : X_1 + X_2 + \cdots + X_n \notin \{-9, -8, \ldots, -1, 0, 1, \ldots, 99\}\}. \quad (5.27)$$

This is a stopping time that can be interpreted as a game where a gambler continues playing until the first time where either ten dollars are lost or one hundred dollars are won.

Stopping times are used to determine the statistics of the first time an event occurs. They typically occur as exceptional conditions in computer systems. For example, in a single server queue with a finite waiting room one stopping time of interest is the first time that the queue overflows and thus a customer is denied entrance to the waiting room. In a computer system this corresponds to a buffer overflow condition and typically invokes procedures to slow down the arrival process to the queue and thus avoid further customer loss. A stopping time for a system with real time constraints might be the first time a message arrives at its destination past its deadline. For example, in an ATM network this might signal that the job stream has not been assigned sufficient bandwidth. In inventory control systems, stopping times can correspond to the first time the inventory drops below a certain threshold. In such cases the system would automatically create orders to replace the missing inventory. In addition to models of this type, stopping times allow us to derive useful distributions, as in our derivation of the geometric distribution considered in Example 4.7. In Chapter 9, we will use stopping times to derive phase distributions that are extensively used in performance models. We are now in a position to derive Wald's equality.

5.1.9
Wald's Equality

We are interested in calculating the expectation of a random variable determined by a stopping time. This is a classical result called *Wald's equality*,[3] and is it proved here.

Theorem 5.18 (Wald's equality) *Let* N*, for* $N \geq 1$ *and such that* $E[N]$ *exists, be a random variable that is a stopping time for the sequence of independent and identically distributed random variables* $X_i, i = 1, 2, \ldots$*, having finite expectation. Let*

$$S_N = X_1 + X_2 + \ldots + X_N. \quad (5.28)$$

Then

$$E[S_N] = E[X] \, E[N].$$

---

[3]Named after Abraham Wald (1902–1950), an American statistician.

Proof: Observe that $X_n, n = 1, 2, \ldots$, is included in the sum (5.28) only if $N \geq n$. Thus we can write

$$
\begin{aligned}
E\left[S_N\right] &= E\left[\sum_{n=1}^{\infty} X_n I_{\{N \geq n\}}\right] \\
&= \sum_{n=1}^{\infty} E\left[X_n I_{\{N \geq n\}}\right], \quad\quad (5.29)
\end{aligned}
$$

where the interchange of sum and expectation in (5.29) follows from the fact that we assume that X_n has finite expectation (we can substitute X_n by $|X_n|$). We now claim that X_n is independent of $I_{\{N \geq n\}}$. To see this we write

$$
I_{\{N \geq n\}} = 1 - \prod_{j=1}^{n-1} I_{\{N = j\}}.
$$

The result then follows from the independence of random variables X_i and from Lemma 5.16, which implies that the event $\{N = j\}$, for $j = 1, 2, \ldots, n-1$, is independent of X_n. We can thus write (5.29) as

$$
\begin{aligned}
E\left[S_N\right] &= E\left[X\right] \sum_{n=1}^{\infty} E\left[I_{\{N \geq n\}}\right] \\
&= E\left[X\right] \sum_{n=1}^{\infty} P\left[N \geq n\right] = E\left[X\right] E\left[N\right].
\end{aligned}
$$

∎

Example 5.19 Let $X_i = 1$ if the ith toss of a coin lands heads up and 0 otherwise. Then $E\left[X\right] = 1/2$ and

$$
N \stackrel{\text{def}}{=} \min\{n : X_1 + X_2 + \cdots + X_n = 10\}
$$

is a stopping time that can be thought of as the total number of plays required until a gambler wins 10 dollars. Wald's equality applies and, since

$$
E\left[X_1 + X_2 + \cdots + X_N\right] = 10,
$$

it yields $E\left[N\right] = 20$, as one expects.

Wald's equality can sometimes yield nonintuitive results, as the following examples show:

Example 5.20 (Gamblers Ruin Problem) Consider the stopping time of Example 5.17 given by (5.27). In this example random variable $X_i, i = 1, 2, \ldots$, corresponds to the ith toss of a fair coin and equals 1 with probability $1/2$ and -1 otherwise. The game stops the first time a gambler loses all of their money (let's be generous and say this equals 10 dollars!) or wins 100 dollars. Intuitively, one believes that since winning and losing a dollar are equally likely events, there is a greater chance of first losing 10 dollars than there is of first winning 100 dollars. This seems to imply that $E\left[X_1 + X_2 + \cdots + X_N\right] < 0$. Wald's equality, however, shows that

$$E\left[X_1 + X_2 + \cdots + X_N\right] = E\left[X\right] E\left[N\right] = 0$$

since $E\left[X\right] = 0$. It is clear that N has a finite expectation and thus satisfies the conditions for Wald's inequality. Let w be the probability that the game stops by the player winning 100 dollars. The expected winnings are given by $100w - 10(1 - w)$, and since Wald's equality implies that this equals 0 we must conclude that $w = 1/11$.

The specification of the problem allows us to be more general. If we define

$$N \stackrel{\text{def}}{=} \min\{n : X_1 + X_2 + \cdots + X_n \notin [-m, k]\}, \quad m, k > 0,$$

then Wald's inequality implies that the probability of first hitting k, denoted by w, satisfies $kw - m(1 - w) = 0$ and thus

$$w = \frac{m}{k + m}.$$

Example 5.21 (First Passage Times) Let $X_i = 1$ if the ith toss of a coin lands heads up and -1 otherwise. Suppose that a player wins one dollar on the ith toss if head lands up and loses one dollar otherwise. Then

$$N \stackrel{\text{def}}{=} \min\{n : X_1 + X_2 + \cdots + X_n = 1\}$$

is a stopping time that corresponds to a first passage time (see Example 5.14). Applying Wald's inequality, however, yields a contradiction since clearly $X_1 + X_2 + \cdots + X_N = 1$ but $E\left[X\right] = 0$. This implies that N cannot have a finite expectation, which is indeed the case (see Problem 5.4).

5.2 Expectations of Functions of Random Variables

We now explore some properties of the expectation of functions of random variables. We call $E\left[X^i\right]$ the ith *moment* of the random variable X. Often one forms the expectation of a function of X, say, $E\left[h(X)\right]$. If we can write h as a power series

$$h(x) = h_0 + h_1 x + h_2 x^2 + \cdots,$$

where $h_i, i = 0, 1, \ldots$, are constant coefficients then

$$E\left[h(X)\right] = h_0 + h_1 E\left[X\right] + h_2 E\left[X^2\right] + \cdots \qquad (5.30)$$

and thus the moments of X define $E\left[h(X)\right]$.

Example 5.22 (Moments of Exponential Random Variables) The nth moment of an exponential random variable X with parameter λ is given by

$$E[X^n] = \int_0^\infty x^n \lambda e^{-\lambda x} dx = \frac{n!}{\lambda^n}$$

where we have used (see (B.12))

$$n! = \int_0^\infty e^{-t} t^x dx$$

to evaluate the integral.

Example 5.23 (Survivor Representation of Moments - Nonnegative Random Variables) For nonnegative random variables (5.23) implies that

$$E[X] = \int_0^\infty \overline{F_X}(x) dx.$$

This can be generalized, and we can write

$$E[X^k] = k \int_0^\infty x^{k-1} \overline{F_X}(x) dx, \qquad k = 1, 2, \dots . \tag{5.31}$$

To show this we perform the integral in (5.31) by parts yielding

$$\int_0^\infty x^{k-1} \overline{F_X}(x) dx = \left. \frac{x^k \overline{F_X}(x)}{k} \right|_0^\infty + \int_0^\infty \frac{x^k}{k} f_X(x) dx$$

$$= \frac{E[X^k]}{k}.$$

Example 5.24 (Expected Number in a Single Server Discrete Time Queue) Consider a single server queue where time is slotted into unit intervals and each customer requires one slot of service. Let N_i be a random variable that denotes the number of customers in the system at the end of slot i, for $i = 1, 2, \dots$, and suppose that the number of customer arrivals in each slot is independent and identically distributed. We denote the number of arrivals in slot i by V_i and assume that customers arriving during a slot are not serviced in that slot. For this model it is clear that

$$N_i = (N_{i-1} - 1)^+ + V_i. \tag{5.32}$$

To interpret (5.32) note that if $N_{i-1} > 0$ then there is a customer in service that leaves at the end of the slot and thus the number of customers in the system is $N_{i-1} - 1$ plus the number of arrivals in the slot, V_i. If $N_{i-1} = 0$ then the number of customers in the ith slot is simply given by V_i. Assume that N_i reaches a limiting distribution as $i \to \infty$. Denote this random variable by

N and denote a generic arrival random variable by V, that is, $V =_{st} V_i, i = 1, 2, \ldots$. Equation (5.32) can then be written as

$$N = \left(\hat{N} - 1\right)^+ + V, \tag{5.33}$$

where $\hat{N}$, for $\hat{N} =_{st} N$, is a random variable that is independent of V. We are interested in determining an expression for the average number of customers in the system.

Let $E[N_S]$ be the average number of customers in the server. It is clear that we can write this as

$$E[N_S] = \sum_{i=1}^{\infty} f_N(i) \tag{5.34}$$

since if $N > 0$ then the server has a customer. (The density function $f_N(i)$ equals the probability that there are $i - 1$ customers in the queue and 1 customer in service.)

To determine the expectation of (5.33), we first calculate

$$
\begin{aligned}
E\left[(N - 1)^+\right] &= \sum_{i=1}^{\infty}(i - 1)f_N(i) \\
&= E[N] - \sum_{i=1}^{\infty} f_N(i) \\
&= E[N] - E[N_S], \tag{5.35}
\end{aligned}
$$

where we have used (5.34) to obtain (5.35). Taking the expectation of (5.33) yields

$$E[N] = E[N] - E[N_S] + E[V],$$

and since the $E[N]$ terms cancel, this implies that

$$E[N_S] = E[V]. \tag{5.36}$$

It is interesting to note that taking the expectation of (5.33) does not provide any information about $E[N]$ since it drops out of the equation!

A "trick" that can be applied in situations like this is to perform a similar calculation for the second moment by squaring (5.33). This yields

$$N^2 = \left[(N - 1)^+\right]^2 + 2V(N - 1)^+ + V^2. \tag{5.37}$$

A calculation like that leading to (5.35) shows that

$$E\left[(N - 1)^+\right]^2 = E\left[N^2\right] - 2E[N] + E[V],$$

where we have used (5.36). Taking the expectation of (5.37) yields

$$E\left[N^2\right] = E\left[N^2\right] - 2E[N] + E[V] + 2E[V]\left(E[N] - E[V]\right) + E\left[V^2\right],$$

and conveniently the second moment of N drops out of this equation (this is why the trick works!). This leaves us with an equation for $E[N]$,

$$E[N] = \frac{E[V] + E[V^2] - 2(E[V])^2}{2(1 - E[V])}$$

$$= E[N_S] + \frac{E[V^2] - E[V]}{2(1 - E[V])}. \tag{5.38}$$

Rewriting the equation in the form (5.38) shows that the second term equals the average number of customers waiting for service, that is,

$$E[N_Q] = \frac{E[V^2] - E[V]}{2(1 - E[V])}.$$

The equation shows that the expected number of customers in the queue increases without bound as $E[V] \to 1$. We will revisit this example when discussing the expected response time for M/G/1 queues in Section 7.2.

5.2.1 Variance

A particularly useful statistic of a random variable is the variance that was first mentioned in Example 5.10. We denote the variance of random variable X equivalently by $\mathrm{Var}(X)$ and σ_X^2 and term $\sqrt{\mathrm{Var}(X)}$ and σ_X the *standard deviation* of X. By definition the variance is given by

$$\sigma_X^2 = \mathrm{Var}(X) \stackrel{\text{def}}{=} E\left[(X - E[X])^2\right]. \tag{5.39}$$

After applying property 2 of expectation, this can be rewritten in terms of moments by

$$\mathrm{Var}(X) = E\left[(X - E[X])^2\right]$$

$$= E\left[X^2 - 2XE[X] + (E[X])^2\right]$$

$$= E\left[X^2\right] - (E[X])^2, \tag{5.40}$$

as given by (5.22). Properties of variance given in Table 5.2 can be easily established (we suggest the reader prove them in Problem 5.11).

The variance for the distributions previously considered can be easily calculated and are listed in Tables 5.9 and 5.10.

Example 5.25 Variance measures the "dispersion" of a random variable about its mean. Consider a Bernoulli random variable with probability p. A simple calculation shows that the variance is $\mathrm{Var}(B_{1,p}) = p(1-p)$. If $p = 0$ or

#	Property
1	$\mathrm{Var}\,(X) = 0$ if and only if X is a degenerate random variable.
2	$\mathrm{Var}\,(aX + b) = a^2 \mathrm{Var}\,(X)$ for a, b constants.
3	$\mathrm{Var}\,(X_1 + X_2 + \cdots + X_n) = \mathrm{Var}\,(X_1) + \mathrm{Var}\,(X_2) + \cdots + \mathrm{Var}\,(X_n)$ if n is a constant and $X_i, i = 1, 2, \ldots, n$ are independent.
4	$\mathrm{Var}\,(X_1 + X_2 + \cdots + X_N) = E\,[N]\mathrm{Var}\,(X) + (E\,[X])^2\,\mathrm{Var}\,(N)$ if $X_i, i = 1, 2, \ldots,$ is a sequence of independent and identically distributed random variables with mean and variance, $E\,[X]$ and $\mathrm{Var}\,(X)$, respectively, and N is a stopping time for the sequence.

TABLE 5.2. Properties of Variance

$p = 1$ then clearly there is no dispersion in the value of the random variable and this is reflected in the fact that the variance for these cases equals 0. The largest dispersion in the Bernoulli random variable occurs when all outcomes are equally likely, that is, when $p = 1/2$ for which the variance is $1/4$.

5.2.2 Relationship Between Expectation and Variance

The expectation and variance of a random variable X provide *independent characterizations* of X. We can think of $E\,[X]$ as a measure of the location or mean of X. Shifts in X, say to $a + X$ simply cause shifts in the location, that is, $E\,[a + X] = a + E\,[X]$. The expectation of X does not provide information about the spread or dispersion of the distribution of X about its mean. Variance is a measure of this dispersion, and the fact that shifting X does not change its dispersion is reflected in the property that $\mathrm{Var}\,(a + X) = \mathrm{Var}\,(X)$.

There is another qualitative description of a random variable that can be termed "variability." Consider a random variable X with $E\,[X] = 1$ and $\mathrm{Var}\,(X) = 1$ and another random variable Y with $E\,[Y] = 10^{100}$ and $\mathrm{Var}\,(Y) = 1$. Both random variables have the same variance but, since $E\,[Y] \gg E\,[X]$, X is more "variable" than Y. This motivates a definition where variance and expectation are compared. The squared *coefficient of variation* of random variable X, denoted by C_X^2, serves this purpose and is defined by

$$C_X^2 \overset{\text{def}}{=} \frac{\text{Var}\,(X)}{(E\,[X])^2}.$$

(5.41)

The variability of X increases with C_X^2.

Example 5.26 (Relationship Among Expectation, Variance, and Coefficient of Variation) Consider a sequence of random variables $X_n, n = 1, 2, \ldots$, that have a distribution given by

$$P\,[X_n = x] = \begin{cases} 1/2, & x = (1+\alpha)n, \\ 1/2, & x = (1-\alpha)n, \end{cases}$$

where α, for $0 < \alpha < 1$, is some fixed constant. Simple calculations show that

$$E\,[X_n] = n, \quad \sigma_{X_n}^2 = (\alpha n)^2, \quad C_{X_n}^2 = \alpha^2.$$

This reflects the fact that the location $(E\,[X_n])$ increases linearly with n, and the dispersion $(\sigma_{X_n}^2)$ increases as n^2, but the variability $(C_{X_n}^2)$ remains fixed for all n.

Both the expectation and variance of a random variable are measures that depend only on the distribution of one random variable. Very often, however, we are concerned with the relationship between several random variables. Such relationships are described by joint distribution functions. Similar to the scalar case it is frequently convenient to characterize joint distributions by some statistical values. In the case of independent random variables, such characterizations reduce to their scalar components. If random variables are not independent, then a useful characterization that provides a measure of the dependency is covariance.

**5.2.3
Covariance**

If X and Y are not independent then $\text{Var}\,(X+Y)$ is, in general, not equal to the sum of the variances of X and Y. We can calculate

$$\text{Var}\,(X+Y) \;=\; E\left[((X+Y) - E\,[X+Y])^2\right]$$

$$=\; \text{Var}\,(X) + \text{Var}\,(Y) + 2\text{Cov}\,(X,Y),$$

where

$$\text{Cov}\,(X,Y) \overset{\text{def}}{=} E\,[(X - E\,[X])\,(Y - E\,[Y])] = E\,[XY] - E\,[X]E\,[Y]$$

(5.42)

is the *covariance* of X and Y.

Covariance is a measure of the "dependency" of two random variables and it is obvious from its definition that if X and Y are independent then $\text{Cov}(X, Y) = 0$. A subtle problem arises with providing semantics for covariance, however, since two random variables can be dependent and yet have a covariance of 0 (see Example 5.27). The strongest statement that can be made is that if the covariance of two random variables is nonzero then the random variables must be dependent. Consider the following two examples:

Example 5.27 Let X equal $-1, 0$, or 1 with equal probability and let $Y = |X|$. Since Y is defined in terms of X, the random variables X and Y are dependent. Simple calculations, however, show that $E[XY] = 0$ and $E[X] = 0$ and thus that $\text{Cov}(X, Y) = 0$.

Example 5.28 Assume that X is nonzero, and let $Y = 1/X$. Then if $E[X]E[Y] = 1$ random variables X and Y are uncorrelated (note that $E[XY] = 1$). As an example, consider

$$
f_X(x) = \begin{cases} 1/9, & x = -1, \\ \\ 4/9, & x = 1/2, \ 2. \end{cases}
$$

A simple calculation shows that X and $1/X$ are uncorrelated and have unit mean.

It is convenient to normalize the covariance of two random variables by defining a *correlation coefficient* given by

$$
\rho(X, Y) \stackrel{\text{def}}{=} \frac{\text{Cov}(X, Y)}{\sqrt{\text{Var}(X)\,\text{Var}(Y)}}. \tag{5.43}
$$

This normalization implies that $-1 \le \rho(X, Y) \le 1$ with equality 1 (resp., -1) if and only if $Y = aX$ (resp., $Y = -aX$) where $a > 0$ (see Problem 5.15). We say that two random variables are positively or negatively correlated if $\rho(X, Y) > 0$ or $\rho(X, Y) < 0$, respectively, and say that they are uncorrelated if $\rho(X, Y) = 0$. A positive (resp., negative) correlation between random variables intuitively implies that if X is large then Y is likely to be large (resp., small). Consider the following example:

Example 5.29 (Negative Correlation) Let X be defined as in Example 5.27, that is, it takes on values $-1, 0, 1$ with equal probability and let $Y = 1/(2 + X)$. Since Y is expressed as a decreasing function of X it is plausible to believe that X and Y are negatively correlated. Simple calculations show that

$$E\left[XY\right] = -2/9, \quad E\left[X\right] = 0, \quad E\left[Y\right] = 11/18,$$

$$E\left[X^2\right] = 2/3, \quad E\left[Y^2\right] = 49/108,$$

and thus

$$\rho\left(X,Y\right) = \frac{-2/9}{\sqrt{\frac{2}{3}\left(\frac{49}{108} - \left(\frac{11}{18}\right)^2\right)}} = -0.96,$$

justifying our intuition.

5.3 Moment Generating Functions

The *moment generating function* is defined by

$$\mathcal{M}_X(\theta) \overset{\text{def}}{=} \int_{-\infty}^{\infty} e^{\theta x} f_X(x) dx$$

$$= E\left[e^{\theta X}\right], \quad \theta > 0, \qquad (5.44)$$

provided the integral exists. Expanding the exponential as a power series and taking the expectation implies that

$$\mathcal{M}_X(\theta) = 1 + \theta E\left[X\right] + \frac{\theta^2 E\left[X^2\right]}{2!} + \frac{\theta^3 E\left[X^3\right]}{3!} + \cdots . \quad (5.45)$$

From (5.45) it is easy to see that

$$\mathcal{M}_X(0) = 1 \qquad (5.46)$$

and the nth moment of the random variable is given by

$$\mathcal{M}_X^{(n)}(0) = E\left[X^n\right], \quad n = 1, 2, \ldots, \qquad (5.47)$$

and hence the name of the function is explained. From (5.40) we have that

$$\sigma_X^2 = \mathcal{M}_X^{(2)}(0) - \left(\mathcal{M}_X^{(1)}(0)\right)^2. \qquad (5.48)$$

We record this and other useful properties of moment generating functions in Table 5.3.

It is often more convenient to determine the moments of a distribution by first calculating its moment generating function and then using (5.47)

#	Property
1	$\mathcal{M}_Y(\theta) = e^{b\theta}\mathcal{M}_X(a\theta)$ if $Y = aX + b$.
2	$\mathcal{M}_{X+Y}(\theta) = \mathcal{M}_X(\theta)\mathcal{M}_Y(\theta)$ if X and Y are independent random variables.
3	$\mathcal{M}_{X_1+X_2+\cdots+X_n}(\theta) = (\mathcal{M}_X(\theta))^n$ if $X_i, i = 1, 2, \ldots, n$, are independent and identically distributed random variables and satisfy $X =_{st} X_i$.
4	$\mathcal{M}_X^{(n)}(0) = E[X^n]$.
5	$\mathcal{M}_{X_1+X_2+\cdots+X_N}(\theta) = \mathcal{G}_N(\mathcal{M}_X(\theta))$ if $X_i, i = 1, 2, \ldots$, is a sequence of independent and identically distributed random variables with moment generating function $\mathcal{M}_X$ and N is a stopping time for the sequence with probability generating function $\mathcal{G}_N$.
6	$\mathcal{M}_X(\theta) = \mathcal{G}_X(e^\theta)$ if X is a discrete random variable.

TABLE 5.3. Properties of Moment Generating Functions

to pick off the moments. As an example, let X be an exponential random variable with parameter λ. The moment generating function is given by

$$\mathcal{M}_X(\theta) = \int_0^\infty e^{\theta x} \lambda e^{-\lambda x} dx = \frac{\lambda}{\lambda - \theta}. \qquad (5.49)$$

Differentiation yields

$$\mathcal{M}_X^{(n)}(\theta) = \frac{\lambda n!}{(\lambda - \theta)^{n+1}} \quad n = 1, 2, \ldots,$$

which, using (5.47), implies that

$$E[X^n] = \frac{n!}{\lambda^n},$$

a result we previously derived using integration in Example 5.22.

Property 5 of moment generating functions follows from a simple conditioning argument. For notational simplicity let $Y \stackrel{\text{def}}{=} X_1 + X_2 + \cdots + X_N$, where N is a stopping time for the sequence $X_1, X_2, \dots$. Conditioning on the value of N and using property 3 of moment generating functions shows that

$$\mathcal{M}_Y(\theta) = \sum_{n=1}^{\infty} (\mathcal{M}_X(\theta))^n \, P[N = n].$$

Comparing this with the definition of a probability generating function (4.24) shows that $\mathcal{M}_Y(\theta) = \mathcal{G}_X(\mathcal{M}_X(\theta))$ as claimed. One can use this relationship to obtain the moments of Y by differentiation. The first differentiation (we are using (5.47)) yields

$$E[Y] = E[X] \, E[N]$$

which is another proof of Wald's theorem, Theorem 5.18 (property 8 of Table 5.1). The second differentiation and some algebra yield

$$\sigma_Y^2 = E[N]\sigma_X^2 + (E[X])^2 \, \sigma_N^2,$$

which we have seen before as the fourth property of variance.

It is interesting to note that if X is a discrete random variable with generating function $\mathcal{G}_X(z)$ then its moment generating function is given by

$$\mathcal{M}_X(\theta) = \sum_{x=0}^{\infty} P[X = x] \, e^{\theta x} = \mathcal{G}_X(e^{\theta}). \tag{5.50}$$

This equals the generating function of X evaluated at $z = \exp[\theta]$. Using this relationship we can state properties of probability generating functions that are analogous to those just given for moment generating functions.

5.3.1 Probability Generating Functions

Recall that from (4.24) the definition of the probability generating function for the discrete random variable X is

$$\mathcal{G}_X(z) \overset{\text{def}}{=} \sum_{i=0}^{\infty} P[X = i] \, z^i, \tag{5.51}$$

which can also be viewed as the expectation

$$\mathcal{G}_X(z) \overset{\text{def}}{=} E\left[z^X\right]. \tag{5.52}$$

Probability generating functions satisfy all the properties of ordinary generating functions (see Table 3.6) and analogous properties to those just listed for moment generating functions (see Table 5.4).

#	Property
1	$\mathcal{G}_Y(z) = z^b \mathcal{G}_X(az)$ if $Y = aX + b$.
2	$\mathcal{G}_{X+Y}(z) = \mathcal{G}_X(z)\mathcal{G}_Y(z)$ if X and Y are independent random variables.
3	$\mathcal{G}_{X_1+X_2+\cdots+X_n}(z) = (\mathcal{G}_X(z))^n$ if $X_i, i = 1, 2, \ldots, n$, are independent and identically distributed random variables and satisfy $X =_{st} X_i$.
4	$\mathcal{G}_X^{(n)}(1) = E[(X)_n] = E[X(X-1)\cdots(X-n+1)]$.
5	$\mathcal{G}_{X_1+X_2+\cdots+X_N}(z) = \mathcal{G}_N(\mathcal{G}_X(z))$ if $X_i, i = 1, 2, \ldots$, is a sequence of independent and identically distributed discrete random variables with probability generating function $\mathcal{G}_X$ and N is a stopping time for the sequence with probability generating function $\mathcal{G}_N$.
6	$\mathcal{G}_X(z) = \mathcal{M}_X(\ln(z))$.

TABLE 5.4. Properties of Probability Generating Functions

To determine the moments of a discrete random variable X we can use the equivalence given in (5.50). Simple manipulations yield

$$\mathcal{M}_X^{(1)}(\theta) = e^\theta \mathcal{G}_X^{(1)}(e^\theta),$$

$$\mathcal{M}_X^{(2)}(\theta) = e^{2\theta} \mathcal{G}_X^{(2)}(e^\theta) + e^\theta \mathcal{G}_X^{(1)}(e^\theta),$$

from which the first two moments can be derived, yielding

$$E[X] = \mathcal{M}_X^{(1)}(0) = \mathcal{G}_X^{(1)}(1),$$

$$E[X^2] = \mathcal{M}_X^{(2)}(0) = \mathcal{G}_X^{(2)}(1) + \mathcal{G}_X^{(1)}(1). \tag{5.53}$$

Moment generating functions for some common densities are given in Table 5.10, and in Table 5.9 we list common probability generating functions.

5.3.2 Moment Generating Functions Determine Densities

We use moment generating functions in applied probability to calculate probability distributions for continuous random variables in much the same way as probability generating functions are used for discrete random variables. The mathematical foundation for this, similar to Theorem 4.12, is that a moment generating function uniquely determines a distribution. We record this in the following theorem:

Theorem 5.30 *Let X and Y be random variables with moment generating functions $\mathcal{M}_X(\theta)$ and $\mathcal{M}_Y(\theta)$, respectively. If $\mathcal{M}_X(\theta) = \mathcal{M}_Y(\theta)$ for some interval around 0 then $X =_{st} Y$.*

It is clear that a distribution function determines a set of moments when these moments exist and it is also clear that the existence of a moment generating function implies the existence of all moments. A natural question that arises is how far a set of moments characterizes a distribution. A somewhat surprising result here is that two different distributions can have the same set of moments. It is also surprising that the existence of all moments does not imply that a moment generating function exists (see [98] for counterexamples to these suppositions).

5.4 The Laplace Transform

A function that is closely related to the moment generating function is the *Laplace transform*.[4] For a given real valued function $h(t), t \geq 0$, the Laplace transform, denoted by $h^*(s)$, is defined by

$$h^*(s) \stackrel{\text{def}}{=} \int_0^\infty e^{-sx} h(x) dx. \tag{5.54}$$

If $h(t), t \geq 0$, is a function corresponding to a density function of a nonnegative random variable X then it is clear from (5.44) that the Laplace transform of the function $h^*(s)$ equals the moment generating function of X evaluated at $-s$, that is,

$$h^*(s) = \mathcal{M}_X(-s). \tag{5.55}$$

To preserve the semantics associated with random variables we use the "$\mathcal{L}_X(s)$ notation" when dealing with the Laplace transform of a random variable X,

$$\mathcal{L}_X(s) \stackrel{\text{def}}{=} \int_0^\infty e^{-sx} f_X(x) dx \tag{5.56}$$

and use the "$h^*(s)$ notation" otherwise. Notice that an interpretation of the transform for the random variable case is that it is the expectation

---

[4]Named after Pierre Simon Laplace (1749–1827).

of $\exp[-sX]$ and thus can be written as

$$\mathcal{L}_X(s) = E\left[e^{-sX}\right].$$

Using (5.55) shows that the following relationship holds between the moment generating function of X and its Laplace transform

$$\mathcal{L}_X(s) = \mathcal{M}_X(-s).$$

A probabilistic interpretation of the Laplace transform is found in Problem 5.21. Like the theorem for moment generating functions, Theorem 5.30, if two random variables have the same Laplace transform then they are stochastically equal.

5.4.1
Common
Transforms

Extensive tables of Laplace transforms can be found in [1, 30, 42] and in Table C.2 we provide a partial list. We write

$$f(t) \iff f^*(s)$$

to indicate transform pairs. The derivation of some elementary transforms is given in the following example:

Example 5.31 (Derivation of Some Simple Transforms) Recall that

$$n! = \int_0^\infty e^{-x} x^n dx.$$

By a simple conversion this implies that

$$\frac{t^{n-1}}{(n-1)!} \iff \frac{1}{s^n}, \quad n = 1, 2, \ldots. \tag{5.57}$$

Let

$$h(t) = \lambda e^{-\lambda t};$$

then

$$h^*(s) = \int_0^\infty e^{-st} \lambda e^{-\lambda t} dt = \frac{\lambda}{\lambda + s}, \tag{5.58}$$

and thus

$$\lambda e^{-\lambda t} \iff \frac{\lambda}{\lambda + s}.$$

5.4.2
Properties of
the Laplace
Transform

The Laplace transform is generally more useful than the moment generating function since it can be applied to functions that are not densities.[5] The Laplace transform is the continuous counterpart of the ordinary generating function defined in Chapter 3. To see the relationship between these two functions assume that the function $h(t)$ takes on nonzero values for integer t, that is, that $h(t) = h_i$, for $t = i$ and $i = 0, 1, \ldots$, and let $H(x)$ be the generating function

$$H(x) = \sum_{i=0}^{\infty} h_i x^i.$$

Then it follows that

$$h^*(s) = \int_0^{\infty} e^{-sx} h(x) dx = \sum_{i=0}^{\infty} e^{-si} h_i = H(e^{-s}) \qquad (5.59)$$

and thus the Laplace transform equals the ordinary generating function evaluated at e^{-s}. If h_i represents a probability density, then equations (5.59) and (5.55) show that we have simply established a different form of the relationship between moment generating functions and their corresponding probability generating functions for discrete random variables given in (5.50).

Since Laplace transforms are equivalent to moment generating functions when the underlying function is a density, it follows that they satisfy analogous properties to those given for moment generating functions in Section 5.3. For example, the analogous moment property to (5.47) is given by

$$\mathcal{L}_X^{(n)}(0) = (-1)^n E\left[X^n\right]. \qquad (5.60)$$

Analogous properties for Laplace transforms of a random variable to those found on the list given for moment generating functions are given in Table 5.5.

In Table 5.6 we provide a more complete list of the most useful properties of Laplace transforms that are needed in applications. This table is similar to Table 3.6 for ordinary generating functions. As with generating functions, Laplace transforms are often used when forming the convolution of two random variables. To derive the Laplace transform

---

[5]Note that if $f(x) \geq 0, 0 \leq x \leq \infty$, and $c = \int_0^{\infty} f(x) dx$ is finite, then we can *normalize* f to make it a density; that is, $f(x)/c$ is a density. If we let X be a random variable with this density then clearly the Laplace transform of f and the moment generating function of X are related by $f^*(s) = c\mathcal{M}_X(-s)$. The Laplace transform, however, is defined even if f cannot be normalized, that is, if $c = \infty$.

#	Property
1	$\mathcal{L}_Y(s) = e^{-bs}\mathcal{L}_X(as)$ if $Y = aX + b$.
2	$\mathcal{L}_{X+Y}(s) = \mathcal{L}_X(s)\mathcal{L}_Y(s)$ if X and Y are independent random variables.
3	$\mathcal{L}_{X_1+X_2+\cdots+X_n}(s) = (\mathcal{L}_X(s))^n$ if $X_i, i = 1, 2, \ldots, n$, are independent and identically distributed random variables and satisfy $X =_{st} X_i$.
4	$\mathcal{L}_X^{(n)}(0) = (-1)^n E[X^n]$.
5	$\mathcal{L}_{X_1+X_2+\cdots+X_N}(s) = \mathcal{G}_N(\mathcal{L}_X(s))$ if $X_i, i = 1, 2, \ldots$, is a sequence of independent and identically distributed random variables with Laplace transform $\mathcal{L}_X$ and N is a stopping time for the sequence with probability generating function $\mathcal{G}_N$.
6	$\mathcal{L}_X(s) = \mathcal{G}_X(e^{-s})$ if X is a discrete integer random variable.

TABLE 5.5.   Properties of Laplace Transforms

for a convolution, let $h(t)$ be the density function that results when we convolve two continuous density functions f and g. The convolution operator *for density functions* is defined by

$$h(t) \stackrel{\text{def}}{=} \int_0^t f(t-x)g(x)dx. \tag{5.61}$$

We equivalently write this using the convolution notation as

$$h(t) = f \circledast g(t). \tag{5.62}$$

These definitions are analogous to the convolution of two discrete sequences as given by (3.55) and (3.56). (The convolution operator for *distribution functions* is defined differently; see (6.26).) To calculate $h^*(s)$ we write

$$h^*(s) \quad = \quad \int_0^\infty e^{-st}h(t)dt$$

Property		Function	Laplace Transform
	Addition	$af(t) + bg(t)$	$af^*(s) + bg^*(s)$
	Convolution	$f \circledast g(t)dx$	$f^*(s)g^*(s)$
Shifting	Linear	$f(t-a)$	$e^{-as}f^*(s)$
	Exponential	$e^{-at}f(t)$	$f^*(s+a)$
Scaling	Power	$t^n f(t)$	$(-1)^n \frac{d^n f^*(s)}{ds^n}$
	Linear	$tf(t)$	$-\frac{d}{ds}f^*(s)$
	Inverse power	$f(t)/t^n$	$\int_{s_1=s}^{\infty} ds_1 \int_{s_2=s_1}^{\infty} ds_2 \cdots \int_{s_n=s_{n-1}}^{\infty} ds_n f^*(s_n)$
	Harmonic	$f(t)/t$	$\int_s^{\infty} f^*(s_1)ds_1$
Integrals	Single	$\int_0^t f(x)dx$	$\frac{f^*(s)}{s}$
	Multiple	$\underbrace{\int_0^{\infty} \cdots \int_0^{\infty}}_{n \text{ times}} f(x)dx^n$	$\frac{f^*(s)}{s^n}$
Derivatives	Single	$\frac{d}{dt}f(t)$	$sf^*(s) - f(0^-)$
	Multiple	$\frac{d^n f^*(s)}{dt^n}$	$s^n f^*(s) - \sum_{i=0}^{n-1} s^{n-1-i} f^{(i)}(0^-)$
Values	Initial	$\lim_{t\to 0} f(t)$	$\lim_{s\to\infty} sf^*(s)$
	Integral	$\int_0^{\infty} f(x)dx$	$f^*(0)$
	Final	$\lim_{t\to\infty} f(t)$	$\lim_{s\to 0} sf^*(s)$

TABLE 5.6. Algebraic Properties of Laplace Transforms

$$= \int_0^\infty e^{-st} dt \int_0^t f(t-x)g(x)dx$$

$$= \int_0^\infty e^{-sx} g(x)dx \int_x^\infty e^{-s(t-x)} f(t-x)dt$$

$$= g^*(s) \ f^*(s).$$

To derive the single derivative property listed in Table 5.6, let

$$h(t) \overset{\text{def}}{=} \frac{d}{dt} f(t),$$

and thus, by definition,

$$h^*(s) = \int_0^\infty e^{-st} h(t)dt = \int_0^\infty e^{-st} \frac{d}{dt} f(t)dt.$$

Integrating this by parts yields

$$h^*(s) = e^{-st} f(t)\big|_0^\infty + s \int_0^\infty e^{-st} f(t)dt$$

$$= \lim_{t\to\infty} e^{-st} f(t) - \lim_{t\to 0} e^{-st} f(t) + s f^*(s). \qquad (5.63)$$

To evaluate the limits in (5.63) we first observe that for $f^*(s)$ to exist it must be the case that

$$\lim_{t\to\infty} e^{-st} f(t) = 0,$$

and thus the first limit is 0. The second limit is given by

$$\lim_{t\to 0} e^{-st} f(t) = f(0^-),$$

where the notation $f(0^-)$ means that the limit is taken from below so as to capture any singularities at the point 0. Thus we have that

$$h^*(s) = s f^*(s) - f(0^-).$$

A typical case of how singularities arise in practice is found in Example 4.11 where we calculated the density function for the waiting time in a single server queue (see (4.23)). In this example an impulse at 0 reflects the case where arriving customers do not have to wait before receiving service.

Example 5.32 (Transform for Erlang Random Variables) Using the convolution property of Laplace transforms and the fact that an Erlang density is the convolution of exponentials shows that

$$\frac{\lambda^n t^{n-1}}{(n-1)!} e^{-\lambda t} \iff \frac{\lambda^n}{(\lambda+s)^n}, \quad n=1,2,\dots . \qquad (5.64)$$

Example 5.33 (Example of Inversion) We wish to invert

$$f^*(s) = \frac{a^2}{s^2(s+a)}. \tag{5.65}$$

To do this we first note that (5.65) is the product of two easily inverted transforms

$$at \iff \frac{a}{s^2}, \qquad ae^{-at} \iff \frac{a}{a+s}.$$

This implies that f is the convolution of a linear and exponential function and hence

$$f(t) = \int_0^t a^2 x e^{-a(t-x)} dx = ae^{-at} \int_0^t ax e^{ax} dx.$$

To evaluate this, we integrate by parts to yield

$$
\begin{aligned}
f(t) &= ae^{-at} \left\{ \left. xe^{ax} \right|_0^t - \int_0^t e^{ax} dx \right\} \\
&= ae^{-at} \left\{ te^{at} - \left. \frac{e^{ax}}{a} \right|_0^t \right\} \\
&= at - (1 - e^{-at}).
\end{aligned}
$$

Hence we have established that

$$at - (1 - e^{-at}) \iff \frac{a^2}{s^2(s+a)}. \tag{5.66}$$

5.4.3 Solving Differential Equations with Laplace Transforms

Similar to ordinary generating functions, there is a one-to-one mapping between functions and their Laplace transforms, and as with generating functions, Laplace transforms can be readily used to solve the continuous counterpart to discrete recurrence relations, that is, differential equations. Consider the following example.

Example 5.34 (Using Transforms to Solve a Simple Differential Equation) Let

$$f(t) = -2f^{(1)}(t) - f^{(2)}(t), \quad t \geq 0, \tag{5.67}$$

with boundary conditions $f(0) = 0$ and $f^{(1)}(0) = 1$. To solve this differential equation we first take the Laplace transform of both sides, yielding

$$
\begin{aligned}
f^*(s) &= -2(sf^*(s) - f(0^-)) - (s^2 f^*(s) - sf(0^-) - f^{(1)}(0^-)) \\
&= \frac{f(0^-)(2+s) + f^{(1)}(0^-)}{1 + 2s + s^2}.
\end{aligned}
$$

Substituting the boundary values and simplifying yield

$$f^*(s) = \frac{1}{(1+s)^2}.$$

This is the product of two transforms, each of which is the transform for an exponential density with parameter 1. Thus we conclude that $f(t)$ must be the density for an Erlang random variable with parameters $(2,1)$, that is, that

$$f(t) = te^{-t},$$

which is easily seen to satisfy (5.67) and the boundary values.

Example 5.35 (Partial Fraction Expansion) We wish to find the function $f(t)$ that satisfies

$$f^*(s) = \frac{1-3s}{(1-2s)(1-4s)^2}.$$

We can save some work by observing that we have created a partial fraction expansion similar to this in Example 3.23. Lifting the results from that section shows that

$$f^*(s) = \frac{-1}{2(1-2s)} + \frac{1}{2(1-4s)^2} + \frac{1}{1-4s}. \tag{5.68}$$

A simple calculation shows that

$$\frac{a}{1-bs} = \frac{a/b}{b(1/b - s)}$$

and thus we see that we can write (5.68) more conveniently as

$$f^*(s) = \frac{-1}{2} \frac{1/2}{1/2 - s} + \frac{1}{2} \left(\frac{1/4}{1/4 - s}\right)^2 + \frac{1}{4} \frac{1/4}{1/4 - s}. \tag{5.69}$$

Equation (5.69) is now easy to invert since each term corresponds to an exponential density or a convolution of an exponential density. Inverting thus yields

$$\begin{aligned}
f(t) &= \frac{-1}{2}(1/2)e^{-t/2} + \frac{1}{2}(1/4)^2 te^{-t/4} + \frac{1}{4}(1/4)e^{-t/4} \\
&= \frac{(t+2)e^{-t/4}}{32} - \frac{e^{-t/2}}{4}.
\end{aligned}$$

5.5 Fundamental Theorems

In this section we derive several theorems that follow immediately from the definition of expectation. Many of the theorems in this section have derivations that are surprisingly easy especially when one considers the depth of the results. This ease in deriving such profound results is a consequence of the inexorable search of mathematics to discover proofs with the most direct and simple expression. For example, it took a mathematician as great as Jacob Bernoulli twenty years of effort to prove a special case of the weak law of large numbers, which we will effortlessly prove within a page. Paul Erdös (1913-), arguably one of the world's greatest mathematicians, has an apt way of summarizing proofs that are so direct and simple. The highest complement he can give to a mathematical proof is to say that it is *straight from the book*. This means that the proof is so elegant and simple and goes so directly to the heart of the matter that it comes straight from God's book of sublime proofs[51].

5.5.1 Markov's Inequality

We can use the definition of expectation to obtain a simple yet powerful, inequality on the survivor function of a random variable called *Markov's inequality*.[6] Let h be a nonnegative and nondecreasing function and let X be a random variable. Observe that if the expectation of $h(X)$ exists then it is given by

$$E[h(X)] = \int_{-\infty}^{\infty} h(z) f_X(z) dz. \tag{5.70}$$

By assumptions on h it easily follows that

$$\int_{-\infty}^{\infty} h(z) f_X(z) dz \geq \int_{t}^{\infty} h(z) f_X(z) dz \geq h(t) \int_{t}^{\infty} f_X(z) dz. \tag{5.71}$$

Combining (5.70) and (5.71) yields Markov's inequality:

$$P[X \geq t] \leq \frac{E[h(X)]}{h(t)}, \qquad \textbf{Markov's Inequality.} \tag{5.72}$$

This simple inequality is surprisingly useful, and various other well-known inequalities are directly derivable from it. We will derive two such inequalities: Chebyshev's inequality and Chernoff's bound.

First we derive the simplest form of Markov's inequality that arises when $h(x) = x^+$. Markov's inequality implies that

$$P[X \geq t] \leq \frac{E[X^+]}{t}, \quad t > 0.$$

If X is nonnegative then this implies that

$$P[X \geq t] \leq \frac{E[X]}{t}, \quad t > 0, \qquad \textbf{Simple Markov Inequality,} \tag{5.73}$$

---

[6]Named after A.A. Markov (1856–1922), a Russian mathematician who made fundamental contributions to probability theory.

which is primarily of interest for large t. This form of Markov's inequality is quite weak but can be used to quickly check statements made about the tail of a distribution of a random variable when the expectation is known. For example, if the expected response time of a computer system is 1 second then one would doubt that many users of the system have experienced response times greater than 10 seconds. The simple Markov inequality shows that $P[X \geq 10] \leq .1$ and thus at most 10% of the response times in the system can be greater than 10 seconds. The simple Markov inequality is a first-order inequality since only knowledge of $E[X]$ is required.

5.5.2 Chebyshev's Inequality

A second-order inequality derivable from Markov's inequality is Chebyshev's inequality,[7] where we assume that the variance σ_X^2 of X is known (and finite). Set $Y \overset{\text{def}}{=} (X - E[X])^2$ and $h(x) = x$. Then Markov's inequality implies that

$$P\left[Y \geq t^2\right] \leq \frac{E[Y]}{t^2}. \tag{5.74}$$

Making the observation that

$$P\left[Y \geq t^2\right] = P\left[(X - E[X])^2 \geq t^2\right] = P\left[|X - E[X]| \geq t\right],$$

and that

$$E[Y] = E\left[(X - E[X])^2\right] = \sigma_X^2.$$

These equalities imply that (5.74) can be rewritten as

$$P\left[|X - E[X]| \geq t\right] \leq \frac{\sigma_X^2}{t^2}, \qquad \textbf{Chebyshev's Inequality.} \tag{5.75}$$

Chebyshev's inequality holds for all distributions and cannot be improved (there are distributions for which it holds as an equality). It provides intuition about the meaning of the variance of a random variable since it shows that wide diversions from $E[X]$ are unlikely if σ_X^2 is small. For example, setting $t = c\sigma_X$ in (5.75) for a positive constant c yields

$$P\left[|X - E[X]| \geq c\sigma_X\right] \leq \frac{1}{c^2}.$$

This shows that *for all distributions* the probability that a random variable is greater than c standard deviations from its mean is less than $1/c^2$.

---

[7]Named after P.L. Chebyshev (1821–1894).

The generality of Chebyshev's inequality does not imply that it is not useful. Later it allows us to establish the weak law of large numbers, which characterizes the extent of variations of a statistical average.

5.5.3 Chernoff's Bound

The moment generating function of a random variable X can be used in a variety of ways to characterize properties of X. The most notable of these is Chernoff's bound, which characterizes the tail of a distribution. Let $h(x) = \exp[\theta x]$ for $\theta \geq 0$. Then Markov's inequality (5.72) implies that

$$P[X \geq t] \leq e^{-\theta t} \mathcal{M}_X(\theta), \quad \theta \geq 0. \tag{5.76}$$

Chernoff's bound arises from (5.76) and from the observation that since the equation is valid for all $\theta \geq 0$, the tightest bound arises for the "best" θ. Because this may be obtained as a limiting value we write this as an "inf" rather than as a "min" and obtain

$$P[X \geq t] \leq \inf_{\theta \geq 0} e^{-\theta t} \mathcal{M}_X(\theta), \quad \textbf{Chernoff's Bound}. \tag{5.77}$$

Recall that the simple Markov inequality and Chebyshev's inequality can be viewed as first-order and second-order bounds, respectively, because they require only knowing the first (simple Markov) or first two (Chebyshev) moments of a random variable X. Chernoff's bound, in contrast, requires knowledge of the moment generating function, which equivalently implies knowing all the moments of X. We thus expect Chernoff's bound to be tighter than either Markov's inequality or Chebyshev's inequality. Chernoff's bound is used when there is no simple representation of $\overline{F_X}(a)$. For example, if X is exponential with parameter λ then (4.82) can be directly used to determine $\overline{F_X}(a)$ and there is no need to use Chernoff's bound. The following example shows a case where explicit representation for the survivor function is not possible.

Example 5.36 (Bounding the Tail of Coin Tossing Experiments)
Let $Y_i, i = 1, 2, \ldots$, be independent Bernoulli random variables with parameter $1/2$ (hence they correspond to fair coin tosses), and let

$$X_n = Y_1 + Y_2 + \ldots + Y_n$$

be the total number of heads obtained in n tosses. Clearly we have that

$$E[X_n] = \frac{n}{2}, \quad \sigma^2_{X_n} = \frac{n}{4}, \quad \mathcal{M}_{X_n}(\theta) = \left[\frac{1 + e^\theta}{2}\right]^n.$$

Our objective is to calculate the relative accuracy of bounding the tail of the distribution of the number of heads obtained in a large number of tosses. Let

$\alpha > 1/2$ and consider the following bounds for the event that there are more than αn heads in n tosses of the coin. The simple form of Markov's inequality (5.73) yields

$$P[X_n \geq \alpha n] \leq \frac{1}{2\alpha},$$

which is independent of n and thus provides a loose bound.

To use Chebyshev's inequality we write $\alpha n = n/2 + (\alpha - 1/2)n$ and observe that $P[X_n \geq \alpha n] = P[X_n - n/2 \geq (\alpha - 1/2)n]$. We thus have

$$P[X_n - n/2 \geq (\alpha - 1/2)n] \quad \leq \quad P[|X_n - n/2| \geq (\alpha - 1/2)n]$$

$$\leq \quad \frac{n/4}{((\alpha - 1/2)n)^2}$$

$$= \quad \frac{1}{4n(\alpha - 1/2)^2}.$$

Even though Chebyshev's inequality equally counts deviations above and below the mean, we see that it yields a substantially better bound than Markov's inequality.

To calculate Chernoff's bound for X_n, we write

$$P[X_n \geq \alpha n] \leq \inf_{\theta \geq 0} e^{-\theta \alpha n} \left[\frac{1 + e^\theta}{2} \right]^n. \tag{5.78}$$

Using calculus we differentiate the right-hand side with respect to θ to derive the optimal value $\theta^\star$ given by

$$\theta^\star = \ln(\alpha/(1 - \alpha)). \tag{5.79}$$

Since $\alpha > 1/2$ equation (5.79) implies that $\theta^\star \geq 0$. Substituting this into (5.78) yields

$$P[X_n \geq \alpha n] \leq \frac{[1/(2(1 - \alpha))]^n}{[\alpha/(1 - \alpha)]^{\alpha n}},$$

which is more conveniently expressed as

$$P[X_n \geq \alpha n] \leq \exp(-n\beta),$$

where

$$\beta \stackrel{\text{def}}{=} \ln(1/(1 - \alpha)) + \alpha \ln(\alpha/(1 - \alpha)).$$

This last form shows that for any fixed α, $P[X_n \geq \alpha n]$ is exponentially decreasing in n. Next we list some representative values, for $n = 100$, that show the relative values obtained in these inequalities.

α	0.55	0.60	0.80
Markov's inequality	0.90	0.83	0.62
Chebyshev's inequality	0.100	0.025	0.002
Chernoff's bound	0.006	1.8×10^{-9}	1.9×10^{-84}

Example 5.37 (Bounding the Tail of an Erlang Random Variable)
In this example we investigate the tail of the Erlang distribution. In contrast to exponential distributions, there is no closed formula for the survivor function of an Erlang distribution, and here we use Chernoff's bound to characterize tail probabilities. Let $Y_i, i = 1, 2, \ldots$, be a sequence of independent exponential random variables with parameter λ and define X_n to be their statistical average

$$X_n \stackrel{\text{def}}{=} \frac{Y_1 + Y_2 + \cdots + Y_n}{n}.$$

Thus X_n is an Erlang random variable with parameters $(n, n\lambda)$. We can use the convolution property of moment generating functions and Chernoff's bound to write

$$P\left[X_n \geq a\right] \leq e^{-\theta a} \mathcal{M}_{X_n}(\theta) = e^{-\theta a} \mathcal{M}_{E_{1,n\lambda}}^n(\theta),$$

where $E_{1,n\lambda}$ is an exponential random variable with parameter $n\lambda$. Using (5.49) we have that

$$P\left[X_n \geq a\right] \leq \inf_{\theta \geq 0} e^{-\theta a} \left[\frac{n\lambda}{n\lambda - \theta}\right]^n. \tag{5.80}$$

Using elementary calculus to derive the value of θ that minimizes (5.80) shows that $\theta^\star$ satisfies

$$\theta^\star = n\lambda - \frac{n}{a},$$

and thus we have the bound

$$P\left[X_n \geq a\right] \leq e^{n(1-\lambda a)} \left(\lambda a\right)^n. \tag{5.81}$$

As a simple application of this result, suppose we wish to determine the probability that X_n is greater than α times its expected value $E\left[X_n\right] = \lambda^{-1}$, for $\alpha \geq 1$. In this case $\theta^\star = n\lambda(1 - 1/\alpha)$ and $a = \alpha/\lambda$. Thus (5.81) yields

$$P\left[X_n \geq \alpha E\left[X_n\right]\right] \leq e^{-n(\alpha-1)} \alpha^n = \left[\frac{\alpha}{e^{\alpha-1}}\right]^n.$$

Name	Function	Random Variable	Inequality
Markov's inequality	h nonnegative, nondecreasing	X	$P[X \geq t] \leq E[h(X)]/t$
Simple Markov	$h(x) = x$	X	$P[X \geq t] \leq E[X]/h(t)$
Chebyshev	$h(x) = x$	$(X - E[X])^2$	$P[|X - E[X]| \geq t] \leq \sigma_X^2/t^2$
Chernoff	$h(x) = \exp[\theta x],\ \theta \geq 0$	X	$P[X \geq t] \leq \inf_\theta \exp[-\theta t]\mathcal{M}_X(\theta)$

TABLE 5.7. Inequalities Derivable from Markov's Inequality

Notice that since $\alpha < e^{\alpha-1}$ for $\alpha > 1$ we have that

$$P[X_n \geq \alpha E[X_n]] \longrightarrow 0, \quad \text{as } n \longrightarrow \infty.$$

We summarize the inequalities derivable from Markov's inequality in Table 5.7.

5.5.4 Weak Law of Large Numbers

Recall that in Section 2.1 we discussed the frequency-based approach to probability pioneered by von Mises. In that section we argued that if an experiment was performed numerous times, then the statistical average of a particular outcome would be close to the probability of that outcome. In other words, one could estimate the "probability" of a particular outcome using its statistical average. In this section we prove a theorem that supports this intuition. Suppose the outcomes of a series of experiments are independent and identically distributed random variables given by $X_i, i = 1, 2, \ldots$, with finite mean and variance $E[X]$ and σ_X^2, respectively. Let S_n be a random variable corresponding to the sum of the first n experiments,

$$S_n \stackrel{\text{def}}{=} \sum_{i=1}^{n} X_i.$$

The *statistical average* of the first n experiments is the value S_n/n. Intuitively, for large n, we expect the statistical average to be close to $E[X]$. The weak law of large numbers provides a foundation for such a belief.

To derive the law we note that

$$E\left[\frac{S_n}{n}\right] = E[X],$$

and, from properties 2 and 3 of variance, that

$$\text{Var}\left(\frac{S_n}{n}\right) = \frac{1}{n}\,\sigma_X^2.$$

We are interested in variations of the statistical average about its mean, and to evaluate this we use Chebyshev's inequality to write

$$P\left[\left|\frac{S_n}{n} - E\left[X\right]\right| \geq \epsilon\right] \leq \frac{\text{Var}\left(S_n/n\right)}{\epsilon^2} = \frac{\sigma_X^2}{n\epsilon^2}. \tag{5.82}$$

Inequality (5.82) shows that as the number of experiments increases it becomes less likely that the statistical average differs from $E\left[X\right]$. Specifically, for any $\epsilon > 0$, (5.82) implies that

$$\lim_{n\to\infty} P\left[\left|\frac{S_n}{n} - E\left[X\right]\right| \geq \epsilon\right] = 0, \quad \text{\textbf{Weak Law of Large Numbers.}} \tag{5.83}$$

Convergence of the random variables S_n/n to $E\left[X\right]$ as stipulated by (5.83) is termed *convergence in probability*.

Definition 5.38 *We say that a sequence of random variables Y_n converges* in probability *to a random variable Y, denoted by $Y_n \xrightarrow{P} Y$, if* $P\left[|Y_n - Y| \geq \epsilon\right] \longrightarrow 0$.

We can thus restate the weak law of large numbers (5.83) as

$$\frac{S_n}{n} \xrightarrow{P} E\left[X\right] \quad \text{as} \quad n \longrightarrow \infty.$$

It is somewhat surprising that the weak law of large numbers can be derived so simply. It follows essentially from the observation that the expectation of Y_n is independent of n whereas the variance of Y_n decreases as $1/n$. This implies that $\text{Var}\left(Y_n\right) \to 0$, which implies, from property 1 of variance, that Y_n becomes "more deterministic" with increasing n. Notice that, when expressed in terms of the coefficient of variations, this implies that $C_{Y_n}^2 \to 0$, again expressing the fact that the variability of Y_n decreases with n.

Mathematically, the weak law of large numbers follows from Markov's inequality, which in turn follows from the definition of expectation. Since expectation is a linear operator applied to functions, the law could be expressed purely algebraically without invoking any notions of probability or random variables. Problem 5.24 presents algebraic versions of Markov's and Chebyshev's inequalities. As often noted in Chapter 3, it is frequently the case that a purely algebraic derivation lacks semantic meaning as to why a particular result holds and the algebraic versions of the above inequalities are, indeed, more cryptic than their probabilistic counterparts. Both of these inequalities, however, originated in probability theory, and for this reason it might seem less natural to express them purely algebraically. We can, however, reverse this procedure and, in Problem 5.29, we show how probability theory can be used to prove a theorem that originated in algebra. In the problem we lead the reader through a proof of a classical result in algebra, the Stone–Weierstrass theorem. Our proof of this theorem is a natural extension of the law of large numbers and follows from a probabilistic interpretation of *Bernstein's polynomials* used in an elegant proof of the theorem due to Bernstein. Again, when couched in a probabilistic framework, the proof seems to flow more intuitively than in its algebraic version.

5.5.5 Strong Law of Large Numbers**

We have been careful in the above discussion to identify (5.83) as the *weak* law of large numbers. It is clear that in so doing we are distinguishing such a law from a companion law called the *strong law of large numbers*. We will state the strong law here but will not prove it (see [14] for a good reference on convergence of random variables and a proof of the strong law under general conditions).

To motivate our discussion we perform a simple coin tossing experiment. Consider a coin that lands heads up with probability p. For the sake of the example let $p = 1/4$ and assign the value of 1 to heads and 0 to tails. An experiment consists of an infinite number of tosses. Let Ω denote the set of outcomes of all possible experiments. For any particular experiment $\omega \in \Omega$ the corresponding sequence of 0s and 1s is known deterministically and is termed the *sample path* of ω. The "randomness" in experiments arises by selecting one experiment from the set. We define $Y_n(\omega)$ to be the statistical average of the first n outcomes of experiment ω. Intuitively, for large n the value of $Y_n(\omega)$ is close to p since heads lands up with probability p. Clearly, however, there are sample paths for which this is not the case, and we find it convenient to list two such paths.

The sample path of experiment ω_1 consists of an alternation of heads and tails (starting by 1,0,1,0, ...). An easy calculation shows that

$$Y_n(\omega_1) = \begin{cases} 1/2, & n = 2k, \\ k/(2k-1), & n = 2k-1, \quad k = 1, 2, \dots . \end{cases}$$

Notice that $Y_n(\omega_1)$ converges to $1/2$ and not to p as expected. Experiment ω_2 has a sample path that consists of a sequence of 2^i heads fol-

---
**Can be skipped on first reading.

lowed by a sequence of 2^{i+1} tails for $i = 0, 2, 4, \ldots$ (the sequence starts with $1, 0, 0, 1, 1, 1, 1, 0, 0, \ldots$). For $n = 2^{2k-1} - 1$, $k = 1, 2, \ldots$, it is clear that there are $2^0 + 2^2 + \cdots + 2^{2(k-1)}$ heads and thus

$$Y_n(\omega_2) = \frac{\sum_{j=0}^{k-1} 2^{2j}}{2^{2k-1} - 1} = \frac{2^{2k} - 1}{3(2^{2k-1} - 1)}, \quad n = 2^{2k-1} - 1, \quad k = 1, 2, \ldots .$$
(5.84)

For $n = 2^{2k} - 1$, $k = 1, 2, \ldots$, there are the same number of heads but a larger number of total tosses and thus

$$Y_n(\omega_2) = \frac{2^{2k} - 1}{3(2^{2k} - 1)}, \quad n = 2^{2k} - 1, \quad k = 1, 2, \ldots .$$
(5.85)

We do not need to specify $Y_n(\omega_2)$ for other values of n (the persistent reader can supply such a derivation) since our purpose is to establish that the sequence does not converge. This follows from (5.84) and (5.85), which, for large values of k, are approximately equal to $2/3$ and $1/3$, respectively. The strong law of large numbers essentially says that experiments such as ω_1 and ω_2 are "rare" and those for which the sequence of values $Y_n(\omega)$ converges to p are almost always encountered (they occur with probability 1).

To state the strong law precisely we need to refine the definitions we gave in Section 4.5.2 concerning almost sure equality. In that section we pointed out that X was contained in some set of states *almost surely* if the probability that X was not in that set has probability 0. Typically, almost sure equality is used in reference to the convergence of a sequence of random variables. To be more precise, let $X_n(\omega), n = 1, 2, \ldots$, be the sample path obtained from a random experiment $\omega \in \Omega$. It is important, as in the coin tossing experiment above, to observe that once ω is specified the sample path $X_n(\omega), n = 1, 2, \ldots$, is a deterministic sequence of values.

Definition 5.39 *Let $\Omega' \subset \Omega$ be a set of random experiments $\omega \in \Omega$ that satisfy*

$$X_n(\omega) \longrightarrow c, \quad \omega \in \Omega'.$$

We say that X_n converges almost surely *(also called* strong convergence*) to the value c, denoted by $X_n \xrightarrow{a.s.} c$, if $P[\Omega'] = 1$. This form of convergence is equivalently termed: X_n converges to c almost certainly or with probability 1 (w.p.1).*

Almost sure convergence intuitively means that the random experiments with sample paths that do not converge to c occur in a set $\Omega - \Omega'$ that has a probability of 0.

Definition 5.39 is an adaptation of the definition of convergence for a deterministic sequence that accounts for the fact that the sequence arises from a random experiment. To see this, recall that by definition of a limit, if a deterministic sequence a_n satisfies $a_n \to a$ then, for any $\epsilon > 0$, there exists a value $n(\epsilon)$ so that $|a_n - a| \leq \epsilon$ as $n \to \infty$ for all $n \geq n(\epsilon)$. There are thus *no occurrences* of $|a_n - a| > \epsilon$ for values of $n \geq n(\epsilon)$. When sequences correspond to random experiments, as in the coin tossing experiment mentioned earlier, this type of convergence is too strong. For example, the sample path of ω_1 converges to a value different from p and the sample path of ω_2 does not converge. There is still a sense of convergence to p, however, since the set of experiments that converge to p have probability 1. Violations of sample paths that do not converge to p have probability of 0. Recall that since a probability of 0 does not imply impossibility (see Section 4.5.2) we can only conclude that violations of this type of convergence are extremely rare but not impossible.

We can now state the strong law of large numbers as

$$P\left[\lim_{n\to\infty} \left|\frac{S_n}{n} - E[X]\right| \geq \epsilon\right] = 0,$$

which can equivalently be stated, using Definition 5.39, as

$$\frac{S_n}{n} \xrightarrow{a.s.} E[X] \quad \text{as} \quad n \longrightarrow \infty, \quad \textbf{Strong Law of Large Numbers.}$$

$$(5.86)$$

The strong law makes a precise statement regarding sample paths obtained in the "typical" experiment, that is, for sufficiently large n there is a large probability that a randomly selected sample path has a statistical average close to $E[X]$. In contrast, the weak law makes a statement regarding the entire ensemble of sample paths, that is, for sufficiently large n there is a large probability that, *averaged over all sample paths*, the statistical average is close to $E[X]$. The weak law does not make any statement regarding particular sample paths of random experiments and, specifically, does not imply that a randomly selected sample path converges to $E[X]$. Conceivably it could be the case that all sample paths either converge to values different from $E[X]$ or do not converge at all (as with experiments ω_1 and ω_2, respectively) and the weak law could still hold. In these cases the strong law would be violated. It is obvious from what we have said that the strong law implies the weak law.

Example 5.40 (The Variability of Erlang Random Variables) Erlang random variables are frequently found in computer models for processes that are less variable than exponential random variables. The fact that an Erlang random variable is the deterministic sum of exponentials implies that such models often retain the tractability inherent in the memoryless property of exponential distributions. This feature explains their prevalence in computer models. Recall that $E_{n,n\lambda}$ denotes an Erlang random variable with parameters $(n, n\lambda)$. We can use the strong law of large numbers to show that $E_{n,n\lambda}$, for large n, is "close to being deterministic." Our analysis here is a stronger version of that presented in Example 5.37, which bounded the tails of Erlang random variables.

Let S_n be defined by

$$S_n \stackrel{\text{def}}{=} E_{1,\lambda}^1 + E_{1,\lambda}^2 + \cdots + E_{1,\lambda}^n,$$

where we recall that $E_{1,\lambda}^i, i = 1, 2, \ldots$, are independent exponential random variables with parameter λ. Define

$$X_n \stackrel{\text{def}}{=} \frac{1}{n} S_n = \frac{E_{1,\lambda}^1 + E_{1,\lambda}^2 + \cdots + E_{1,\lambda}^n}{n}. \tag{5.87}$$

Note that, from (4.88), S_n is an Erlang random variable with parameters (n, λ) and thus from (5.9) has an expectation of $n\lambda^{-1}$. The random variable X_n is the statistical average of S_n (we do not currently know its distribution) and thus has expectation λ^{-1}. By the strong law of large numbers we know that

$$X_n \stackrel{a.s.}{\longrightarrow} \lambda^{-1} \text{ as } n \longrightarrow \infty \tag{5.88}$$

and thus X_n becomes "more deterministic" for large n.

To make this result useful in modeling, we recall from Example 4.17 that exponential random variables do not change distributions when scaled. In particular, writing $E_{1,\lambda}^i$ as $E_{1,n\lambda/n}^i$ and using (4.84) allows us to write

$$\frac{1}{n} E_{1,\lambda}^i = \frac{1}{n} E_{1,n\lambda/n}^i =_{st} E_{1,n\lambda}^i.$$

Thus, we can write X_n of (5.87) as

$$X_n = E_{1,n\lambda}^1 + E_{1,n\lambda}^2 + \cdots + E_{1,n\lambda}^n.$$

This form of X_n is suitable for modeling because, rather than being a statistical average as in (5.87), it is the sum of independent exponential random variables. This implies, from (4.88), that $X_n =_{st} E_{n,n\lambda}$ and thus X_n is Erlang with parameters $(n, n\lambda)$. Of course (5.88) still holds, and this provides the basis for the common assumption that for large n, the random variable $E_{n,\lambda}$ becomes less "variable" and it explains why Erlang distributions are used to model processes (such as service processes) that are "more deterministic" than exponential.

5.5.6 The Central Limit Theorem

The strong law of large numbers shows that if S_n is the sum of n independent and identically distributed random variables with the same distribution as X, then the statistical average S_n/n converges in a strong (sample path) sense to $E[X]$. The law does not, however, provide information about variations of the statistical average about its mean (we call this the *normalized average*) for large values of n. To describe precisely what we mean by this, define the normalized average by

$$Y_n \stackrel{\text{def}}{=} \frac{S_n}{n} - E[X]. \tag{5.89}$$

From the strong law of large numbers we know that $Y_n \stackrel{a.s.}{\longrightarrow} 0$ and thus, for large n, the value of Y_n is almost certainly close to 0. To analyze variations of the normalized average away from 0 we need to "magnify" the values of Y_n. To do this we multiply Y_n by an increasing function of n; a power of n, say n^α. It is necessary to determine the value of α that leads to a nontrival limit. For example, if α is selected too large then $n^\alpha Y_n \to \infty$ whereas if it is selected too small then $n^\alpha Y_n \to 0$. To obtain a nontrival limit for the normalized average the scaling factor must satisfy $\alpha = 1/2$ (see Problem 5.31) and this leads us to define the scaled random variable Z_n,

$$Z_n \stackrel{\text{def}}{=} \sqrt{n}\, Y_n = \frac{S_n - nE[X]}{\sqrt{n}}.$$

Our object in this section is to derive the limiting distribution of Z_n as $n \to \infty$. The limiting distribution of Z_n is one of the most important distributions in probability: the *normal distribution*.[8]

To calculate the limiting distribution of Z_n we assume that X has a moment generating function (the central limit theorem holds under less restrictive assumptions) and we determine the asymptotic properties of the moment generating function of Z_n. Using properties 1 and 3 of moment generating functions (Table 5.3) implies that

$$\mathcal{M}_{Z_n}(\theta) = \exp(-E[X]\theta\sqrt{n})\ \left(\mathcal{M}_X(\theta/\sqrt{n})\right)^n.$$

Some algebra shows that this can be rewritten as

$$\mathcal{M}_{Z_n}(\theta) = \exp\left(n\left\{\ln\left(\mathcal{M}_X(\theta/\sqrt{n})\right) - E[X]\theta/\sqrt{n}\right\}\right). \tag{5.90}$$

We are interested in evaluating (5.90) as $n \to \infty$. For notational convenience, define $\varepsilon_n \stackrel{\text{def}}{=} \theta/\sqrt{n}$ and

$$\Psi_n(\theta) \stackrel{\text{def}}{=} n\left\{\ln\left(\mathcal{M}_X(\varepsilon_n)\right) - E[X]\varepsilon_n\right\}.$$

---

[8] An entire family of distributions, called *stable distributions*, arises by scaling normalized averages by an appropriate power of n. See [103] for a thorough exposition of this family, which includes the Cauchy and normal distributions.

Substituting these definitions into (5.90) shows that

$$\mathcal{M}_{Z_n}(\theta) = \exp\left(\Psi_n(\theta)\right) \tag{5.91}$$

and thus to calculate the limit of $\mathcal{M}_{Z_n}$ it suffices to determine the limit of Ψ_n.

Calculating the limiting value of Ψ_n requires some algebraic manipulation. We first find it convenient to write Ψ_n in the form

$$\Psi_n(\theta) = \theta^2\left(\frac{\ln\left(\mathcal{M}_X(\varepsilon_n)\right) - E\left[X\right]\varepsilon_n}{\varepsilon_n^2}\right).$$

Using the fact that $\varepsilon_n \to 0$ as $n \to 0$ implies that

$$\lim_{n\to\infty} \Psi_n(\theta) = \theta^2 \lim_{\varepsilon_n\to 0} \frac{\ln\left(\mathcal{M}_X(\varepsilon_n)\right) - E\left[X\right]\varepsilon_n}{\varepsilon_n^2}.$$

As this yields an indeterminate form we use l'Hospital's rule to yield

$$\lim_{n\to\infty} \Psi_n(\theta) = \theta^2 \lim_{\varepsilon_n\to 0} \frac{\mathcal{M}_X^{(1)}(\varepsilon_n)/\mathcal{M}_X(\varepsilon_n) - E\left[X\right]}{2\varepsilon_n}.$$

This again is indeterminate and using l'Hospital's rule again yields

$$\lim_{n\to\infty} \Psi_n(\theta) = \theta^2 \lim_{\varepsilon_n\to 0} \frac{\mathcal{M}_X^{(2)}(\varepsilon_n)\mathcal{M}_X(\varepsilon_n) - \left(\mathcal{M}_X^{(1)}(\varepsilon_n)\right)^2}{2\left(\mathcal{M}_X(\varepsilon_n)\right)^2}.$$

From (5.46) and (5.48) we find that

$$\lim_{n\to\infty} \Psi_n(\theta) = \frac{\theta^2\sigma_X^2}{2}$$

and thus from (5.91) we finally conclude

$$\lim_{n\to\infty} \mathcal{M}_{Z_n}(\theta) = e^{(\theta\sigma x)^2/2}. \tag{5.92}$$

We now need to invert this moment generating function to determine its corresponding density function which we denote by f_Z. This is left to the reader in Problem 5.32; here we state the result:

$$f_Z(x) = \frac{1}{\sigma_X\sqrt{2\pi}}\, e^{-x^2/(2\sigma_X^2)}, \qquad -\infty < x < \infty, \tag{5.93}$$

which is a special case of the *normal density*. In general, the density of a normal random variable X has two parameters (μ, σ^2) (these correspond to the mean and variance of X) and is written as

$$f_X(x) = \frac{1}{\sigma\sqrt{2\pi}}\, e^{-(x-\mu)^2/(2\,\sigma^2)}, \qquad -\infty < x < \infty.$$

The distribution function, for which there is no simple expression, is defined by

$$\Phi(x; \mu, \sigma^2) \stackrel{\text{def}}{=} \frac{1}{\sigma\sqrt{2\pi}} \int_{-\infty}^{x} e^{-(z-\mu)^2/(2\sigma^2)} dz.$$

A commonly used function, called the *error function* is defined by

$$\text{erf } x \stackrel{\text{def}}{=} \frac{2}{\sqrt{\pi}} \int_{0}^{x} e^{-u^2} du. \tag{5.94}$$

The error function is frequently used to express results for normal distributions and is used in Table C.2 when listing the Laplace transform for various functions. We let N_{μ,σ^2} denote a normal random variable with parameters (μ, σ^2).

The preceding derivation shows that the moment generating function Z_n converges to the moment generating function of a normal random variable with parameters $(0, \sigma_X^2)$. Since moment generating functions uniquely determine distributions, we can also say that the distribution of Z_n converges to the normal distribution with the above parameters. This establishes the central limit theorem which is stated as

$$\lim_{n \to \infty} P\left[\frac{S_n - nE[X]}{\sqrt{n}} \leq x\right] = \Phi(x; 0, \sigma^2),$$

or equivalently as

$$\sqrt{n}\left(\frac{S_n}{n} - E[X]\right) \xrightarrow{D} N_{0,\sigma_X^2}, \qquad \textbf{Central Limit Theorem.} \tag{5.95}$$

Sometimes the central limit theorem is stated as

$$\frac{S_n - nE[X]}{\sqrt{n}} \xrightarrow{D} N_{0,\sigma_X^2},$$

but this last form does not preserve the sense that to obtain the theorem we scale a sequence that converges to 0. In Figure 5.4 we plot the density function of the normal for various parameters.

The central limit theorem is a result through which we view into the depths of mathematics.[9] It is perhaps the first result that we have seen in this book that cannot be anticipated merely by intuition since the form of the result must be obtained through mathematical analysis. It

---

[9]The first version of this theorem appeared in 1718 in a paper by Abraham de Moivre (1667–1754). It was all but forgotten until Pierre-Simon Laplace (1749–1827) rescued it from oblivion in his treatise, *Théorie Analytique des Probabilités*, published in 1812. For this reason the theorem is often called the de Moivre–Laplace integral theorem.

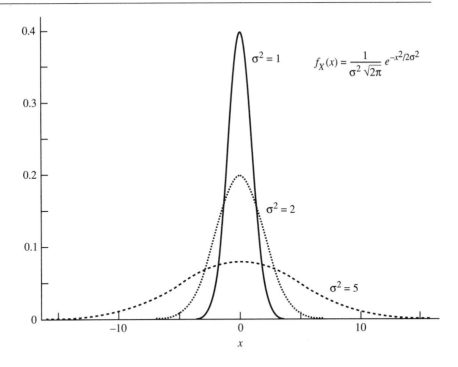

FIGURE 5.4. Density of the Normal Random Variable

is probably safe to say that the normal distribution is one of the most important distributions in probability theory, and this is why it stands in such a prominent position in the family of distributions derived in previous chapters (see Figure 4.4). The central limit theorem plays a fundamental role in the fields of simulation, reliability, estimation, and many other areas of applied probability.

5.6 Summary of Chapter 5

We established several fundamental theorems of probability in this chapter. Markov's inequality, Chebyshev's inequality, and Chernoff's bound, which are derivable from it, provide useful information on the tail of a distribution and allow us to establish the weak law of large numbers. This, together with the strong law, closes the gap between the frequency-based approach and the axiomatic approach to probability, both of which were discussed in Chapter 2. The Central limit theorem is perhaps the most profound theorem encountered thus far in the book; it plays a prominent role in probability. It is thus fitting that it serves as an anchor in Figure 4.4, closing our development of distribution functions. The figure shows that starting from classical algebraic variables (degen-

Law	Representation $(S_n \stackrel{\text{def}}{=} X_1 + X_2 + \cdots + X_n)$
Jensen's inequality	$E[h(X)] \geq h(E[X])$, h convex
Wald's equality	$E[S_N] = E[X]E[N]$, N a stopping time
Markov's inequality	$P[X \geq t] \leq E[h(X)]/h(t)$, h nonnegative and nondecreasing
Simple Markov inequality	$P[X \geq t] \leq E[X]/t$, $X \geq 0$
Chebyshev's inequality	$P[|X - E[X]| \geq t] \leq \sigma_X^2/t^2$
Weak law of large numbers	$\lim_{n \to \infty} P[|S_n/n - E[X]| \geq \epsilon] = 0$, $S_n/n \xrightarrow{P} E[X]$
Strong law of large numbers	$P[\lim_{n \to \infty} |S_n/n - E[X]| \geq \epsilon] = 0$, $S_n/n \xrightarrow{a.s.} E[X]$
Chernoff's bound	$P[X \geq a] \leq \inf_{\theta \geq 0} e^{-\theta a} \mathcal{M}_X(\theta)$
The central limit theorem	$\lim_{n \to \infty} P\left[\frac{S_n - nE[X]}{\sqrt{n}} \leq x\right] = \Phi(x; 0, \sigma^2)$, $\sqrt{n}\left(\frac{S_n}{n} - E[X]\right) \xrightarrow{D} N_{0,\sigma_X^2}$

TABLE 5.8. Fundamental Laws of Probability

erate random variables) by simple operations we can develop common discrete distributions - the Bernoulli, geometric, binomial, discrete uniform, and negative binomial - and their corresponding limiting forms - the exponential, Poisson, continuous uniform, and Erlang - and end with a distribution that is central to probability theory - the normal distribution.

The fundamental theorems established in the chapter are summarized in Table 5.8.

5.6.1
Ending
Historical Notes

With this chapter we finish our investigations into the foundations of probability and then proceed to build upon these results in the study of stochastic processes, a modern branch of mathematics. Throughout the previous three chapters we have indicated in the text and footnotes some aspects of the history that led to the founding of the field of probability. It is thus appropriate to finish this chapter with a final glimpse at this history. As in all human endeavors, this history is replete with its personal intrigues.

The custom of mathematical writing of the 17th and 18th centuries was to intersperse new results with praise and criticisms of fellow mathematicians. The result of such a practice is that many controversies find permanent archival within the catalogue of mathematical discoveries. Questions of priority were commonly raised in technical publications as were differences of opinion that sometimes led to bitter grievances. One revisits these differences of opinion when reading the original sources; some of these differences include: Liebnitz versus Newton about the discovery of calculus, Montmort versus de Moivre regarding the originality of *De Mensura Sortis*, and Jacob Bernoulli versus Jacques Bernoulli (his brother) on the discovery of isoperimetric curves. Such arguments diminish in significance with the detachment of historical perspective, and it is not without some irony that the probabilistic results of the era that sometimes caused turmoil can now be condensed into a few chapters of a book introducing the subject. Over the course of time the quest of mathematics for simplification has unified the results of this great age into a handful of theorems.

Mathematicians of this time frequently had to make great personal sacrifices to pursue their careers, often acting against the best wishes of their parents to pursue a more traditional and secure vocation. Some of the greatest probabilists left lasting contributions to the field at the cost of living their lives in poverty. Secure positions were difficult to obtain, and mathematicians often supported themselves by tutoring or making financial arrangements to sell mathematical results. These results were typically related to games of chance and sold to gamblers seeking to gain an edge on opponents. It is not surprising that the three great texts of the day - *Deratiocinniis in Ludo Aleae* by Huygens (1657), *Ars Conjectandi* by Jacob Bernoulli (1713), and *The Doctrine of Chance* by de Moivre (1718) - express their results in terms of games of chance.

Abraham de Moivre (1667–1754) may have contributed more to the field of probability than any other mathematician of his era and, although he was elected a Fellow of the Royal Society in 1697, he lived most of his life in the manner just described - tutoring mathematics and when he was too old and feeble to tutor, selling mathematical results on games of chance. In a letter to Jacques Bernoulli dated December 2, 1707, he described the sorry state of his financial condition:[10]

> *I am obliged to work almost from morning to night, that is, teaching my pupils and walking. But since the city is very large, a considerable part of my time is employed solely in walking. That is what reduces the amount of profit I can make, and cuts into my leisure for study. Moreover one is obliged here more than anywhere else in the world to maintain a certain appearance; that is almost as essential as scientific merit. What remains to me at the end of the*

---

[10]See Helen Walker's biography in Todhunter [115].

*year is that I have indeed lived, and with some pleasure in my con-
nection with the Society, but without having anything to spare. If
there were some position where I could live tranquilly and where I
could save something, I should accept it with all my heart.*

One of the world's greatest mathematicians, a mathematician of such
caliber that Isaac Newton deflected mathematical questions by reply-
ing, "Ask Mr. de Moivre, he knows all that better than I do," thus spends
most of his precious time walking the rainy streets of London between
appointments to teach elementary mathematics!

De Moivre was close friends with many of the mathematicians of his
time, including those of the Bernoulli family. We have previously mentioned some of the work of Jacob Bernoulli (1654–1705) (sometimes
called Jacques Bernoulli), for whom the Bernoulli distribution is named.
Jacob is also credited with *Bernoulli numbers* used in number theory,
the *Bernoulli equation* of differential equations, and the *lemniscate of
Bernoulli* encountered in most first courses of calculus. Jacob Bernoulli
is not to be confused with his younger brother Johann Bernoulli (1667–
1748) (also referred to as Jean I Bernoulli), who was his active and often
bitter rival. Johann's work included contributions to calculus, analysis
of optical phenomena associated with refraction and reflection, and the
derivation of the equation for the *brachistochrone*, the curve of quick-
est descent of a weighted particle in a gravitational field. Johann had
three sons Nicolaus III (1695–1726), Daniel (1700–1782), and Johann
II (1710–1790) - all of whom became mathematicians. Nicolaus III and
Daniel are responsible for results in probability and later became known
for a problem posed while working in St. Petersburg, now referred to as
the *St. Petersburg paradox* (you can tackle this paradox in Problem 5.1).

The Bernoullis were an active mathematical family; eight notable mathematicians can be counted among the descendants of Nicolaus Bernoulli
(1623–1708), a successful merchant who, ironically, did not encourage
his sons to study mathematics. Jacob's motto was, *Invito patre sidera
verso*, "Against my father's will I study the stars," an indication of the
resistance he met in pursuing his avocation. Nicolaus wanted his sons to
pursue more practical careers and had definite plans for Jacob to go into
the ministry and Johann to be a merchant or a physician. This partially
explains the curious financial arrangement that Johann made with Marquis l'Hospital to receive a regular stipend in exchange for mathematical
results. As previously mentioned (see footnote 5 of Chapter 4), this led
to a major result of Johann's that determined the value of the indeterminate form 0/0 to be named "l'Hospital's rule" rather than Bernoulli's
Rule! Other mathematicians in the Bernoulli family include Nicolaus
II (1687–1759), Johann III (1744–1807), and Jacob II (1759–1789). We

Random Variable	X	$E[X]$	σ_X^2	$\mathcal{G}_X(z)$
Discrete uniform	U_n	$(n+1)/2$	$(n^2-1)/12$	$\frac{z(1-z^n)}{n(1-z)}$
Bernoulli	$B_{1,p}$	p	$p(1-p)$	$1-p(1-z)$
Binomial	$B_{n,p}$	np	$np(1-p)$	$(1-p(1-z))^n$
Geometric	$G_{1,p}$	$1/p$	$(1-p)/p^2$	$\frac{pz}{1-(1-p)z}$
Negative binomial	$G_{n,p}$	n/p	$n(1-p)/p^2$	$\left(\frac{pz}{1-(1-p)z}\right)$
Poisson	P_λ	λ	λ	$\exp[-\lambda(1-z)]$

TABLE 5.9. Properties of Common Discrete Distributions

can only imagine the chagrin it must have caused the elder Nicolaus to see so many of his descendants follow such an impractical vocation as mathematics when they could have had successful careers as merchants. It is almost impossible to measure how much mathematics gained from Nicolaus's loss!

The Poisson distribution is named after Siméon Poisson (1781–1840), a French mathematician born before the Revolution, who first discovered it in the same way as we derived it in the text (it can be found in his treatise, *Recherches sur la Probabilité des Jugements*, published in 1837). Poisson averred that there were only two good things in life: discovering

Random Variable	X	$E[X]$	σ_X^2	$\mathcal{M}_X(\theta)$	$\mathcal{L}_X(s)$
Continuous uniform	$U_{[a,b]}$	$(a+b)/2$	$(b-a)^2/12$	$\frac{e^{b\theta}-e^{a\theta}}{\theta(b-a)}$	$\frac{e^{-as}-e^{-bs}}{s(b-a)}$
Exponential	$E_{1,\lambda}$	$1/\lambda$	$1/\lambda^2$	$\frac{\lambda}{\lambda-\theta}$	$\frac{\lambda}{\lambda+s}$
Erlang	$E_{n,\lambda}$	n/λ	n/λ^2	$\left(\frac{\lambda}{\lambda-\theta}\right)^n$	$\left(\frac{\lambda}{\lambda+s}\right)^n$
Normal	$N_{\mu,\sigma}$	μ	σ^2	$\exp[\mu\theta+\theta^2\sigma^2/2]$	$\exp[-\mu s+s^2\sigma^2/2]$

TABLE 5.10. Properties of Common Continuous Distributions

mathematics and teaching mathematics! He evidently excelled at both, publishing more than 400 works and enjoying a reputation of being a superb teacher.

5.7
Problems for
Chapter 5

5.1 **[R1] The St. Petersburg Paradox.** This paradox was first posed by Nicolaus III Bernoulli (1695–1726) and his brother Daniel Bernoulli (1700–1782) in 1725 while Nicolaus was on the mathematical faculty of the Academy of St. Petersburg (he was there for only eight months and showed great promise before tragically dying by drowning). The problem goes as follows: Assume that a fair coin is tossed and a player receives one dollar if a head appears on the first toss, two dollars if a head does not appear until the second toss, four dollars if the first occurrence of a head is on the third toss, and so on. Hence the player receives 2^{n-1} dollars if the first occurrence of heads is on the nth toss. A straightforward calculation,

$$\sum_{n=1}^{\infty} 2^{n-1} \left(\frac{1}{2}\right)^n = \infty,$$

shows that the player's expected winnings is infinite. This implies that the game is fair only if the player is willing to stake an infinite amount of money. In relation to common experience, this seems to be a paradoxical result. The problem so puzzled mathematicians of the 18th century that Georges Louis Leclerc, Comte de Buffon (1707–1788) (who is famous for his random needle estimation of π), made an empirical test of the problem and found that in 2,084 games a total of 10,057 dollars were paid to the player, less than an average of 5 dollars per game! How do you explain the paradox?

Problems Related to Section 5.1

5.2 **[E1]** Provide proofs for the properties of expectation given in Table 5.1.

5.3 **[M3]** Random variables often have nonintuitive properties. It would seem strange, for example, for the sum of identically distributed random variables to have the same distribution as any one of them. Let $X_i, i = 1, 2, \ldots$, be a sequence of independent Cauchy random variables (see (5.12) for the density function). Show that

$$X_1 =_{st} X_1 + X_2 + \cdots + X_n$$

for any finite value of n.

5.4 **[M2]First Passage Times.** Random variables that do not have an expectation often occur in a nonintuitive manner. Recall that in Section 3.3.5 we analyzed the statistics of a first passage problem for a fair coin experiment. In that problem we derived an expression for f_n, $n = 1, 2, \ldots$, the probability that the total winnings of a given player equals one dollar after the nth toss. We showed that $f_{2k} = 0, k = 1, 2, \ldots$, and

$$f_{2k-1} = \frac{1}{k} \frac{1}{2^{2k-1}} \binom{2(k-1)}{k-1}, \quad k = 1, 2, \ldots .$$

Show that the expectation of the first passage time is undefined. (Hint: Use the bounds on Stirling's approximation given in (B.8)).

5.5 [M3] Show that (5.15) and (5.16) follow from the definition of convexity given in Definition 5.5. Additionally show that all convex functions are continuous.

5.6 [M2] The picture in Figure 5.2 graphically suggests that $g(x,a) \leq h(x)$ where $g(x,a)$ is defined in (5.17). Analytically prove this. (Hint: Create a Taylor expansion of h about the point $a + \epsilon$.)

5.7 [E2] Show that the assumption that the expectations of X and Y exist is necessary in property 1 of expectation by constructing two random variables where $E[X + Y]$ exists but $E[X]$ and $E[Y]$ do not.

5.8 [E1] What is the relationship between $1/E[X]$ and $E[1/X]$?

5.9 [E1] Establish the survivor representation of expectation given in (5.23) for continuous random variables given by

$$E[X] = \int_0^\infty \overline{F_X}(x)dx.$$

5.10 [M3] Consider a single-lane road containing an infinite number of cars. Cars select positive speeds from a continuous distribution independently of each other. A particular car travels along the road at the minimum of its selected speed and the speed of the car ahead of it. It is clear that cars form platoons that travel together at the speed of the lead car of the platoon. What is the expected number of cars in a platoon?

Problems Related to Section 5.2

5.11 [E1] Provide proofs for the properties of variance given in Table 5.2.

5.12 [M2] Let h be any function satisfying

$$h(x) = x, \quad x = 0, \ 1/2, \ 1,$$

$$h(x) > x, \quad 0 < x < 1/2,$$

$$h(x) < x, \quad 1/2 < x < 1.$$

Show that all possible relationships can hold (e.g., $\leq, =$, and $\geq$) between $E[h(X)]$ and $h(E[X])$ where X is a Bernoulli random variable with parameter p.

5.13 [E2] Let $Y = aX + b$ where a, b are constants and X is a random variable. Suppose that $C_Y^2 = C_X^2$. Write an expression for $E[Y]$. What is its maximum value?

5.14 [E3] In Example 5.24 we showed that to obtain an equation for $E[N]$ from the equation given in (5.33),

$$N = (N - 1)^+ + V,$$

we had to take the expectation of the second power of the equation. Show that to obtain an equation for the ith moment one has to consider the $i + 1$st power. Provide an equation for the third moment.

5.15 [**M3**] In this problem we prove the Cauchy–Schwarz inequality given by

$$(E[XY])^2 \le E[X^2] E[Y^2].\qquad(5.96)$$

In (5.96) equality holds if and only if either $P[Y=0]=1$ or $P[X=aY]=1$ for some constant a. We show that the Cauchy–Schwarz inequality implies that $-1 \le \rho(X_1, X_2) \le 1$ with equality 1 (resp., -1) if and only if $X_2 = aX_1$ (resp., $X_2 = -aX_1$) where $a > 0$. To prove (5.96) argue as follows:

1. Show that if $P[Y=0]=1$ then equality trivially holds.

2. Define $h(a) = E\left[(X - aY)^2\right]$ and show that $h(a)$ is nonnegative and reaches a minimum at

$$a^\star = \frac{E[XY]}{E[Y^2]}.$$

3. Show that $h(a^\star) \ge 0$ implies the Cauchy–Schwarz inequality with equality if $P[X - aY = 0] = 1$.

4. Use (5.96) to show that $-1 \le \rho(X_1, X_2) \le 1$ with equality 1 (resp., -1) if and only if $X_2 = aX_1$ (resp., $X_2 = -aX_1$) where $a > 0$.

5.16 [**R1**] In Example 5.29 we considered a random variable Y that was a function of X and calculated the covariance between the two random variables. Consider the following generalization. Let X be defined as in the example, that is, it takes on the values $-1, 0, 1$ with equal probability and let $Y_a = 1/(a+X)$ for $a > 1$. In Example 5.29 $a = 2$ and $\text{Cov}(X, Y_2) = -0.96$. Without doing any calculation describe the relationship between $\text{Cov}(X, Y_a)$ and $\text{Cov}(X, Y_2)$ for $a > 2$. Derive an expression for the covariance that justifies your conclusion. Without doing any calculation, hypothesize what happens as $a \to \infty$. Justify your conclusion.

Problems Related to Section 5.3

5.17 [**M3**] Derive an expression for the probability generating function of equation (5.33) and provide a method to determine all the moments of N.

5.18 [**E3**] Let X_i, $i = 1, 2, \ldots$, be a sequence of independent and identically distributed discrete random variables and let N be a stopping time for this sequence. Prove that

$$\mathcal{M}_{X_1 + X_2 + \cdots + X_N}(\theta) = \mathcal{M}_N(\ln \mathcal{M}_X(\theta)).\qquad(5.97)$$

Apply (5.97) for the case where X_i geometrically distributed with parameter p, for $i = 1, 2, \ldots$, and where N is geometrically distributed with parameter α to prove that the geometric sum of geometric random variables is geometric (4.42). Express (5.97) in terms of generating functions.

5.19 [**M1**] Let $S_n = X_1 + X_2 + \cdots + X_n$ where X_i are independent and identically distributed random variables. Is the coefficient of variation $C_{S_n}^2$ increasing or decreasing in n?

5.20 [**M2**] Show that

$$\text{Var}\left(\sum_{i=1}^{n} X_i\right) = \sum_{i=1}^{n} \text{Var}\left(X_i\right) + 2\sum_{i=1}^{n-1}\sum_{j=i+1}^{n} \text{Cov}\left(X_i, X_j\right).$$

Problems Related to Section 5.4

5.21 [**M2**] **Probabilistic Interpretation of Laplace Transforms.** In this problem we derive a probabilistic interpretation of $\mathcal{L}_X(s)$. Let Z be an exponential random variable with parameter s. Show that

$$\mathcal{L}_X(s) = P\left[Z > X\right].$$

Derive a similar interpretation for $\mathcal{G}_X(z)$.

5.22 [**E2**] Invert the following Laplace transform,

$$f^*(s) = \frac{s^3 + 8s^2 + 22s + 16}{4(s^3 + 5s^2 + 8s + 4)}.$$

Is f a density function?

5.23 [**M2**] For $f(t) \iff f^*(s)$ show that

$$\int_0^t \frac{f(x)}{x}\,dx \quad\iff\quad \frac{1}{s}\int_s^\infty f^*(y)dy$$

$$\int_t^\infty \frac{f(x)}{x}\,dx \quad\iff\quad \frac{1}{s}\int_0^s f^*(y)dy$$

Problems Related to Section 5.5

5.24 [**M2**] **Algebraic Version of the Simple Markov and Chebyshev Inequalities.** We show that Markov's (the simple version as given by (5.73)) and Chebyshev's inequalities can be expressed in a purely algebraic fashion without recourse to probability theory. This shows that these inequalities are fundamentally algebraic in nature. The derivations of the inequalities here are less intuitive than when couched within a probabilistic framework. Let $f(x)$ be a nonnegative function defined over the interval $[0, \infty)$ that integrates to 1 (note that f satisfies the definition of a density function). Define

$$\alpha^{(i)} \stackrel{\text{def}}{=} \int_0^\infty x^i f(x)dx, \quad i = 0, 1, \ldots,$$

and note that if X is a random variable with density f then

$$E\left[X\right] = \alpha^{(1)}, \qquad \sigma_X^2 = \alpha^{(2)} - (\alpha^{(1)})^2,$$

and

$$P[X > t] = \int_t^\infty f(x)dx$$

and

$$P[|X - \beta| > t] = \int_0^{t-\beta} f(x)dx + \int_{t+\beta}^\infty f(x)dx, \quad \beta \geq 0.$$

1. Prove the algebraic version of the simple Markov inequality:

$$\int_t^\infty f(x)dx \leq \alpha^{(1)}/t.$$

2. Prove for any values $\beta, t > 0$ that

$$\int_0^{\beta-t} f(x)dx + \int_{\beta+t}^\infty f(x)dx \leq \int_0^\infty \left(\frac{x - \beta}{t}\right)^2 f(x)dx. \tag{5.98}$$

Hint: Show that

$$\left(\frac{x - \beta}{t}\right)^2 \geq 1, \quad 0 \leq x \leq \beta - t, \quad \beta + t \leq x.$$

3. Show that (5.98) yields the algebraic version of Chebyshev's inequality when $\beta = \alpha^{(1)}$.

5.25 [M2] **Derivation of a Special Case of Markov's Inequality.** In this problem we derive the simple case of Markov's inequality using a different approach than given in the text. For a given nonnegative random variable X let $\mathcal{A}_t$ be the event that X is greater than or equal to t, $\mathcal{A}_t = \{X \geq t\}$ and let Y be a random variable defined by $Y = tI_{\mathcal{A}_t}$, $t > 0$. Use the fact that $X \geq Y$ to derive the simple Markov inequality (5.73).

5.26 [M2] **Different Derivation of Chernoff's Bound.** In this problem we derive Chernoff's bound using a method that suggests several generalizations. Let h be a function defined by

$$h(x) = \begin{cases} 0, & x \leq a, \\ 1, & x > a. \end{cases}$$

Use the fact that $h(x) \leq \exp[\theta(x - a)]$ to derive Chernoff's bound (5.77).

5.27 [M2] **Bernstein's Inequality.** Show that Markov's inequality (5.72) implies Bernstein's inequality given by

$$P[|X| \geq x] \leq \inf_{\theta > 0} e^{-\theta x} \mathcal{M}_X(\theta) + \inf_{\alpha > 0} e^{-\alpha x} \mathcal{L}_X(\alpha).$$

5.28 [M3] **Generalization of the Weak Law of Large Numbers.** In this exercise we generalize the weak law of large numbers. The derivation in Section 5.5.4 assumed that the random variables $X_i, i = 1, 2, \ldots,$ were independent and identically distributed. We relax this constraint by considering the case where the variance of this sequence of random variables is uniformly bounded. Suppose that there is a constant C that satisfies $\text{Var}(X_i) \leq C$, $i = 1, 2, \ldots$. We then claim that

$$\lim_{n \to \infty} P\left[\left|\frac{1}{n}\sum_{i=1}^{n} X_i - \frac{1}{n}\sum_{i=0}^{n} E[X_i]\right| < \epsilon\right] = 1. \tag{5.99}$$

Notice that (5.99) is a version of a weak law of large numbers where we have not assumed that the random variables have the same expectation or variance. To prove (5.99), argue as follows:

1. Show that Chebyshev's inequality can be expressed as

$$P[|X - E[X]| < t] \geq 1 - \frac{\sigma_X^2}{t^2}. \tag{5.100}$$

2. Show that by assumption we can write the following bound:

$$\text{Var}\left(\frac{1}{n}\sum_{i=1}^{n} X_i\right) \leq \frac{C}{n}. \tag{5.101}$$

3. Show that (5.101) implies that (5.100) is given by

$$P\left[\left|\frac{1}{n}\sum_{i=1}^{n} X_i - \frac{1}{n}\sum_{i=0}^{n} E[X_i]\right| < \epsilon\right] \geq 1 - \frac{C}{n\epsilon^2}. \tag{5.102}$$

4. Show that (5.99) follows from (5.102) as $n \to \infty$.

5.29 [R2] Probabilistic Proof of the Stone–Weierstrass Theorem. Problem 5.24 showed that we could view Markov's and Chebyshev's inequalities algebraically. In this problem we conversely show that an important algebraic theorem can be proved probabilistically. As in Chapter 3 such arguments often provide intuition as to why the property holds that is not evident in algebraic proofs. Let f be a real valued function defined over the interval $[0, 1]$. The Stone–Weierstrass theorem states that there exists a polynomial p such that, for any value $\epsilon > 0$, $|p(x) - f(x)| < \epsilon$ uniformly in $x \in [0, 1]$. An elegant proof of this theorem was given by Bernstein using what are now called Bernstein polynomials. The nth Bernstein polynomial associated with f is given by

$$B_n(x) = \sum_{k=0}^{n} f\left(\frac{k}{n}\right)\binom{n}{k} x^k (1-x)^{n-k}, \quad 0 \leq x \leq 1. \tag{5.103}$$

Bernstein showed that

$$\lim_{n \to \infty} B_n(x) = f(x) \tag{5.104}$$

uniformly in x, thus providing a constructive proof of the Stone–Weierstrass theorem. In this problem we show that (5.104) follows from the strong law of large numbers (the following derivation establishes only pointwise convergence). In the problem we need a fact not given in the text.

Fact. If $x_n \xrightarrow{a.s.} x$ then $f(x_n) \xrightarrow{a.s.} f(x)$ for all bounded continuous functions.

1. To couch (5.103) in a probabilistic framework show that

$$B_n(x) = E\left[f\left(B_{n,x}/n\right)\right],$$

where $B_{n,x}$ is a binomial random variable with parameters (n, p).

2. Show that

$$\frac{B_{n,x}}{n} \xrightarrow{a.s.} x.$$

3. Use the fact stated above to complete the proof.

5.30 **[M2]** Let $M_n = \max\{U^1_{[0,1]}, U^2_{[0,1]}, \dots, U^n_{[0,1]}\}$ where $U^i_{[0,1]}$ are independent uniform random variables over the interval $[0, 1]$. Show that

$$\lim_{n \to \infty} P\left[n(1 - M_n) < x\right] = 1 - e^{-x},$$

and thus that

$$\lim_{n \to \infty} n(1 - M_n) =_{st} E_{1,x},$$

where $E_{1,x}$ is an exponential random variable with parameter x.

5.31 **[R2]** In deriving the central limit theorem we scaled the normalized average, defined by (5.89), by n^α for $\alpha = 1/2$. Show, by attempting to carry out the derivation leading to (5.92), that this is the only scaling factor that leads to nontrivial limits (i.e., larger values of α diverge to ∞ and smaller values converge to 0).

5.32 **[E2]** Use

$$\int_{-\infty}^{\infty} e^{-x^2/2}dx = \sqrt{2\pi}$$

to show that

$$\mathcal{M}_{N_{\mu,\sigma^2}}(\theta) = \exp\left(\mu\theta + \theta^2\sigma^2/2\right),$$

where N_{μ,σ^2} is a normal random variable with parameters (μ, σ^2).

Part II

Stochastic Processes

6

The Poisson Process
and Renewal Theory

In this chapter we analyze a stochastic process termed a *renewal process*. A *stochastic process* is a set of random variables $\{X(t) : t \in \mathcal{T}\}$ defined on a common probability space. It is typical to think of t as time, $\mathcal{T}$ as a set of points in time, and $X(t)$ as the value or state of the stochastic process at time t. We classify processes according to time and say that they are discrete time or continuous time depending on whether $\mathcal{T}$ is discrete (finite or countably infinite) or continuous. If $\mathcal{T}$ is discrete then we typically index the stochastic process with integers, that is, $\{X_1, X_2, \ldots\}$. In Part I of the book all random variables that we considered either were independent or had a very simple form of dependency (e.g., they were correlated). Stochastic processes are introduced as a way to capture more complex forms of dependency between sets of random variables.

An example of a stochastic process that arises in many applications is a single server queue. The state of the system is the number of customers in the queue and the arrival (resp., departure) of a customer increases (resp., decreases) the state by 1 unit. For such a process the average number of customers in the system can be determined using results developed in the following chapters. Little's result (first mentioned in Chapter 1) allows one to derive the average customer response time. Another stochastic process can be defined for this model. Let $X(t)$ be the total number of customers that have arrived to the system over the time period $[0, t]$. Clearly, $X(t_i)$ and $X(t_j)$, for $0 < t_j < t_i$, are related since $X(t_i)$ equals $X(t_j)$ plus the number of arrivals that take place in the interval $(t_j, t_i]$. The process $X(t)$ is called a *counting process*, and a special type of such a process is a renewal process, the topic of this chapter.

A renewal process, $N(t)$, is created by forming the partial sums of independent and identically distributed random variables, $X_1 + X_2 + \cdots$. Each random variable X_i corresponds to an event, and $N(t)$ is a random variable that counts the number of completed events in the time interval $[0, t]$. Such processes are widely used to model arrival streams and service executions in computer systems. If a process is used to model customer arrivals then $N(t) = n$ means that n customer arrivals occurred in the first t time units.

The Poisson process is an important special case of a renewal process. Like the Poisson distribution to which it is related, it is characterized by several unique and desirable mathematical properties. These properties not only lead to tractable mathematical models but also explain why Poisson processes are prevalent in natural and man-made systems. Generally, a Poisson process arises whenever there is a large number of sources that produce events independently with small probability. In the context of a computer system, these events can correspond to customer arrivals from a large number of independent sources. Such models include a set of automatic teller machines being served by a central facility, a set of workstations issuing requests to a data server, or requests for a certain memory module over an interconnection network. The common feature of such systems is that the total rate of requests to the common destination over a given time interval equals the sum of the requests coming from numerous sources each of which issues requests with a small rate.

To model such arrivals we break time into infinitesimal slots, ϵ, and model the requests from each of the sources in a given slot as a Bernoulli random variable with a small parameter. Let n be the number of sources and let $p_n \epsilon$ be the probability that a given source generates a request in each slot. If $p_n \to 0$ as $n \to \infty$ in such a way that

$$\lim_{n \to \infty} n p_n = \lambda,$$

then the arrival stream of requests to the central facility is a Poisson process. We call λ the *rate* of the process. For this process the distribution of the number of requests contained in any collection of disjoint time periods having length t is Poisson with parameter λt.

To derive properties of the Poisson process, we find it convenient to start from the simpler binomial process. A binomial process can be thought of as an infinite sequence of coin tosses where an event corresponds to heads landing up. Properties of binomial processes are easily derived due to the simplicity of the process. The Poisson process is a limiting form of a binomial process where time is scaled by a factor of $1/n$, the number of coin tosses is scaled by a factor of n, and the probability of heads landing up is scaled by a factor of $1/n$. The limiting process as $n \to \infty$ is a Poisson process.

Renewal theory can be used to analyze service executions in a computer system. For service executions we are typically interested in the remaining service time that is seen by an arriving customer to the system (in the terminology of renewal theory this is called the forward recurrence time or residual time). If customer J arrives to a busy first-come-first-served queue at time t, for example, then before it can receive service it must wait for all the other customers that are ahead of it in the system to finish service. For customers in the queue, this time corresponds to their service demand, but for the customer already in service it corresponds to the time remaining in its service demand since it has already received some portion of service. If the time of J's arrival t is "randomly selected" with respect to the service process (a notion we make precise in this chapter) then the remaining service time is stochastically equal to the forward recurrence time of the service process. Thus renewal theory plays a key role in analyzing queueing systems.

Section 6.1 considers the simple discrete time binomial process and establishes some of its properties. This process is generalized to continuous time, yielding the Poisson process in Section 6.2, and properties analogous to the binomial case are established. Properties of the Poisson process for modeling computer systems are found in Section 6.2.7. Our discussion of general renewal processes starts in Section 6.3 and therein we establish all of the properties of renewal sequences that we will need later in the book when analyzing models of systems. Conclusions are presented in Section 6.8. We will sometimes adopt the convention to suppress the dependency on time in our notation of a renewal process; we sometimes write X instead of $X(t)$.

6.1 Binomial Processes

For discrete time systems one often models arrival processes to a system as arising from a binomial source. In such models one assumes that time is slotted into unit intervals and at most one arrival can occur in any slot. Arrivals are assumed to be random and occur in each slot independently with probability p. We can think of the process taking place in discrete time and being equivalent to an infinite series of coin tosses where arrivals correspond to heads landing up with probability p. In models of computer systems, arrivals correspond to the aggregate request of a group of users and the slot size in the model is selected so that there is a small probability of having more than one arrival in any slot. This arrival process has the following properties, which can easily be inferred from Figure 6.1.

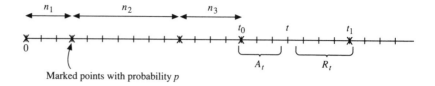

FIGURE 6.1. Binomial Processes

6.1.1
Basic Properties
of a Binomial
Process

1. **Independent Increments**: Consider slots $n_i, i = 1, 2, \ldots$, that satisfy $0 \leq n_1 < n_2 < n_3 \cdots$, and let $N_i, i = 1, 2, \ldots$, be the total number of arrivals found in the interval $[n_{i-1}, n_i - 1]$. We call the random variables N_i *increments*. In a binomial process the increments are independent.

2. **Stationary Increments**: The distribution of N_i depends only on the value of the difference $n_i - n_{i-1}$ and not on the value of n_i.

3. **Geometric Interarrival Times**: Let m_i be the number of slots between the ith and $i-1$st arrival to the system. Then the distribution of $m_i, i = 1, 2, \ldots$, is geometric with parameter p and starts at 0. Furthermore, the set of random variables $\{m_1, m_2, \ldots\}$ are mutually independent.

4. **Binomial Distribution**: Let $\mathcal{S}$ be any set of n slots. Then it follows that the number of arrivals found in $\mathcal{S}$ is distributed as a binomial random variable with parameters (n, p).

The above properties essentially follow from the fact that, in a binomial process, each slot contains an arrival independent of all other slots with the same probability. In general, *none of the above properties hold* for an arbitrary discrete time process. A simple example where all of these properties are violated is a discrete degenerate process in which arrivals occur every m slots. For this case increments are neither independent nor stationary since slot n contains an arrival only if m evenly divides n. Obviously the interarrival times are not geometric and, additionally, the number of arrivals in a collection of n slots depends on the particular collection of slots.

It is important here to distinguish a "process" from a "random variable." We can think of a process as an infinite sequence of events or "epochs" that are separated by "inter-epoch" times. In the binomial process above, inter-epoch times are independent geometric random variables. The name "binomial process" indicates that the number of epochs contained in any selection of n slots is binomially distributed. This relates to the binomial process with the binomial random variable. It is tempting to believe that for any random variable we can define a process that satisfies the above analogy. This, however, is not the case. For example, there is *no process* for which the number of epochs contained in any selection of n slots is distributed as a geometric random variable.

6.1.2 Residual and Age Distributions for Binomial Processes

In this section we derive distributions for the age and residual life random variables of a binomial process. These random variables are shown in Figure 6.1. Consider a particular point t (it can be marked or unmarked). We define the *age* at time t, a random variable denoted by A_t, to be the number of leftmost points between t and the first marked point (point t_0 in the figure). The *residual time* associated with point t, a random variable denoted by R_t, is defined to be the number of rightmost points between t and the first marked point (point t_1). If the values of $X_i, i = 1, 2, \ldots$, correspond to the lifetime of a component that is replaced upon wearing out, then the age of the component associated with point t is the amount of time the component has been in operation and the residual time is the amount of time the component will continue to be in operation until it needs to be replaced. The age (resp., residual life) is also called the *backward* (resp., *forward*) *recurrence time*. In our calculations we assume that point 0 is marked.

6.1.3 Randomly Selected Times

We are interested in determining the distribution of the age and residual life when t is "randomly selected." The notion of random selection is easily understood when there is a finite set of objects. In the study of combinatorics we made extensive use of this notion in many variations of ball and urn problems and, for these cases, a random selection was defined as selecting an object from a set of n items with uniform probability (4.31). When there are an infinite number of possible selections, however, the notion of random selection is not as easily defined. We cannot simply select t uniformly over $\{1, 2, \ldots, n\}$ and then let $n \to \infty$ since the limiting distribution is not defined. By definition, t is a random selection with respect to a renewal sequence $X_1, X_2, \ldots$ if t satisfies two criteria: it is independent of the renewal sequence and it falls in a "stationary portion" of the sequence. For our purposes we say that the process is *stationary* if the distributions of A_t and R_t, do not depend on the value t.

It is important to realize that independence and stationarity of a randomly selected point are two different notions. For example, assume that point 0 is marked and suppose that t is selected independently of the renewal sequence but is finite. Clearly then t does not fall within a stationary portion of the sequence since A_t cannot have a value greater than t and thus is not independent of t. To avoid this problem, we calculate the distributions of the age and residual time at time t and then let $t \to \infty$. Intuitively, for large t, the influence of the boundary at 0 has negligible effect on the age and residual life.[1] We will also discuss a different approach in Section 6.3, where we create a renewal process that is stationary for all values of time. This method, however, requires knowing the limiting distributions of the age and residual life.

6.1.4 Derivation of the Joint Distribution

In this section we derive the joint distribution of (A_t, R_t) for a fixed value of t and then let $t \to \infty$. The discussion and derivation are simplified by the use of Figure 6.1. Let us first use intuition to anticipate the result. Consider a point t. It is clear that all points to the right of t are marked independently with probability p and thus the number of slots to the first rightmost marked point from t has the same distribution as the number of tosses required before the first head in a coin tossing experiment where the coin lands heads up with probability p. Thus the distribution of the residual life is the same as the distribution of the stopping time of a coin tossing experiment, that is, it is geometric with parameter p.

The same argument does not hold for the age because of boundary effects. For example, if $t = 1$ then it is clear that either $A_t = 0$, if point 1 is marked, or $A_t = 1$ otherwise. Boundary effects become negligible, however, as $t \to \infty$ and an argument similar to that for the residual life but occurring *backwards in time* (i.e., leftwards on the time axis) suggests that the first leftmost marked point is also distributed as a geometric random variable with parameter p. Intuitively, we expect the age and residual life distributions to be geometric with parameter p. We can extract a bit more from these arguments by observing that the value of the age does not affect the value of the residual life. This suggests that these distributions are also independent of each other. We now prove these assertions.

We calculate the joint distribution of (A_t, R_t) by considering two cases.

Case 1. $(A_t = t)$ If $A_t = t$ then the first renewal occurs after or at point t. There are two different subcases to consider: point t is either unmarked or marked. If t is unmarked then it must be the case that the first renewal takes place at time $t + j$, for some j, which implies that $X_1 = t + j$ and therefore that $P[A_t = t, R_t = j] =$

---

[1]For the general case we must be more careful and impose a technical condition for this to be strictly correct. If X is a degenerate random variable with value α, for example, the fact that there is a renewal at point 0 implies that there is also a renewal at $\alpha \lfloor t/\alpha \rfloor$ and so $A_t = t - \alpha \lfloor t/\alpha \rfloor$ and $R_t = \alpha - A_t$. For this case the boundary always has an influence on the values of the age and residual life. Precise conditions for A_t and R_t to be independent of the boundary are given in Section 6.3.

$P[X_1 = t + j]$. If, on the other hand, point t is marked then it corresponds to the time of the first renewal and the event $\{A_t = t, R_t = j\}$ is equivalent to $X_1 = t$ and $X_2 = j$. This implies that $P[A_t = t, R_t = j] = P[X_1 = t, X_2 = j]$. Combining these two subcases yields

$$
\begin{aligned}
P[A_t = t, \ R_t = j] \ &= \ P[X_1 = t + j] + P[X_1 = t, X_2 = j] \\
\\
&= \ p(1-p)^{t+j-1} + p^2(1-p)^{t+j-2} \\
\\
&= \ p(1-p)^{t+j-2}.
\end{aligned}
\tag{6.1}
$$

Case 2. $(A_t = i < t)$ The second case arises when t occurs after the first renewal. We can account for all possibilities by partitioning events according to the number of renewals that have occurred prior to t. Since each renewal is at least one unit, there are at most $t - i$ renewals before time t. As above, there are two subcases to consider: point t is either unmarked or marked. If t is unmarked then it falls within a renewal epoch that occurs after the first renewal. Hence, using the law of total probability and letting $S_n \overset{\text{def}}{=} X_1 + X_2 + \cdots + X_n$, we can write these events as

$$
\{A_t = i, \ R_t = j, \ t \text{ unmarked}\} = \{S_n = t - i, \ X_{n+1} = i + j\}_{n=1}^{t-i}
\tag{6.2}
$$

(recall that the notation on the right-hand side of this equation is a shorthand for the union of sets $\{S_n = t - i, \ X_{n+1} = i + j\}$ over $n = 1, 2, \ldots, t-i$). If t is marked then the length of the renewal epoch just before time t equals i and that of the renewal epoch just after time t equals j. Hence we can write

$$
\{A_t = i, R_t = j, \ t \text{ marked}\} = \{S_n = t - i, \ X_{n+1} = i, \ X_{n+2} = j\}_{n=1}^{t-i}.
$$

Using the independence of renewal epochs and the fact that the sum of geometric random variables is distributed as a negative binomial (4.53) allows us to write

$$
P[A_t = i, \ R_t = j]
$$

$$
= \ \sum_{n=1}^{t-i} \{ P[S_n = t - i, \ X_{n+1} = i + j]
$$

$$
+ \ P[S_n = t - i, \ X_{n+1} = i, \ X_{n+2} = j] \}
$$

$$
= \ \sum_{n=1}^{t-i} p^n \binom{t-i-1}{t-i-n} (1-p)^{t-i-n} \{ p(1-p)^{i+j-1}
$$

$$+p^2(1-p)^{i+j-2}\}$$

$$= \sum_{n=1}^{t-i} p^n \begin{pmatrix} t-i-1 \\ t-i-n \end{pmatrix} p(1-p)^{t+j-n-2}, \quad i=1,2,\ldots,t-1.$$

We use the identity (see Problem 6.1)

$$\sum_{n=1}^{k} \alpha^n \begin{pmatrix} k-1 \\ k-n \end{pmatrix} = (1+\alpha)^{k-1}\alpha \tag{6.3}$$

to simplify the above equation to yield

$$P\left[A_t = i,\ R_t = j\right] = p^2(1-p)^{i+j-2}, \quad i=1,2,\ldots,t-1. \tag{6.4}$$

It is interesting to note that (6.4) depends on the value of t only through the range of variable i.

Combining both cases yields

$$P\left[A_t=i, R_t=j\right] = \begin{cases} p^2(1-p)^{i+j-2}, & i=1,2,\ldots,t-1,\ j=1,2,\ldots, \\ \\ p(1-p)^{t+j-2}, & i=t,\ j=1,2,\ldots\ . \end{cases}$$
$$\tag{6.5}$$

It can easily be checked that the summation of (6.5) over $i = 1, 2, \ldots$, and $j = 1, 2, \ldots$, equals 1 and thus (6.5) is a density. As $t \to \infty$ the limiting form of equation (6.5) is a product of two geometric distributions with parameter p. Thus the limiting form of random variables A_t and R_t are independent and geometrically distributed with parameter p, a result our intuitive remarks above anticipated. We record this result in the following proposition (the definition of convergence in distribution is found in Definition 4.14).

Proposition 6.1 *Let N be a binomial process with parameter p and let A_t and R_t be random variables corresponding to the age and residual life associated with a point t. Let A and R denote random variables that satisfy $A_t \xrightarrow{D} A$, and $R_t \xrightarrow{D} R$ as $t \to \infty$. Then A and R are independent geometric random variables with parameter p.*

6.1.5
Length Biasing

Proposition 6.1 shows that

$$A =_{st} R =_{st} G_{1,p}, \tag{6.6}$$

where $G_{1,p}$ is a geometric random variable with parameter p. It also follows that A and R are independent random variables. Clearly, since both A and R have distributions that are equal to the interarrival distribution of the binomial process, it must be the case that a selected point in the sequence does not fall in a "typical" interval. In other words, the distribution of the interval containing the randomly selected point is not the same as the distribution of the underlying sequence. This is indeed the case and follows from the observation that a randomly selected point is more likely to fall within a larger interval of time than within a smaller one. In fact, since both A and R are independent and identically distributed, the distribution of the interval containing the randomly selected point, a random variable denoted by L, is given by

$$f_L(i) = (i - 1)p^2(1 - p)^{i-2}, \quad i = 2, 3, \ldots . \tag{6.7}$$

Hence L has a negative binomial distribution (this follows from the fact that the sum of two geometric random variables is distributed as a negative binomial (4.53)).

The bias of having a randomly selected point fall within a large interval is called *length biasing* and is a feature of all renewal sequences. It is clear, however, that the distributional equalities (6.6) arise because the geometric distribution is memoryless (see Problem 6.3). Naturally these conclusions are strongly based on the fact that successive marked points are independent of previously marked points, and we would not expect all of the equalities in (6.6) to hold generally. For example, in the degenerate case considered previously, if $X = X_i = n$ with probability 1, then it is easy to show that the age and residual life are uniformly distributed over $(1, n)$ and are *not independent*. In this case $A =_{st} R$, $A \neq_{st} X$, and A and R satisfy $A = n - R$. It is interesting to note that for this and the above example $A =_{st} R$. This equality does hold generally, and we will prove it by deriving the density functions of A and R in Section 6.5. Calculations involving the age and residual life of a random variable are frequently used to analyze queueing systems (see Section 7.2).

Example 6.2 (Residual Service Times) Consider a first-come-first-served single server queue that operates in discrete time. Assume that time is slotted into unit intervals and that the service requirement of the customer in the queue is geometric with probability p. Consider a time t (assumed to fall at the beginning of a time slot) and let N_t be the total number of customers in the system at time t. Let V_t be a random variable that denotes the total time needed to completely service all these N_t customers. We call V_t the *virtual waiting time* for time t since it equals the time spent waiting if a customer arrived at time t. We are interested in calculating the distribution of V_t given the value of N_t.

Clearly if $N_t = 0$ then $V_t = 0$. If $N_t = n$ for $n \geq 1$ then there is one customer receiving service and $n-1$ customers waiting in the queue. The virtual waiting time equals the residual service time of the customer in service, which we denote by R_t, plus the sum of the service requirements of the $n - 1$ waiting customers. Hence we have that

$$V_t = R_t + X_1 + X_2 + \cdots + X_{N_t - 1},$$

where $X_i, i = 1, 2, \ldots, N_t - 1$ are independent, geometric, random variables with parameter p. If we suppose that the queue has a stationary distribution, that is, that $\lim_{t \to \infty} N_t = N$ exists, then our comments above show that R_t also achieves a limiting distribution that is geometric with parameter p. The limiting random variable for the virtual waiting time, denoted by $V = \lim_{t \to \infty} V_t$, thus has a distribution given by

$$P[V = i] = \begin{cases} P[N = 0], & i = 0, \\ \sum_{n=1}^{i} p^n \begin{pmatrix} i - 1 \\ i - n \end{pmatrix} (1 - p)^{i-n}, & i = 1, 2, \ldots, \end{cases} \tag{6.8}$$

where we have used the fact that the sum of geometric random variables is negative binomial (4.53) in the second case of (6.8). In Chapter 7 we establish techniques that are used to determine the limiting distribution of N.

Like the analogy between binomial and Poisson random variables established in Chapter 4 (see Section 4.5.3) there is a continuous limiting form of the binomial process - the Poisson process. This process is most frequently used to model arrival processes because mathematical expressions associated with it are more tractable than corresponding expressions for binomial processes. The reasons for this tractability are similar to those delineated in Section 4.5.5. In that section we showed that, unlike binomial random variables, the sum of Poisson random variables is always Poisson regardless of their parameter values. Additionally, we established a tractable Poisson approximation to binomial density functions and showed that the Poisson distribution is obtained as a general limit of sums of Bernoulli random variables that does not depend on the exact values of the Bernoulli probabilities. All of these properties can be extended to Poisson processes; in the next section we will explore these and several generalizations of the Poisson process that considerably enhance its modeling capabilities.

6.2 Poisson Processes

Similar to the approach of Section 4.5.3, we derive the Poisson process as a limiting form of a binomial process. Consider a binomial arrival process with slot size ϵ and let $N_\epsilon(t)$ denote the number of arrivals in t time intervals (this corresponds to t/ϵ slots). To avoid trivialities assume that t/ϵ is integer. The probability of having an arrival in each slot is given by $\lambda\epsilon$ and hence the density of $N_\epsilon(t)$ is binomial with parameters $(t/\epsilon, \lambda\epsilon)$,

$$P[N_\epsilon(t) = i] = \begin{pmatrix} t/\epsilon \\ i \end{pmatrix} (\lambda\epsilon)^i (1 - \lambda\epsilon)^{t/\epsilon - i}, \quad t \geq 0. \tag{6.9}$$

To simplify this and determine the limiting distribution, let $n_\epsilon \stackrel{\text{def}}{=} t/\epsilon$. Similar to the calculations starting from (4.69), we multiply by 1 in the form of $(n_\epsilon)^i/(n_\epsilon)^i$, yielding

$$P[N_\epsilon(t) = i] = \frac{(n_\epsilon)_i}{n_\epsilon^i} \frac{(n_\epsilon \lambda \epsilon)^i}{i!} \frac{(1 - \lambda\epsilon)^{n_\epsilon}}{(1 - \lambda\epsilon)^i}, \quad t \geq 0. \quad (6.10)$$

Equation (3.16) shows that, for any fixed i, $(n_\epsilon)_i/n_\epsilon^i \to 1$ as $\epsilon \to 0$ and, similar to (4.70), equation (6.10) reaches a limiting value of

$$P[N(t) = i] = \frac{(\lambda t)^i}{i!} e^{-\lambda t}, \quad t \geq 0, \quad (6.11)$$

where $N(t)$ is the limiting version of $N_\epsilon(t)$, that is, $N_\epsilon(t) \xrightarrow{D} N(t)$ as $\epsilon \to 0$. We say that N is a *Poisson process* with parameter λ.

6.2.1
Basic Properties
of a Poisson
Process

Since the Poisson process is a limiting form of a binomial process it shares similar properties.

1. **Independent Increments**: Consider time epochs that satisfy $0 \leq t_1 < t_2 \cdots < t_k$ and let $N_i, i \geq 1$, be the number of arrivals found in the time interval $[t_{i-1}, t_i)$. We call the random variables N_i *increments* and note that Poisson processes have independent increments.

2. **Stationary Increments**: The distribution of N_i depends only on the size of the intervals, $t_i - t_{i-1}$, and does not depend on the value of t_i. This implies that the arrival rate, defined to be the expected number of arrivals per unit time, is constant over time. To see this observe that for any small interval of time, Δt, the probability that an arrival occurs in Δt equals $\lambda \Delta t$. This follows from

$$P[N(\Delta t) = 1] = \lambda \Delta t e^{-\lambda \Delta t}$$

$$= \lambda \Delta t (1 - \lambda \Delta t + \frac{(\lambda \Delta t)^2}{2!} - \cdots)$$

$$= \lambda \Delta t + o(\Delta t). \quad (6.12)$$

3. **Single Arrivals and Exponential Interarrival Times:** Let τ_i be the time between the ith and $i - 1$st arrivals. Then the distribution of τ_i is exponentially distributed with parameter λ, for example, $P[\tau_1 > t] = P[N(t) = 0] = \exp(-\lambda t)$. Furthermore, the set of interarrival times $\{\tau_1, \tau_2, \ldots\}$ consist of mutually independent random variables.

The fact that there are single arrivals results directly from the binomial limiting argument given above. Formally, let $N(\Delta t)$ be a random variable that counts the number of arrivals in Δt time units. Then an argument similar to that leading to (6.12) implies that

$$P\left[N(\Delta t) = i\right] = o(\Delta t^i), \quad i = 1, 2, \ldots .$$

This shows that the leading order term occurs for $i = 1$ and thus the probability that there is more than one arrival in Δt is negligible as $\Delta t \to 0$.

4. **Poisson Distribution:** Let $\mathcal{S}$ be any set of time intervals having a total length of t time units. The distribution of the number of arrivals found in set $\mathcal{S}$ is Poisson with parameter λt.

Variations of the properties listed above lead to other types of Poisson processes (similar generalizations can be considered for binomial arrival processes but they do not typically prove to be mathematically tractable). In Section 6.3.7 we show that the first two properties listed above are sufficient to define a Poisson process.

6.2.2 Relaxing Stationarity: Nonhomogenous Poisson Processes

If one relaxes the stationary property then the resultant process is called a *nonhomogenous Poisson process* and is used to model time varying arrival processes. To precisely define this process, suppose that the arrival rate is a function of t, denoted by $\lambda(t)$, and let $N(t)$ be the number of arrivals in the interval $(0, t)$. An argument similar to that leading to (6.11) (see Problem 6.5) shows that the arrival process has a Poisson type distribution given by

$$P\left[N(t) = i\right] = \frac{\Lambda(t)^i e^{-\Lambda(t)}}{i!}, \tag{6.13}$$

where

$$\Lambda(t) \overset{\text{def}}{=} \int_0^t \lambda(x)dx. \tag{6.14}$$

Example 6.3 Suppose we wish to model an arrival stream into a computer facility and observe that before and after lunch time arrivals occur at rate λ_0 and in the middle of lunch they occur at rate λ_1, $\lambda_1 < \lambda_0$. If lunch is from time t_0 to t_1 we could model this arrival stream using a time varying Poisson process with rates that linearly decrease to λ_1 over the interval $(t_0, (t_0 + t_1)/2)$

and then linearly increase to λ_0 over $((t_0+t_1)/2, t_1)$. Some algebra shows that for this model we can write $\Lambda(t)$ as

$$\Lambda(t) = \begin{cases} \lambda_0, & t \le t_0, \ t \ge t_1, \\ \lambda_1 + m\,|t - (t_1 + t_0)/2|, & t_0 \le t \le t_1, \end{cases}$$

where the slope m is given by

$$m = \frac{2(\lambda_0 - \lambda_1)}{t_1 - t_0}.$$

The distribution for the number of arrivals in $[0, t]$ can thus be obtained from (6.13) and (6.14).

6.2.3 Relaxing Single Arrivals: Compound Poisson Processes

Relaxing the single arrival assumption yields a *compound Poisson process*. For processes of this type each arrival corresponds to a random number of customers and the total number of customers arriving in $(0, t)$ equals the number of customers brought to the system by $N(t)$ arrivals, where $N(t)$ is a Poisson process. More precisely, this corresponds to the random sum operator where N is Poisson. We set

$$X(t) = X_1 + X_2 + \cdots + X_{N(t)}$$

and are interested in the distribution of $X(t)$. Clearly $X(t)$ is Poisson if $X_i, i \ge 1$, are degenerate random variables with unit values. In general, however, determining the distribution of $X(t)$ is difficult.

Example 6.4 (Batch Arrival Process) Consider the case of a parallel-processing system consisting of a single queue and multiple processors. Jobs arriving to the system consist of a possibly random number of tasks, each of which must be serviced to completion before we consider the job completed. Since each job consists of a random number of tasks, we can model the arrival process as a compound Poisson process, which is also called a *batch arrival process*. Let B be a random variable that denotes the number of tasks comprising a job. Using the random sum property of probability generating functions (see the fifth property of probability generating functions given in Section 5.3) shows that

$$\mathcal{G}_{X(t)}(z) = \mathcal{G}_{N(t)}\left(\mathcal{G}_B(z)\right). \tag{6.15}$$

The random variable $N(t)$ is Poisson with parameter λt and thus has a probability generating function given by (see Table 5.9)

$$\mathcal{G}_{N(t)}(z) = e^{-\lambda t(1-z)}.$$

If B is geometric with parameter p, then

$$\mathcal{G}_B(z) = \frac{pz}{1 - (1-p)z},$$

and thus for this case (6.15) yields

$$\mathcal{G}_{X(t)}(z) = \exp\left[\frac{-\lambda t(1-z)}{1-(1-p)z}\right].$$

The mean and variance can be obtained from this by differentiation. Note that if B is a Bernoulli random variable with parameter p then it has a probability generating function given by $\mathcal{G}_B(z) = 1 - p(1-z)$ and thus for this case (6.15) yields

$$\mathcal{G}_{X(t)}(z) = e^{-\lambda t p(1-z)},$$

which shows that it is Poisson with parameter $\lambda p t$ as would be expected.

6.2.4 Relaxing Independent Increments: Modulated Poisson Processes

Relaxing the independent increment property yields a *modulated Poisson process*. This process can be expressed more generally when we develop Markov process theory, and here we content ourselves with a special case of the process. Assume that there are two states, 0 and 1, for a "modulating process." When the state of the modulating process equals 0 then the arrival rate of customers is given by λ_0, and when it equals 1 then the arrival rate is λ_1. The initial state of the modulating process is randomly selected and is equally likely to be state 0 or 1. The residence time in a particular modulating state is exponentially distributed with parameter μ and, after expiration of this time, the modulating process changes state. Thus there is an alternating sequence of states $\dots, 1, 0, 1, 0, \dots$, and the times in each state are exponentially distributed. For a given period of time $(0, t)$, let Υ be a random variable that indicates the total amount of time that the modulating process has been in state 0. Let $X(t)$ be the number of arrivals in $(0, t)$. It is clear then that, given Υ, the value of $X(t)$ is distributed as a nonhomogenous Poisson process and thus

$$P\left[X(t) = n \mid \Upsilon = \tau\right] = \frac{\left(\lambda_0 \tau + \lambda_1(t - \tau)\right)^n e^{-(\lambda_0 \tau + \lambda_1(t - \tau))}}{n!}. \tag{6.16}$$

The difficulty in determining the distribution for $X(t)$ is to calculate the density of Υ. In Problem 6.6 we lead the reader through a derivation of this density. There are some limiting cases that are of interest. As $\mu \to 0$ the probability that the modulating process makes no transitions within t seconds converges to 1, and thus the initial state determines the arrival rate at time t. We expect for this case that

$$P\left[X(t) = n\right] = \frac{1}{2}\left\{\frac{(\lambda_0 t)^n e^{-\lambda_0 t}}{n!} + \frac{(\lambda_1 t)^n e^{-\lambda_1 t}}{n!}\right\}. \tag{6.17}$$

On the other hand, if $\mu \to \infty$ then the modulating process makes an infinite number of transitions within t seconds and we expect the modulating process to spend an equal amount of time in each state yielding

$$P\left[X(t) = n\right] = \frac{(\beta t)^n e^{-\beta t}}{n!}, \tag{6.18}$$

where

$$\beta \stackrel{\text{def}}{=} \frac{\lambda_0 + \lambda_1}{2}.$$

It is clear that we can generalize modulating processes so that they have more than two states and an arbitrary transition structure. Such processes are called Markov processes and will be studied in Chapter 8.

Example 6.5 (Modeling Voice) To transmit voice over a computer network it must first be digitized. A basic feature of speech is that it comprises an alternation of silent periods and non-silent periods and this explains why it is frequently modeled as a modulated process. During talk spurts, voice is coded and sent as packets and silent periods can be modeled as producing no packets. Assume that the arrival rate of packets during a talk spurt period is Poisson with rate λ_1 and silent periods produce a Poisson rate with $\lambda_0 \approx 0$. If the duration of times for talk and silent periods are exponentially distributed with parameters μ_1 and μ_0, respectively, then the model of the arrival stream of packets is given by a modulated Poisson process.

6.2.5 Generalized Limit that Leads to the Poisson Process

Recall that in Section 4.5.3 we showed that a Poisson random variable is the limiting case of the sum of a large number of Bernoulli random variables if the sum of their, possibly different, parameter values equals λ (see condition (4.77)). Analogously, the superposition of a large number of binomial processes, each with small parameter, yields a Poisson process if conditions on their parameter values, similar to those of (4.77), are satisfied. More precisely, consider a set of binomial processes, $N_i, i = 1, 2, \ldots, m$, with parameter values $\lambda_i(m), i = 1, 2, \ldots, m$, where $\lambda_i(m) \to 0$ as $m \to \infty$ in a way that satisfies

$$\lambda_1(m) + \lambda_2(m) + \cdots + \lambda_m(m) \longrightarrow \lambda \quad \text{as} \quad m \longrightarrow \infty, \tag{6.19}$$

where λ is a positive constant. Then the superposition of these processes, given by $N(t) = N_1(t) + N_2(t) + \cdots + N_m(t)$, is a Poisson process with rate λ as $m \to \infty$. The derivation of this closely follows that leading to equation (4.80) of Section 4.5.3 and that given in the beginning of this section.

This provides insight into why the Poisson process appears so frequently in mathematical models of computer systems since it shows that the superposition of many binomial sources of traffic, each of which has a small arrival rate, form a Poisson stream of traffic. In computer systems, there are often many potential sources of traffic, each of which only infrequently issues requests. Regardless of the individual rates of request generation, the above limiting argument shows that the aggregate traffic presented to the system forms a Poisson source provided that (6.19) is satisfied.

We close this section by mentioning that, although we have only considered the superposition of binomial processes, a Poisson process limit is obtained more generally. The *Palm–Khintchine theorem* (see [49], page 158) shows that a Poisson process arises when a large number of independent renewal processes are superimposed, provided that a condition like (6.19) holds on the rates of each process.

6.2.6 Residual and Age Distributions for Poisson Processes

An important property of the exponential distribution that follows from its relationship with the geometric distribution concerns the distributions of the age and residual times. Assume that $X_i, i = 1, 2, \ldots$, is a renewal sequence starting at time 0 consisting of independent, exponential, random variables with parameter λ. Let t be a point in time and define A_t and R_t to be the age (equiv., the backwards recurrence time) and residual life (equiv., the forward recurrence time), respectively. The analogy with the binomial case suggests that, as $t \to \infty$, the age and residual life of a Poisson process are independent exponential variables with parameter λ. Our derivation of this result is abbreviated since it closely follows that given for the binomial case and will be generalized in Section 6.5.

For a given value of t, there are two cases that must be considered: either $A_t = t$ if t occurs before the first renewal or $A_t = x, 0 \leq x < t$, otherwise. To simplify the derivation, let $Z_t \overset{\text{def}}{=} (A_t, R_t)$ be a random vector corresponding to the age and residual life random variables and let X be an exponential random variable with parameter λ. To calculate the joint density of Z, denoted by $f_{Z_t}(x, y)$, we consider the two cases outlined above. The first case where $A_t = t$ occurs if t lies within the first renewal interval.[2] Analogous to the derivation in Section 6.1.4 leading to (6.1) we have

$$f_{Z_t}(t, y) = f_{X_1}(t + y) = \lambda e^{-\lambda(t+y)}, \quad 0 \leq y. \tag{6.20}$$

---

[2]In contrast to the geometric case we do not need to consider the event that $\{X_1 = t\}$ since this occurs with zero probability.

The second case where $A_t < t$ occurs when t occurs after the first renewal. Analogous to the derivation leading to (6.4) we have

$$
\begin{aligned}
f_{Z_t}(x, y) &= \sum_{n=1}^{\infty} \frac{\lambda^n (t-x)^{n-1} e^{-\lambda(t-x)}}{(n-1)!} \, \lambda e^{-\lambda(x+y)} \\
&= \lambda^2 e^{-\lambda(x+y)}, \quad 0 \leq x < t, \, 0 \leq y. \quad (6.21)
\end{aligned}
$$

Combining the two cases (6.20) and (6.21) yields

$$
f_{Z_t}(x, y) = \begin{cases} \lambda^2 e^{-\lambda(x+y)}, & 0 \leq x < t, \, 0 \leq y, \\ \lambda^2 e^{-\lambda(t+y)}, & x = t, \; 0 \leq y, \end{cases} \quad (6.22)
$$

where, as in (6.5), the dependency on t in the first equation of (6.22) only occurs in the domain of x. As $t \to \infty$ the density of Z_t approaches the product of two exponential densities with parameter λ. Thus, for the limiting case, the age and residual life random variables are exponential and independent, a result we record here.

Proposition 6.6 *Let $X_i, i = 1, 2, \ldots$, be a renewal sequence of independent exponential random variables with parameter λ. Let A_t and R_t be random variables corresponding to the age and residual life associated with a point t and let $A_t \xrightarrow{D} A$ and $R_t \xrightarrow{D} R$ as $t \to \infty$. Then A and R are independent exponential random variables with parameter λ.*

The fact that the age and residual life have the same distribution as a renewal epoch follows from the memoryless property of exponential random variables. The renewal epoch intersected by a randomly selected point is distributed as an Erlang random variable with parameters $(2, \lambda)$ and hence has an expected length twice that of a typical renewal epoch. These results are not surprising when the Poisson process is viewed as the limiting form of a binomial process that is equivalent to a series of independent coin tosses. When the Poisson process is derived along different lines, however, the fact that the age and residual life have the same distribution as a renewal epoch seems counterintuitive and paradoxical. In fact in the literature this result is often called the *inspection paradox* or the *residual life paradox*. The source of the "paradox" lies in the fact that selected intervals are biased by their lengths, and in Section 6.5 we will present a derivation of the age and residual life for the general case. This derivation shows that if renewal epochs are not deterministic then the expected value of the length of an intersected renewal epoch is *strictly larger* than the expected value of a renewal epoch. With the previous derivation in mind, this is an intuitive rather than paradoxical result.

6.2.7 Superposition and Splitting of Poisson Processes

The Poisson process is by far the most important process in modeling, and in this section we examine some of its unique properties. These properties are readily apparent when one considers that the Poisson process is derived from the binomial processes, which, as seen in Section 6.1, can be viewed in terms of coin tosses.

Suppose one superimposes Poisson processes $N_i, i = 1, 2, \ldots, m$ with rates λ_i. The resultant process is a Poisson process with rate $\lambda = \lambda_1 + \lambda_2 + \cdots + \lambda_m$. This essentially follows from the fact that the sum of Poisson random variables is Poisson as established in Section 4.5.5 (see (4.76)). It is also the case that if one splits a Poisson process then each of the split processes is also Poisson. To precisely describe what we mean by splitting a Poisson process (see Figure 6.2 for a pictorial description) assume that N is a Poisson process with parameter λ.

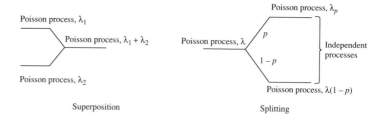

FIGURE 6.2. Superposition and Splitting of a Poisson Process

Suppose the system consists of two subsystems (the results that follow hold for any finite number of subsystems). With probability p we route a customer to subsystem 1 and with probability $1-p$ we route the customer to subsystem 2. Let $N_1(t)$ and $N_2(t)$ be random variables that denote the number of customers that arrive to subsystem 1 and 2, respectively, over t units of time. We are interested in determining the distribution of these random variables. We determine this using conditional probabilities. It is clear that if a total of n customers arrive to the system over t units of time then the number that are routed to the first subsystem is binomially distributed with parameters (n, p). Since the total number of customer arrivals is a Poisson process with parameter λ we have

$$
\begin{aligned}
P\left[N_1(t) = j\right] &= \sum_{n=j}^{\infty} P\left[N_1(t) = j \mid N(t) = n\right] P\left[N(t) = n\right] \\
&= \sum_{n=j}^{\infty} \binom{n}{j} p^j (1-p)^{n-j} \frac{(\lambda t)^n}{n!} e^{-\lambda t} \\
&= \frac{(\lambda p t)^j}{j!} e^{-\lambda t} \sum_{n=j}^{\infty} \frac{(\lambda(1-p))^{n-j}}{(n-j)!}
\end{aligned}
$$

$$= \frac{(\lambda pt)^j e^{-\lambda pt}}{j!}.$$

Thus the distribution of $N_1(t)$ is Poisson with parameter λpt. The distribution of N_2 is also Poisson but with parameter $\lambda(1-p)t$. We have shown that the distribution of a split Poisson process is a Poisson process.

It also follows that N_1 and N_2 are independent Poisson processes. To establish independence, consider two sets of time intervals $\mathcal{S}_i, i = 1, 2$, and assume that their intersection has length s and is denoted by $\mathcal{S}$. By definition of the split process and of the Poisson process, it is clear that the number of arrivals to the subsystems are independent over disjoint time intervals, that is, that $N_1(\mathcal{S}_1 - \mathcal{S})$ and $N_2(\mathcal{S}_2 - \mathcal{S})$ are independent random variables. To establish independence over all time intervals it thus suffices to show that $N_1(\mathcal{S})$ and $N_2(\mathcal{S})$ are also independent. This follows from a direct calculation, which shows that

$$P\left[N_1(\mathcal{S}) = j, \ N_2(\mathcal{S}) = n - j\right] = P\left[N_1(\mathcal{S}) = j\right] P\left[N_2(\mathcal{S}) = n - j\right].$$
(6.23)

To establish (6.23) note that the left-hand side can be written as

$$P\left[N_1(\mathcal{S}) = j, \ N_2(\mathcal{S}) = n - j\right] = \binom{n}{j} (1-p)^j p^{n-j} \frac{e^{-\lambda s}(\lambda s)^n}{n!}$$

and the right-hand side can be written as

$$P\left[N_1(\mathcal{S}){=}j\right] P\left[N_2(\mathcal{S}){=}n{-}j\right] = \frac{e^{-\lambda ps}(\lambda ps)^j}{j!} \ \frac{e^{-\lambda(1-p)s}(\lambda(1-p)s)^{(n-j)}}{(n-j)!}.$$

These equations show that (6.23) is satisfied and thus establishes the independence of N_1 and N_2 over $\mathcal{S}$. Combining this with the independence over the disjoint intervals completes the argument that N_1 and N_2 are independent processes. We record this in the following proposition.

Proposition 6.7 *Let N be a Poisson process with rate λ and let $N_i, i = 1, 2, \ldots, m$, be m processes obtained from N by splitting N according to probabilities $p_i, i = 1, 2, \ldots, m$, where $p_1 + p_2 + \cdots + p_m = 1$. Then $N_i, i = 1, 2, \ldots, m$ are independent Poisson processes with rates λp_i.*

One can actually establish stronger results along these lines. For example, it can be shown that if N and N_1 are Poisson with parameters λ and λ_1 then N_2 is necessarily Poisson with parameter $\lambda - \lambda_1$ (see [57], page 228)). Properties of the Poisson process have many ramifications in models of computer systems. Consider a set of queues that have the property that when the arrival stream of customers is Poisson its output stream is also Poisson. Suppose we join such queues together in a network that randomly routes customers in a feed-forward fashion so that queue i can route traffic to queue j only if $j > i$. Then Proposition 6.7 implies that all of the flows in the network are Poisson (we analyze such networks in Examples 10.9 and 10.10).

6.3 Renewal Processes

The binomial and Poisson processes are special cases of a *renewal process*. We introduced the notion of a renewal process in Section 4.4 and there we established that the renewal counting process for geometric (resp., exponential) random variables is binomial (resp., Poisson). In this section we develop further results along these lines that will be needed in later chapters. These results unify those of the previous sections. We begin by recalling the definition a renewal process.

Let $\{X_n,\ n = 1, 2, \ldots\}$ be a sequence of nonnegative, independent, and identically distributed random variables and let X be a generic random variable of the sequence (i.e., $X =_{st} X_n$). In this section we assume that $P[X > 0] > 0$ and $P[X < \infty] = 1$ and that X has bounded moments and thus a moment generating function. To avoid technicalities we assume that X is a continuous random variable and we simplify notation by defining

$$\gamma \stackrel{\text{def}}{=} 1/E[X]. \tag{6.24}$$

We define $S_0 = 0$ and

$$S_n \stackrel{\text{def}}{=} X_1 + X_2 + \cdots + X_n, \quad n \geq 1,$$

and let

$$N(t) \stackrel{\text{def}}{=} \max\{n \,:\, S_n \leq t\}.$$

Recall that S_n is the time of the nth renewal and $N(t)$ is the number of renewals that occur within the interval $(0, t]$. We let $F_n,\ n = 1, 2, \ldots,$ be the distribution of the sum of n independent random variables distributed as X. It is convenient to write this using the convolution notation previously introduced (see (5.62)). To do this we first define the convolution operator as applied to distribution functions.

Let $H(t)$ be the distribution function that results when we sum two random variables having distributions F and G and densities f and g, respectively. The convolution operator for distribution functions is denoted by

$$H(t) = F \circledast G(t), \quad 0 \leq t < \infty, \tag{6.25}$$

and is defined by

$$H(t) \stackrel{\text{def}}{=} \int_0^t F(t-x)g(x)dx \tag{6.26}$$

(equiv., $H(t) \stackrel{\text{def}}{=} \int_0^t G(t-x)f(x)dx$). Comparing (5.61), which is the definition of convolution of density functions, and (6.26) shows that the form of the convolution operator depends on whether the functions being convolved are distributions or densities. This difference in the definitions arises from the fact that the convolution of two distribution functions using the form given by (5.61) does not necessarily lead to a distribution function (see Problem 6.10).

Continuing our analysis, we write F_n, $n = 1, 2, \ldots$, as the nth-fold convolution of F_X, that is,

$$F_n(x) \stackrel{\text{def}}{=} \underbrace{F_X \circledast F_X \circledast \cdots \circledast F_X}_{n \text{ times}} \tag{6.27}$$

and let f_n be the corresponding density function. The density (resp., distribution) of S_n is given by f_n (resp., F_n).

Renewal theory is concerned with properties of $N(t)$, and we recall that the distribution of $N(t)$ is related to the survivor of S_{n+1} by (see (4.96))

$$P[N(t) \le n] = P[S_{n+1} > t], \quad t \ge 0. \tag{6.28}$$

Using this we can derive the density of $N(t)$ as

$$
\begin{aligned}
P[N(t) = n] &= P[N(t) \le n] - P[N(t) \le n-1] \\[2mm]
&= P[S_{n+1} > t] - P[S_n > t] \\[2mm]
&= F_n(t) - F_{n+1}(t). \tag{6.29}
\end{aligned}
$$

Example 6.8 (Density of N(t) for Poisson Process) Let N be a Poisson process with rate λ. Recall that the sum of independent and identically distributed exponential random variables is an Erlang random variable (see (4.88)) and thus, from (4.87),

$$F_n(t) = \int_0^t \frac{\lambda^n t^{n-1} e^{-\lambda t}}{(n-1)!} dt.$$

This integral can be evaluated using the identity given in (4.97), which upon substitution into (6.29) yields

$$P[N(t) = n] = \int_0^t \frac{\lambda^n t^{n-1} e^{-\lambda t}}{(n-1)!} dt - \int_0^t \frac{\lambda^{n+1} t^n e^{-\lambda t}}{n!} dt$$

$$= \sum_{k=0}^{n} \frac{(\lambda t)^k e^{-\lambda t}}{k!} - \sum_{k=0}^{n-1} \frac{(\lambda t)^k e^{-\lambda t}}{k!}$$

$$= \frac{(\lambda t)^n e^{-\lambda t}}{n!},$$

as we already know from (6.11). In Problem 6.12 we generalize this result to Erlang distributions.

We can now obtain our first result for general renewal processes, which shows that[3] $N(t)/t \xrightarrow{a.s.} \gamma$, or equivalently written, that

$$P\left[\lim_{t\to\infty} \frac{N(t)}{t} = \gamma\right] = 1.$$

We first, however, show that $N(t)$ grows without bound as $t \to \infty$ but that for any finite t, $N(t)$ is bounded.

Lemma 6.9 *The number of renewals in t time units satisfies*

$$P[N(t) < \infty] = 1, \quad t < \infty, \tag{6.30}$$

and

$$N(t) \xrightarrow{a.s.} \infty, \quad \text{as } t \longrightarrow \infty. \tag{6.31}$$

Proof: From the strong law of large numbers we know that

$$\frac{S_n}{n} \xrightarrow{a.s.} E[X], \quad \text{as } n \longrightarrow \infty,$$

and since $E[X] > 0$ this implies that S_n must grow to infinity as $n \to \infty$. Therefore the event $\{S_n \le t\}$ can hold for at most a finite number of values of n. This establishes (6.30). To establish (6.31) we argue that (6.28) implies that

$$P\left[\lim_{t\to\infty} N(t) > n\right] = P[S_{n+1} < \infty].$$

The only way that $S_{n+1} = \infty$, however, is for at least one of the values of the first $n+1$ renewal epochs to have an infinite value. By assumption this occurs with probability 0. ∎

---

[3]See Definition 5.39 for the definition of almost sure convergence.

**6.3.1
Limiting
Number of
Renewals per
Unit Time**

In this section we establish the following proposition.

Proposition 6.10 *The number of renewal epochs in t units of time satisfies*

$$\frac{N(t)}{t} \xrightarrow{a.s.} \gamma, \quad \text{as} \quad t \longrightarrow \infty. \tag{6.32}$$

Proof: To prove this, observe that $S_{N(t)} \leq t < S_{N(t)+1}$ and thus it must be the case that

$$\frac{S_{N(t)}}{N(t)} \leq \frac{t}{N(t)} < \frac{S_{N(t)+1}}{N(t)}. \tag{6.33}$$

Since $N(t) \xrightarrow{a.s.} \infty$ as $t \to \infty$ it follows from the strong law of large numbers that $S_{N(t)}/N(t) \xrightarrow{a.s.} E[X]$ as $t \to \infty$. Combining these results shows that $S_{N(t)}/N(t) \xrightarrow{a.s.} E[X]$ as $t \to \infty$. The same limiting value holds for $S_{N(t)+1}/N(t)$. This can be easily seen by writing

$$\frac{S_{N(t)+1}}{N(t)} = \frac{S_{N(t)+1}}{N(t)+1} \frac{N(t)+1}{N(t)}$$

and by observing that $(N(t)+1)/N(t) \xrightarrow{a.s.} 1$ as $t \to \infty$. Equation (6.32) is thus satisfied since (6.33) shows that it is bounded by two sequences, each of which converges almost surely to $E[X]$. ∎

**6.3.2
Expected
Number of
Renewals**

We are often interested in the expected number of arrivals from a renewal source over a given period of time. To derive an equation for this quantity, define

$$M(t) \stackrel{\text{def}}{=} E[N(t)]. \tag{6.34}$$

It is clear that $M(t)$ is a nondecreasing function of t. We can obtain an equation for $M(t)$ as follows:

$$
\begin{aligned}
M(t) &= \sum_{n=1}^{\infty} nP[N(t) = n] \\
&= \sum_{n=1}^{\infty} n\left(F_n(t) - F_{n+1}(t)\right) \\
&= F_1(t) + \sum_{n=1}^{\infty}(n+1)F_{n+1}(t) - \sum_{n=1}^{\infty} nF_{n+1}(t) \\
&= \sum_{n=1}^{\infty} F_n(t). \tag{6.35}
\end{aligned}
$$

Equation (6.35) is difficult to manipulate since it involves convolutions of the distribution function F, and this suggests that we derive the equivalent Laplace transform equation. To do this, let

$$M^*(s) \stackrel{\text{def}}{=} \int_0^\infty e^{-st} M(t) dt.$$

(Note that we do not write the Laplace transform of M as $\mathcal{L}_M(s)$ since M is not a probability density function.) Take the transform of both sides of (6.35) (using the single integral and the convolution properties of Laplace transforms; see Table 5.6) to yield

$$M^*(s) = \frac{\mathcal{L}_X(s)}{s(1 - \mathcal{L}_X(s))}. \tag{6.36}$$

For some densities (6.36) can be inverted to explicitly obtain an expression for $M(t)$. The following lemma parallels Lemma 6.9 and shows that $M(t)$ is finite for all $t < \infty$ but has a limiting value of infinity.

Lemma 6.11 *The expected number of renewals satisfies*

$$M(t) < \infty, \quad t < \infty, \tag{6.37}$$

and

$$M(t) \longrightarrow \infty, \quad \text{as } t \longrightarrow \infty.$$

Proof: Using the notation defined by (6.27) it follows that

$$F_n(t) = F_{n-m} \circledast F_m(t), \quad 0 \leq m \leq n.$$

We can thus derive the inequality

$$
\begin{aligned}
F_n(t) &= \int_0^t F_{n-m}(t-x) f_m(x) dx \\
&\leq F_{n-m}(t) \int_0^t f_m(x) dx = F_{n-m}(t) F_m(t).
\end{aligned}
$$

Let k be a value such that $F_k(t) < 1$ (our assumption that $P[X > 0] > 0$ assures that there exists such a k). Writing

$$F_{nk+i}(t) \leq (F_k(t))^n F_i(t)$$

and substituting this into (6.35) yields

$$
\begin{aligned}
M(t) &= \sum_{n=0}^\infty \sum_{i=1}^k F_{nk+i}(t) \\
&\leq \sum_{n=0}^\infty (F_k(t))^n \sum_{i=1}^k F_i(t) \\
&= \frac{\sum_{i=1}^k F_i(t)}{1 - F_k(t)}.
\end{aligned}
$$

This establishes (6.37). To show that $M(t)$ grows without bound as $t \to \infty$ we can use the final value property of Laplace transforms to write

$$\lim_{t \to \infty} M(t) = \lim_{s \to 0} \frac{\mathcal{L}_X(s)}{1 - \mathcal{L}_X(s)}.$$

The result then follows from the fact that $\lim_{s \to 0} \mathcal{L}_X(s) = 1$. ∎

Another property of the expected number of renewals that parallels Proposition 6.10 is that $M(t)/t \to \gamma$ as $t \to \infty$. Since $M(t) = E[N(t)]$, it is plausible to believe that the limit of $M(t)/t$ is the same as that given in (6.32). In contrast to the Lemma 6.11, however, this is *not a consequence* of Proposition 6.10, as seen in the following example. The proof that $M(t)/t$ reaches γ in the limit will be given in Proposition 6.17.

Example 6.12 ($Y_n \xrightarrow{a.s.} Y \nRightarrow E[Y_n] \longrightarrow E[Y]$) Consider the two point distribution

$$P[Y_n = y] = \begin{cases} n/(n+1), & y = 0, \\ 1/(n+1), & y = n+1. \end{cases}$$

It is clear that for all n, $E[Y_n] = 1$ and thus

$$\lim_{n \to \infty} E[Y_n] = 1.$$

It is also the case that $Y_n \xrightarrow{a.s.} 0$ or, written equivalently, that

$$P\left[\lim_{n \to \infty} Y_n = 0\right] = 1.$$

Thus the value of the random variable and its expectation differ in the limit.

For some densities we can write an explicit expression for $M(t)$. Consider the following example.

Example 6.13 (M(t) for Poisson Processes) For a Poisson process with parameter λ we have already shown that interarrival times are exponentially distributed and from (5.58) have derived that if X is exponential with parameter λ then $\mathcal{L}_X(s) = \lambda/(\lambda + s)$. Thus we have that

$$M^*(s) = \frac{\lambda/(\lambda + s)}{s(1 - \lambda/(\lambda + s))}$$

$$= \frac{\lambda}{s^2}.$$

Inverting this (see (5.57)) yields

$$M(t) = \lambda t. \tag{6.38}$$

Equation (6.38) is intuitive since each renewal epoch has an average length of $E[X] = 1/\lambda$ and thus t time units, on average, contain $t/E[X] = \gamma t$ renewals (note that $\lambda = \gamma$ for the Poisson process). This line of argument, however, leads to the plausible conclusion that

$$M(t) = \gamma t \tag{6.39}$$

holds for all renewal processes. Renewal theory is known for results that run counter to one's intuition and, in fact, we will later show that the only continuous renewal process for which (6.39) holds is the Poisson process. As a simple counterexample, consider the following example of renewal processes.

Example 6.14 (M(t) for Erlang Renewal Processes) Let X be an Erlang random variable with parameters $(2, 2\lambda)$. Notice that (5.9) implies that $E[X] = 1/\lambda$ and from (5.64) that

$$\mathcal{L}_X(s) = \left(\frac{2\lambda}{2\lambda + s}\right)^2.$$

Substituting this into (6.36) and simplifying yields

$$M^*(s) = \frac{4\lambda^2}{s^2(4\lambda + s)}.$$

To invert this we observe that in Example 5.33 we encountered a similar transform. Using the inversion derived there (5.66) yields

$$\frac{1}{4}\left[4\lambda t - (1 - e^{-4\lambda t})\right] \iff \frac{1}{4}\left[\frac{(4\lambda)^2}{s^2(4\lambda + s)}\right]$$

and thus

$$M(t) = \lambda t - \frac{1 - e^{-4\lambda t}}{4}. \tag{6.40}$$

Observe that in this example $M(t) \neq \gamma t$ (note that $\gamma = \lambda$ for the example) and thus the supposition that (6.39) always holds is false. To gain insight into why this is the case we will consider the renewal density function.

6.3.3
Renewal Density

Equation (6.39) is satisfied if the renewal process has a constant *arrival rate*, which mathematically means that the probability of having a renewal in the interval $(t, t + \Delta t)$ is the same for all t. Hence the Poisson process of Example 6.13 satisfies (6.39) but the Erlang process of Example 6.14 does not.

To make this more precise define the *renewal density* as

$$m(t) \stackrel{\text{def}}{=} \frac{dM(t)}{dt}. \tag{6.41}$$

The name "renewal density" is not to be confused with a probability density function but rather should be interpreted in physical terms where density is a measure of "mass per unit space." In this interpretation mass is analogous to renewals and space is analogous to time. This interpretation is clear when we write (6.41) as

$$m(t) = \lim_{\Delta t \to 0} \frac{M(t + \Delta t) - M(t)}{\Delta t}. \tag{6.42}$$

For small Δt we can write $M(t + \Delta t) - M(t) = E\left[I_{(t, t+\Delta t)}\right]$, where

$$I_{(t, t+\Delta t)} = \begin{cases} 1, & \text{if } (t, t + \Delta t) \text{ contains a renewal,} \\ 0, & \text{if } (t, t + \Delta t) \text{ does not contain a renewal.} \end{cases}$$

Writing (6.41) as (6.42) shows that $m(t)\Delta t$ is the probability of a renewal occurring in the interval $(t, t + \Delta t)$. This explains why we call $m(t)$ a density; it can be thought of as a measure of the denseness of renewals on the time axis. We can now more precisely define what we mean when we say that a renewal process is stationary. We will call a renewal process *stationary* if the value of the renewal density $m(t)$ is independent of the value of t. In Proposition 6.15 we will prove that this implies that $m(t) = \gamma$ for all t if X is continuous. As with the derivation of the Poisson process, given in Section 6.2, a stationary limit results when $t \to \infty$.

6.3.4 Limiting Renewal Density

An important result can now be stated for the asymptotic value of $m(t)$.

Proposition 6.15 *The renewal density defined by (6.41) satisfies*

$$\lim_{t \to \infty} m(t) = \gamma.$$

Proof: Using the single derivative property of Laplace transforms (see Table 5.6) and (6.36) shows that

$$m^*(s) = \frac{\mathcal{L}_X(s)}{1 - \mathcal{L}_X(s)}.$$

Using the final value property of Laplace transforms shows that

$$\lim_{t\to\infty} m(t) = \lim_{s\to 0} sm^*(s)$$

$$= \lim_{s\to 0} \frac{s\mathcal{L}_X(s)}{1 - \mathcal{L}_X(s)}. \tag{6.43}$$

Since X is a random variable and thus has a density that integrates to 1 we have that $\mathcal{L}_X(0) = 1$ and thus the limit in (6.43) yields the indeterminate form $0/0$. Using l'Hospital's rule yields

$$\lim_{t\to\infty} m(t) = \lim_{s\to 0} \frac{s\frac{d\mathcal{L}_X(s)}{ds} + \mathcal{L}_X(s)}{-\frac{d\mathcal{L}_X(s)}{ds}} = \gamma.$$

■

6.3.5 Limiting Expected Number of Renewals per Unit Time

We can now establish that the limiting rate of renewals equals γ for all renewal processes. To do this we must first prove a technical lemma:

Lemma 6.16 *Suppose that $\lim_{\tau\to\infty} g(\tau)$ exists and that $\int_0^\tau g(t)dt < \infty$ for all finite values of τ. Then*

$$\lim_{\tau\to\infty} g(\tau) = \lim_{\tau\to\infty} \frac{1}{\tau}\int_0^\tau g(t)dt.$$

Proof: Let $\lim_{\tau\to\infty} g(\tau) = g$ and thus for any value $\epsilon > 0$ there exists a value $\tau(\epsilon)$ such that $|g(\tau) - g| < \epsilon$ for $\tau > \tau(\epsilon)$. We can write

$$\int_0^\tau g(t)dt - \int_0^{\tau(\epsilon)} g(t)dt = \int_{\tau(\epsilon)}^\tau g(t)dt. \tag{6.44}$$

For $\tau > \tau(\epsilon)$ we can write

$$(\tau - \tau(\epsilon))(g - \epsilon) \le \int_{\tau(\epsilon)}^\tau g(t)dt \le (\tau - \tau(\epsilon))(g + \epsilon).$$

Substituting (6.44) into this and simplifying yields

$$g - \epsilon \le \frac{\int_0^\tau g(t)dt - \int_0^{\tau(\epsilon)} g(t)dt}{\tau - \tau(\epsilon)} \le g + \epsilon.$$

Since ϵ is arbitrary, letting $\tau \to \infty$ proves the lemma. ■

Proposition 6.17 *The expected rate of renewals, $M(t)/t$ satisfies*

$$\frac{M(t)}{t} \longrightarrow \gamma, \quad \text{as} \quad t \longrightarrow \infty.$$

Proof: The proof follows from an application of Lemma 6.16. The conditions for the lemma follow from the definition of the renewal density function, (6.41), Proposition 6.15, and Lemma 6.11, which shows that $\int_0^t M(t)dt < \infty$ for any finite value of t. ∎

Proposition 6.17 is an extremely important result, since it shows that γ is the expected rate of renewals for all renewal processes. In terms of applications it establishes the intuitive fact that the average arrival rate of a renewal source is the inverse of the expected interarrival time.

6.3.6 Nonstationary and Nonindependent Increments

If a renewal process has a constant rate then it satisfies $m(t) = \gamma$. This equation, however, holds only for Poisson processes. In the next example we provide insight into why this is true by considering an Erlang renewal process.

Example 6.18 (Nonindependent and Nonstationary Increments)
Consider the Erlang process of Example 6.14. Differentiation of (6.40) yields

$$m(t) = \lambda(1 - e^{-4\lambda t}) \tag{6.45}$$

and thus $m(t) \neq \gamma = \lambda$, for finite t. Consider increments $[0, t)$ and $[t, 2t)$ and let N_1 and N_2 denote the number of renewals occurring in these respective increments. The fact that the renewal density function (6.45) depends on time implies that the distribution of the number of renewals found in the two increments differ and thus the process has nonstationary increments.

We can also show that N_1 and N_2 are dependent random variables. To establish this let X be an Erlang random variable with parameters $(2, 2\lambda)$. The distribution function of X is given by

$$F_X(t) = \int_0^t 4\lambda^2 x e^{-2\lambda x} dx = 1 - e^{-2\lambda t}\{1 + 2\lambda t\}. \tag{6.46}$$

We can establish dependency between N_1 and N_2 by showing that the events $\{N_1 = 0\}$ and $\{N_2 = 0\}$ are not independent.

It is clear that $\{N_1 = 0\}$ and $X > t$ are equivalent events and thus, using (6.46), that

$$P[N_1 = 0] = P[X > t] = (1 + 2\lambda t)e^{-2\lambda t}.$$

Let Υ, $0 \leq \Upsilon < t$, be a random variable that denotes the time of the last renewal epoch in the first increment $[0, t)$. We set $\Upsilon = 0$ if there are no renewals in this first epoch, that is, if $N_1 = 0$. Observe that the event $\{N_1 \neq 0\}$ implies that Υ lies in the range $0 < \Upsilon < t$. Using (6.46) we can write

$$P[N_2 = 0] \quad = \quad P[X > t + (t - \Upsilon)]$$

$$= \quad E_\Upsilon\left[(1 + 2\lambda(2t - \Upsilon))e^{-2\lambda(2t - \Upsilon)}\right]. \qquad (6.47)$$

From (6.47) it is clear that the probability of the event $\{N_2 = 0\}$ depends on the value of Υ, which implies that it depends on the value of N_1. In particular

$$P[N_2 = 0 \mid N_1 = 0] \neq P[N_2 = 0 \mid N_1 \neq 0],$$

and thus N_1 and N_2 are dependent random variables.

The above example shows that the Erlang renewal process has nonstationary increments (as implied by the time dependency of its renewal density function) and does not have independent increments. It is important to make such distinctions when analyzing renewal processes, and we record these observations as follows:

Nonstationary Increments: Renewal processes do not, in general, have stationary increments. In Section 6.4 we show how to construct a stationary renewal process.

Nonindependent Increments: In general, the number of renewals found in nonoverlapping time intervals are not independent random variables. This is true regardless of whether the process is stationary.

6.3.7 Stationary and Independent Increments $\Longrightarrow$ Poisson Process

In this section we show that the Poisson process is the only renewal process having independent and stationary increments. To establish this result, assume that N is a stationary renewal process and suppose that there is a renewal epoch at time t and no further renewals in the interval $(t, t + \tau_1)$. The probability that the first renewal after $t + \tau_1$ occurs after time $t + \tau_1 + \tau_2$ equals

$$P[X > \tau_1 + \tau_2 \mid X > \tau_1] = \frac{\overline{F_X}(\tau_1 + \tau_2)}{\overline{F_X}(\tau_1)}. \qquad (6.48)$$

If the renewal process has independent increments, it necessarily follows that the event of no arrivals in $(t, t + \tau_1)$ is independent of the event of no arrivals in $(t + \tau_1, t + \tau_1 + \tau_2)$. Hence it must be the case that

$$P[X > \tau_1 + \tau_2 \mid X > \tau_1] = \overline{F_X}(\tau_2)$$

and thus, from (6.48), that

$$\overline{F_X}(\tau_1 + \tau_2) = \overline{F_X}(\tau_1) \ \overline{F_X}(\tau_2).$$

Comparing this with (4.86) and the arguments in Section 4.5.9 shows that the only continuous distribution that satisfies this is the exponential distribution. Thus for increments to be independent it must be the case that the renewal random variables are exponential. This provides the intuition why the Poisson process is the only continuous renewal process that has independent increments. Next we discuss how one can dispense with letting $t \to \infty$ as a mechanism to achieve a stationary process.

6.4 Equilibrium Renewal Processes

Consider a renewal process where the first renewal has a different distribution than subsequent renewals. Let the first renewal time be given by random variable X_e and let subsequent renewal times be distributed as X, $X_i =_{st} X, i = 2, 3, \ldots$. As before set $\gamma \stackrel{\text{def}}{=} 1/E[X]$. Such a renewal process is called a *delayed renewal process*, and we are interested in a particular case where the density of X_e is given by[4]

$$f_{X_e}(x) = \gamma \overline{F_X}(x). \tag{6.49}$$

Observe that from the generalized form of the survivor representation of expectation (see (5.31)) it follows that

$$E\left[X_e^k\right] = \gamma \frac{E\left[X^{k+1}\right]}{k+1},$$

and thus the assumption that X has finite moments implies that X_e also has finite moments. The Laplace transform of X_e is given by

$$f_{X_e}{}^*(s) = \gamma \frac{1 - F_X{}^*(s)}{s}. \tag{6.50}$$

For reasons that will soon become obvious we term the process just defined an *equilibrium renewal process*. In the following we subscript all renewal quantities associated with the equilibrium process with an "*e*." Thus $N_e(t)$ is the renewal process just defined and $N(t)$ is the usual renewal process where the first renewal has the same distribution as X.

Intuitively, since $N_e(t)$ and $N(t)$ differ only in the distribution of the first renewal epoch, it is plausible that they share similar properties. In

---

[4]We will later show that this is the density for the residual life of a randomly selected point t (see (6.58)) for a stationary process (previously obtained by letting $t \to \infty$). By definition, an equilibrium process starts out with a renewal length equal in distribution to the residual life. Intuitively, we can imagine that such a process is created by letting t grow extremely large so that the stationary distribution of the residual life is achieved and then reindexing t from 0 so that the residual life is the first renewal epoch. Viewed in this manner it is not surprising that an equilibrium process is stationary from the start of the process.

particular, we expect that $N_e(t)/t$ reaches the same limiting value as $N(t)/t$ and thus that

$$\frac{N_e(t) - N(t)}{t} \xrightarrow{a.s.} 0, \quad \text{as } t \longrightarrow \infty,$$

which implies that $N_e(t)/t \xrightarrow{a.s.} \gamma$. This result can be established similarly to the proof of Proposition 6.10. We also expect that $M_e(t)/t$ and $M(t)/t$ reach the same limiting value. From Proposition 6.17 this implies that $M_e(t)/t \to \gamma$ as $t \to \infty$. In fact, we can prove a stronger result of this type for equilibrium renewal processes.

Proposition 6.19 *Consider an equilibrium renewal process. Then*

$$M_e(t) = \gamma t, \quad t \geq 0.$$

Proof: To prove this we first establish an equation for $M_e(t)$. Let $\hat{F}_n$ be the convolution of F_{X_e} with the $n - 1$st-fold convolution of F_X, that is,

$$\hat{F}_n(x) \stackrel{\text{def}}{=} F_{X_e} \circledast \underbrace{F_X \circledast F_X \circledast \cdots \circledast F_X}_{n - 1 \text{ times}}, \quad n = 2, 3 \ldots,$$

and let $\hat{F}_1(x) = F_{X_e}(x)$. Similar to the derivation of (6.35) we have that

$$M_e(t) = \sum_{n=1}^{\infty} \hat{F}_n(t),$$

which, on using (6.50), yields

$$\begin{aligned}
M_e^*(s) &= \frac{\gamma(1 - \mathcal{L}_X(s))}{s^2} \sum_{n=1}^{\infty} \{\mathcal{L}_X(s)\}^{n-1} \\
&= \frac{\gamma}{s^2}.
\end{aligned}$$

This can be easily inverted, yielding

$$M_e(t) = \gamma t, \quad t \geq 0. \tag{6.51}$$

∎

It is clear from (6.41) and Proposition 6.19 that the renewal density function of equilibrium renewal processes, $m_e(t)$, is simply a constant,

$$m_e(t) = \gamma. \tag{6.52}$$

Recall that in Example 6.13 we showed that (6.51) was satisfied for Poisson processes (see (6.38)) but not for Erlang processes (see (6.40) of Example 6.14). We are now in the position to precisely explain the disparity between these two results: the Poisson process is an equilibrium process whereas the Erlang process is not. This is readily manifest in the fact that for the Poisson process $R =_{st} X_1 =_{st} X$ (see Proposition 6.6).

6.5 Joint Distribution of the Age and Residual Life

The derivation of the joint distribution of the age and residual life of the binomial process was relatively straightforward because the probability of having a renewal in all slots was equivalent to a coin toss with probability p. We implicitly used this uniformity in deriving the Poisson process since this process results as the limit of a binomial process. In terms of renewal functions, this uniformity is manifest in Proposition 6.15, which shows that the renewal density equals γ for large t, and thus the probability of having an arrival in an interval of Δt is given by $\gamma \Delta t$ for all renewal processes. We now use this observation to derive the joint distribution of the age and residual life.

Our derivation here will be abbreviated since it closely mimics the derivation given previously for the Poisson process. Recall that we denote the joint random variable of the age and residual time for point t by $Z_t \stackrel{\text{def}}{=} (A_t, R_t)$. As before, there are two cases to consider; $A_t = t$ and $A_t < t$. If $A_t = t$ then the first renewal occurs after time t. Analogous to the similar case for the Poisson process, (6.20), we can write

$$f_{Z_t}(t, y) = f_X(t + y), \quad 0 \le y. \tag{6.53}$$

To calculate the second case, where $A_t < t$, we observe that $A_t = x$ and $R_t = y$ occur if there is a renewal epoch at $t - x$ that has length $x + y$. The probability that there is a renewal point in $(x, x + dx)$ is approximated by $m(t - x)dx$. Similarly the probability that the renewal epoch has length contained in the interval $(x + y, x + y + dy)$ is approximated by $f_X(x + y)dy$. We can thus write

$$f_{Z_t}(x, y)dx dy \approx m(t - x)dx f_X(x + y)dy. \tag{6.54}$$

As both dx and dy approach zero we have

$$f_{Z_t}(x, y) = m(t - x)f_X(x + y). \tag{6.55}$$

We are interested in the random vector Z_t as $t \to \infty$, denoted by Z. The first case given by (6.53) vanishes in the limit, and application of Proposition 6.15 to the second case (6.55) yields the main result

$$f_Z(x, y) = \gamma f_X(x + y). \tag{6.56}$$

The density for the age at time t can be found by integrating (6.54) over y and hence we obtain

$$f_{A_t}(x) = m(t - x) \int_0^\infty f_X(x + y)dy = m(t - x)\overline{F_X}(x), \quad x < t.$$

Similarly the stationary density, obtained from (6.56), is given by

$$f_A(x) = \gamma \int_0^\infty f_X(x + y)dy = \gamma\overline{F_X}(x). \tag{6.57}$$

The joint distribution $f_Z(x, y)$ is a symmetric function, and thus the density function of the stationary residual life must be the same as (6.57),

$$f_R(x) = \gamma\overline{F_X}(x). \tag{6.58}$$

It is thus clear that $A =_{st} R$ as intimated in Section 6.1.2. It is also clear that the stationary joint distribution of an equilibrium renewal process, Z_e, is given by

$$f_{Z_e}(x, y) = \gamma f_X(x + y).$$

Example 6.20 (Age for Uniform Renewal Process) Consider a renewal process where X is uniformly distributed over $[0, 1]$ and thus $E[X] = 1/2$ and $F(x) = x$. The age of a randomly selected point has density

$$f_A(x) = 2(1 - x)$$

and therefore has expectation

$$E[A] = 2 \int_0^1 x(1 - x) = \frac{1}{3}.$$

6.5.1 Stationary Process for all Time

Comparing (6.49) with (6.58) shows that the first renewal of an equilibrium process has the same distribution as the residual life. Consider now creating a renewal process, N_s, over the interval $(-\infty, \infty)$ defined so that time 0 corresponds to a point within renewal epoch X_0, which starts at a point distributed as $-A$ and ends at a point distributed as R. The remaining renewal epochs, $X_{-1}, X_{-2}, \ldots$, and $X_1, X_2, \ldots$, are distributed as the generic random variable X.

In such a construction, point 0 is equivalent to a randomly selected point! The age and residual life of a point t that is selected independently of the renewal process is independent of t (see Problem 6.14). Specifically, $A_t =_{st} A$ and $R_t =_{st} R$ for any t that is selected independently of the process. Thus N_s is a stationary process for all time.

**6.5.2
Moments for
the Age and
Residual Life**

Calculation of the moments of the age and residual life distributions is facilitated by use of the survivor representation of expectation (5.31). We only have to consider the age since the residual life is identically distributed. Straightforward use of (5.31) yields

$$E\left[A^k\right] = E\left[R^k\right] = \frac{E\left[X^{k+1}\right]}{(k+1)E\left[X\right]}. \tag{6.59}$$

The first moment can also be written as

$$E\left[A\right] = E\left[R\right] = \frac{1}{2}\left(E\left[X\right] + \frac{\sigma_X^2}{E\left[X\right]}\right) = \frac{E\left[X\right]}{2}(1 + C_X^2). \tag{6.60}$$

It is interesting to observe from (6.59) the kth moment of the age depends on the $k+1$st moment of the generic renewal random variable and, in particular, that (6.60) shows that $E\left[A\right]$ increases with the variance of the generic random variable. This increase in the "order" of the dependency for the age and residual life is an extremely important property of renewal processes and will be used to study queueing systems (e.g., see (7.25)).

**6.5.3
Length Biasing
for General
Renewal
Processes**

Suppose we consider the distribution of the length of the renewal epoch intersected by the point t. Denote this random variable by L_t and denote the limiting random variable as $t \to \infty$ by L. Naively one might believe that, since the interval is made up of an age and residual period, that $L_t =_{st} A_t + R_t$. This is incorrect because, as previously mentioned, A_t and R_t are not necessarily independent random variables. It is the case, however, that $E\left[L_t\right] = E\left[A_t\right] + E\left[R_t\right]$ and thus, using (6.60), that the expected length of the renewal epoch intersected by a randomly selected point is

$$
\begin{aligned}
E\left[L\right] &= E\left[X\right] + \frac{\sigma_X^2}{E\left[X\right]} & (6.61) \\[2mm]
&= E\left[X\right](1 + C_X^2) & (6.62) \\[2mm]
&= \frac{E\left[X^2\right]}{E\left[X\right]}. & (6.63)
\end{aligned}
$$

This is a key result since, as indicated in Section 6.2.6, it shows that the expected length of the intersected interval for continuous renewal processes is at least $E\left[X\right]$ (6.61), increases with the variability of X (6.62), and depends only on the first two moments of X (6.63).

The density of L can be obtained from (6.56) by integrating over all (x, y) such that $x + y$ is a constant, that is,

$$f_L(\ell) = \gamma \int_0^\ell f_Z(x, \ell - x)dx$$

$$= \gamma \ell f_X(\ell). \tag{6.64}$$

The bias toward intersecting larger length intervals is readily apparent in the "$\ell f_X(\ell)$" factor of (6.64), which can be interpreted as the product of the size of the interval, ℓ, by its relative frequency, $f_X(\ell)$. The moments of L are given by

$$E\left[L^k\right] = \frac{E\left[X^{k+1}\right]}{E\left[X\right]}, \quad k = 1, 2, \ldots, \tag{6.65}$$

and comparison to (6.59) shows that

$$E\left[L^k\right] = (k+1)E\left[A^k\right] = (k+1)E\left[R^k\right], \quad k = 1, 2, \ldots \,.$$

Example 6.21 (Examples of Length Distributions) For X exponential with parameter λ we obtain

$$f_L(x) = \lambda^2 x e^{-\lambda x},$$

and thus the length is an Erlang random variable with parameters $(2, \lambda)$. This is readily apparent from Proposition 6.6, which for this case shows that the age and residual life are exponentially distributed and independent.

For the Erlang case considered in Example 6.14, we have that

$$f_L(x) = 4\lambda^3 x^2 e^{-2\lambda x},$$

and for the uniform random variable case considered in Example 6.20, we find that

$$f_L(x) = 2x, \quad 0 \le x \le 1.$$

6.6 Alternating Renewal Processes

A useful generalization of a renewal process arises when renewal intervals consist of the sums of other random variables. It suffices to consider the simple case where $X_n = Y_n + Z_n, n = 1, 2, \ldots$, where we call Y_n the first and Z_n the second phase of the nth renewal interval (see Problem 6.17). The sequence X_n is independent and identically distributed, but we allow the random variables (Y_n, Z_n) to have any joint distribution as long as $E\left[Y_n\right] + E\left[Z_n\right]$ is finite. For alternating renewal processes we are typically interested in the probability that the process is in a certain phase at a particular point in time.

To derive an expression for this, let $\pi_1(t)$ be the probability that at time t the process is in the first phase. Consider the following two cases: the first renewal takes place after time t and the first renewal takes place prior to time t. In the first case it is clear that the process is still in phase 1 at time t and this occurs with probability $\overline{F_Y}(t)$. In the second case a renewal point occurs in the interval $(t - x, t - x + dx)$ and the length of the first phase is larger than x. For a particular value of x, $x < t$, such an event occurs with probability $m(t - x)dx\overline{F_Y}(x)$. Combining the two cases yields

$$\pi_1(t) = \overline{F_Y}(t) + \int_0^t m(t - x)\overline{F_Y}(x)dx. \tag{6.66}$$

Example 6.22 (Alternating Poisson Process) Let Y_n and Z_n be independent and exponentially distributed random variables with parameters λ and μ, respectively. Such a process is called an *alternating Poisson process*. To determine an explicit expression for $\pi_1(t)$ we first determine an expression for the renewal density function. It is clear that

$$\mathcal{L}_X(s) = \frac{\lambda\mu}{(\lambda + s)(\mu + s)}.$$

Using this in (6.36) and simplifying yields

$$M^*(s) = \frac{\lambda\mu}{s^2(\lambda + \mu + s)}.$$

To invert this, first write

$$\frac{\lambda\mu}{\lambda + \mu}t \iff \frac{\lambda\mu}{(\lambda + \mu)s^2}, \qquad (\lambda + \mu)e^{-(\lambda+\mu)t} \iff \frac{\lambda + \mu}{\lambda + \mu + s},$$

and thus

$$\begin{aligned}
M(t) &= \frac{\lambda\mu e^{-(\lambda+\mu)t}}{\lambda + \mu} \int_0^t (\lambda + \mu)xe^{(\lambda+\mu)x}dx \\
&= \frac{\lambda\mu}{\lambda + \mu}t - \frac{\lambda\mu}{(\lambda + \mu)^2}\left(1 - e^{-(\lambda+\mu)t}\right),
\end{aligned}$$

where we have used the derivation in Example 5.33 to evaluate the integral. Differentiation yields

$$m(t) = \frac{\lambda\mu}{\lambda + \mu}\left(1 - e^{-(\lambda+\mu)t}\right).$$

Substituting this into (6.66) and simplifying yields

$$\begin{aligned}
\pi_1(t) &= e^{-\lambda t} + \frac{\lambda\mu}{\lambda + \mu}\int_0^t \left(1 - e^{-(\lambda+\mu)(t-x)}\right)e^{-\lambda x}dx \\
&= \frac{\mu}{\lambda + \mu} + \frac{\mu}{\lambda + \mu}e^{-(\lambda+\mu)t}.
\end{aligned}$$

It is interesting to observe in the above example that

$$\pi_1 \overset{\text{def}}{=} \lim_{t\to\infty} \pi_1(t) \tag{6.67}$$

has the value

$$\pi_1 = \frac{\mu}{\lambda+\mu}.$$

To gain an intuition for this result, observe that we can rewrite this limit as

$$\begin{aligned}
\pi_1 &= \frac{\mu}{\lambda+\mu} \\
&= \frac{1/\lambda}{1/\lambda + 1\mu}. \tag{6.68}
\end{aligned}$$

Equation (6.68) shows that the limiting probability of being in the first phase of an alternating renewal process equals the average length of the first phase divided by the expected length of a renewal interval. This makes intuitive sense and, in fact, holds generally for any alternating renewal process as the following proposition shows.

Proposition 6.23 *Let N be an alternating renewal process as defined above where $X_n = Y_n + Z_n$. Then*

$$\pi_1 = \frac{E[Y]}{E[Y]+E[Z]},$$

where π_1 is defined in (6.67).

Proof: Observe that Proposition 6.15 shows that

$$\lim_{t\to\infty} m(t) = \frac{1}{E[Y]+E[Z]}.$$

Taking the limit of (6.66) and using the fact that the expected value of Y is the integral of its survivor function (5.31) yields the proposition. ■

Example 6.24 (Loss Systems) Consider a single server queue that does not have a waiting room. Customers that arrive to the system when the server is busy are assumed to be lost. Assume that arrivals come from a Poisson source with rate λ and that the expected service time is μ. We claim that this system is an alternating renewal process. The system alternates between being busy and idle and the times when the server is busy are independent and identically distributed. For the process to be an alternating renewal process it must also be the case that the times when the server is idle are independent and identically distributed. Consider a point in time immediately after a service completion, say t_s, and let $t_a < t_s$ be the time of the last customer arrival to the system (this can be either the customer just served or one that was lost). Let X be an exponential random variable that corresponds to the interarrival time of the first arrival past t_a. The service time acts as an interruption of X and, since X is memoryless, this implies that the time until the next arrival after service completion is exponential with parameter λ. Clearly the set of idle times is mutually independent and hence the process is an alternating renewal process. Proposition 6.23 thus implies that if we look at the system at a random point in time then the probability of seeing the system in phase 1 is given by $\mu/(\lambda + \mu)$. It is important to note that if we relax the assumption of Poisson arrivals then the process is not an alternating renewal process because renewal intervals are no longer independent.

6.6.1 Time Averages

In this section we show that Proposition 6.23 permits a different interpretation in terms of time averages. To establish this, let $P_1(t)$ be the fraction of time over t time units that the renewal process is in the first phase of an alternating renewal process. Define $\ell_1(t)$ by

$$\ell_1(t) \overset{\text{def}}{=} \min\left\{Y_{N(t)+1}, t - S_{N(t)}\right\}.$$

In words, $\ell_1(t)$ is the total time spent in phase 1 over the interval $(S_{N(t)}, t)$. The total time spent in phase 1 up to time t, denoted by $L_1(t)$, is thus given by

$$L_1(t) = Y_1 + Y_2 + \cdots + Y_{N(t)} + \ell_1(t).$$

Similarly $t - S_{N(t)} - \ell_1(t)$ is the total time spent in phase 2 over the interval $(S_{N(t)}, t)$ and the total time spent in phase 2 up to time t is given by

$$L_2(t) = Z_1 + Z_2 + \cdots + Z_{N(t)} + t - S_{N(t)} - \ell_1(t).$$

We can write the total time as

$$t = L_1(t) + L_2(t).$$

The average amount of time spent in phase 1 can be thus written as

$$P_1(t) = \frac{L_1(t)}{L_1(t) + L_2(t)}.$$

Similar to the proof of Proposition 6.10, it follows from the strong law of large numbers that

$$\frac{L_1(t)}{N(t)} \xrightarrow{a.s.} E[Y], \quad \text{as } t \longrightarrow \infty,$$

and a similar limit holds for $L_2(t)$. Thus we conclude that

$$P_1(t) \xrightarrow{a.s.} \frac{E[Y]}{E[Y] + E[Z]}, \quad \text{as } t \longrightarrow \infty.$$

Comparing this to Proposition 6.23 shows that the limiting time average of being in phase 1 is the same as the probability of finding the renewal process in the first phase at a randomly selected time. This is an important point, and it provides additional intuition as to why Proposition 6.23 is valid.

6.7 Generalizations for Lattice Random Variables

All of the preceding results assume that X is a continuous random variable. In this section we briefly describe how results generalize when this is not the case. To illustrate differences that arise when X is not continuous, consider a degenerate renewal process where $X = E[X]$. It is clear that

$$N(t) = M(t) = \lfloor t/E[X] \rfloor, \tag{6.69}$$

and the renewal density function defined by (6.41) is not defined because M is not differentiable. Equation (6.69) implies that

$$\lim_{t \to \infty} \frac{N(t)}{t} = \lim_{t \to \infty} \frac{M(t)}{t} = \gamma$$

and thus Propositions 6.10 and 6.17 continue to hold. The age and residual life distributions for the degenerate process are given by

$$A_t = t - E[X] \lfloor t/E[X] \rfloor$$

and

$$R_t = E[X] \lceil t/E[X] \rceil - t$$

and are related by $A_t + R_t = E[X]$. Since the age and residual life depend on t for all t the process is not stationary.

The degenerate process is a special case of a process where X has a *lattice* distribution, so named because all the probability mass lies on a lattice set of points. Precisely, a distribution is lattice if there exists a value d so that $\sum_{n=0}^{\infty} P[X = nd] = 1$. The value of the maximum d that satisfies this is called the *period*. For such distributions $M(t)$ is not differentiable anywhere. As with the degenerate case, however, for lattice processes the limiting values of $N(t)/t$ and $M(t)/t$ given in Propositions 6.10 and 6.17, respectively, continue to hold. Our proof of Proposition 6.10 for the limiting value of $N(t)/t$ is actually valid for lattice distributions but our proof of Proposition 6.17 for the limiting value of $M(t)/t$ is not since it assumes that $M(t)$ is differentiable (Problem 6.18 proves the proposition more generally). As with the degenerate process, lattice processes generally do not have stationary versions since for all times random variables A_t and R_t are not independent of t. For such processes $A_t \geq t - d\lfloor t/d \rfloor$, which clearly depends on t.

We should point out that discrete time processes are a special type of lattice process where renewals can take place only on integer values. Recall that in Section 6.1 we studied the binomial process in which time was slotted into unit intervals. This is a discrete time process that is examined in interval slots and has a period of 1. If we define $m(t)$ to equal the difference $M(t) - M(t-1)$ for such processes then it can be shown that

$$\lim_{t \to \infty} m(t) = \gamma \tag{6.70}$$

(see Problem 6.19). Furthermore, a stationary process is obtained when $t \to \infty$. Expressed in terms of our previous definitions, we can now see that the binomial process is a discrete time equilibrium process and as $t \to \infty$ it is stationary and has independent age and residual life distributions that are geometric.

6.8 Summary of Chapter 6

In this chapter we established several properties of Poisson processes that are frequently used to model arrival sources. As shown in the chapter, a Poisson process arises as a limiting form of a binomial process and, more generally, is obtained from the superposition of many renewal processes, each with small rate. We will frequently make use of the properties of Poisson processes in the rest of the text. For example, in Chapter 10 we will derive properties of queueing networks that consist of queues that have the property that they produce Poisson departure processes when their arrival streams are Poisson processes. Such queues are called quasi-reversible queues; they play a key role in establishing results of product form networks.

In Table 6.1 we summarize some fundamental results that we derived for the Poisson, general renewal and equilibrium renewal processes. These results will be used later when studying queueing systems. Expected values of basic renewal random variables are given in Table 6.2. We can summarize our results by considering two renewal processes with generic renewal epochs given by X and Y, respectively. If $E[X] = E[Y]$ then we have shown that both processes have the same asymptotic rate of renewals γ and have the same asymptotic renewal density given by γ. This last result implies that for both processes the probability of finding a renewal in the period $(t, t + \Delta t)$ is $\gamma \Delta t$ for sufficiently large t and small Δt. This constant arrival rate implies that both processes have stationary increments for large t but it does not imply that they have

Property/Value	Poisson Process	General Renewal	Equilibrium Renewal
$N(t)/t$	$\xrightarrow{a.s.} \lambda$	$\xrightarrow{a.s.} \gamma$	$\xrightarrow{a.s.} \gamma$
$M(t)/t$	λ	$\longrightarrow \gamma$	γ
$m(t)$	λ	$\longrightarrow \gamma$	γ
Independent increments	Yes	No	No
Stationary increments	Yes	No	Yes
Splitting	Invariant	Not invariant	Not Invariant

TABLE 6.1. Summary of Results for Renewal Processes

independent increments. Increments, for continuous renewal processes, are independent only for Poisson processes. If, furthermore, $E[X^2] = E[Y^2]$ then, regardless of higher moments, we have shown that both processes have the same expected values for the age, residual life, and length of a renewal intersected by a randomly selected point in time. The fact that higher moments do not enter into calculations of this type is an important result in terms of modeling.

Variable	Approximation/Expected Value	Density/Transform
$M(t)$	$\approx \gamma t$	$\mathcal{L}_X(s)/s(1 - \mathcal{L}_X(s))$
$m(t)$	$\approx \gamma$	$\mathcal{L}_X(s)/(1 - \mathcal{L}_X(s))$
Age A	$E[X^2]/2E[X]$	$\gamma \overline{F_X}(x)$
Residual life R	$E[X^2]/2E[X]$	$\gamma \overline{F_X}(x)$
Length L	$E[X^2]/E[X]$	$\gamma x f_X(x)$

TABLE 6.2. Values of Renewal Random Variables

Bibliographic Notes

Çinclar [12] has a thorough exposition of properties of the Poisson process (see Chapter 4). An approach to renewal theory based on the key renewal theorem can be found in Ross [101]. An exposition based on the renewal density function is found in Cox [23]. Other books on stochastic processes with sections on renewal theory include Heyman and Sobel [49], Karlin and Taylor [57], Medhi [81], Trivedi [118] and Wolff [125]. Our presentation of renewal theory was greatly influenced by Cox [23] and Ross [101].

6.9
Problems for Chapter 6

Problem Related to Section 6.1

6.1 [M2] Establish the identity used in equation (6.3) of Section 6.1.4

$$\sum_{n=1}^{k} \alpha^n \binom{k-1}{k-n} = (1+\alpha)^{k-1}\alpha. \tag{6.71}$$

Problems Related to Section 6.2

6.2 [M3] Consider a renewal sequence given by $U_n^i, i = 1, 2, \ldots$, where U_n^i are independent and uniformly distributed random variables over the set $\{1, 2, \ldots, n\}$. Calculate the joint distribution of the age and residual life random variables (A_t, R_t) for the point t. What are their limiting values?

6.3 [R2] Let X be a generic random variable of a stationary renewal process and let A be a random variable that denotes the age of a randomly selected point. Show that $A =_{st} X$ if and only if X is memoryless.

6.4 [M2] Let $N(t)$ count the number of arrivals in t time units from a Poisson source. Show that the distribution of $N(s)$ conditioned on $\{N(t) = n\}$, for $s < t$, is binomial with parameters $(n, s/t)$.

6.5 [M3] Derive (6.13) and (6.14) by determining the limiting process obtained from a nonhomogenous binomial process where slot $n_\epsilon = t/\epsilon$ is marked with probability $\lambda(t)\epsilon$.

6.6 [M3] In this exercise we derive an expression for the density of Ψ so that the density of $X(t)$ of (6.16) can be evaluated. Recall that $X(t)$ is a random variable that denotes the number of arrivals in t seconds from a Poisson process that is modulated by random variable Ψ. Let M be a random variable that denotes the number of state transitions of the modulating process that occur within t time units.

1. Show that the density of M is given by

$$P[M = m] = \begin{cases} (\mu t)^{m-1} e^{-\mu t} / (m-1)!, & m \geq 1, \\ e^{-\mu t}, & m = 0. \end{cases} \tag{6.72}$$

2. Use the fact that the initial state of the modulating process is selected with uniform probability to argue that, for $\Upsilon = t$, the density has an impulse at t and thus can be written as

$$f_\Upsilon(x) = \frac{1}{2} e^{-\mu x} \delta_0(x - t) + g_\Upsilon(x), \quad 0 \leq x \leq t, \tag{6.73}$$

where the first (resp., second) term corresponds to the case where there are no (resp., some) transitions of the modulating process.

3. Show that if there are an even number of transitions then g_Υ is given by

$$g_\Upsilon(\tau \mid M = 2\ell) = \frac{\mu^{2\ell} \left(\tau(t-\tau)\right)^{\ell-1}}{\left((\ell-1)!\right)^2} e^{-\mu t}, \quad \ell \geq 1, \;\; 0 < \tau < t, \qquad (6.74)$$

4. Show that if there are an odd number of transitions then

$$g_\Upsilon(\tau \mid M = 2\ell + 1) = \frac{\mu^{2\ell+1} \left(\tau(t-\tau)\right)^{\ell-1} t}{2\ell!(\ell-1)!} e^{\mu t}, \quad \ell \geq 1, \;\; 0 < \tau < t. \qquad (6.75)$$

5. Use (6.72) to uncondition (6.74) and (6.75) and thus, for $0 < \tau < t$,

$$g_\Upsilon(\tau) = \sum_{\ell=1}^{\infty} \{g_\Upsilon(\tau \mid M = 2\ell + 1) P\left[M = 2\ell + 1\right]$$

$$+ \, g_\Upsilon(\tau \mid M = 2\ell) P\left[M = 2\ell\right]\}. \qquad (6.76)$$

6. Use this and (6.73) to uncondition (6.16) and write

$$P\left[X(t) = n\right] = \int_0^t P\left[X(t) = n \mid \Upsilon = \tau\right] f_\Upsilon(\tau). \qquad (6.77)$$

Notice that although (6.77) cannot be written in closed form, it can easily be evaluated numerically by truncating the summation in (6.76) and using numerical integration.

6.7 [R2] Show that the limiting expressions of (6.17) and (6.18) are correct.

6.8 [M3] Provide the details leading to equations (6.20) and (6.21).

Problems Related to Section 6.3

6.9 [M2] Derive the joint density of (A_t, R_t) using the same technique as that used in the derivation of the distribution of L_t in Section 6.5.

6.10 [E2] We were careful in the text when calculating convolutions to distinguish the cases where densities and distributions were being convolved. Show that if two distribution functions F_1 and F_2 are convolved using the form given by (6.26), that is,

$$F(t) = \int_0^\infty F_1(t - x) F_2(x) dx,$$

then the resultant function is not necessarily a distribution function.

6.11 [M2] Using the notation defined in (6.27), show that

$$M(t) \leq \inf_k \frac{k}{1 - F_k(t)}.$$

6.12 [**M2**] Let N be a renewal process where renewal epochs are Erlang with parameters (m, λ). Show that

$$P\left[N(t) = n\right] = \sum_{i=nm}^{nm+m-1} \frac{(\lambda t)^i e^{-\lambda t}}{i!}.$$

6.13 [**M3**] Consider a renewal process with a renewal point at time 0. Is $M(t)/t$ a monotone function in t?

Problems Related to Section 6.4

6.14 [**M2**] Let X denote a generic renewal epoch and let A and R denote generic random variables distributed as the age and residual life of X. Consider the following process defined over the infinite time interval $(-\infty, \infty)$. The epoch that contains time 0, denoted by X_0, is constructed to start at a random point in time which is distributed as $-A$ and to end at a random point, that is distributed as R. Epochs that occur prior to and after epoch X_0 are distributed as X. Show that the process so constructed is a stationary renewal process. Do this by showing that the distributions of A_t and R_t are independent of t for all possible values of t.

6.15 [**M3**] (**Continuation**) Let N_e be the equilibrium Erlang renewal process of Problem 6.12. Show that

$$P\left[N_e(t) = n\right] = \frac{e^{-\lambda t}}{m} \left\{ \sum_{i=nm}^{nm+m-1} \frac{(nm + m - i)(\lambda t)^i}{i!} \right.$$

$$\left. - \sum_{i=nm-m}^{nm-1} \frac{(i - nm + m)(\lambda t)^i}{i!} \right\}, \qquad n = 1, 2, \ldots$$

and

$$P\left[N_e(t) = 0\right] = \frac{e^{-\lambda t}}{m} \sum_{i=0}^{m-1} \frac{(m - i)(\lambda t)^i}{i!}.$$

Hint: Show that

$$P\left[N_e(t) = n\right] = \gamma \int_0^t \left(P\left[N(x) = n - 1\right] - P\left[N(t) = n\right] \right) dx.$$

Problem Related to Section 6.5

6.16 [**R2**] Let X be a random variable denoting the length of a generic renewal epoch and let L be the length of a renewal epoch intersected by a randomly selected point in time. Prove or disprove the relationship $C_L^2 \leq C_X^2$.

Problem Related to Section 6.6

6.17 [**M3**] In this problem we consider a generalization of the alternating renewal process considered in Section 6.6. Let N be a renewal process where $X_i, i = 1, 2 \ldots,$ are independent and identically distributed random variables. Suppose that $X_i = Y_{1,i} + Y_{2,i} + \cdots + Y_{\ell,i}$ where $(Y_{1,i}, Y_{2,i}, \ldots, Y_{\ell,i})$ have some joint distribution but are independent over i. Derive an expression for $\pi_i, i = 1, 2, \ldots, \ell$, the distribution of finding the process in phase i as $t \to \infty$.

Problems Related to Section 6.7

6.18 [**M3**] In this problem we provide a proof that

$$\lim_{t \to \infty} \frac{M(t)}{t} = \gamma,$$

for all distributions of X, the generic renewal epoch.

1. Show that $N(t) + 1$ is a stopping time for the sequence S_n, $n = 1, 2, \ldots$.
2. Show that this implies that

$$E\left[X_1 + X_2 + \cdots + X_{N(t)+1}\right] = E\left[X\right](M(t) + 1).$$

   (Hint: Use Wald's inequality.)
3. Show that

$$\frac{M(t)}{t} > \gamma$$

   and thus that

$$\liminf_{t \to \infty} \frac{M(t)}{t} \geq \gamma.$$

   (Hint: Show that $X_1 + X_2 + \cdots + X_{N(t)+1} = S_{N(t)+1} > t$.)
4. Let renewal epochs be determined by modified random variables $X_n' = \min\{X_n, C\}$ for some constant C (quantities that are primed correspond to the modified renewal process). Show that

$$t + C \geq E\left[X'\right](M'(t) + 1)$$

   and thus that

$$\limsup_{t \to \infty} \frac{M(t)}{t} \leq \frac{1}{E\left[X'\right]}.$$

5. Let $C \to \infty$ to complete the proof.

6.19 [**M2**] Consider a discrete time renewal process defined on the integers with period 1. Show that the limit in (6.70) is satisfied.

7

The M/G/1 Queue

In this chapter we analyze a simple single server queue that is frequently used to model components of computer systems. This queue is termed the *M/G/1* queue. This is standard "queueing notation," first introduced by Kendall. Typically a queue is described by four variables

$$A/S/k/c,$$

which have the following interpretation:

A,S - The arrival (A) or service (S) process where M means Poisson arrivals and exponential service times, G means the process is generally distributed, E_ℓ denotes an ℓ-stage Erlang distribution, and D denotes a deterministic distribution.

k - The number of servers.

c - The buffer size of the system. This is the total number of customers the system can hold. Arrivals to the system already containing c customers are assumed to be lost. If not specified, c is assumed to be infinite.

For the M/G/1 queue, the "M" indicates Poisson arrivals, the "G" indicates general service times (i.e., service times that are not necessarily exponential), and the "1" indicates a single server. An M/M/k/c queue has Poisson arrivals, exponential service times, k servers, and space for a maximum of c customers.

All the mathematical tools required in this analysis of the M/G/1 queue have been established in the previous chapters, and this chapter illustrates the power of these tools to derive a wide spectrum of important and useful results. We start with some elementary systems theory and establish three fundamental properties of systems.

7.1 Elementary Systems Theory

In this section we establish three important properties or laws of systems that we will apply to the M/G/1 queue analyzed in this chapter. It is important to note that the laws established here hold for systems more general than the M/G/1 queue. We view a system in terms of its input and output properties; customers arrive to the system, spend time there, and then leave. We do not need to specify how the customers are managed within the system. A requirement we have is that the system does not create or destroy customers. Customers that arrived to the system before a given time t must either have departed at some prior time $t' \leq t$ or must still reside in the system.

Basic quantities that are of interest include the average time spent in the system by a customer and the average number of customers in the system. To relate these and other quantities, we first introduce some notation and then establish two fundamental laws, Little's law and the level crossing law. These laws hold for both physical and probabilistic systems. We then establish a third law that holds only for probabilistic systems and requires the arrival process to be Poisson. Our derivations of these laws is at an intuitive level. Precise proofs for some of the results of this section are delicate and require arguments that lie outside the scope of the text.

It is convenient to express performance measures in terms of time, and we let $A(t)$ be the number of customer arrivals to the system in the time interval $[0, t]$, $D(t)$ to be the number of customer departures in $[0, t]$, and $N(t)$ to be the number of customers resident in the system at time t. This last quantity can be expressed as

$$N(t) = A(t) - D(t) + N(0),$$

where $N(0)$ is the number of customers in the system at time 0. If not explicitly specified we assume that the system is initially empty, that is, that $N(0) = 0$. To avoid some technical difficulties later, we also assume that the system operates in a way so that $N(t) = o(A(t))$. Hence even if the number in the system grows without bound it does so in a manner such that

$$\lim_{t \to \infty} \frac{N(t)}{A(t)} = 0.$$

If we start with an empty system then clearly $A(t) \geq D(t)$ for all t. We define the average arrival rate of customers by time t, $\overline{A}(t)$, by

$$\overline{A}(t) \stackrel{\text{def}}{=} \frac{A(t)}{t}, \tag{7.1}$$

and let T_i denote the response time (the total amount of time spent in the system) of the ith customer.

7.1.1
Time Averages
and Customer
Averages

We wish to distinguish between two types of averages: time averages and customer averages. The *time average* of the number of customers in the system until time t is the quantity

$$\overline{N}(t) \overset{\text{def}}{=} \frac{1}{t} \int_0^t N(x)dx.$$

If we think of each customer as accumulating one second of residence time for each second they spend in the system, then $\overline{N}(t)$ is the average residence time of customers per unit time until time t. This average is taken with respect to *time*.

The *customer average* response time until time t is the quantity

$$\overline{T}(t) \overset{\text{def}}{=} \frac{1}{A(t)} \sum_{i=1}^{A(t)} T_i.$$

The value of $\overline{T}(t)$ is the average response time of customers that have arrived to the system by time t (some of these customers may still be resident in the system at time t, but we still count their total response times in $\overline{T}(t)$). This average is taken with respect to *customers*.

7.1.2
Almost Sure
Limits

The results derived in this section are expressed in terms of limits and, as Table 4.3 shows, there are different types of limits that are defined for random variables. Since the first two laws that we will define are applicable to both physical and probabilistic systems it is natural to express limits for the probabilistic case in terms of almost sure convergence. This type of convergence most closely corresponds to the convergence of real numbers, which is used to express the laws for the physical case. You may recall that almost sure convergence was defined in Definition 5.39. We write $X_n \overset{a.s.}{\longrightarrow} c$ if $P\left[\lim_{n\to\infty} |X_n(\omega) - c| \geq \epsilon\right] = 0$ for all $\epsilon > 0$. Intuitively, this implies that the only random experiments ω for which the sequence of random values $X_n(\omega)$ does not converge to c have a collective probability of 0. Such sequences are not necessarily impossible but are so rare that they might as well be ignored in practice. For a particular random experiment ω, the sample path $X_n(\omega)$ for $n = 0, 1, \ldots$ is a sequence of real numbers. Although these values depend on the particular experiment ω, practically all experiments yield a sequence that converges to c. In this sense the convergence of $X_n(\omega)$ to c is analogous to the convergence of a sequence of real numbers.

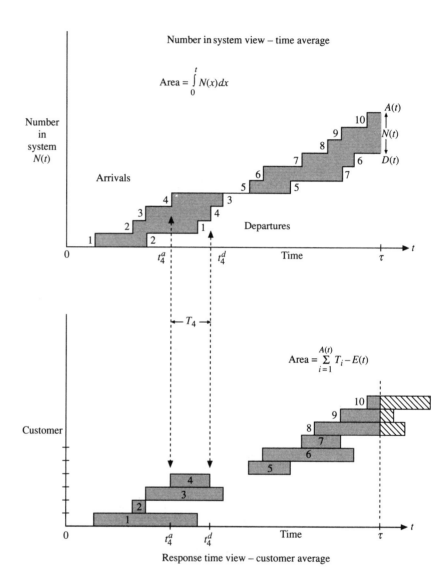

FIGURE 7.1. Little's Law

We assume that arrivals and departures of customers occur throughout time and that $A(t) \xrightarrow{a.s.} \infty$ and $D(t) \xrightarrow{a.s.} \infty$ as $t \to \infty$. We define the following limits, which are assumed to exist. The average arrival rate of customers to the system, λ, is defined by

$$\overline{A}(t) \xrightarrow{a.s.} \lambda, \quad \text{as } t \longrightarrow \infty. \tag{7.2}$$

The average number of customers in the system, $\overline{N}$, is defined by

$$\overline{N}(t) \xrightarrow{a.s.} \overline{N}, \quad \text{as } t \longrightarrow \infty, \tag{7.3}$$

and similarly the average customer response time, $\overline{T}$, is defined by

$$\overline{T}(t) \xrightarrow{a.s.} \overline{T}, \quad \text{as } t \longrightarrow \infty. \tag{7.4}$$

Little's law, our first result, mathematically relates the time average $\overline{N}(t)$ and customer average $\overline{T}(t)$.

7.1.3
Little's Law

In the upper plot of Figure 7.1, we plot a particular sequence of arrivals and departures for 10 customers that arrive to a hypothetical system. The arrival times of these customers, which are labeled 1 through 10, correspond to the upward jumps on the $A(t)$ curve, and the departure times for the customers are the labeled upward jumps on the $D(t)$ curve. Notice that departures are not necessarily in the same order as arrivals, for example, customer 4 departs prior to customer 3. Also notice that at time τ, labeled in the figure, customers 8, 9, and 10 are still resident in the system. From the figure, it is clear that the area until a given time t that is bounded by the two curves is given by

$$\text{Area}(t) = \int_0^t (A(t) - D(t))dt = \int_0^t N(t)dt. \tag{7.5}$$

The upper plot represents a view of the number of customers in the system. We can equivalently plot the activity of the system from a customer's point of view.

The lower plot in Figure 7.1 is the view as seen by a customer in terms of its response time. To explain this plot, observe that the box corresponding to the forth customer starts at the time of its arrival and ends when it leaves the system. The length of the box equals T_4 and, by definition, it is 1 unit high. The boxes associated with customers 8, 9, and 10 are depicted differently in the figure to show that these customers are still in the system at time τ. It is clear that the total area of boxes that strictly lie before a given time t equals $\text{Area}(t)$. We can equivalently write this in terms of customer response times as

$$\text{Area}(t) = \sum_{i=1}^{A(t)} T_i - E(t), \tag{7.6}$$

where $E(t)$ accounts for the portion of a customer's response time that lies past time t. For the example given in Figure 7.1, $E(\tau)$ is the "hatched" area of customers 8, 9, and 10 that corresponds to the portion of their residence times that fall past time τ. The value of $E(t)$ equals 0 if there are no customers in the system at time t. For example, if we evaluate $E(t)$ after the departure of customer 3 but prior to the arrival of customer 5 then the error term would equal 0.

Equating (7.5) and (7.6), and using (7.1) and some minor rearrangement implies that

$$\overline{N}(t) = \overline{A}(t)\overline{T}(t) - \frac{E(t)}{t}. \qquad (7.7)$$

If customers have finite response time, that is, if they eventually leave the system, then we expect $E(t)$ to be small with respect to t for large t. In other words, we expect that $E(t)/t \to 0$ as $t \to \infty$. Assume that this is the case. Then taking the limit $t \to \infty$ in (7.7) and using (7.2), (7.3), and (7.4) yields an important relationship called Little's law [80]:

$$\overline{N} = \lambda\overline{T}, \qquad \textbf{Little's Law}. \qquad (7.8)$$

As stated in (7.8), Little's law is a relationship that holds for a deterministic sequence of values. Specifically, the values of $\overline{A}(t)$, $\overline{N}(t)$, and $\overline{T}(t)$ and their limiting values λ, $\overline{N}$, and $\overline{T}$, respectively are deterministic and depend on a particular sequence of arrivals and departures to the system. Typically we are interested in applying Little's law to a system that is governed by distributions that determine the random arrival and departure sequences. Consider such a system with random variables N and T corresponding to the number of customers in the system and the response time of a typical customer. Performance measures can be expressed in terms of the expected values of these random variables, and it is desirable to use Little's result to establish a relationship between $E[N]$ and $E[T]$. Little's law, as given by (7.8), suggests that

$$E[N] = \lambda E[T], \qquad (7.9)$$

where, as before, λ is the arrival rate of customers to the system. To establish (7.9) we must establish that $E[N] = \overline{N}$ and $E[T] = \overline{T}$. Conditions for when these equalities are satisfied lie outside our scope. We can, however, provide a simple intuitive argument that establishes the equalities. We will sketch such an argument for the first equality.

We can write

$$E[N] = \sum_{n=0}^{\infty} nP[N = n].$$

Let $\ell_n(t)$ be the total amount of time in the interval $[0,t]$ that $N(t)$ equals n, that is, it is the total time in the first t time units that the system contains n customers. It is clear that $t = \sum_{n=0}^{\infty} \ell_n(t)$ and that we can write

$$\int_0^t N(x)dx = \sum_{n=0}^{\infty} n\ell_n(t).$$

Suppose that $\ell_n(t)/t$ reaches a limit (in an almost sure sense) as $t \to \infty$. This limit corresponds to the fraction of time the system has n customers and intuitively equals $P[N = n]$, that is, one expects that

$$\frac{\ell_n(t)}{t} \xrightarrow{a.s.} P[N = n] \quad \text{as} \quad t \to \infty. \tag{7.10}$$

If (7.10) holds then we can write

$$
\begin{aligned}
\lim_{t\to\infty} \frac{1}{t} \int_0^t N(x)dx &= \lim_{t\to\infty} \sum_{n=0}^{\infty} n \frac{\ell_n(t)}{t} \\
&= \sum_{n=0}^{\infty} n \lim_{t\to\infty} \frac{\ell_n(t)}{t} \\
&= \sum_{n=0}^{\infty} nP[N = n] = E[N].
\end{aligned}
$$

A similar argument establishes that $E[T] = \overline{T}$ and thus (7.9) follows from Little's law given in the form (7.8). It is typical to refer to both (7.9) and (7.8) as "Little's law," since for computer performance models both equalities typically hold.

Little's law was used as a folk theorem long before there was a formal proof (see [122] for a more rigorous proof and a recent summary of its characteristics). The law requires that customers are not created or lost within the system. Arrivals could arrive from external sources and depart to an external sink (called an *open* system) or, after leaving the system, recirculate back as new arrivals (called a *closed* system). Loosely speaking, if the number of customers in the system does not grow without bound so that all customers that arrive to the system eventually leave the system then Little's law applies. In the case of an open queueing system, it is usual to assume that the law holds if the average arrival rate to the system is less than the average rate at which customers can be served. For a single server queue this implies that the average time between successive arrivals, the expected *interarrival time*, is greater than the average service time and thus, intuitively, an infinite queue does not build up (there are pathological counterexamples to this supposition).

In such cases Little's law is applicable. Consider the following examples of Little's law as applied to queueing networks.

Example 7.1 (An Open Queueing Network and Little's Law) Consider the open queueing network, depicted in Figure 1.3, that has three service centers corresponding to a CPU, a floppy disk drive, and a hard disk drive. The average arrival rate into the system is λ. We index performance measures for the CPU and the floppy and disk drive by the indices 0, 1, and 2, respectively, and assume that Little's law holds for the system and for each queue. Applying Little's result to the entire system implies that

$$E[N] = \lambda E[T], \tag{7.11}$$

where $E[N]$ is the average number of customers resident in the system and $E[T]$ is the average customer response time. Let λ_i and $E[N_i]$, for $i = 0, 1, 2$, be the average arrival rate and average number in queue i, respectively, and let $E[R_i], i = 0, 1, 2$, be the average time spent in queue i before departing from the system. Because some customers are fed back through the system multiple times (this is shown by the CPU–I/O loop in the figure) the average arrival rate into the CPU is larger than that the exogenous arrival rate, that is, $\lambda_0 > \lambda$ and the total time spent in a particular queue by a customer is composed of multiple visits. After customers are served by the CPU they are routed to either the floppy disk or the hard disk. If we assume that the arrival rate of customers into the CPU is less than its service rate and, similarly, assume that the arrival rate into each of these I/O servers is smaller than their respective service rates, then it must be the case that $\lambda_0 = \lambda_1 + \lambda_2$. Little's law, applied to each of the service centers, implies that

$$E[N_i] = \lambda_i E[R_i], \quad i = 0, 1, 2. \tag{7.12}$$

Since a customer that is resident in the system is resident in one of the queues we have that

$$E[N] = E[N_0] + E[N_1] + E[N_2].$$

Thus using (7.11) and (7.12) implies that

$$E[T] = \frac{\lambda_0}{\lambda} E[R_0] + \frac{\lambda_1}{\lambda} E[R_1] + \frac{\lambda_2}{\lambda} E[R_2]. \tag{7.13}$$

A customer's expected response time is determined by the arrival rate into each queue and the expected amount of time it spends in each of the queues prior to departure from the system. In Chapter 10 we establish techniques for calculating the response times of (7.13).

Example 7.2 (A Multiclass Closed Queueing Network and Little's Law) Consider an extension of the model analyzed in Example 7.1 consisting of a two-class closed queueing network with a CPU, a floppy disk, and a hard disk drive. To analyze the system we extend the notation presented in Example 7.1 to include a class index. Hence $\lambda_i^j, i = 0, 1, 2, j = 1, 2$, is the average arrival rate of class j customers into service center i (where 0, 1, and 2 correspond to the CPU, the floppy drive, and the hard disk drive, respectively). The average arrival rate of customers into the terminal center is denoted by γ. Applying Little's law to the entire system implies that

$$N = \gamma E\left[T\right], \tag{7.14}$$

where N is the total number of customers in the system (this is a fixed quantity and thus we do not write it as an expectation). In (7.14), $E\left[T\right]$ is the average customer response time (this is averaged over both classes of customers). If we define γ^i, for $i = 1, 2$, to be the average arrival rate of customers of class i into the terminals and N^i and $E\left[T^i\right]$ to be the number and average class i response time then it is clear that Little's law implies that

$$N^i = \gamma^i E\left[T^i\right],$$

and

$$N = N^1 + N^2, \qquad \gamma = \gamma^1 + \gamma^2.$$

These expressions imply that

$$E\left[T\right] = \frac{\gamma^1}{\gamma} E\left[T^1\right] + \frac{\gamma^2}{\gamma} E\left[T^2\right].$$

7.1.4
Level Crossing
Law

Consider a system in which customers arrive to and leave from the system one at a time. Upon arrival, a customer records the number of customers already in the system (not including itself), and likewise at departure a customer records the number of customers left behind in the system (again not including itself). We are interested in calculating the fraction of arriving and departing customers that "see" i customers in the system, variables denoted by a_i and d_i, respectively. To calculate an expression relating these quantities, let $A_i(t)$ be the number of arriving customers in $[0, t]$ that find the system with i customers and, similarly, let $D_i(t)$ be the number of departing customers that leave behind i other customers. The fraction of customers that see i customers in the system equals $A_i(t)/A(t)$ and the fraction that leave i customers behind equals $D_i(t)/D(t)$. We are interested in determining the limiting values of a_i and d_i defined by (we assume these limits exist)

$$\frac{A_i(t)}{A(t)} \xrightarrow{a.s.} a_i, \quad \text{as } t \longrightarrow \infty, \quad i = 0, 1, \ldots,$$

and

$$\frac{D_i(t)}{D(t)} \xrightarrow{a.s.} d_i, \quad \text{as } t \longrightarrow \infty, \quad i = 0, 1, \ldots.$$

Observe that

$$1 = \sum_{i=0}^{\infty} a_i = \sum_{i=0}^{\infty} d_i$$

and thus it is also appropriate to think of a_i and d_i as being probabilities.

An arrival to the system with i customers, for $i = 0, 1, \ldots$, causes a transition in the system from i to $i + 1$ customers. We denote such a transition by $i \mapsto i + 1$ and say that the transition *crosses level i in the upward direction*. Similarly a departure from state $i + 1$, for $i = 0, 1, \ldots$, leading to transition $i + 1 \mapsto i$ is said to *cross level i in the downward direction*. Since arrivals and departures occur one at a time, the number of times level i is crossed in the upward and downward directions can differ by at most 1, that is, the number of $i \mapsto i + 1$ transitions can differ by at most one from the number of $i + 1 \mapsto i$ transitions. Crossing level i in the upward direction corresponds to an arrival seeing i customers in the system and, similarly, crossing it in the downward direction corresponds to a departing customer seeing i customers. It thus follows that

$$|A_i(t) - D_i(t)| \leq 1, \quad i = 0, 1, \ldots . \tag{7.15}$$

Simple algebra shows that

$$\frac{A_i(t)}{A(t)} - \frac{D_i(t)}{D(t)} = \frac{A_i(t) - D_i(t)}{A(t)} - \frac{D_i(t)}{D(t)} \frac{N(t)}{A(t)}. \tag{7.16}$$

In light of (7.15), the limiting value of the first term of the right-hand side of (7.16) is zero. This follows from the assumptions that $A(t) \xrightarrow{a.s.} \infty$ and $D(t) \xrightarrow{a.s.} \infty$ as $t \to \infty$ and $A(t) \geq D(t)$. The second term has a limit of zero since, by assumption, $N(t) = o(A(t))$ and $D_i(t)/D(t)$ is bounded. Hence we conclude that

$$a_i = d_i, \quad i = 0, 1, \ldots, \quad \textbf{Level Crossing Law.} \tag{7.17}$$

Note that the level crossing law does not imply that the fraction of *time* that the system has i customers equals a_i. The fractions taken in this law are with respect to the *number of customers* rather than to time. It is clear that a time average generally differs from a customer average. As an example, consider a system that processes customers extremely fast when there is more than one customer in the system and not at all when there is exactly one customer. It would then be the case that the *fraction of time* the system has i customers is close to 0 for $i \neq 1$ and close to 1 for $i = 1$. The level crossing law would hold, however, for all i.

The level crossing law can be used to simplify the derivation of queue length statistics for single server first-come-first-served queues. Analysis is often more straightforward if one calculates the number of customers that are seen by a departing customer. These customers are exactly those that have arrived during the departing customer's time in the system (also called its *sojourn time*). The level crossing law shows that the distribution seen by a departing customer is the same as that seen by an arriving customer. Since the customers already resident in the system are those that delay the arriving customer, one can use the distribution seen by an arrival to calculate response time statistics. We will use this technique when deriving queue length and response time statistics for the M/G/1 queue.

**7.1.5
Poisson Arrivals
See Time
Averages**

In this section we compare two different views of the system state: the view as seen by an arriving customer and that as seen by a random point in time. Our objective is to show that these views are statistically identical if arrivals are Poisson. In general, these different viewpoints lead to different statistical outcomes as illustrated by the following example.

Example 7.3 (Arrival Times are Not Random Times) Consider a single server queue with arrivals at each integer time unit. Assume that customers require exactly $1/2$ time unit of service. Let τ be a selected point in time and let $M(\tau)$ denote the number of customers in the system just prior to time τ. Note that the fractional portion of τ can be written as $\tau - \lfloor \tau \rfloor$, where $\lfloor x \rfloor$ is the largest integer that is less than or equal to x. The value of $M(\tau)$ is given by

$$M(\tau) = \begin{cases} 0, & \tau - \lfloor \tau \rfloor > 1/2, \\ 1, & \tau - \lfloor \tau \rfloor \leq 1/2. \end{cases}$$

We are interested in calculating the expected number of customers in the system at time τ and to do this we must specify how τ is selected. It is important to note that the selection of τ is under our control.

Let τ_a be a randomly selected arrival point. It is clear that the number of customers already in the system (we do not include the new arrival) is equal to 0 since the model implies that $\tau_a - \lfloor \tau_a \rfloor = 0$. Thus the expected number of customers in the system as "seen" by an arriving customer equals 0.

On the other hand, select τ_r to be a random point in time in the sense specified in Chapter 6. Then the expected number of customers in the system at time τ_r is greater than 0. This follows from the fact that there is a nonzero probability that τ_r occurs during the service of a customer. We conclude that the expected values of $M(\tau)$ differ if τ is selected as an arrival time or a randomly selected time.

Example 7.3 shows that, generally, the statistics seen by an arriving customer differ from those seen by a random point in time. An important result that we derive next shows that these two viewpoints are identical if the arrival process is Poisson; *the statistics of a system as seen by an arriving customer are identical to those seen at a randomly selected time if the arrival process is Poisson.* This is an important property of Poisson arrivals, and it plays a key role in performance modeling.

Consider a system with a Poisson arrival process, Z, with rate λ. We suppose that arrivals occur infinitely far in the past so that for any particular time τ there is a previous arrival with probability 1. Let $M(\tau)$ be a random variable whose value depends on the times of previous arrivals. We assume that $M(\tau)$ is evaluated *just before* the instant τ. Like Example 7.3, we are interested in the statistics of $M(\tau)$ as seen at a customer arrival epoch and by a randomly selected point in time. We argue that if customer arrivals are Poisson, then these two viewpoints are stochastically equivalent, that is, they yield the same expected value for M. As Example 7.3 points out, this equivalence cannot be taken for granted and therefore must arise from special properties of Poisson arrivals. Implicit in our assumptions is that the system under investigation is stationary and thus the statistics of M do not depend on the absolute value of time.

Let τ_a be a randomly selected arrival point and let τ_r be a randomly selected point in time. We first argue that both τ_a and τ_r fall in a stationary portion of the Poisson process Z. This follows by definition of the random point in time, τ_r, and for the random arrival point, τ_a, because a Poisson process is an equilibrium renewal process.

We now argue that the arrival process prior to time τ_a is statistically identical to the arrival process prior to time τ_r. The distribution of the time until the first arrival before τ_a is exponential with rate λ. Similarly, Proposition 6.6 shows that this also is true for the distribution of time until the first arrival before the randomly selected point, τ_r. It is easy to see that equivalence of these first prior arrivals implies that this equivalence holds for all arrivals before these times. We conclude that the arrival process before τ_a is stochastically identical to the arrival process prior to time τ_r. (The Poisson process is the only process for which this holds.)

Since the value of the measure M only depends on the arrival times before time τ, we conclude that Poisson arrivals and random points in time see values of M that are stochastically equal. This fact is expressed by the acronym **PASTA**, which stands for Poisson Arrivals See Time Averages. The name doesn't yet fit with the result we have just derived since we have shown that the statistics of M as seen by a Poisson arrival are the same as those seen by a random point in time rather than as those obtained from a time average. To bridge this gap we must show that the statistics seen by a random point in time and that of a time average are identical. To do this let $m(\tau) \stackrel{\text{def}}{=} E\left[I_{\{M \le m\}}(\tau)\right]$ be the probability that $\{M \le m\}$ at time τ. As in Chapter 6 the value of $m(\tau)$ at a "random point in time" is the value obtained as $\tau \to \infty$. Assume that

$$\lim_{\tau \to \infty} m(\tau) = m.$$

The time average of m is given by

$$m_{\text{avg}} \stackrel{\text{def}}{=} \lim_{\tau \to \infty} \frac{1}{\tau} \int_0^{\tau} m(t)dt.$$

We thus need to show that $m = m_{\text{avg}}$. This, however, follows from an application of Lemma 6.16.

PASTA is one of the most important results in queueing theory. A typical application of it is to follow a customer through the system starting from its arrival all the way to its departure from a queue. By summing up the delays along the customer's route we can derive expressions for the customer's response time. To do this we use PASTA to argue that on arrival the statistics of the queue seen by the customer are identical to the stationary statistics of the queue.

7.2 Expected Values for the M/G/1 Queue

In this section we derive expected values for the M/G/1 queue. These values include the expected waiting and response times as well as the expected number in the queue and in the system. It is perhaps surprising that we can obtain these measures so easily, and a quick review of the following arguments will show that all that is required are the three properties of systems derived in the previous section and the expected value of the residual life of a renewal process as derived in Section 6.5.

We first make some definitions that will be used throughout the text. Let the Poisson arrival rate be denoted by λ, and let customer service times be generally distributed according to generic random variable S. The service rate of the queue is denoted by μ, and this implies that $\mu = 1/E[S]$. We let W and T denote random variables for the waiting time and response time of a randomly selected customer, and as usual, $E[W]$ and $E[T]$ are their expected values. The waiting time is the time spent in the queue before receiving service, and the response time includes this and the service time. Clearly we have

$$E[T] = E[W] + E[S]. \tag{7.18}$$

The number in the queue is denoted by N_q, the number in the server by N_s, and the number in the system by N. The expected values of these random variables satisfies

$$E[N] = E[N_q] + E[N_s]. \tag{7.19}$$

Our object in this section is to derive expressions for all of these expected values. We start by calculating the utilization of the queue.

7.2.1 Utilization of the M/G/1 Queue

The *utilization* of the queue, denoted by ρ, is the fraction of time that the server is busy. To derive an expression for this, let $I_{\{N_s>0\}}$ be an indicator random variable that equals 1 if the server is busy. We can write

$$\rho = E\left[I_{\{N_s>0\}}\right], \tag{7.20}$$

which using property 7 of expectation, can equivalently be written as

$$\rho = P\left[N_s > 0\right]. \tag{7.21}$$

Another expression for this is

$$\rho = E\left[N_s\right], \tag{7.22}$$

which follows from the definition of expectation. In words, the utilization equals the probability that the server is busy or equivalently equals the expected number of customers in service.

To find an expression for this we use Little's result where the "system" consists of the server and hence is limited to have at most one customer. Applying Little's law implies that the average number of customers in the server is given by

$$E\left[N_s\right] = \min\left\{1, \lambda/\mu\right\}. \tag{7.23}$$

If $\rho = 1$ then we say that the queue is *saturated*. Typically for such systems this implies that the number of customers in the queue is unbounded.

7.2.2 Expected Waiting and Response times for the M/G/1 Queue

To derive an expression for the expected waiting and response times for an M/G/1 queue we "tag" a customer and follow it through the queue as it accumulates its expected time in the system. This technique is called *the method of moments* and is widely used to calculate expected response times. Consider a randomly selected arriving customer. We argue that this customer's arrival time can be viewed as a randomly selected time. This follows from the PASTA property of Poisson arrival streams. We thus can write the tagged customer's expected waiting time as

$$E\left[W\right] = E\left[N_q\right]E\left[S\right] + \rho E\left[R\right]. \tag{7.24}$$

The first term in (7.24) accounts for expected total time for the customers in the queue, and the second term, $E\left[R\right]$, is the expected amount of time remaining for the customer in service conditioned on the server being busy. The probability that the server is busy is given by ρ.

Since the arrival time is equivalent to a randomly selected time, the remaining service time can be viewed as that obtained for a renewal sequence consisting of generic random variables S. Hence we recognize $E[R]$ as the expected residual time of this renewal process. With (6.60) we obtain

$$E[R] = \frac{E[S^2]}{2E[S]} = \frac{E[S]}{2}(1 + C_S^2). \tag{7.25}$$

Using Little's theorem (7.9) we have that $E[N_q] = \lambda E[W]$, and thus substituting these equations in (7.24) and using (7.25) shows that

$$E[W] = \frac{\rho}{1-\rho} E[R] \tag{7.26}$$

$$= \frac{\rho E[S](1 + C_S^2)}{2(1-\rho)} \tag{7.27}$$

$$= \frac{\lambda E[S^2]}{2(1-\rho)}. \tag{7.28}$$

Each of these equivalent expressions provides insight into waiting times found in queueing systems. Equation (7.26) shows that the expected residual life of the service process determines the expected waiting time and thus, as evident in (7.28), the expected waiting time depends only on the first two moments of service time. Equation (7.27) shows that the expected waiting time increases with the variability of the service process. Equation (7.18) implies that the expected response time for such an M/G/1 is given by

$$E[T] = E[S] + \frac{\rho E[S](1 + C_S^2)}{2(1-\rho)} \tag{7.29}$$

$$= \frac{E[S]}{1-\rho} \left(1 - \frac{\rho(1 - C_S^2)}{2} \right). \tag{7.30}$$

Little's result immediately implies that

$$E[N_q] = \frac{\rho^2(1 + C_S^2)}{2(1-\rho)}, \tag{7.31}$$

and

$$E[N] = \rho + \frac{\rho^2(1 + C_S^2)}{2(1-\rho)}. \tag{7.32}$$

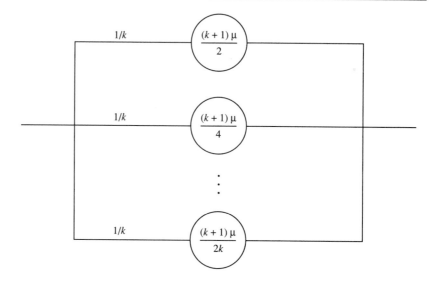

FIGURE 7.2. The Hyperexponential Distribution $H_2(k)$

This last equation is referred to as the *Pollaczek–Khinchin (P–K) mean value formula.*

In Table 7.1 we evaluate these expectations for common queues, the M/D/1 queue that has a coefficient of variation of service time equal to zero, the M/M/1 queue for which $C_S^2 = 1$, and the M/E_k/1 queue with $C_S^2 = 1/k$. We also define a queue with hyperexponential service times, denoted by $H_2(k)$, which is depicted in Figure 7.2. The service time is given by

$$S = \begin{cases} E_{1,k\mu}, & \text{with probability } \frac{k}{k+1}, \\[2ex] E_{1,\mu/k}, & \text{with probability } \frac{1}{k+1}, \end{cases} \tag{7.33}$$

where $E_{1,\alpha}$ denotes an exponential random variable with parameter α. A simple calculation shows that the expectation of this hyperexponential random variable is independent of k and equals

$$E\left[S\right] = \frac{1}{\mu}.$$

The variance of $H_2(k)$ is given by

$$\sigma_S^2 = \frac{k}{k+1}\frac{1}{(k\mu)^2} + \frac{1}{k+1}\left(\frac{k}{\mu}\right)^2$$

Queue	$E[W]$	$E[T]$	$E[N_q]$	$E[N]$
M/G/1	$\frac{\rho E[S](1+c_S^2)}{2(1-\rho)}$	$E[S] + \frac{\rho E[S](1+c_S^2)}{2(1-\rho)}$	$\frac{\rho^2(1+c_S^2)}{2(1-\rho)}$	$\rho + \frac{\rho^2(1+c_S^2)}{2(1-\rho)}$
M/D/1	$\frac{E[S]\rho}{2(1-\rho)}$	$\frac{E[S](2-\rho)}{2(1-\rho)}$	$\frac{\rho^2}{2(1-\rho)}$	$\frac{\rho(2-\rho)}{2(1-\rho)}$
M/M/1	$\frac{E[S]\rho}{1-\rho}$	$\frac{E[S]}{1-\rho}$	$\frac{\rho^2}{1-\rho}$	$\frac{\rho}{1-\rho}$
M/E_k/1	$\frac{E[S]\rho(1+k)}{2k(1-\rho)}$	$\frac{E[S](\rho+k(2-\rho))}{2k(1-\rho)}$	$\frac{\rho^2(k+1)}{2k(1-\rho)}$	$\frac{\rho(\rho+k(2-\rho))}{2k(1-\rho)}$
M/$H_2(k)$/1	$\frac{E[S]\rho(k^2+1)}{2k(1-\rho)}$	$\frac{E[S](\rho(k-1)^2+2k)}{2k(1-\rho)}$	$\frac{\rho^2(k^2+1)}{2k(1-\rho)}$	$\frac{\rho(\rho(k-1)^2+2k)}{2k(1-\rho)}$

TABLE 7.1. Expected Values for Common M/G/1 Queues

$$= \frac{k^3+1}{k(k+1)\mu^2}$$

$$= \frac{k(k-1)+1}{k\mu^2}.$$

Thus the coefficient of variation is given by

$$C_S^2 = \frac{k(k-1)+1}{k}.$$

For $k = 1$ we recover an exponential random variable, and this yields the M/M/1 queue whereas the coefficient of variation increases without bound as $k \to \infty$. Hence between the distributions for E_k and $H_2(k)$ we can cover M/G/1 systems where the coefficient of variation is less than or equal to 1 (M/E_k/1 queues) and greater than or equal to 1 (M/$H_2(k)$/1 queues).

Observe in this table that waiting and response times are multipliers of $E[S]$, whereas there is no such multiplier for the number in the queue or system. This points out a form of "dimensional analysis" where the units of an equation conform to its semantic interpretation. Waiting times and response times are expressed in units of time where a unit corresponds to an average service time, $E[S]$. The number of customers in the queue and in the system, however, are *dimensionless* quantities. Note that these quantities can be viewed in terms of the expected number of customers found in the server, $E[N_s]$. Equation (7.22) shows that this equals ρ.

It is also important to observe the common term, $1/(1 - \rho)$, found in all of the expressions. This term expresses the rate of growth of the queue as a function of its utilization and is a salient feature of open queueing systems. This explains the similarity of the expected response time curves depicted in Figure 7.3. For $\rho > 1$ the equations are meaningless since the queue grows without bound. All of the measures for this case are infinity. Next we make some general observations about M/G/1 queueing systems that provide intuition regarding their response time behavior.

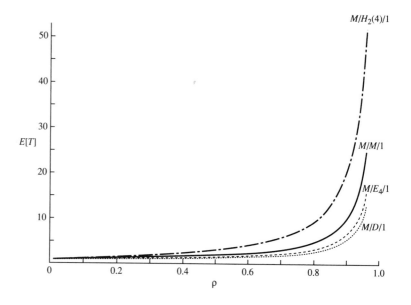

FIGURE 7.3. Expected Response Times for Common M/G/1 Queues

7.2.3 General Observations on Queueing Systems

The M/M/1 queue and the M/D/1 queue are clearly special cases of the $M/E_k/1$ given in the table. When $k = 1$ we obtain the M/M/1 queue whereas Example 5.40 shows that M/D/1 arises as $k \to \infty$. The expected waiting times for $M/E_k/1$ monotonically decrease with k. The extremes have a ratio of

$$\frac{E\left[W_{M/M/1}\right]}{E\left[W_{M/D/1}\right]} = 2 \tag{7.34}$$

and thus an average customer spends twice as much time waiting for service in an M/M/1 queue than in an M/D/1 queue. For the $M/H_2(k)/1$

queue, the expected waiting time increases monotonically with k and, for any fixed utilization ρ, the expected waiting time grows without bound as $k \to \infty$. To see this, form the ratio

$$\frac{E\left[W_{\mathrm{M}/H_2(k)/1}\right]}{E\left[W_{\mathrm{M}/\mathrm{M}/1}\right]} = \frac{k^2 + 1}{2k}. \tag{7.35}$$

It is important to gain an insight into why the variation in service time plays such a key role in M/G/1 queues; to do this we must first understand why customers must wait for service.

Consider the M/D/1 queue. Since the arrival rate, λ, is smaller than the service rate, μ, one might believe there should be no queueing since the server operates faster than the average rate the server receives work. This would certainly be the case if arrivals and service times were both deterministic, that is, in a D/D/1 queue. For Poisson arrivals, however, arrivals occasionally "bunch up" so that the server cannot process them as fast as they come in. This is what causes queueing delays. For example, if service times are one time unit and the system is empty, then the first customer to arrive immediately goes into service. The number of arrivals during the first customer's service time is Poisson with parameter λ. These arrivals, and subsequent arrivals while the server is busy, must wait before receiving service. This process generates a *busy period* that starts when the server becomes busy and ends when it goes idle. Every customer served in the busy period, besides the first one that generated the period, must wait before receiving service.

For an M/M/1 queue, this bunching up is increased by the occurrence of long service times. This increases the probability of many arrivals during a customer's service time and increases the waiting times of future arrivals. This increase can be substantial because waiting times are cumulative. For example, if there are n customers in the queue and the service time of the customer in service is increased by ϵ time units, then the cumulative effect on the sum of waiting times is $n\epsilon$, that is, each customer is delayed for an additional ϵ time units. Therefore, as the queue length increases, the effects of having a large service time are magnified.

This, essentially, is the role of the variability term $1 + C_S^2$ that appears in (7.27). If one fixes $E[S]$ so that the utilization is fixed, then increasing C_S^2 increases the chance of having a large service time hold up many queued customers. As an example, consider a two point distribution where S equals 0 with probability q and equals $1/(1-q)$ with probability $1-q$. In this distribution $E[S] = 1$ independently of q but $C_S^2 = q/(1-q)$. Substituting this into (7.27) shows that

$$E[W] = \frac{\rho}{2(1 - \rho)(1 - q)}. \tag{7.36}$$

For this system the effect of changing the coefficient of variation is dramatic since it essentially has the same influence as that of changing the utilization of the system. As $q \to 1$, a large number of customers have zero service times and, if there is no queue, these customers immediately pass through the system. Occasionally, however, there is a customer with a very long service time, that is, with probability $1 - q$ a customer requires a service time of $1/(1 - q)$ units. This causes arrivals to build up in the queue, and the cumulative effect of waiting times takes over. This effect swamps the benefits received by those customers that enter and immediately leave the system. It is interesting to note here that, as in the $M/H_2(k)/1$ queue, expected waiting times can be infinite even if the utilization of the system is close to zero.

7.2.4 Busy Period for M/G/1 Queues

As just mentioned a busy period starts when the server becomes busy and ends when it goes idle. The queue alternates between idle and busy periods and, in fact, this is an alternating renewal process. To see this, consider the end of a busy period at time t. Since arrivals are independent of the service process, the memoryless property of exponential random variables implies that the time until the first arrival past t is exponentially distributed with rate λ and does not depend on previous arrivals. Hence idle times are independent and exponentially distributed. Similarly busy periods are also independent and identically distributed since each is generated by the first arrival after an idle period. Thus busy and idle periods alternate, and successive alternations for each are independent and identically distributed. This forms an alternating renewal process.

Let $E[B]$ denote the expected busy period and $E[I]$ denote the expected idle time. Our previous comments imply that

$$E[I] = \frac{1}{\lambda}. \tag{7.37}$$

It is also clear that the fraction of time the system is busy is also equal to the fraction of time the server is busy, and thus we can write

$$\rho = \frac{E[B]}{E[B] + E[I]}. \tag{7.38}$$

From (7.21), however, ρ is also equal to the probability that a randomly selected point in time lands in a busy period. Substituting (7.37) into (7.38) yields the following equation for the expected length of a busy period:

$$E[B] = \frac{1}{\lambda} \frac{\rho}{1 - \rho}. \tag{7.39}$$

Using $\rho = \lambda E[S]$ in (7.39) yields

$$E[B] = \frac{E[S]}{1 - \rho}. \qquad (7.40)$$

It is interesting to note here that the expected length of a busy period for the M/G/1 queue does not depend on higher moments of the service time (its distribution does, as will be later shown) and, comparison with Table 7.1 shows that it is identical to the expected response time of an M/M/1 queue! This is a surprising result, but a more surprising result is obtained by comparing (7.40) with (7.30). This shows that if $C_S^2 > 1$ then the expected length of a busy period is *smaller* than the expected customer response time, that is, $E[B] < E[T]$. This is counter-intuitive since it must be the case that the length of a given busy period is *larger* than the response times of all the customers served in that busy period. This would seem to imply that the expected length of a busy period must be at least equal to the expected customer response time since there is at least one customer served in a busy period. The above mathematics shows that this line of reasoning is false and highlights the need for having a formal way to reason about probabilistic systems.

The fallacy in the reasoning lies in the fact that the length of a busy period and the response times of customers served in it are *dependent random variables*. We cannot thus argue about relationships between these random variables separately. Consider, for example, busy periods consisting of one customer. For such busy periods, it is likely that the customer service time comprising the busy period is smaller than a typical service time. This follows from the fact that, by assumption, no arrivals occurred during the customer service time (See Problem 7.8). If there is a large fraction of these busy periods, then the expected length of a busy period will be shorter than the expected customer response time even though there may occasionally be a large busy period. In Section 7.4 we calculate the distribution for the number of customers served in a busy period. This shows that there is a large fraction of busy periods with few customers and that this fraction increases with the coefficient of variation.

To determine the expected number of customers served in a busy period we can argue as follows. Let $N_{b,i}$ be the number of customers in the ith busy period and let $E[N_b]$ be the average number of customers served in a busy period. This is given by

$$E[N_b] = \lim_{n \to \infty} \frac{1}{n} \sum_{i=1}^{n} N_{b,i}. \qquad (7.41)$$

Clearly only the first customer sees the server idle; all subsequent customers in the busy period see the customer busy. Thus, from n busy periods, the probability that a randomly selected customer sees the system idle is given by $n/\sum_{i=1}^{n} N_{b,i}$. From PASTA, the probability that an arriving customer sees the server idle is given by $1 - \rho$. Thus we can write

$$\lim_{n \to \infty} \frac{n}{\sum_{i=1}^{n} N_{b,i}} = 1 - \rho,$$

which using (7.41) implies that

$$E[N_b] = \frac{1}{1 - \rho}. \tag{7.42}$$

We could have argued more directly from (7.40) since each customer has an average service time of $E[S]$ but the above argument provides an interesting alternative derivation. In Section 7.4 we derive an expression for the distribution for the number of customers served in a busy period.

7.3 Distribution of Queue Length for the M/G/1 Queue

The previous derivations of the expected values for the M/G/1 queue used the method of moments to trace a customer through the system while adding the expected delays encountered along its way. The method works because, as the first property of expectation in Table 5.1 shows, the expectation of the sum of random variables is the sum of their expectations. This line of approach can also be used to determine the variance of performance measures *provided that all random delays encountered are independent random variables*. If this is the case, then property 3 of variance (given in Table 5.2) shows that the variance of the sum of random variables is the sum of their variances, and thus the method of moments is applicable. In general, however, the delays encountered by the tagged customer in the system are not independent. For example, in the M/G/1 queue the number of customers in the queue and the residual service time are not independent random variables. A simple way to see this is to assume that the customer found in service has an extremely large service time. This implies that the expected residual service time (note that this is conditioned on the length of the service interval) is also large. The expected queue length, conditioned on the service interval is also large since many customers could have arrived during the service interval. Hence a large residual lifetime implies that the queue length is also large and so the corresponding delays are not independent. To determine the variance of queue length we use a different method, which yields the probability generating function for the number of customers in the queue.

To do this, define V_i to be the number of arrivals to the system during the ith service interval and denote a generic service time by S. Because interarrival times are exponential, and thus memoryless, the time from the end of a service epoch until the next arrival epoch is exponentially distributed. This implies that idle periods are independent and identically distributed and thus that random variables $\{V_1, V_2, \ldots\}$ are independent and identically distributed. (It is important to observe that these statements do not hold for the more general class of G/G/1 queues. Generally in such a queue, idle periods are not independent and identically distributed nor is the set of random variables $\{V_1, V_2, \ldots\}$ independent or identically distributed.) It is clear that the number of customers in the queue at the end of the ith service completion, a random variable denoted by N_i, satisfies[1]

$$N_i = (N_{i-1} - 1)^+ + V_i, \quad i = 1, 2, \ldots, \tag{7.43}$$

where we set $N_0 = 0$.

Assume as $i \to \infty$ that random variable N_i converges almost surely to a random variable N. We can think of N as denoting the number of customers in the system at the end of a service epoch after the system has run for a long time. Let V be a generic random variable for the number of arrivals during a service interval. Let $\hat{N} =_{st} N$ be a random variable that is independent of V. We can then write a "stationary" version of (7.43) as (it is stationary since the equation does not depend on i)

$$N =_{st} \left(\hat{N} - 1 \right)^+ + V.$$

We wish to use this equation to determine the density function of N, and it is convenient to do this using probability generating functions.

Before proceeding with this we make an important observation about the equation. Let N_a and N_d be random variables that denote the number of customers seen by an arriving and a departing customer. Furthermore, let N_r denote the number of customers in the system as seen at a random point in time. It is clear that by definition $N =_{st} N_d$. Because customers arrive and depart from an M/G/1 queue one at a time, the level crossing law implies that $N_a =_{st} N_d$. Since arrivals are Poisson, however, PASTA implies that $N_a =_{st} N_r$. Therefore we conclude

$$N =_{st} N_d =_{st} N_a =_{st} N_r.$$

We can now derive the probability generating function for N.

To calculate the probability generating function for N we use the interpretation of (5.52) to write

$$\mathcal{G}_N(z) \stackrel{\text{def}}{=} E\left[z^N\right].$$

---

[1]Observe that we have seen an equation similar to this in Example 5.24, equation (5.32), where we considered a discrete time single server queue. In Problem 7.9 we lead the student through an alternative derivation of the Pollaczek–Khinchin equation based on this observation.

This implies that

$$E\left[z^N\right] \;=\; E\left[z^{(\hat{N}-1)^{+}+V}\right]$$

$$=\; E\left[z^{(\hat{N}-1)^{+}}\right] E\left[z^V\right],$$

and thus we have that

$$\mathcal{G}_N(z) = \mathcal{G}_{(\hat{N}-1)^+}(z)\,\mathcal{G}_V(z). \tag{7.44}$$

The problem in evaluating (7.44) is to determine an expression for the probability generating functions for random variables $\left(\hat{N}-1\right)^{+}$ and V.

We determine both of these functions by direct calculations starting with $\left(\hat{N}-1\right)^{+}$. From the definition of the probability generating function given in (5.51) we can write

$$\mathcal{G}_{(\hat{N}-1)^+}(z) = \sum_{i=0}^{\infty} P\left[\left(\hat{N}-1\right)^{+} = i\right] z^i. \tag{7.45}$$

To determine an expression for the right-hand side of this equation observe that

$$P\left[\left(\hat{N}-1\right)^{+} = i\right] = \begin{cases} P\left[\hat{N}=0\right] + P\left[\hat{N}=1\right], & i=0, \\[2mm] P\left[\hat{N}=i+1\right], & i=1,2\ldots\,. \end{cases}$$

Substituting this into (7.45) yields

$$\sum_{i=0}^{\infty} P\left[\left(\hat{N}-1\right)^{+} = i\right] z^i \;=\; P\left[\hat{N}=0\right] + \frac{1}{z}\{P\left[\hat{N}=1\right] z$$

$$+ P\left[\hat{N}=2\right] z^2 + \cdots\}$$

$$=\; P\left[\hat{N}=0\right] + \frac{1}{z}\left\{\mathcal{G}_N(z) - P\left[\hat{N}=0\right]\right\}.$$

Since there are no customers in the system if and only if there are no customers in the server we can write

$$P\left[\hat{N}=0\right] = P\left[N_s = 0\right] = 1 - \rho.$$

Substituting this into the preceding equation yields

$$\mathcal{G}_{(\hat{N}-1)^+}(z) \;=\; 1 - \rho + \frac{1}{z}\left\{\mathcal{G}_N(z) - (1-\rho)\right\}$$

$$=\; \frac{\mathcal{G}_N(z) - (1-\rho)(1-z)}{z}. \tag{7.46}$$

Having derived an expression for the first probability generating function on the right-hand side of (7.44) we now derive the second generating function, that for V. Given that a service interval has length x, the distribution of V is Poisson with parameter λx. The probability that a service time is in the interval $(x, x + dx)$, however, is given by $f_S(x)dx$. Thus, using conditional probability, we have

$$
\begin{aligned}
\mathcal{G}_V(z) &= \sum_{i=0}^{\infty} \int_0^{\infty} \frac{(\lambda x)^i}{i!} e^{-\lambda x} f_S(x) dx z^k \\
&= \int_0^{\infty} e^{-\lambda x} \left(\sum_{i=0}^{\infty} \frac{(\lambda x z)^i}{i!} \right) f_S(x) dx \\
&= \int_0^{\infty} e^{-\lambda x (1-z)} f_S(x) dx \\
&= \mathcal{L}_S(\lambda(1-z)).
\end{aligned}
\tag{7.47}
$$

Therefore we can determine the probability generating function of V from the Laplace transform of S.

Substituting (7.46) and (7.47) into (7.44) yields

$$
\begin{aligned}
\mathcal{G}_N(z) &= \mathcal{L}_S(\lambda(1-z)) \frac{\mathcal{G}_{\hat{N}}(z) - (1-\rho)(1-z)}{z} \\
&= \mathcal{L}_S(\lambda(1-z)) \frac{(1-\rho)(1-z)}{\mathcal{L}_S(\lambda(1-z)) - z},
\end{aligned}
\tag{7.48}
$$

which is termed the *Pollaczek–Khinchin (P–K) transform equation*. The Laplace transform of S thus yields the probability generating function of N, the number of customers in the system.

7.3.1 General Observations on the Pollaczek–Khinchin Equation

We can make some simple, but invaluable, observations about the form of (7.48). First, by setting $z = 0$ in this equation, the initial value property of probability generating functions shows that the probability that the system is idle is given by $1 - \rho$. The complete summation property of probability generating functions implies that $\mathcal{G}_N(1) = 1$. Since the numerator of (7.48) has the factor $1 - z$, this implies that the denominator also must have this factor, otherwise at $z = 1$ the expression for $\mathcal{G}_N(z)$ would not be defined.[2] Therefore this implies that we can write

$$
\mathcal{L}_S(\lambda(1-z)) - z = (1-z)Q(z),
$$

---

[2]This follows because the probability generating function for a random variable is an analytic function; see Appendix C.

where $Q(z)$ is a function satisfying $Q(1) \neq 0$. In the following section we will use these observations to invert $\mathcal{G}_N(z)$ for the $M/H_2(k)/1$ system.

Inverting (7.48) yields an expression for $P[N = n], n = 0, 1, \ldots$. We can determine the inversion in two ways. In the first we invert the transform by table lookup or by creating a partial fraction expansion. We will use this technique to find the queue length probabilities of the $M/M/1$ and $M/H_2(k)/1$ queues of Table 7.1. Often it is difficult, if not impossible, to invert the transform using these techniques. In these cases it is convenient to create a power series expansion of $\mathcal{G}_N(z)$ and then equate $P[N = n]$ with the coefficient of z^n. We will use this technique to determine queue length probabilities for the $M/D/1$ and $M/E_k/1$ queue. To do this we express (7.48) in an equivalent form:

$$\mathcal{G}_N(z) = (1 - \rho)(1 - z) \sum_{n=0}^{\infty} \left(\frac{z}{\mathcal{L}_S(\lambda(1 - z))} \right)^n, \qquad (7.49)$$

where we have used the equality $(1 - x)^{-1} = 1 + x + x^2 + \cdots$.

One final observation deals with our derivation of (7.47), which establishes a relationship between V and S. Random variable V is the number of Poisson arrivals at rate λ that occur during a service time distributed as S. Clearly there is nothing special about a service time and, if X is any continuous random variable with a Laplace transform and U is a Poisson process with rate γ, then the number of customer arrivals from U that occur during a sample from X satisfies

$$\mathcal{G}_U(z) = \mathcal{L}_X(\gamma(1 - z)).$$

An important use of this equation is to establish a relationship between the response time and number of customers in an $M/G/1$ queue. Consider a random customer J that is tagged on its arrival to the queue. The customers in the queue at J's departure are those customers that arrived to the queue during J's sojourn time. From the level crossing property this number of customers is stochastically equal to the number of customers in the system at J's arrival time, which by PASTA is stochastically equal to the stationary number of customers in the queue at a random point in time. Denoting this last random variable by N and the generic customer response time by T yields the following fundamental relationship between these random variables:

$$\mathcal{G}_N(z) = \mathcal{L}_T(\lambda(1 - z)). \qquad (7.50)$$

Hence, if one can determine $\mathcal{G}_N(z)$ then one immediately also obtains the Laplace transform of T. This is exactly the method we use to determine the distribution of T. (Observe that Little's law for the $M/G/1$ queue is obtained by differentiating (7.50); see Problem 7.5 for a generalization.)

**7.3.2
Distribution for
Common
M/G/1 Queues**

In this section we derive the distributions of the number in the system for the M/G/1 queues given in Table 7.1. We provide several different solution techniques in this section. These techniques can be used as a guide for possible approches when attempting to determine the stationary distribution of queueing systems encountered in practice.

**7.3.3
The M/M/1
Queue**

If S is exponentially distributed with parameter μ then

$$\mathcal{L}_S(s) = \frac{\mu}{s + \mu}.$$

This implies that

$$\mathcal{L}_S(\lambda(1 - z)) = \frac{\mu}{\lambda(1 - z) + \mu} = \frac{1}{\rho(1 - z) + 1}.$$

Substituting this into (7.48) and simplifying yields

$$\mathcal{G}_N(z) = \frac{1 - \rho}{1 - \rho z}. \tag{7.51}$$

This can be easily inverted since it corresponds to a geometric random variable with parameter $1 - \rho$ and hence

$$P[N = n] = (1 - \rho)\rho^n, \quad n = 0, 1, \ldots . \tag{7.52}$$

To illustrate how one uses a power series expansion to obtain this result, we first observe that

$$\frac{1}{\mathcal{L}_S(\lambda(1 - z))} = (1 + \rho(1 - z)).$$

Substituting this into (7.49) yields

$$
\begin{aligned}
\mathcal{G}_N(z) &= (1 - \rho)(1 - z) \sum_{j=0}^{\infty} z^j (1 + \rho(1 - z))^j \\
&= (1 - \rho)(1 - z) \sum_{j=0}^{\infty} z^j \sum_{n=0}^{j} \binom{j}{n} \rho^n (1 - z)^n \\
&= (1 - \rho) \sum_{n=0}^{\infty} \rho^n (1 - z)^{n+1} \sum_{j=n}^{\infty} \binom{j}{n} z^j \\
&= (1 - \rho) \sum_{n=0}^{\infty} \rho^n z^n, \tag{7.53}
\end{aligned}
$$

where we have used the equality (see Table C.5)

$$\sum_{j=n}^{\infty} \binom{j}{n} z^j = \frac{z^n}{(1-z)^{n+1}}.$$

From (7.53) it is clear that the coefficient of z^n is $(1-\rho)\rho^n$ and thus we obtain (7.52).

7.3.4 The M/$H_2(k)$/1 Queue

Most M/G/1 distributions are not so easily obtained as with the M/M/1 queue and it is instructive to consider the M/$H_2(k)$/1 queue defined earlier. The algebraic steps required to determine its distribution are similar to those typically needed for M/G/1 systems. We start by calculating the Laplace transform for the random variable defined in (7.33):

$$\mathcal{L}_S(s) = \frac{k}{k+1}\frac{k\mu}{s+k\mu} + \frac{1}{k+1}\frac{\mu/k}{s+\mu/k},$$

and thus some algebra yields

$$\mathcal{L}_S(\lambda(1-z)) = \frac{(k^2-k+1)(1-z)\rho+k}{(\rho(1-z)+k)(k\rho(1-z)+1)}. \tag{7.54}$$

It is useful at this point to check when $z = 1$ that $\mathcal{L}_S(0) = 1$, which is easily seen to be the case for (7.54).

Substituting this into (7.48) yields

$$\mathcal{G}_N(z) = \frac{\left\{(k^2-k+1)(1-z)\rho+k\right\}(1-\rho)(1-z)}{(k^2-k+1)(1-z)\rho+k-z\left\{(\rho(1-z)+k)(k\rho(1-z)+1)\right\}}. \tag{7.55}$$

Again, a simple check on the calculation is to substitute $z = 1$ into (7.55) to verify that this yields

$$\mathcal{G}_N(1) = 1. \tag{7.56}$$

As mentioned earlier, since the numerator of (7.55) has a factor of $1-z$, this implies that the denominator must also have this factor, otherwise it would violate (7.56). Simple algebra shows that

$$(k^2-k+1)(1-z)\rho+k-z\left\{(\rho(1-z)+k)(k\rho(1-z)+1)\right\}$$

$$= (1-z)\left\{k\rho^2 z^2 - \rho(k^2+k\rho+1)z + (k^2-k+1)\rho+k\right\},$$

and thus we can write

$$\mathcal{G}_N(z) = \frac{\left\{(k^2-k+1)(1-z)\rho+k\right\}(1-\rho)}{k\rho^2 z^2 - \rho(k^2+k\rho+1)z + (k^2-k+1)\rho+k}. \tag{7.57}$$

We can check this equation by setting $z = 1$ in (7.57). As an additional check that the algebra is correct, we use the initial value property of probability generating functions by setting $z = 0$ that yields $\mathcal{G}_N(0) = 1 - \rho$ as expected.

To invert (7.57) we create a partial fraction expansion of the ratio; this requires determining the roots of the polynomial in the denominator of (7.57) (we are following the recipe given in Section 3.4.2 for partial fraction expansions). Using the quadratic formula with some algebra yields the roots

$$r_i = \frac{k^2 + k\rho + 1 + (-1)^i \sqrt{(k^2 + k\rho + 1)^2 - 4k\left\{(k^2 - k + 1)\rho + k\right\}}}{2k\rho},$$

(7.58)

for $i = 1, 2$. For k integer, both roots are distinct and thus with $\alpha_i \stackrel{\text{def}}{=} 1/r_i$ we can write

$$\mathcal{G}_N(z) = \frac{\left\{(k^2 - k + 1)\rho(1 - z) + k\right\}(1 - \rho)}{\left\{(k^2 - k + 1)\rho + k\right\}(1 - \alpha_1 z)(1 - \alpha_2 z)}$$

$$= a_1 \frac{1 - \alpha_1}{1 - \alpha_1 z} + a_2 \frac{1 - \alpha_2}{1 - \alpha_2 z},$$

(7.59)

where $a_i, i = 1, 2$ are constants that can be determined by (3.83). This yields

$$a_1 = \frac{\left\{k\alpha_1 - (k^2 - k + 1)\rho(1 - \alpha_1)\right\}(1 - \rho)}{\left\{(k^2 - k + 1)\rho + k\right\}(\alpha_1 - \alpha_2)(1 - \alpha_1)}$$

(7.60)

and

$$a_2 = \frac{\left\{(k^2 - k + 1)\rho(1 - \alpha_2) - k\alpha_2\right\}(1 - \rho)}{\left\{(k^2 - k + 1)\rho + k\right\}(\alpha_1 - \alpha_2)(1 - \alpha_2)}.$$

Equation (7.59) can now be inverted by inspection; it yields

$$P[N = n] = a_1(1 - \alpha_1)\alpha_1^n + a_2(1 - \alpha_2)\alpha_2^n, \quad n = 0, 1, \ldots,$$

(7.61)

and thus the queue length is distributed as the weighted sum of two geometric random variables (observe that $a_1 + a_2 = 1$). We can write (7.61) more conveniently as the convex sum of two geometric random variables:

$$P[N = n] = q(1 - \alpha_1)\alpha_1^n + (1 - q)(1 - \alpha_2)\alpha_2^n, \quad n = 0, 1, \ldots,$$

(7.62)

where we have set

$$q \stackrel{\text{def}}{=} a_1.$$

(7.63)

We can check this result by setting $k = 1$, which should yield the result for an M/M/1 queue. For this value of k the roots given by (7.58) yield

$$r_1 = \frac{1}{\rho}, \quad r_2 = \frac{1+\rho}{\rho},$$

which implies that

$$\alpha_1 = \rho, \quad \alpha_2 = \frac{\rho}{1+\rho}.$$

These values imply that $q = 1$ and thus (7.62) yields the expression for the M/M/1 queue given in (7.52).

7.3.5 The M/D/1 Queue

For the M/D/1 queue, we must use the power series method to determine the queue length probabilities. For the M/D/1 queue,

$$\mathcal{L}_S(\lambda(1-z)) = e^{-\rho(1-z)},$$

and thus substituting this into (7.49) yields

$$\mathcal{G}_N(z) = (1-\rho)(1-z) \sum_{j=0}^{\infty} z^j e^{\rho j(1-z)}. \tag{7.64}$$

We need to determine the coefficients of z^n in this expression. To do this we rewrite the summand in the expression as

$$(1-z)z^j e^{j\rho(1-z)} = e^{j\rho}\left(z^j - z^{j+1}\right) \sum_{k=0}^{\infty} (-1)^k \frac{(j\rho z)^k}{k!}$$

$$= e^{j\rho} \sum_{k=0}^{\infty} \left(z^{j+k} - z^{j+k+1}\right)(-1)^k \frac{(j\rho)^k}{k!}. \tag{7.65}$$

The coefficient of z^n is obtained from (7.65) by factoring all j and k values that sum to n (first term of z^{j+k}) and sum to $n-1$ (second term of z^{j+k+1}). Collecting these terms implies that the coefficient of z^n can be written as

$$\sum_{k=0}^{n} e^{k\rho} \left\{ (-1)^{n-k} \frac{(k\rho)^{n-k}}{(n-k)!} - (-1)^{n-k-1} \frac{(k\rho)^{n-k-1}}{(n-k-1)!} \right\}$$

$$= \sum_{k=0}^{n} e^{k\rho}(-1)^{n-k} \frac{(k\rho + n - k)(k\rho)^{n-k-1}}{(n-k)!}.$$

Thus, collecting the above terms implies that the queue length probabilities for the M/D/1 queue are given by

$$P[N = n] = (1 - \rho) \sum_{k=0}^{n} e^{k\rho}(-1)^{n-k} \frac{(k\rho + n - k)(k\rho)^{n-k-1}}{(n-k)!}. \qquad (7.66)$$

**7.3.6
The M/E_k/1
Queue**

We again use the power series method and note that for the M/E_k/1 queue

$$\mathcal{L}_S(\lambda(1-z)) = \left(\frac{k\mu}{\lambda(1-z) + k\mu} \right)^k.$$

It thus follows from (7.49) that

$$\mathcal{G}_N(z) = (1 - \rho)(1 - z) \sum_{n=0}^{\infty} z^n \left[1 + \frac{\rho(1-z)}{k} \right]^{kn}. \qquad (7.67)$$

Using the binomial theorem to expand the power and collecting the coefficients of z^n yields

$$P[N{=}n] = (1 - \rho) \sum_{j=0}^{n} (-1)^{n-j} \frac{\alpha^{n-j-1}}{(1-\alpha)^{kj}} \left[\binom{kj}{n-j} \alpha + \binom{kj}{n-j-1} \right], \qquad (7.68)$$

where $\alpha \overset{\text{def}}{=} \rho/(\rho + k)$. In Table 7.2 we summarize the distributions for the queue lengths just derived for these common M/G/1 queues, and in Figure 7.4 we compare the queue length probabilities for the common M/G/1 queues derived earlier.

**7.3.7
Distribution of
Response Time
for M/G/1
Queues**

The Laplace transform of the response time is obtained from (7.50) and (7.48). After a simple change of variables this yields

$$\mathcal{L}_T(s) = \mathcal{L}_S(s) \frac{s(1 - \rho)}{s - \lambda + \lambda \mathcal{L}_S(s)}. \qquad (7.69)$$

It is clear from the way we have decomposed the transforms in (7.69) that the convolution property of Laplace transforms implies that the transform for the waiting time W is given by

$$\mathcal{L}_W(s) = \frac{s(1 - \rho)}{s - \lambda + \lambda \mathcal{L}_S(s)}. \qquad (7.70)$$

Queue	Queue Length Distribution $P[N = n]$	Equation #
M/M/1	$(1-\rho)\rho^n$	(7.52)
$M/H_2(k)/1$	$q(1-\alpha_1)\alpha_1^n + (1-q)(1-\alpha_2)\alpha_2^n$	(7.62)
M/D/1	$(1-\rho)\sum_{k=0}^{n} e^{k\rho}(-1)^{n-k}\frac{(k\rho+n-k)(k\rho)^{n-k-1}}{(n-k)!}$	(7.66)
$M/E_k/1$	$(1-\rho)\sum_{j=0}^{n}(-1)^{n-j}\frac{\alpha^{n-j-1}}{(1-\alpha)^{kj}}\left[\begin{pmatrix} kj \\ n-j \end{pmatrix}\alpha + \begin{pmatrix} kj \\ n-j-1 \end{pmatrix}\right]$	(7.68)

TABLE 7.2. Queue Length Distributions for Common M/G/1 Queues

We can rewrite (7.70) by arranging it as follows:

$$\mathcal{L}_W(s) = \frac{1-\rho}{1-\rho\left[\frac{1-\mathcal{L}_S(s)}{sE[S]}\right]}. \qquad (7.71)$$

To interpret this expression observe that

$$\frac{1-F_S(t)}{E[S]} \iff \frac{1-\mathcal{L}_S(s)}{sE[S]},$$

and hence the bracketed expression in the denominator of (7.71) is the transform for a random variable with density function given by $\overline{F}(t)/E[S]$. We recognize this as the density function of the residual service time (see (6.58)), R, and thus we can write (7.71) as

$$\mathcal{L}_W(s) = \frac{1-\rho}{1-\rho\mathcal{L}_R(s)}. \qquad (7.72)$$

Similarly, we can express the transform for the response time as

$$\mathcal{L}_T(s) = \mathcal{L}_S(s)\frac{1-\rho}{1-\rho\mathcal{L}_R(s)}. \qquad (7.73)$$

We could have guessed that the residual service time would have come into play in these equations since our derivation of the expected waiting time is expressible in these terms as shown by (7.26). The M/M/1 queue lends itself readily to inversion. Since R is exponential with parameter

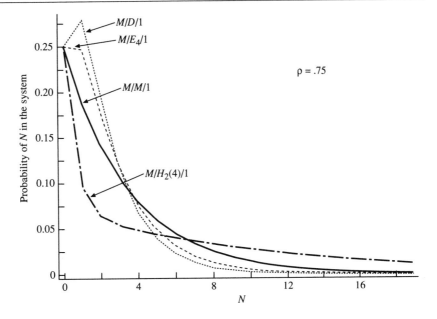

FIGURE 7.4. Queue Length Probabilities for Common M/G/1 Queues

μ we have

$$\mathcal{L}_T(s) \;=\; \frac{\mu}{s+\mu}\,\frac{1-\rho}{1-\rho\left[\frac{\mu}{s+\mu}\right]}$$

$$=\; \frac{\mu-\lambda}{s+\mu-\lambda}.$$

This is easily inverted, yielding

$$f_T(x) = (\mu - \lambda)e^{-(\mu-\lambda)x}. \tag{7.74}$$

We conclude that the response time for the M/M/1 queue is exponentially distributed with parameter $\mu - \lambda$.

We can argue this differently. Let $X_i, i = 1, 2, \ldots$, be independent and identically distributed exponential random variables with parameter μ. If an arriving customer sees N customers in the system, then, since the residual service time is also exponential, the customer's response time can be written as the random sum

$$T = X_1 + X_2 + \cdots + X_{N+1}.$$

Since N is geometric with parameter $1 - \rho$ and starts from 0 it thus follows that T is exponential with parameter $\mu(1-\rho)$. This follows from the property established in Section 4.5.10 (see (4.89)) that a geometric sum of exponential random variables is exponentially distributed.

7.4 Distribution of the Number of Customers in a Busy Period

We can derive the distribution for the number served in a busy period, a random variable denoted by N_b using a simple, but ingenious, argument. Consider the first customer, say J, that arrives to an empty system. This customer initiates a busy period that consists of all future arrivals that occur before the server becomes idle. Recall that V_1 is a random variable that denotes the number of customers that arrive during J's service time. We can imagine that each of these V_1 customers generates sub-busy periods that are generated from the customers that arrive during their service time. The number of customers served in these sub-busy periods is distributed identically to that of the initial busy period. Thus we can write

$$N_b = 1 + N_b^1 + N_b^2 + \cdots + N_b^{V_1},$$

where $N_b^i, i = 1, 2, \ldots, V_1$, are independent random variables with identical distributions to N_b, that is, $N_b =_{st} N_b^i, i = 1, 2, \ldots, V_1$. In words, the "1" counts customer J, and N_b^i counts the ith arrival during J's service time, as well as all customers that arrive during the service time of this arrival.

It is now straightforward to derive the probability generating function for N_b. Clearly

$$\mathcal{G}_{N_b}(z) = E\left[z^{N_b}\right] = E\left[z^{1+N_b^1+N_b^2+\cdots+N_b^{V_1}}\right],$$

and we can calculate

$$
\begin{aligned}
\mathcal{G}_{N_b}(z) &= \sum_{i=0}^{\infty} E\left[z^{1+N_b^1+N_b^2+\cdots+N_b^{V_1}} \mid V_1 = i\right] P\left[V_1 = i\right] \\
&= z \sum_{i=0}^{\infty} \mathcal{G}_{N_b}^i(z) P\left[V_1 = i\right] \\
&= z \mathcal{G}_V\left(\mathcal{G}_{N_b}(z)\right).
\end{aligned}
\tag{7.75}
$$

We can simplify (7.75) by using the relationship between V and S given in (7.47) to finally yield

$$\mathcal{G}_{N_b}(z) = z\mathcal{L}_S(\lambda(1 - \mathcal{G}_{N_b}(z))). \tag{7.76}$$

The recursive nature of the busy period is apparent from the fact that the generating function of N_b is expressed in terms of itself.

The moments for the number of customers served can be easily obtained from (7.76). The chain rule for differentiation yields

$$E\left[N_b\right] = \mathcal{G}_{N_b}^{(1)}(1)$$

$$= -\lambda \mathcal{L}_S^{(1)}(0)\mathcal{G}_{N_b}^{(1)}(1) + \mathcal{L}_S(0)$$

$$= \frac{1}{1-\rho}, \tag{7.77}$$

which corroborates our previous derivation of (7.42). Continuing (see Problem 7.14) results in

$$E\left[N_b^2\right] = \frac{2\rho(1-\rho) + \rho^2(1+C_S^2)}{(1-\rho)^3} + \frac{1}{1-\rho},$$

and

$$\sigma_{N_b}^2 = \frac{\rho(1-\rho) + \rho^2(1+C_S^2)}{(1-\rho)^3}. \tag{7.78}$$

There is a large variation in the number of customers served in a busy period as reflected in the $(1-\rho)^{-3}$ term of (7.78). To better see this note that the coefficient of variation is given by

$$C_{N_b}^2 = \rho + \frac{\rho^2(1+C_S^2)}{1-\rho}.$$

This shows that $C_{N_b}^2 \to \infty$ as $\rho \to \infty$. It is typical for a set of busy periods to have many that are small and a few that are extremely large. The large periods result from a positive feedback characteristic of busy periods. As the number of arrivals to a busy period increase, the chance that subsequent arrivals also will occur in the busy period increases as well. Typically, (7.76) is difficult to invert but some special cases can be obtained. As usual, the M/M/1 queue is amenable to inversion, and we will provide the density of N_b for this system in the following subsection.

7.4.1 Distribution of N_b for the M/M/1 Queue

The distribution of the number of customers served in a busy period for the M/M/1 queue can be found by inverting (7.76). It is instructive, however, to proceed along a different approach to this distribution since it is similar to a problem addressed in Chapter 3. Recall that in Section 3.3.5 we analyzed a coin tossing experiment where a fair coin is tossed and if it lands heads up then player 1 wins a dollar, otherwise the player loses a dollar. We used generating functions to calculate the distribution of the number of tosses that were necessary for the first player to win exactly one dollar, which is given by (3.70). This experiment is a *first passage problem* and a moment's reflection shows that the busy period for an M/M/1 queue is also a first passage problem since the busy period ends the first time the queue length drops to 0. The derivation of the number of customers served in an M/M/1 busy period that we present here closely follows that of the derivation of the first time the player is up one dollar in the coin tossing game.

To analyze the busy period, consider the first customer served in an M/M/1 busy period, say customer J. The probability that J is the only customer served in the busy period equals the probability that J's service completes before there are any additional arrivals to the system. This probability, which we denote by p, is given by

$$p = \frac{\mu}{\lambda + \mu}. \tag{7.79}$$

Suppose that a customer, say customer J', arrives during J's service time. The remaining service time for J is exponentially distributed with parameter μ, and thus both J and J' can be thought of as generating independent and identically distributed sub-busy periods. The total number of customers in the entire busy period is thus equal to the sum of the number of customers served in each sub-busy period.

To express this precisely, let b_n be the probability that there are n customers served in a busy period,

$$b_n \stackrel{\text{def}}{=} P[N_b = n], n = 1, 2, \ldots,$$

where we set $b_0 = 0$. Our earlier comments show that

$$b_1 = p.$$

With probability $1 - p$ the arrival J' occurs and the probability that the busy period has n customers equals all possible ways the sub-busy periods of J and J' sum up to n customers. Since each sub-busy period has at least one customer we have

$$b_n = (1 - p)\left[b_1 b_{n-1} + b_2 b_{n-2} + \cdots + b_{n-2} b_2 + b_{n-1} b_1\right], \quad n = 2, 3, \ldots . \tag{7.80}$$

Define the probability generating function of N_b by

$$\mathcal{G}_{N_b}(z) = \sum_{n=0}^{\infty} b_n z^n.$$

Using the initial condition and (7.80) we have

$$\mathcal{G}_{N_b}(z) = pz + (1 - p)\sum_{n=2}^{\infty} b_n z^n$$

$$= pz + (1-p)\sum_{n=2}^{\infty} z^n \sum_{i=1}^{n-1} b_i b_{n-i}$$

$$= pz + (1-p)\sum_{i=1}^{\infty} b_i z^i \sum_{n=i+1}^{\infty} b_{n-i} z^{n-i}$$

$$= pz + (1-p)\sum_{i=1}^{\infty} b_i z^i \sum_{j=1}^{\infty} b_j z^j$$

$$= pz + (1-p)\mathcal{G}_{N_b}^2(z). \qquad (7.81)$$

Using (7.79) and the fact that $p = 1/(1+\rho)$ shows that (7.81) is equivalent to

$$\rho \mathcal{G}_{N_b}^2(z) - (1+\rho)\mathcal{G}_{N_b}(z) + z = 0$$

and thus the quadratic equation implies that

$$\mathcal{G}_{N_b}(z) = \frac{1+\rho \pm \sqrt{(1+\rho)^2 - 4\rho z}}{2\rho}.$$

The root corresponding to the positive radical can be eliminated since it does not satisfy the property that the sum of the probabilities must equal 1 (i.e., $\mathcal{G}_{N_b}(1) = 1$). Therefore we have

$$\mathcal{G}_{N_b}(z) = \frac{1+\rho - \sqrt{(1+\rho)^2 - 4\rho z}}{2\rho}$$

$$= \frac{(1+\rho)}{2\rho}\left[1 - \sqrt{1 - \frac{4\rho z}{(1+\rho)^2}}\right], \qquad (7.82)$$

and it can be easily checked that this is what one obtains from (7.76) for the M/M/1 queue.

To invert (7.82) we use the same procedure as in Section 3.3.3 that uses the power series expansion

$$\sqrt{1-x^2} = 1 - \binom{1/2}{1} x^2 + \binom{1/2}{2} x^4 - \binom{1/2}{3} x^6 + \cdots .$$

With this expansion we can write (7.82) as

$$\mathcal{G}_{N_b}(z) = \frac{1+\rho}{2\rho}\sum_{i=1}^{\infty}(-1)^{i-1}\binom{1/2}{i}\left(\frac{2}{1+\rho}\right)^{2i}\rho^i z^i.$$

To determine the values of b_n we pick off the coefficient of z^n, which implies that

$$b_n = (-1)^{n-1}\binom{1/2}{n} 2^{2n-1}\rho^{n-1}(1+\rho)^{1-2n}, \qquad n = 1, 2, \ldots . \qquad (7.83)$$

We can write (7.83) more conveniently using the identity given in (B.2), which shows that b_n is expressed in terms of Catalan numbers as

$$b_n = \frac{1}{n} \left(\begin{array}{c} 2(n-1) \\ n-1 \end{array} \right) \rho^{n-1}(1+\rho)^{1-2n}, \quad n = 1, 2, \dots . \quad (7.84)$$

In Figure 7.5 we plot the distribution for the number of customers for an M/M/1 busy period.

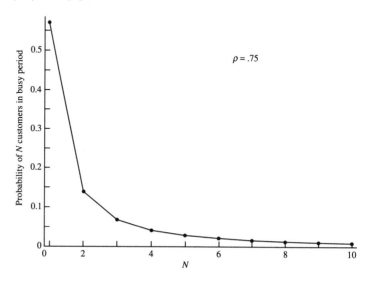

FIGURE 7.5. Distribution of Number of Customers in an M/M/1 Busy Period

7.5
Summary of
Chapter 7

In this chapter we derived performance measures for the M/G/1 queue. These results are important in modeling since components of a computer system are frequently modeled as M/G/1 queueing systems. After establishing four basic properties of systems, we used simple derivations to determine distributions and expected values for the queue length and response time of M/G/1 systems. For the special case of an M/M/1 queue all distributions can be derived in closed form. We summarize the results of this chapter in Tables 7.3 and 7.4.

It is important to point out what we have *not* derived in this chapter to highlight the necessity of the assumptions that were used to derive M/G/1 statistics.

1. Our results are not applicable to queues where the arrival stream is not Poisson. Consider a G/G/1 queue with an arrival stream that is

Statistical Measures for the M/G/1 Queue		
	Density/Transform	Expectation
N_q	$\frac{(1-\rho)(1-z)}{\mathcal{L}_S(\lambda(1-z))-z}$	$\frac{\rho^2\left(1+c_S^2\right)}{2(1-\rho)}$
N	$\mathcal{L}_S(\lambda(1-z))\,\frac{(1-\rho)(1-z)}{\mathcal{L}_S(\lambda(1-z))-z}$	$\rho+\frac{\rho^2(1+c_S^2)}{2(1-\rho)}$
N_b	$z\mathcal{L}_S(\lambda(1-\mathcal{G}_{N_b}(z)))$	$\frac{1}{1-\rho}$
W	$\frac{s(1-\rho)}{s-\lambda+\lambda\mathcal{L}_S(s)}$	$\frac{\rho E[S](1+c_S^2)}{2(1-\rho)}$
T	$\mathcal{L}_S(s)\frac{s(1-\rho)}{s-\lambda+\lambda\mathcal{L}_S(s)}$	$E[S]+\frac{\rho E[S](1+c_S^2)}{2(1-\rho)}$
B	$\mathcal{L}_S[s+\lambda-\lambda\mathcal{L}_B(s)]$	$\frac{E[S]}{1-\rho}$

TABLE 7.3. Summary of M/G/1 Measures

a renewal process with interarrival times that have a *general* (non-exponential) distribution. Recall that to derive the expected response time as in (7.24) we tagged a customer and followed it through the system summing up all components of its expected delay in the system, that is, the method of moments. The first component of customer delay is the remaining service time for the customer in the server. For Poisson arrivals we could argue that this was equivalent to what would be obtained when we sampled service intervals randomly in time. This followed because of the PASTA property of Poisson arrivals. If the arrival process is not Poisson, however, then an arrival does not see the server at a random point in time. As Example 7.3 shows, arrival times are not generally equivalent to random times. Calculating the corresponding term to $E[R]$ in (7.24) is not straightforward, and we will revisit this problem in Chapter 9 for some special cases. In Chapter 8 we will derive the stationary statistics for an M/G/1 queue using a different technique, that of the embedded Markov process. This will also allow us to analyze the G/M/1 queue.

2. For similar reasons our results are not applicable to queues with more than one server, that is, the M/G/k queue. A difficulty arises in calculating the first component of a customer's delay - the time after

	Statistical Measures for the M/M/1 Queue		
	Density/Transform	Expectation	Variance
N_q	$\delta_0(1-\rho)(1+\rho)+(1-\delta_0)(1-\rho)\rho^{n+1}$	$\frac{\rho^2}{1-\rho}$	$\frac{\rho^2(1+\rho(1-\rho))}{(1-\rho)^2}$
N	$(1-\rho)\rho^n$	$\frac{\rho}{1-\rho}$	$\frac{\rho}{(1-\rho)^2}$
N_b	$\frac{1}{n}\begin{pmatrix} 2(n-1) \\ n-1 \end{pmatrix}\rho^{n-1}(1+\rho)^{1-2n}$	$\frac{1}{1-\rho}$	$\frac{\rho(1+\rho)}{(1-\rho)^3}$
W	$(1-\rho)\delta_0(x)+\lambda(1-\rho)e^{-\mu(1-\rho)x}$	$\frac{\rho}{\mu-\lambda}$	$\frac{\rho(2-\rho)}{(\mu-\lambda)^2}$
T	$\mu(1-\rho)e^{-\mu(1-\rho)x}$	$\frac{1}{\mu-\lambda}$	$\frac{1}{(\mu-\lambda)^2}$
B	$\frac{1+\rho+s/\mu-\sqrt{(1+\rho+s/\mu)^2-4\rho}}{2\rho}$	$\frac{1}{\mu-\lambda}$	$\frac{\rho(1+\rho)}{(\mu-\lambda)^2(1-\rho)}$

TABLE 7.4. Summary of M/M/1 Measures

the arrival of the tagged customer of the first service completion. To calculate this time, let $R_i, i = 1, 2, \ldots, k$, be the remaining service time for each server. Then clearly the first component of a tagged customer's delay is given by

$$\tau = \min\{R_1, R_2, \ldots, R_k\}.$$

Two difficulties, however, are encountered.

(a) The first is that the random variables R_i are not mutually independent, and thus calculation of the distribution for the minimum requires knowing their joint distribution. An example that points out the dependency between these random variables is a two server case where service times equal 0 with probability q and equal $1/(1-q)$ with probability $1-q$. Suppose that q is close to 1 and call customers with service times of $1/(1-q)$ *large customers*. The probability that a server is busy is identical to the probability that it is serving a large customer. Let p be the unconditioned probability that a given server is busy.

We claim that the conditional probability that server 2 is busy given that server 1 is busy is greater than p. This follows from the fact that during the time server 1 is busy, all customers that leave the system must have passed through server 2. During this time the system essentially acts as a one server queue, and this increases the probability that server 2 will encounter a large customer and thus become busy. Since the probability that server 2 is busy depends on the state of server 1, it follows that the residual times found by a tagged customer are not independent random variables.

(b) The second reason the M/G/k queue poses mathematical difficulties lies in the fact that the residual service times are still correlated after the first service completion from the system. The time for the next service completion, which moves the tagged customer one step closer in the queue, also depends on the joint distribution of residual service times $(R'_1, R'_2, \ldots, R'_k)$ where $R'_i = R_i - \tau$ if server i is not the first service completion and equals S, otherwise. The time of the next service completion is thus given by $\tau' = \min\{R'_1, R'_2, \ldots, R'_k\}$, and this is not generally computable.

3. We have also not addressed the case where the arrival process, queue length, and service process are not independent of each other. Many possible types of dependency can arise in practice. For example, a natural wish of customers is to avoid long lines, and thus it is not uncommon for arrival rates to decrease as the queue length increases. For computer systems this occurs when users, frustrated by long response times, decrease the rate of their requests to the system. We will discuss extensions along these lines in Chapter 10 when we discuss queueing networks.

4. We have also not addressed the batch arrival case where each Poisson arrival consists of a random number of customers.

Bibliographic Notes

The most referenced proof of Little's law is, not surprisingly, found in Little [80]. A recent review of literature regarding Little's law can be found in Whitt [122]. See Wolff [126] for further results on PASTA. The level crossing law can be found in Kleinrock [66].

The M/G/1 queue is frequently found in performance models of computer systems, and it is therefore not a surprise that there are many texts that contain section devoted to is exposition. I have a special regard for the elegant treatment found in Kleinrock [66]. Other excellent books that can be consulted include Beneš [6], Cohen [19], Cooper [21], Cox and Smith [24], Gross and Harris [44], Kobayashi [70], Lavenberg [76], Saaty [102], Takács [112], Trivedi [118], and Wolff [125].

7.6
Problems for
Chapter 7

Problems Related to Section 7.1

7.1 [M1] Consider the following k server queue. Arrivals to the queue are exponential with rate λ and the average service time in any server is exponentially distributed with mean $1/\mu$. Customers are selected by the servers from the head of the queue when they become idle. If more than one server is idle, then a customer at the head of the queue is allocated to one of the free servers with equal probability. What is the utilization of a server? What is the expected number of customers found in service?

7.2 [R2](Continuation) Assume in Problem 7-1 that the service time of the ith server is exponentially distributed with mean $1/\mu_i$. What is the utilization of the ith server? What is the expected number of customers found in service? What is the probability that a tagged customer will be served by server i?

7.3 [E1] What is the probability that a departing customer from an M/G/1 queue leaves behind an empty queue?

Problems Related to Section 7.2

7.4 [M1] Show that the expected number of customer arrivals during a service time in an M/G/1 queue is ρ.

7.5 [M2] For an M/G/1 queue:

1. Show that

$$\frac{d^i \mathcal{G}_N(z)}{dz^i} = (-\lambda)^i \frac{d^i \mathcal{L}_T(y)}{dy^i} \ . \ \ i = 1, 2, \ldots,$$

2. Use this to show

$$E\left[N(N-1)\cdots(N-i+1)\right] = \lambda^i E\left[T^i\right], \quad i = 1, 2, \ldots \ . \tag{7.85}$$

Observe that (7.85) is Little's formula when $i = 1$.

7.6 [E3](Modeling Disk Drives as M/G/1 Queues) The service time of a disk drive consists of three independent components. First the head must be moved over the track (called the *seek time* and denoted by D_s), then there is a *rotational delay* denoted by D_r for the sector to pass under the head, and finally the data can be *transferred*, a random variable we denote by D_t. Hence the service time is given by

$$S = D_s + D_r + D_t.$$

Suppose that the disk rotates at ω revolutions per second, that there are n sectors per track, and that there are m tracks. Assume that data are transferred from the head to the computer system at the same rate it rotates under the head, that is, it takes $1/(n\omega)$ seconds to transfer a sector. We assume that all of the sectors on the disk are accessed uniformly and thus the rotational delay is uniformly distributed over the time it takes for the disk to make one rotation. We also assume that the time needed to move the head from track i to track j is given by $f(|i - j|)$. The mechanics for a disk arm consists of automatic gearing, which allows the head to accelerate as it moves, and thus f is a sublinear function. Assume that f is given by

$$f(i) = ci \left(2 - \frac{i}{m-1}\right), \quad i = 0, 1, \dots, m - 1,$$

where c is some constant. Assume that arrivals to the disk are Poisson with rate λ and that requests are served in first-come-first-served order. Write an expression for the expected disk response time as a function of ω, n, c, and m.

7.7 [**M2**] A computer facility can measure the time that customers spend waiting before receiving service but cannot measure how long customers receive service. Arrivals to the facility are Poisson and the system administrator has accurate estimates of the expected response time for two arrival rates $\lambda_i, i = 1, 2$, which yield expected waiting times of $E[W_i], i = 1, 2$, respectively. What is the expected waiting time and expected length of a busy period for an arrival rate of λ?

7.8 [**E3**] Consider M/G/1 busy periods that consist of only one customer and let S' be a random variable that denotes the service time of customers served in such busy periods. Calculate the distribution of S' and calculate the mean and variance of S' in terms of an unconditioned service time S.

Problems Related to Section 7.3

7.9 [**M2**](**Derivation of the Pollaczek–Khinchin Equation**) In this problem we provide an alternative derivation of the Pollaczek–Khinchin mean value equation based on the recurrence of (7.43),

$$N_i = (N_{i-1} - 1)^+ + V_i.$$

1. Show that

$$E[N] = E[V] + \frac{E[V^2] - E[V]}{2(1 - E[V])}.$$

2. Use this and (7.85) of Problem 7.5 to derive the Pollaczek–Khinchin (P–K) mean value formulas.

7.10 [**M3**] Consider an M/G/1 system that starts with n customers and assume that arrivals are Poisson with rate λ. Let S be a generic service time and assume that $\lambda E[S] < 1$. At time 0 we start serving the first customer in the queue. We are interested in the total number of customers served before the system become idle for the first time. What is the mean and variance of the number of customers that are served during this first busy period? Derive an expression for its probability generating function.

7.11 [M1] Let D be the time between two successive customer departures from an M/G/1 queue. Derive the Laplace transform of D. (Hint: Condition on a departure leaving behind an empty/non-empty queue).

7.12 [M3](Continuation) Let (D_1, D_2) be random variables denoting two successive interdeparture times for three customer departures from an M/G/1 system. Derive the Laplace transform for the joint distribution of (D_1, D_2). Are they independent random variables?

7.13 [M3] Consider an M/G/1 system where the Laplace transform of service time is given by

$$\mathcal{L}_S(s) = \frac{c}{Q(s)},$$

where

$$Q(s) = q_0 + q_1 s + q_2 s^2 + \cdots .$$

Derive an expression for $P[N = n]$ in terms of the constants c and $q_i, i = 0, 1, \ldots$.

7.14 [E1] Perform the calculation needed to derive the moments of the busy period given in (7.78).

7.15 [M1] Let N be the queue length of an M/G/1 queue. Show that the corresponding service time S satisfies

$$\mathcal{L}_S(s) = \frac{(\lambda - s)\mathcal{G}_N(1 - s/\lambda)}{\lambda \mathcal{G}_N(1 - s/\lambda) - (1 - \rho)s}.$$

7.16 [R1] Prove or disprove: For any nonnegative discrete random variable X there exists a service time Y such that X is the queue length for an M/G/1 queue.

7.17 [M3] A computer system has Poisson arrivals at rate λ and consists of two classes of jobs. With probability p an arriving job is of class 1 and otherwise it is of class 2. Class i has service times that are distributed as $S_i, i = 1, 2$. We index queue statistics by i and thus $E[T_i]$ is the response time of the ith class whereas $E[T]$ is the average response time.

1. Derive expressions for $E[W_i]$, $E[T_i]$, $E[W]$, and $E[T]$.

2. Suppose at an arrival instant there are N_q customers in the queue. What is the probability that k of these customers are of class i?

3. What is the probability that an arriving customer sees a class i customer in service?

7.18 [M3] Two M/G/1 queues have the same expected waiting times. Derive an equation that expresses the relationship between the expected service time and the coefficient of variation of service time that must hold between the two systems.

Problems Related to Section 7.4

7.19 [**M3**] Consider an M/G/1 queue that serves customers in last-come-first-served order where we do not allow preemptions of the customer in service. Derive the expected customer response time and an expression for the probability generating function of the number of customers served in a busy period. Derive similar expressions for random order service (here, if there are n customers in the queue at a service completion, then customers are selected uniformly with probability $1/n$).

7.20 [**M3**] Consider the following modification of an M/G/1 queueing system. Service intervals in busy periods alternate between two classes of service. The first customer served in a busy period always has a service time distributed as random variable S_1. Subsequent customers served in the busy period alternate between S_2 and S_1, that is, the second customer has a service time of S_2, the third S_1, and so on. Calculate an expression for the distribution of the number of customers served in a busy period. What is the average number of customers served in a busy period?

8

Markov Processes

In this chapter we consider an important type of stochastic process called the *Markov process*. A Markov process[1] is a stochastic process that has a limited form of "historical" dependency. To precisely define this dependency, let $\{X(t) : t \in \mathcal{T}\}$ be a stochastic process defined on the parameter set $\mathcal{T}$. We will think of $\mathcal{T}$ in terms of time, and the values that $X(t)$ can assume are called *states* which are elements of a *state space* $\mathcal{S}$. A stochastic process is called a Markov process if it satisfies

$$P\left[X(t_0 + t_1) \leq x \mid X(t_0) = x_0, X(\tau), -\infty < \tau < t_0\right]$$

$$= P\left[X(t_0 + t_1) \leq x \mid X(t_0) = x_0\right], \qquad (8.1)$$

for any value of t_0 and for $t_1 > 0$. To interpret (8.1), we think of t_0 as being the present time. Equation (8.1) states that the evolution of a Markov process at a future time, conditioned on its present and past values, depends only on its present value. Expressed differently, the present value of $X(t_0)$ contains all the information about the past evolution of the process that is needed to determine the future distribution of the process. The condition (8.1) that defines a Markov process is sometimes termed the *Markov property*.

Markov processes are classified by whether sets $\mathcal{T}$ and $\mathcal{S}$ are *discrete* (containing a countable number of elements) or *continuous* (containing an uncountable number of elements). A literature review shows that there is a divergence of opinion regarding this classification. Some authors classify Markov processes according to the state space $\mathcal{S}$ and use

---

[1]Named after A.A. Markov (1856–1922), who also first established Markov's inequality from Chapter 5.

the term "Markov chain" to refer to discrete state spaces and "Markov process" to refer to continuous state spaces (see [57, 66, 101, 118, 121] for example). The nomenclature induced by this classification for discrete state spaces leads to the terms: *discrete time Markov chains* and *continuous time Markov chains*. Other authors classify Markov processes according to parameter set $\mathcal{T}$: a Markov process with a discrete set $\mathcal{T}$ is called a "Markov chain" and, if $\mathcal{T}$ is continuous, it is called a "Markov process." This nomenclature is consistent with a variety of authors dating back to Feller (see [34, 61, 70, 81] for example). In this text all examples of Markov processes that we consider have discrete state spaces, and we find it more convenient to adopt this last classification. Throughout this text a *Markov chain* has a discrete parameter set $\mathcal{T}$ and a *Markov process* has a continuous parameter set.

We adopt the convention for Markov chains to index time using integers. The defining equation of a Markov process (8.1) as applied to Markov chains implies that a Markov chain is a set of random variables that satisfy

$$P[X_n = i_n \mid X_{n-1} = i_{n-1}, X_{n-2} = i_{n-2}, \ldots,]$$

$$= P[X_n = i_n \mid X_{n-1} = i_{n-1}]. \tag{8.2}$$

In words, (8.2) shows that the value of X_n, conditioned on the value of X_{n-1}, is independent of the values of random variables X_m for $m < n-1$. It is important to note that there is no restriction in this "one-step" description of a Markov chain. For example, suppose that the value of Y_n, conditioned on the values of Y_{n-2} and Y_{n-1}, is independent of Y_m, for $m < n - 2$. This is still a Markov chain, and it is easy to see that the bivariate chain, $X_n \stackrel{\text{def}}{=} (Y_n, Y_{n-1})$, is a one-step Markov chain. It is clear that one can create a one-step chain for all chains that have conditional dependencies on only a finite number of their previous states. Thus a Markov chain is characterized by the property that it has a limited historical dependency. A similar conclusion holds for Markov processes, and this limited historical dependency explains why so many modeling applications are Markovian. We can now provide some examples of Markov chains.

Example 8.1 (Discrete Time Single Server Queue) Consider a discrete time single server queue where time is divided into slots. Suppose that customers require service equal to one slot, and let X_n be the number of customers in the system at the beginning of slot n. Customers are assumed to arrive to the system at the end of a slot and the number of such arrivals is random and independent in each slot. These assumptions imply that X_n is a Markov chain since we can write

$$X_n = (X_{n-1} - 1)^+ + A_n, \tag{8.3}$$

where A_n is the number of customer arrivals during the nth slot (see (7.43)). Equation (8.3) shows that the value of X_n, conditioned on the value of X_{n-1}, does not depend on the values of random variables X_m for $m < n - 1$. Thus the chain $\{X_0, X_1, \ldots, X_n, \ldots\}$ is Markovian.

Example 8.2 (Statistical Average of Independent Random Variables) Consider the statistical average of a set of independent and identically distributed random variables $Y_i, i = 1, 2, \ldots,$. This is defined by

$$X_n \overset{\text{def}}{=} \frac{Y_1 + Y_2 + \cdots + Y_n}{n}, \quad n = 1, 2, \ldots .$$

Simple algebra shows that

$$\begin{aligned} X_n &= \frac{Y_1 + Y_2 + \cdots + Y_n}{n} = \frac{n-1}{n} \frac{Y_1 + Y_2 + \cdots + Y_{n-1}}{n-1} + \frac{Y_n}{n} \\ &= \frac{1}{n} \left[(n-1)X_{n-1} + Y_n \right]. \end{aligned}$$

This equation shows that the conditional distribution of X_n, given X_{n-1}, is independent of the values of X_m for $m < n - 1$, and thus the sequence of random variables $X_n, n = 1, 2, \ldots,$ is a Markov chain.

To provide examples of Markov processes we must first establish a result that is derivable from the Markov property. Suppose that a Markov process enters state i at time 0. Let H_i denote the amount of time, termed the *holding time*, that the process remains in state i. Suppose at time t the process is still in state i. We wish to determine the probability that it will remain in state i for at least τ more time units. Mathematically we wish to determine the conditional probability $P[H_i > t + \tau \mid H_i > t]$. The Markov property implies that this conditional probability is independent of the past and thus is independent of how long the process has already been in state i. Hence this conditional probability equals the probability that H_i is greater than τ, that is,

$$P[H_i > t + \tau \mid H_i > t] = P[H_i > \tau]. \tag{8.4}$$

Equation (8.4) shows that holding times are memoryless. The results of Section 4.5.9 (see (4.86) and the derivation after this) establish that a continuous random variable that is memoryless is necessarily exponentially distributed. Therefore the Markov property implies that state holding times of Markov processes are exponentially distributed.

Example 8.3 (The M/M/1 Queue) Let $X(t)$ be the number of customers in an M/M/1 queue that has Poisson arrivals at rate λ and exponential service times with expectation $1/\mu$. To show that $X(t)$ is a Markov process we must show that $X(t)$ satisfies the Markov property. To do this we first show that state holding times for the M/M/1 queue are exponentially distributed.

It is clear that the holding time of a state where the system is not empty equals the time until the next arrival or departure from the system. This corresponds to the minimum of two exponential random variables with rates λ and μ, that is, it is exponential with rate $\lambda + \mu$ (see Example 4.20). The holding time of the system when it is empty is the time until the next customer arrival; thus it is exponentially distributed with rate λ. Thus all state holding times are exponentially distributed. To establish that $X(t)$ satisfies the Markov property, it suffices to show that state changes conditioned on the present state are independent of the past.

The state of the system increases by one unit if there is an arrival to the system and decreases by one unit if there is a departure (provided that the system is not empty). If the system is not empty at time t, that is, if $X(t) > 0$, then it is clear that the next state change increases by one unit if the first event after t is an arrival rather than a departure. This probability is given by $\lambda/(\lambda + \mu)$ (see Example 4.21) and is independent of $X(\tau)$ for $\tau < t$. Similarly the next state change decreases by one unit if the first event after t is a departure event. This occurs with probability $\mu/(\lambda + \mu)$ and is independent of $X(\tau)$ for $\tau < t$. If $X(t) = 0$ then the first event after time t must be an arrival causing the state to increase by one unit. This event is clearly independent of $X(\tau)$ for $\tau < t$. We have thus shown that $X(t)$ satisfies the Markov property and is a Markov process.

Example 8.3 presents an informal proof that the stochastic process associated with the number of customers in an M/M/1 queue is a Markov process. It is unusual in practice to formally establish that a stochastic process is Markovian. Most often, this fact is obvious from the statement of the problem.

In Section 8.1 we develop the theory of Markov chains where we classify different types of states and establish several fundamental properties of such chains. The underlying mathematical foundations of Markov chains is renewal theory, and we will make frequent use of the results of Chapter 6 to derive results. In Section 8.2, for example, we use renewal theoretic results to show that for chains of a certain type, the values of the limiting state probabilities satisfy a set of linear equations called global balance equations. The main objective of many Markovian performance models is to solve these sets of equations. A more general form of process, called the *semi-Markov* process, is analyzed in Section 8.3. This process adds a time component to states, and a natural view of a Markov chain is where state holding times are geometrically distributed. This viewpoint leads naturally to Markov processes which, as we have discussed, operate in continuous time and have exponential holding times. In Section 8.4 we derive the set of linear equations that stationary state probabilities satisfy for Markov processes. As discussed in Chapter 7, queueing systems are typically used to model computer systems. In Sections 8.5 and 8.6 we present solutions for simple queues that are widely used in modeling: variations of the M/M/1 queue and the M/G/1 queue, respectively. Section 8.7 considers an important relationship between Markov chains and Markov processes, termed uniformization, and we present some results about transient analysis of Markov processes in Section 8.8.

8.1 Markov Chains

We first concentrate on discrete time Markov processes. As mentioned in the introduction, these processes are termed *Markov chains* to distinguish them from continuous time Markov processes. Without loss of generality, suppose that the state space of the chain, $\mathcal{S}$, is the set of nonnegative integers. The chain is characterized by its evolution among the states of $\mathcal{S}$ over time. To be precise, define the probability $P[X_n = j \mid X_{n-1} = i]$ to be the *transition probability* of state i to state j at time $n-1$. We say that the chain is *time homogeneous* if,

$$P[X_n = j \mid X_{n-1} = i] = P[X_{n+m} = j \mid X_{n+m-1} = i],$$

$$n = 1, 2, \ldots, \quad m \geq 0, \quad i, j \in \mathcal{S}.$$

Throughout this chapter we assume that chains are time homogeneous. With this assumption we can write

$$p_{i,j} \stackrel{\text{def}}{=} P[X_n = j \mid X_{n-1} = i]$$

and we can define the matrix of transition probabilities by

$$\boldsymbol{P} \stackrel{\text{def}}{=} \begin{bmatrix} p_{0,0} & p_{0,1} & \cdots & p_{0,j} & \cdots \\ p_{1,0} & p_{1,1} & \cdots & p_{1,j} & \cdots \\ \vdots & \vdots & \cdots & \vdots & \cdots \\ p_{i,0} & p_{i,1} & \cdots & p_{i,j} & \cdots \\ \vdots & \vdots & \cdots & \vdots & \cdots \end{bmatrix}. \tag{8.5}$$

If $\mathcal{S}$ is finite then $\boldsymbol{P}$ has finite dimension. We write $i \mapsto j$ if $p_{i,j} > 0$, which means that the chain can jump directly from i to j.

We can think of the operation of the chain as follows. The chain starts at time 0 in some state, say $i_0 \in \mathcal{S}$. At the next time unit or step, the chain jumps to a neighboring state i_1 with probability p_{i_0,i_1} provided that $i_0 \mapsto i_1$. It may be the case that this jump is immediately back to the state itself, that is, $i_1 = i_0$. We call such an occurrence a *self-loop*. This procedure is repeated so that, at step n, the chain is in some state i_n, where

$$i_0 \mapsto i_1 \mapsto i_2 \mapsto \cdots \mapsto i_{n-1} \mapsto i_n. \tag{8.6}$$

A sequence of states satisfying (8.6) is called a *path*. We write $i \overset{n}{\rightsquigarrow} j$ if there exists a path of n steps between i and j and $i \rightsquigarrow j$ if there exists a path (possibly an infinite sequence of states) from state i to j. Given an initial state there is a set of possible paths that can be taken by the Markov chain. This is called the set of *sample paths*. The particular sample path taken by the chain is denoted by

$$i_0, i_1, i_2, \ldots, i_{k-1}, i_k, \ldots . \tag{8.7}$$

This sample path shows that the process is found in state i_k at step k, for $k = 0, 1, \ldots$. In the next section we analyze properties of Markov chains in terms of properties of their sample paths.

8.1.1
The Chapman–Kolmogorov Equations

The Markov property allows us to write an expression for the probability that a particular sample path is taken by the chain by multiplying the transition probabilities along the sample path. In particular, the probability of sample path (8.6) is given by

$$p_{i_0,i_1} p_{i_1,i_2} \cdots p_{i_{n-2},i_{n-1}} p_{i_{n-1},i_n}.$$

It is clear from the law of total probability that the probability that the chain is in state j at step n equals the sum of the probabilities of all n-step sample paths from state i to j. For $n = 1$ this probability is simply $p_{i,j}$ and is given by the entry in position (i,j) of matrix $\boldsymbol{P}$ (8.5). For $n = 2$ the probability of all sample paths $i \overset{2}{\rightsquigarrow} j$ can be written as $\sum_{k \in \mathcal{S}} p_{i,k} p_{k,j}$, which is simply the entry in position (i,j) of matrix $\boldsymbol{P}^2$. Let $p_{i,j}^n$ denote the (i,j) entry of matrix $\boldsymbol{P}^n$ (we define $\boldsymbol{P}^0 = \boldsymbol{I}$). Then it is easy to see that $p_{i,j}^n$ is the probability that the process is in state j, n steps after being in state i. The rules for matrix multiplication imply that

$$\boldsymbol{P}^n = \boldsymbol{P}^m \, \boldsymbol{P}^{n-m}, \quad 0 \le m \le n, \tag{8.8}$$

or, written in terms of the matrix entries, that

$$p_{i,j}^n = \sum_{k \in \mathcal{S}} p_{i,k}^m p_{k,j}^{n-m}, \quad 0 \le m \le n. \tag{8.9}$$

Equations (8.9) are called the *Chapman–Kolmogorov equations*.

It is clear that if the Markov chain starts in state 1 then the probability that the chain is in state i at step 1 is given by the $(1,i)$ entry of $\boldsymbol{P}$. More generally, define the vector

$$\boldsymbol{\pi}^0 \overset{\text{def}}{=} (\pi_0^0, \pi_1^0, \ldots),$$

where π_i^0 is the probability that the initial state is i at time 0. It is then clear that the probability that the process is in state i at step 1, denoted by π_i^1, is given by

$$\pi_i^1 = \sum_{k=0}^{\infty} \pi_k^0 p_{k,i}, \quad i = 0, 1, \dots . \tag{8.10}$$

Define $\boldsymbol{\pi}^n \stackrel{\text{def}}{=} (\pi_1^n, \pi_2^n, \dots)$, for $n = 0, 1, \dots$, to be the vector of state probabilities for the nth step of the chain. Using this notation and some elementary matrix algebra allows us to write (8.10) in a simpler form:

$$\boldsymbol{\pi}^1 = \boldsymbol{\pi}^0 \boldsymbol{P}.$$

Because the process is Markovian, we can consider $\boldsymbol{\pi}^1$ to be initial probabilities for the next step of a one-step Markov chain and thus can write

$$\boldsymbol{\pi}^2 = \boldsymbol{\pi}^1 \boldsymbol{P}.$$

More generally we have

$$\boldsymbol{\pi}^n = \boldsymbol{\pi}^{n-1} \boldsymbol{P} \quad n = 1, 2, \dots . \tag{8.11}$$

Iterating this recurrence shows that we can write (8.11) as

$$\boldsymbol{\pi}^n = \boldsymbol{\pi}^0 \boldsymbol{P}^n, \quad n = 1, 2, \dots . \tag{8.12}$$

This equation shows that the probability of state i at step n is simply the sum of the probabilities along all sample paths $j \stackrel{n}{\rightsquigarrow} i$ weighted by the probability of starting in state j. Observe that (8.12) implies that, conditioned on starting in state i, the value of $\boldsymbol{\pi}^n$ is simply the ith row of $\boldsymbol{P}^n$, that is,

$$[p_{i,0}^n, p_{i,1}^n, \dots] = \underbrace{[0, \dots, 0, 1, 0, \dots]}_{\text{1 in } i\text{th position}} \begin{bmatrix} p_{0,0}^n & p_{0,1}^n & \cdots & p_{0,j}^n & \cdots \\ p_{1,0}^n & p_{1,1}^n & \cdots & p_{1,j}^n & \cdots \\ \vdots & \vdots & \cdots & \vdots & \cdots \\ p_{i,0}^n & p_{i,1}^n & \cdots & p_{i,j}^n & \cdots \\ \vdots & \vdots & \cdots & \vdots & \cdots \end{bmatrix}. \tag{8.13}$$

A simple example best illustrates how these equations can be used.

Example 8.4 (Finite Discrete Time Queue) Consider a discrete time single server queue that can hold at most two customers. Each customer requires a service time equal to one time slot, and arrivals are assumed to enter the queue just before the end of a slot. Assume that the number of arrivals in each slot is an independent geometric random variable starting from 0 with parameter $1 - p$. Thus i customers arrive in any given slot with probability $(1 - p)p^i$ for $i = 0, 1, \ldots$. If the number of customers arriving to the queue is larger than the number of available positions then all surplus customers are lost. The state of the system is the number of customers in the system at the beginning of a slot, and a convenient representation of the process is through its state transition diagram as shown in Figure 8.1.

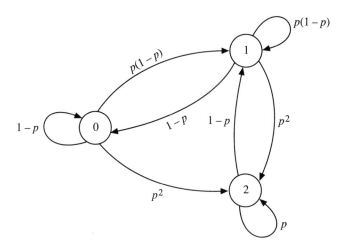

FIGURE 8.1. State Transition Diagram for a Finite Capacity Queue

The transition matrix for the chain is given by

$$P = \begin{bmatrix} 1 - p & p(1 - p) & p^2 \\ 1 - p & p(1 - p) & p^2 \\ 0 & 1 - p & p \end{bmatrix}, \tag{8.14}$$

and it is clear that $i \rightsquigarrow j$ for all i and j. For $p = 1/3$ the matrix P has entries given by

$$P = \frac{1}{9} \begin{bmatrix} 6 & 2 & 1 \\ 6 & 2 & 1 \\ 0 & 6 & 3 \end{bmatrix} = \begin{bmatrix} 0.6666 & 0.2222 & 0.1111 \\ 0.6666 & 0.2222 & 0.1111 \\ 0 & 0.6666 & 0.3333 \end{bmatrix}.$$

Suppose that we start the system in state 3 and thus that $\pi^0 = (0, 0, 1)$. Equation (8.12) implies that $\pi^1 = (0, 0.6666, 0.3333)$. Matrix multiplication shows that

$$P^2 = \left(\frac{1}{9}\right)^2 \begin{bmatrix} 48 & 22 & 11 \\ 48 & 22 & 11 \\ 36 & 30 & 15 \end{bmatrix} = \begin{bmatrix} 0.5925 & 0.2716 & 0.1358 \\ 0.5925 & 0.2716 & 0.1358 \\ 0.4444 & 0.3703 & 0.1851 \end{bmatrix}$$

and so (8.12) implies that $\pi^2 = (0.4444, 0.3703, 0.1851)$. Continuing to the third step shows that

$$P^3 = \left(\frac{1}{9}\right)^3 \begin{bmatrix} 420 & 206 & 103 \\ 420 & 206 & 103 \\ 396 & 222 & 111 \end{bmatrix} = \begin{bmatrix} 0.5761 & 0.2825 & 0.1412 \\ 0.5761 & 0.2825 & 0.1412 \\ 0.5432 & 0.3045 & 0.1522 \end{bmatrix} \tag{8.15}$$

and thus the state probabilities at the third step are given by

$$\pi^3 = (0.5432, 0.3045, 0.1522).$$

It is interesting to note that, since the rows of (8.15) do not substantially differ, the state probabilities after three steps are approximately equal for all initial states. Continuing on shows that, for P^8, all rows have the same values up to four decimal places,

$$P^8 = \begin{bmatrix} 0.5714, 0.2557, 0.1428 \\ 0.5714, 0.2557, 0.1428 \\ 0.5714, 0.2557, 0.1428 \end{bmatrix} \tag{8.16}$$

and implies that $\pi^8 = (0.5714, 0.2557, 0.1428)$. This vector is accurate to four decimal places for *all possible initial states* π^0.

Two closely related hypotheses emerge from Example 8.4:

Hypothesis 1. Matrix P^n converges as n increases. This limiting matrix has identical rows, each of which equals $\lim_{n \to \infty} \pi^n$.

Hypothesis 2. The probability of being in a given state at step n, for n sufficiently large, does not depend on the initial state. Hence the Markov chain loses the "memory" of its starting state.

In the following sections we show that these hypotheses are satisfied by a particular type of Markov chain called an ergodic chain. To do this we first consider more general chains and classify their states.

8.1.2 Classification of States

The classification of states depends on the path structure of the Markov chain. Consider Figure 8.2 which graphically summarizes all of the different possible state classifications. In this figure all arcs that are unlabeled are assumed to have a probability strictly greater than 0 and strictly less than 1.

Observe that state 0 does not satisfy $0 \rightsquigarrow i$ for any $i \neq 0$ and thus, once entered, the system stays in state 0 forever. Such a state is called an *absorbing state*. One can cast first passage problems into absorption time problems of Markov chains by assuming that the first time the chain enters a given set of states it is absorbed. First passage problems previously considered include the coin tossing experiment of Section 3.3.5 and the busy period analysis of Section 7.4.

States 1 and 2 are examples of *transient states*. A state i is transient if, starting from it, there is a positive probability that the chain will never

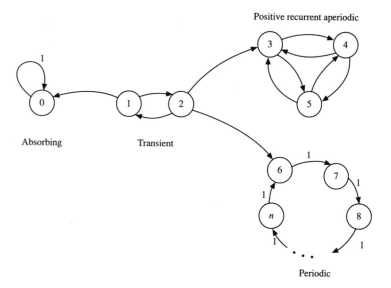

FIGURE 8.2. Classification of States

return. As an example, if there exists a state j satisfying $i \rightsquigarrow j$ but $j \not\rightsquigarrow i$ then i is transient. A transient state can be visited only a finite number of times with probability one (see Problem 8.2). If the chain is started in a transient state then, after a finite number of return visits, the state will be left forever.

States 3, 4, and 5 are examples of *recurrent states*. In contrast to transient states, if the chain is started in a recurrent state i then the probability of eventually returning to i is 1. It is clear that $i \rightsquigarrow j$ implies that $j \rightsquigarrow i$ and we write $i \leftrightarrow j$ and say that states i and j *communicate*. We say that a Markov chain is *irreducible* if all states communicate with each other; otherwise, a chain is *reducible*.

We call a sequence of states starting and ending at i, an *i-cycle* and distinguish two types of recurrent states depending on the expected number of steps in such a cycle. If the expected number of steps in an *i-cycle* is finite then we call state i *positive recurrent* otherwise state i is said to be *null recurrent*. A consequence of positive recurrence is that if i is positive recurrent and if $i \rightsquigarrow j$, then j is also positive recurrent. (In the problems at the end of the chapter we ask you to provide proofs for consequences of the above definitions.)

The last classification of states concerns periodicity. If the *i-cycle*, $i \overset{n}{\rightsquigarrow} i$, exists only when $n = kd$ for some values of k and a fixed value of $d > 1$ then state i is said to be *periodic* with period d. This indicates that the state can only return to itself after some multiple of d steps. For example, if $7 \leq n < \infty$ then states $\{6, 7, \ldots, n\}$ are periodic with period $d = n - 5$. We call states that are not periodic *aperiodic*, and states $\{3, 4, 5\}$ are examples of such states.

A positive recurrent aperiodic state is called an *ergodic* state, and most of the Markov models that arise in applications are irreducible and ergodic, that is, they have state spaces consisting of one set of ergodic states where $i \leftrightarrow j$ for all i, j in the set. In this case it makes sense to determine the limiting state probabilities as alluded to in the hypotheses based on the finite capacity queue problem considered in Example 8.4, which is both irreducible and ergodic.

It is important to note that there can be more than one ergodic set of states. For example, suppose arcs were added in Figure 8.2 so that states in $\{6, 7, \ldots\}$ formed an ergodic set (that is, we add arcs so that the states in this set are not periodic). Then the chain consists of two ergodic sets, $\{3, 4, 5\}$ and $\{6, 7, \ldots, n\}$. Such a chain is not irreducible. The stationary distribution of a particular state in one of these sets clearly depends on the sample path taken by the chain. Before developing the theory required to determine the limiting state probabilities for a system with one ergodic set of states we first derive mathematical criteria that determine the above state classifications.

8.1.3 Mathematical Classification

Let $f_{i,j}^n$ be the probability that, starting in state i, the *first transition* into state j occurs at step n, for $n = 1, 2, \ldots$. By definition we set $f_{i,j}^0 = 0$ and formally define

$$f_{i,j}^n \overset{\text{def}}{=} P[X_0 = i, X_1 \neq j, \ldots, X_{n-1} \neq j, X_n = j]. \quad (8.17)$$

The probability that the Markov chain starting in state i eventually enters state j, a quantity denoted by $f_{i,j}$, can be determined by using the law of total probability. Conditioning on the *first entrance* to state j yields a partition of all the possibilities and therefore

$$f_{i,j} = \sum_{n=1}^{\infty} f_{i,j}^n.$$

It is clear that $f_{i,j} = 0$ if $i \not\rightsquigarrow j$ whereas if $i \rightsquigarrow j$ then $f_{i,j} > 0$. Observe that $f_{i,i}$ can be interpreted as the probability that the chain eventually returns to state i having started there, that is, that the chain contains an i-cycle. We can thus characterize transient and recurrent states as

$$f_{i,i} = 1, \qquad \text{recurrent state}, \qquad (8.18)$$

$$f_{i,i} < 1, \qquad \text{transient state}. \qquad (8.19)$$

Equation (8.18) implies that i-cycles occur with certainty if the state is recurrent, and thus state i will be visited an infinite number of times if the chain is started in state i. Equation (8.19), on the other hand, implies that there is a positive probability that the chain will never return to state i if the chain is transient. It necessarily follows that state i will be visited only a finite number of times. Table 8.1 summarizes the classification of states.

Classification	# Times Visited	Expected # Steps/Visit	Formula
Transient	$< \infty$	—	$f_{i,i} < 1$
Positive recurrent	∞	$< \infty$	$f_{i,i} = 1$
Null recurrent	∞	∞	$f_{i,i} = 1$

TABLE 8.1. Classification of States for Markov Chains

8.1.4 The Stationary Distribution of Ergodic Chains

Many results can be obtained from the simple observation that the Markov property implies that at each return to state i the chain can be viewed as starting over again and thus the sequence of i-cycles forms a renewal sequence. The density of the number of steps required to return to state i is given by the values $f_{i,i}^n, i = 1, 2, \ldots$. More precisely, let $N_{i,j}(n)$ be the number of visits to state j in n steps of the chain given that $X_0 = i$. If the chain starts in state i and if i is recurrent, then $N_{i,i}(n)$ is a renewal process. Otherwise, if j is recurrent and the chain starts in state i then $N_{i,j}(n)$ is a delayed renewal process (provided that $i \leftrightarrow j$). The expected number of visits to state j up to and including n, denoted by $M_{i,j}(n)$, is given by

$$M_{i,j}(n) \stackrel{\text{def}}{=} \sum_{k=1}^{n} p_{i,j}^k. \tag{8.20}$$

We can use the propositions proved in Chapter 6 to establish properties of the recurrent states of Markov chains. (The renewal process $N_{i,j}(n)$ is defined on a lattice but, as mentioned in Section 6.7, Propositions 6.10 and 6.17 are still valid for such distributions.) The first property uses Proposition 6.10 and shows that if j is recurrent and if $i \leftrightarrow j$ then

$$\frac{N_{i,j}(n)}{n} \xrightarrow{a.s.} \pi_j, \quad \text{as} \quad n \longrightarrow \infty, \tag{8.21}$$

where

$$\pi_j \overset{\text{def}}{=} \left[\sum_{n=1}^{\infty} n f_{j,j}^n \right]^{-1}. \tag{8.22}$$

Similarly, Proposition 6.17 implies that if $i \leftrightarrow j$ then

$$\frac{M_{i,j}(n)}{n} \longrightarrow \pi_j, \quad \text{as} \quad n \longrightarrow \infty. \tag{8.23}$$

Equation (6.70) of Section 6.7 shows that if i is aperiodic then

$$\lim_{n \to \infty} p_{i,j}^n = \pi_j \tag{8.24}$$

provided that j is recurrent and $i \leftrightarrow j$. Hence, if the limit exists, the limiting probability of finding the chain in state j is given by π_j. The values of $\pi_i, i = 0, 1, \ldots$, are called the *stationary probabilities* of the Markov chain, and the above equations show that they permit several interpretations:

1. Equation (8.21) shows that π_j equals the proportion of time the Markov chain spends in state j for almost all sample paths of the chain. That is, the set of sample paths where π_j is not equal to the proportion of the number of steps spent in state j collectively has a probability of 0.

2. Equation (8.22) shows that the expected number of steps between successive visits to state j is $1/\pi_j$.

3. Equation (8.23) shows that π_j equals the expected proportion of the number of steps the Markov chain spends in state j over the set of all sample paths of the chain. Example 6.12 shows that this is *not implied* by item 1 above.

4. Equation (8.24) shows that the probability of finding the chain in state j at a random step equals π_j. A random step means, as in Chapter 6, a step n taken as $n \to \infty$. The equation also shows that this probability is independent of the initial starting state i. Hence the Markov chain does not retain memory of its initial state.[2]

---

[2]We are dealing with aperiodic chains in this section. Several of these interpretations remain valid for periodic chains. For periodic chains it does not make sense to speak of limiting probabilities since such limits do not exist. Unlike ergodic chains, it is not true that irreducible periodic chains lose memory of their initial state. To see this, suppose the chain is started in state i, which has period d. Then the probability of finding the system in state i at time step n is equal to zero if n is not a multiple of d.

We will now determine the stationary state probabilities, $\pi_j, j = 0, 1, \ldots,$ for an irreducible Markov chain where all states are ergodic, that is, a Markov chain where all states are positive recurrent, aperiodic, and contained in a single set and where $i \leftrightarrow j$ for all states i, j. This is the case of most interest in modeling and is termed an *ergodic Markov chain*.

The different interpretations of the values of π_j discussed above are important because they lend physical insight into the meaning of the stationary probabilities. These interpretations also show that the stationary distribution of an ergodic Markov chain is independent of its initial state and thus establishes Hypothesis 2 of Section 8.1. Recall that equation (8.12) specifies the state probabilities at the nth step of the chain,

$$\pi^n = \pi^0 P^n, \; n = 1, 2, \ldots,$$

where π^0 is the initial probability vector. The fact that the distribution of the Markov chain does not depend on π^0 implies, as in (8.13), that all rows of $\lim_{n \to \infty} P^n$ are identical and thus establishes Hypothesis 1. We record this in the following proposition:

Proposition 8.5 *Let P be a transition matrix of an irreducible Markov chain consisting of one ergodic set. Then the limiting matrix $\lim_{n \to \infty} P^n$ has identical rows that are equal to the stationary probability vector π.*

Proposition 8.5 provides a reason for why we term π_j the stationary probability of state j. Suppose that we start the chain in state j with probability π_j. That is, we start with the stationary probability vector π. Then the state probabilities after the first step are given by πP. This is again equal to the stationary probability vector π since

$$\pi = \pi P.$$

Hence, the probability vector π is *stationary* in time. In terms of the Markov chain this implies that the joint distribution of the chain defined by

$$P[X_1 = x_1, X_2 = x_2, \ldots, X_k = x_k]$$

is invariant to shifts of time. That is, the chain is stationary if

$$P[X_1 = x_1, \ldots, X_k = x_k] = P[X_{1+j} = x_1, \ldots, X_{k+j} = x_k],$$

for any value j.

In the next section we formally establish these results. Before doing this we note that we distinguish the stationary state distribution from that of a *transient distribution*. The transient distribution π^n, as given by (8.12) is the state probability distribution obtained after n steps starting from a given state π^0. As exemplified in Example 8.4, the value of π^n generally differs from the stationary distribution π and, as we show in the next section, $\pi^n \to \pi$ as $n \to \infty$. By starting with the stationary distribution, all successive "transient" distributions equal the stationary distribution.

There are similarities between the transient and stationary state probabilities of a Markov chain and a renewal process and its corresponding equilibrium renewal process. Like the transient distribution of a Markov chain, an equilibrium renewal process is obtained as a limiting form of a renewal process. Furthermore, if one starts a renewal process with a "special renewal epoch" (with the residual life distribution) then it is an equilibrium renewal process.

8.1.5 Balance Equations for Stationary Probabilities

We can use the Chapman–Kolmogorov equations to determine an equation for the stationary probabilities. These equations imply that

$$p_{i,j}^{n+1} = \sum_{k=0}^{\infty} p_{i,k}^n p_{k,j}$$

$$\geq \sum_{k=0}^{K} p_{i,k}^n p_{k,j}, \quad j = 0, 1, \ldots, \tag{8.25}$$

for any positive integer K. Letting $n \to \infty$ in (8.25) implies that

$$\pi_j \geq \sum_{k=0}^{K} \pi_k p_{k,j}, \quad j = 0, 1, \ldots .$$

Since the left-hand side of this equation is independent of K we have that

$$\pi_j \geq \sum_{k=0}^{\infty} \pi_k p_{k,j} \quad j = 0, 1, \ldots . \tag{8.26}$$

It is easy to argue that (8.26) is satisfied as an equality. Suppose, for some index j, the inequality is strict. By summing (8.26) we obtain

$$\sum_{j=0}^{\infty} \pi_j > \sum_{j=0}^{\infty} \sum_{k=0}^{\infty} \pi_k p_{k,j} = \sum_{k=0}^{\infty} \pi_k \sum_{j=0}^{\infty} p_{k,j} = \sum_{k=0}^{\infty} \pi_k,$$

which is a contradiction. Hence, it necessarily follows that

$$\pi_j = \sum_{k=0}^{\infty} \pi_k p_{k,j} \quad j = 0, 1, \ldots . \tag{8.27}$$

Intuitively, it is clear that the stationary probabilities and, in fact, all transient state probabilities, sum to 1. This follows from the simple observation that $\boldsymbol{P}^n$ is a probability transition matrix, and thus $\boldsymbol{P}^n$ has row sums that equal 1. Consider the transient state distribution $\boldsymbol{\pi}^n$ as given by (8.12). The sum of its components is given by

$$
\begin{aligned}
\sum_{i=0}^{\infty} \pi_i^n &= \sum_{i=0}^{\infty}\sum_{j=0}^{\infty} \pi_j^0 p_{j,i}^n \\
&= \sum_{j=0}^{\infty} \pi_j^0 \sum_{i=0}^{\infty} p_{j,i}^n \\
&= \sum_{j=0}^{\infty} \pi_j^0 = 1.
\end{aligned}
\tag{8.28}
$$

8.1.6 Uniqueness of Stationary Probabilities

Equations (8.27) and (8.28) permit only one solution. To establish this suppose that $\beta_j, j = 0, 1, \ldots$ also satisfies (8.27) and (8.28). It then follows that

$$
\begin{aligned}
\beta_\ell = \sum_{j=0}^{\infty} \beta_j p_{j,\ell} &= \sum_{j=0}^{\infty}\left\{\sum_{k=0}^{\infty} \beta_k p_{k,j}\right\} p_{j,\ell} \\
&= \sum_{k=0}^{\infty} \beta_k \sum_{j=0}^{\infty} p_{k,j} p_{j,\ell} \\
&= \sum_{k=0}^{\infty} \beta_k p_{k,\ell}^2.
\end{aligned}
$$

This is simply a version of (8.12) for the elements of β obtained after the first two steps of the Markov chain. Continuing in this fashion shows that we can write

$$
\begin{aligned}
\beta_j &= \sum_{k=0}^{\infty} \beta_k p_{k,j}^m \\
&\geq \sum_{k=0}^{K} \beta_k p_{k,j}^m,
\end{aligned}
$$

for any positive integer K. Letting $m \to \infty$ implies that

$$
\beta_j \geq \sum_{k=0}^{K} \beta_k \pi_j.
$$

Since the left-hand side is independent of K we have

$$\beta_j \geq \sum_{k=0}^{\infty} \beta_k \pi_j = \pi_j, \quad j = 0, 1, \ldots . \tag{8.29}$$

To establish uniqueness of the stationary probabilities, assume that at least one of the inequalities in (8.29) is strict. This implies that

$$1 = \sum_{j=0}^{\infty} \beta_j \quad > \quad \sum_{j=0}^{\infty} \pi_j = 1,$$

which is a contradiction. Thus the stationary probabilities $\pi_i, i = 0, 1, \ldots$ are unique. We record these important results in the following proposition.

Proposition 8.6 *The stationary probabilities of an irreducible Markov chain with one ergodic set are unique and satisfy*

$$\pi_j = \sum_{k=0}^{\infty} \pi_k p_{k,j}, \quad j = 0, 1, \ldots, \tag{8.30}$$

and

$$\sum_{j=0}^{\infty} \pi_j = 1. \tag{8.31}$$

Written in matrix form these equations are given by

$$\boldsymbol{\pi} = \boldsymbol{\pi} \boldsymbol{P} \tag{8.32}$$

and

$$|\boldsymbol{\pi}| = 1 \tag{8.33}$$

where $|\boldsymbol{x}|$ is the sum of the elements of vector x.

To apply this result we continue the problem of the finite capacity queue considered in Example 8.4.

Example 8.7 (Finite Discrete Time Queue-Continued) The transition matrix for the finite discrete time queue considered in Example 8.4 has a transition matrix given by (8.14). We can write (8.32) as

$$\boldsymbol{\pi} = \boldsymbol{\pi} \begin{bmatrix} 1-p & p(1-p) & p^2 \\ 1-p & p(1-p) & p^2 \\ 0 & 1-p & p \end{bmatrix} . \tag{8.34}$$

The first equation of (8.34) yields

$$\pi_0 = (1-p)\pi_0 + (1-p)\pi_1$$

$$= \frac{1-p}{p}\pi_1.$$

The second equation yields

$$\pi_1 = p(1-p)\pi_0 + p(1-p)\pi_1 + (1-p)\pi_2$$

$$= (1-p)^2\pi_1 + p(1-p)\pi_1 + (1-p)\pi_2$$

$$= \frac{1-p}{p}\pi_2.$$

Using these equations and the normalization (8.33) yields

$$\pi_0 = \frac{(1-p)^2}{1-p(1-p)}, \quad \pi_1 = \frac{p(1-p)}{1-p(1-p)}, \quad \pi_2 = \frac{p^2}{1-p(1-p)}.$$

Substituting $p = 1/3$ into these equations yields

$$\pi_0 = 0.5714, \quad \pi_1 = 0.2857, \quad \pi_2 = 0.1428,$$

as suggested by (8.16).

The fact that there are four equations (three from (8.34) and one for the normalization (8.33)) and three unknown variables implies that one equation of (8.34) is redundant. For this case the third column of the matrix in (8.34) is $p/(1-p)$ times the second column. This implies that the third equation is a multiple of the second equation given above. Such a redundancy is always the case, and one must use the normalization condition to obtain stationary solutions for Markov chains.

We have now established both hypotheses raised by the example of Section 8.1. To determine the unique stationary probabilities of a Markov chain we have to solve a set of linear equations. We call equations (8.32) *global balance equations*. This name immediately raises the question of what is being "balanced," a question we will address in Section 8.2.

Example 8.8 (A Simple Random Walk) Consider a coin tossing experiment consisting of an infinite series of coin tosses. Each coin toss lands heads up with probability p and causes a player to win one dollar. With probability $1-p$ the player loses a dollar with the proviso that the net worth of the player cannot drop below 0. We can think of the game as being played by a "benign adversary" who does not debit the player one dollar when tails lands up and the player has a stake of zero. If we let $X_n, n = 0, 1, \ldots$, be the total amount of money that the player has at the nth game, where we set $X_0 = 0$, then X_n is a Markov chain. The state space of the chain is given by $\mathcal{S} = \{0, 1, \ldots\}$, and the transition matrix is given by

$$
\boldsymbol{P} = \begin{bmatrix}
1-p & p & 0 & 0 & 0 & \cdots \\
1-p & 0 & p & 0 & 0 & \cdots \\
0 & 1-p & 0 & p & 0 & \cdots \\
0 & 0 & 1-p & 0 & p & \cdots \\
\vdots & \vdots & \vdots & \vdots & \vdots & \ddots
\end{bmatrix}. \tag{8.35}
$$

This matrix is tridiagonal and, although it is infinite, it can be easily solved. To solve it, first note that the balance equation for $i = 0$ is given by

$$
\begin{aligned}
\pi_0 &= (1-p)\pi_0 + (1-p)\pi_1 \\
\\
&= \frac{1-p}{p}\pi_1. \tag{8.36}
\end{aligned}
$$

The ith equation of (8.30) is given by

$$
\pi_i = p\pi_{i-1} + (1-p)\pi_{i+1}, i = 1, 2, \ldots . \tag{8.37}
$$

A simple method to solve these equations is to express state probabilities in terms of π_0 and then to determine the value of π_0 by using the fact that the sum of the probabilities equals 1. Hence (8.36) yields

$$
\pi_1 = \frac{p}{1-p}\pi_0. \tag{8.38}
$$

Rewrite (8.37) as follows

$$
\pi_{i+1} = \frac{\pi_i}{1-p} - \frac{p}{1-p}\pi_{i-1}. \tag{8.39}
$$

Setting $i = 1$ in this equation and using (8.38) yields

$$
\begin{aligned}
\pi_2 &= \frac{\pi_1}{1-p} - \frac{p}{1-p}\pi_0 \\
\\
&= \frac{p\pi_0}{(1-p)^2} - \frac{p\pi_0}{1-p} \\
\\
&= \frac{p^2}{(1-p)^2}\pi_0. \tag{8.40}
\end{aligned}
$$

Comparison of (8.38) and (8.40) suggests that

$$
\pi_i = \alpha^i\pi_0, \quad i = 0, 1, \ldots, \tag{8.41}
$$

where $\alpha \overset{\text{def}}{=} p/(1-p)$. To check that this is correct we substitute (8.41) into (8.39) yielding

$$
\alpha^{i+1}\pi_0 = \frac{\alpha^i\pi_0}{1-p} - \frac{p}{1-p}\alpha^{i-1}\pi_0.
$$

Canceling common terms shows that this is satisfied if

$$\alpha^2(1-p) - \alpha + p = 0.$$

The quadratic formula implies that

$$\alpha = \frac{1 \pm \sqrt{1 - 4p(1-p)}}{2(1-p)} = \frac{1 \pm (1-2p)}{2(1-p)}.$$

The positive radical yields $\alpha = 1$, which is eliminated since there is no way the probabilities can sum to 1 with this solution. Taking the negative radical implies that $\alpha = p/(1-p)$ as hypothesized.

Summing (8.41) yields

$$1 = \pi_0 \sum_{i=0}^{\infty} \alpha^i = \frac{\pi_0}{1-\alpha}$$

and thus the final solution is given by

$$\pi_i = (1-\alpha)\alpha^i, \quad i = 0, 1, \ldots . \tag{8.42}$$

The stationary probabilities are thus geometric with parameter $1 - \alpha$.

8.1.7 Guessing Stationary State Probabilities

Because stationary probabilities are unique, another method to determine them is to *guess their values* and then show that they satisfy (8.30) and (8.31). This might not seem to be a reasonable way to determine stationary probabilities but it is surprisingly useful. In Chapter 10 we will determine the stationary distributions for complex Markov processes that arise in networks of queues in exactly this manner. We record this as a corollary to Proposition 8.6.

Corollary 8.9 *Consider an irreducible Markov chain with one set of ergodic states and with transition matrix $\boldsymbol{P}$. Let $\beta_j, j = 0, 1, \ldots$, be a set of nonnegative numbers. If the values of β_i satisfy*

$$\beta_j = \sum_{k=0}^{\infty} \beta_k p_{k,j} \tag{8.43}$$

and

$$\sum_{j=0}^{\infty} \beta_k = 1, \tag{8.44}$$

then $\beta_j, j = 0, 1, \ldots$, are the stationary probabilities of the Markov chain.

The fact that we can guess a solution of a Markov chain is similar to that of guessing the solution of a recurrence relationship. We can apply this technique to the Markov chain considered in Example 8.8. Suppose we guess that

$$\pi_j = (1 - x)x^j$$

for some value of x. Equation (8.37) implies that

$$(1 - x)x^i = p(1 - x)x^{i-1} + (1 - p)(1 - x)x^{i+1}, \quad i = 1, 2, \ldots .$$

Simplifying this yields

$$(1 - p)^2 x^2 - x + p = 0.$$

The quadratic formula implies that

$$x = \frac{1 \pm \sqrt{1 - 4p(1 - p)}}{2(1 - p)} = \frac{1 \pm (1 - 2p)}{2(1 - p)}.$$

The solution corresponding to the positive radical yields $x = 1$, which is eliminated since it does not yield a stationary solution to the Markov chain. Hence the value of x is given by

$$x = \frac{p}{1 - p}$$

and a simple calculation shows that with this value the probabilities also sum to 1. Hence (8.42) is the solution.

It is easy to correct a guess that satisfies the balance equations (8.43) but not the normalization condition (8.44) provided that the sum of the guessed values is finite. Suppose, for example, that $\beta'_j, j = 0, 1, \ldots,$ satisfy the balance equations (8.43). Then it is clear that

$$\beta_j = b\beta'_j \quad j = 0, 1, \ldots,$$

where

$$b = \left[\sum_{k=0}^{\infty} \beta'_k \right]^{-1} \tag{8.45}$$

satisfies both (8.43) and (8.44). The constant b of (8.45) is called a *normalization constant* and the procedure used above is called *normalizing a solution*.

8.1.8
A Note on Periodic Chains

Proposition 8.6 and Corollary 8.9 are valid for irreducible Markov chains with one ergodic set where states are periodic (it is easy to show that

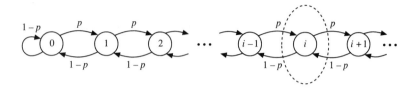

FIGURE 8.3. State Transition Diagram for a Simple Random Walk

if one state is periodic with period d then all states are periodic with the same period; see Problem 8.3). For periodic chains it makes sense to speak of stationary probabilities but not in terms of limiting values of the state transition probabilities. This follows from the fact that the limit of $p_{i,j}^n$, as $n \to \infty$ as given in (8.24), is not defined. For most chains, state transitions can occur from a state back to itself and thus periodicity is not an issue. We will encounter a periodic chain, however, in Section 8.3 when discussing semi-Markov processes where, for the embedded chain, self-loops are explicitly prohibited.

8.2 Global Balance Equations

As illustrated in Example 8.4, the state transition matrix of (8.35) can be conveniently represented by a *probability transition diagram*, which is depicted in Figure 8.3. In this diagram, each circle represents a state of the Markov chain with a value given by the label of the circle. The arcs correspond to the transition probabilities. There is a simple interpretation of the ith equation, as given by (8.37), that is illustrated in the figure. We call the quantity $\pi_i p$ the *probability flux*[3] that flows from state i to state $i+1$. Similarly, let $\pi_i(1-p)$ be the flow of probability flux from state i to state $i-1$. The total probability flux out of state i equals $\pi_i p + \pi_i(1-p) = \pi_i$ and the total probability flux into state i equals $\pi_{i-1}p + \pi_{i+1}(1-p)$. Thus the ith equation of (8.37) shows that the probability flux into state i equals the probability flux out of state i. In other words, probability flux is "balanced" and this explains why we term equations (8.32) global balance equations.

The interpretation of balancing the probability flux into and out of a particular state can be generalized to that of balancing probability flux between sets of states. In Figure 8.3, for example, if we let a set $\mathcal{U} = \{i-1, i, i+1\}, i > 0$, then the probability flux out of set $\mathcal{U}$ equals $\pi_{i-1}(1-p) + \pi_{i+1}p$ and that into set $\mathcal{U}$ equals $\pi_{i-2}p + \pi_{i+2}(1-p)$.

---

[3]The term *flux* comes from physics, where one speaks of magnetism in terms of magnetic flux. Similar to physics where the magnetic flux into a region is balanced by the magnetic flux out of the region, we shall see that the global balance equations can be seen as a balancing of probability flux.

Equating these implies that if probability flux is balanced for set $\mathcal{U}$ then

$$\pi_{i-1}(1-p) + \pi_{i+1}p = \pi_{i-2}p + \pi_{i+2}(1-p). \tag{8.46}$$

Substituting (8.42) into (8.46) shows that this, indeed, is satisfied and thus the probability flux into and out of $\mathcal{U}$ is balanced. Another way to see that (8.46) holds is to observe that this equation arises by summing (8.37) over the indices contained in set $\mathcal{U}$. The sum of (8.37) over indices $i-1, i, i+1$ is given by

$$\begin{aligned}
\pi_{i-1} + \pi_i + \pi_{i+1} \;=\;& p\pi_{i-2} + (1-p)\pi_i \\
&+\; p\pi_{i-1} + (1-p)\pi_{i+1} + p\pi_i + (1-p)\pi_{i+2}.
\end{aligned}$$

Simplification of this equation yields (8.46). This implies that the balancing of probability flux holds for any arbitrary set of states $\mathcal{U}$.

These are important results, and we can express them more generally. The probability flux from state i to state j for Markov chains, denoted by $\tilde{\mathbf{F}}(i,j)$, is defined by

$$\tilde{\mathbf{F}}(i,j) \stackrel{\text{def}}{=} \pi_i p_{i,j}, \quad i,j \in \mathcal{S}. \tag{8.47}$$

We define the probability flux between two subsets of $\mathcal{S}$, say $\mathcal{U}$ and $\mathcal{V}$, as

$$\tilde{\mathbf{F}}(\mathcal{U}, \mathcal{V}) = \sum_{u \in \mathcal{U}, v \in \mathcal{V}} \tilde{\mathbf{F}}(u,v), \quad \mathcal{U}, \mathcal{V} \in \mathcal{S}. \tag{8.48}$$

The global balance equations given by (8.32) can be conveniently restated as

$$\tilde{\mathbf{F}}(\mathcal{U}, \mathcal{U}^c) = \tilde{\mathbf{F}}(\mathcal{U}^c, \mathcal{U}), \quad \text{for all} \quad \mathcal{U} \in \mathcal{S}. \tag{8.49}$$

A way to interpret global balance is that of an "averaging" operation. To explain this, observe that we can couch the dynamic operation of the chain as follows. State i in the system has a value at the nth step given by π_i^n. We will think of this value as being an amount of probability flux. At the $n+1$st step node i receives flux from its neighboring nodes according to the state transition probabilities. That is, if $j \mapsto i$ then node i receives the quantity of flux $\pi_j^n p_{j,i}$ from state j. The value of state i at the $n+1$st step is given by

$$\pi_i^{n+1} = \sum_{j=0}^{\infty} \pi_j^n p_{j,i}, \quad i = 0, 1, \ldots .$$

This represents an averaging of probability flux according to the state transition probabilities, that is, the value of state i at the $n + 1$st step is the average of the values of all of its neighboring states with weights given by the state transition probabilities.

Eventually the probability flux of the states settle down, and the result of the averaging operation is that all states have values equal to their stationary probabilities. This type of dynamic averaging is found in many physical systems. For example, heat flow on a planar object operates in such a manner. For such a system the temperature of a particular point is determined by averaging the temperatures of its neighboring points. Equations that satisfy averaging operations of this type are called *harmonic equations*, and many physical systems are governed by such equations (see [64] for an interesting book describing how one can use electrical analogies to analyze Markov processes).

8.3 Semi-Markov Processes

In this section we consider a generalization of a Markov chain, termed a *semi-Markov process*. In such a process we augment the specification of the process by including a state holding time. The *holding time* of state i, denoted by H_i, is the amount of time that passes before making a state transition from i. The value of H_i, in general, depends on the next state transition, and we let $H_{i,j}$ be the holding time of state i given that the next state transition is to state j. The law of total probability implies that

$$H_i = \sum_{j=0}^{\infty} I_{i,j} H_{i,j}, \qquad (8.50)$$

where $I_{i,j}$ is an indicator random variable that equals 1 with probability $p_{i,j}$. Markov chains have state holding times that are equal to a unit time (i.e., a step) and that are independent of the next state transition. For these processes $H_i = H_{i,j} = 1$.

More generally, holding times do not have to be one time unit nor do they have to be independent of the next state transition. Relaxing these constraints yields a semi-Markov process that operates in the following fashion.

1. After entering state i the process randomly selects its next state $j \neq i$ according to a probability transition matrix.

2. Given that state j is selected, the time spent in state i before jumping to state j is given by a random variable $H_{i,j}$.

A semi-Markov process as just described first selects its next state and then selects a value for its holding time. We could have defined a semi-Markov process in the reverse order: first select a state holding time and then select the next state. Such a formulation is restrictive, however, since it implies that state holding times are necessarily independent of the next transition. Such a formulation corresponds to the special case of a semi-Markov process where $H_{i,j} = H_i$, for all j.

The sample paths for semi-Markov processes are timed sequences of state transitions. We can write a sample path by a sequence of pairs,

$$(i_0, t_0), (i_1, t_1), (i_2, t_2), \ldots, (i_{k-1}, t_{k-1}), (i_k, t_k), \ldots$$

which denotes a sample path where the process is in state i_k over the time period $[t_k, t_{k+1})$. If we view the process only at the times of state transition then the sample paths are identical to those of a Markov chain, as given in (8.7), with transition matrix P^e with entries

$$p^e_{i,j} = \begin{cases} 0, & i = j, \\ p_{i,j}/(1 - p_{i,i}), & i \neq j. \end{cases}$$

This is called the *embedded Markov chain* (the "e" in our notation stands for "embedded"). The stationary probabilities of the embedded Markov chain satisfy global balance. If we let π^e_i be the stationary probability of the embedded chain then

$$\pi^e_i = \sum_{j \neq i} \pi^e_j p^e_{j,i}. \tag{8.51}$$

Clearly π^e_i is the probability that, *at the embedded points*, the process is found in state i. The probability of finding the semi-Markov process in state i at a randomly selected time is *not* given by π^e_i. As a simple counterexample, suppose that the holding time of state 1 is extremely large with respect to the holding times of other states. Then, with a high probability, a random point in time finds the process in state 1 *regardless of the value of* π^e_1 (provided, of course, that π^e_i is positive).

Let π_i denote the stationary probability of the semi-Markov process. It is clear that the value of π_i depends on the stationary probabilities of the embedded chain as well as on the holding time of the state. Intuition suggests that the value of π_i should be proportional to the expected holding time of state i. A plausible conjecture is that π_i is proportional to the product $\pi^e_i E[H_i]$, which, when normalized, yields

$$\pi_i = \frac{\pi^e_i E[H_i]}{\sum_{j=0}^{\infty} \pi^e_j E[H_j]}. \tag{8.52}$$

The following proposition shows that this conjecture is correct.

Proposition 8.10 *Consider a semi-Markov process with state holding times H_i and embedded state transition probabilities $\boldsymbol{P}$. Assume that the embedded Markov chain is positive recurrent and that π_i^e are its stationary probabilities. Then the stationary distribution of the semi-Markov process is given by (8.52).*

Proof: It suffices to show that the fraction of time spent in state i is given by π_i. To do this, let $f_i(n)$ be the fraction of time spent in state i over the first n steps of the embedded process and let $N_i(n)$ be the number of times state i is visited during these n steps. For $j = 1, 2, \ldots$, define $L_i(j)$ to be a random variable that denotes the length of time the process spends in its jth visit to state i. With these definitions we can write

$$f_i(n) \;=\; \frac{\sum_{j=1}^{N_i(n)} L_i(j)}{\sum_{k=0}^{\infty} \sum_{j=1}^{N_k(n)} L_k(j)}$$

$$=\; \frac{\frac{N_i(n)}{n} \sum_{j=1}^{N_i(n)} \frac{L_i(j)}{N_i(n)}}{\sum_{k=0}^{\infty} \frac{N_k(n)}{n} \sum_{j=1}^{N_k(n)} \frac{L_k(j)}{N_k(n)}}. \tag{8.53}$$

Because the embedded chain is positive recurrent $N_i(n) \xrightarrow{a.s.} \infty$ as $n \to \infty$, and the strong law of large numbers implies that

$$\sum_{j=1}^{N_i(n)} \frac{L_i(j)}{N_i(n)} \xrightarrow{a.s.} E\left[H_i\right], \quad \text{as} \quad n \longrightarrow \infty. \tag{8.54}$$

From the Markov property, we can view successive visits to state i as a renewal sequence. The expected number of steps between visits is given by $1/\sum_{n=1}^{\infty} n f_{i,i}^n$ and thus Proposition 6.10 shows that

$$\frac{N_i(n)}{n} \xrightarrow{a.s.} \frac{1}{\sum_{n=1}^{\infty} n f_{i,i}^n}, \quad \text{as} \quad n \longrightarrow \infty. \tag{8.55}$$

Observe that (8.22) implies that

$$\pi_i^e = \frac{1}{\sum_{n=1}^{\infty} n f_{i,i}^n}. \tag{8.56}$$

Substituting (8.54), (8.55), and (8.56) into (8.53) shows that

$$\pi_i = \frac{\pi_i^e E\left[H_i\right]}{\sum_{j=1}^{\infty} \pi_j^e E\left[H_j\right]}$$

and proves the proposition.   ∎

8.3.1
Elimination of
Self-Loops in
Markov Chains

By combining the notion of a state holding in the description of a semi-Markov process we can equivalently view the operation of the chain without self-loops. Recall that a self-loop is a transition from a state immediately back to itself (for state i this occurs with probability $p_{i,i}$). We distinguish the following *equivalent ways* to model the operation of a Markov chain:

Model 1: Self-loops and unit holding times. This is the viewpoint of Section 8.1 where the Markov chain changes state at each step and has self-loops. The holding time in each state is one unit of time.

Model 2: No self-loops and geometric holding times. We view the chain only at epochs where the chain makes transitions that are not self-loops. The operation of the Markov chain is that of a semi-Markov process. Upon entering state i, the chain remains there until it jumps to another state j according to the transition matrix of the embedded Markov chain. The state holding time is independent of j, and is geometrically distributed with an average of

$$E[H_i] = \frac{1}{1 - p_{i,i}}, \quad i = 0, 1, \dots .$$

Denote the stationary probability of the semi-Markov process described in Model 2 by $\tilde{\pi}_i, i = 0, 1, \dots$. Equation (8.52) implies that this can be written as

$$\tilde{\pi}_i = \frac{\pi_i^e / (1 - p_{i,i})}{\sum_{j=0}^{\infty} \pi_j^e / (1 - p_{j,j})}. \tag{8.57}$$

The stationary probabilities of the Markov chain described in Model 1 has stationary probabilities given by π_i of (8.30). It is clear that the two models of the Markov chain given above have the same stationary distribution, and thus $\tilde{\pi}_i = \pi_i$, for $i = 0, 1, \dots$. To analytically establish this, rewrite the global balance equation (8.30) for π_i as

$$\pi_i = \frac{1}{1 - p_{i,i}} \sum_{j \neq i} \pi_j p_{j,i}. \tag{8.58}$$

Also rewrite (8.51) as

$$\pi_i^e = \sum_{j \neq i} \pi_j^e \frac{p_{j,i}}{1 - p_{j,j}}.$$

Equation (8.57) implies that

$$\tilde{\pi}_i = \frac{1}{1 - p_{i,i}} \sum_{j \neq i} \tilde{\pi}_j p_{j,i}. \tag{8.59}$$

Comparing (8.58) and (8.59) shows that $\boldsymbol{\pi}$ and $\tilde{\boldsymbol{\pi}}$ satisfy the same set of linear equations. This, with the observation that both vectors sum to unity implies that $\tilde{\pi}_i = \pi_i, i = 0, 1, \dots$, as required.

8.3.2 Semi-Markov Processes are not Markovian

A moment's reflection shows that semi-Markov processes do not, in general, satisfy the Markov property since the future evolution of the process depends on the current state of the process *and* on the length of time that it has been in that state. To see this, suppose that a semi-Markov process can jump directly to state j from state i. Upon entrance to state i, the probability that the next state is j is given by $p_{i,j}^e$. Suppose that no transition occurs from state i in t time units and let $q_{i,j}(t)$ be the conditional probability that j is the next state. If the Markov property holds, then it must be the case that $q_{i,j}(t) = p_{i,j}^e$. Consider the case where $H_{i,j}$ is a bounded random variable that is less than t and thus that $q_{i,j}(t) = 0 \neq p_{i,j}^e$. In this case the state of the process does not suffice to determine its conditional transition probabilities and thus the semi-Markov process, so described, does not satisfy the Markov property.

There are two cases where a semi-Markov process is Markovian. The first is if state holding times are exponentially distributed. In this case, the memoryless property of exponentials shows that $q_{i,j}(t) = p_{i,j}^e$ for all t. In the second case, we expand the state space of the semi-Markov process to include a component that records the amount of time already spent in the current state. This additional information in the state description makes the process Markovian. Such a modification, however, leads to a continuous state process, which lies outside our scope.

8.4 Markov Processes

Markov processes are natural generalizations of semi-Markov processes with exponentially distributed holding times. Markov processes have no self-loops and their state transitions are characterized by a *generator matrix*, which is analogous to a probability transition matrix. The classification of states and Proposition 8.5 for Markov chains have analogous statements for Markov processes where the probability transition matrix is replaced by a generator matrix.

8.4.1 The Generator Matrix

The generator matrix of a Markov process, denoted by $\boldsymbol{Q}$, has entries that are the rates at which the process jumps from state to state. These entries are defined by

$$q_{i,j} = \lim_{\tau \to 0} \frac{P[X(t+\tau) = j \mid X(t) = i]}{\tau}, \quad i \neq j. \tag{8.60}$$

(We assume that the Markov process is time homogeneous and thus that (8.60) is independent of t.) The total rate out of state i is denoted by q_i and equals

$$q_i = \sum_{j \neq i}^{\infty} q_{i,j}. \tag{8.61}$$

The holding time of state i is exponentially distributed with rate q_i. By definition, we set the diagonal entries of $\boldsymbol{Q}$ equal to minus the total rate,

$$q_{i,i} = -q_i. \tag{8.62}$$

This implies that the row sums of matrix $\boldsymbol{Q}$ equal 0. We next show that the stationary probabilities of a Markov process satisfy a set of linear equations that are expressed in terms of the generator matrix.

8.4.2 The Stationary Distribution of a Markov Process

Markov processes are special cases of semi-Markov processes, and thus we can use (8.52) to determine their stationary distribution. It is clear that the average holding time for state i is given by

$$E[H_i] = \frac{1}{q_i}.$$

To determine the stationary distribution, we must calculate the values $\pi_i^e, i = 0, 1, \ldots$. To do this we first find expressions for $p_{i,j}^e$, the embedded Markov chain. Observe that, by definition, a Markov process has no self-loops and that all transitions occur at exponential rates. These transition rates are given by independent exponential random variables with parameters $q_{i,j}, j \neq i$. The probability of an embedded transition jumping from i to j equals the probability that an exponential random variable with rate $q_{i,j}$ is the minimum of a set of independent exponential random variables with rates $\{q_{i,1}, q_{i,2}, \ldots, q_{i,i-1}, q_{i,i+1}, \ldots\}$. In Example 4.21 we calculated the probability of such an event and (4.94) showed that

$$p_{i,j}^e = \frac{q_{i,j}}{\sum_{j \neq i} q_{i,j}} = \frac{q_{i,j}}{q_i}.$$

From (8.51) it follows that the embedded probabilities satisfy

$$\pi_i^e = \sum_{j \neq i} \pi_j^e \frac{q_{i,j}}{q_i}. \tag{8.63}$$

The stationary distribution of the Markov process using (8.52) is thus given by

$$\pi_i = \frac{\pi_i^e / q_i}{\sum_{j=0}^{\infty} \pi_j^e / q_j}. \tag{8.64}$$

We can write a simple expression for the stationary probabilities in terms of the generator matrix. Multiplying (8.64) by $q_{i,j}$ and summing yields

$$
\begin{aligned}
\sum_{i=0}^{\infty} \pi_i q_{i,j} &= \frac{\sum_{i=0}^{\infty} \pi_i^e q_{i,j}/q_i}{\sum_{j=0}^{\infty} \pi_j^e/q_j} \\[2mm]
&= \frac{\sum_{i \neq j} \pi_i^e q_{i,j}/q_i + \pi_j^e q_{j,j}/q_j}{\sum_{j=0}^{\infty} \pi_j^e/q_j} \\[2mm]
&= \frac{\pi_j^e - \pi_j^e}{\sum_{j=0}^{\infty} \pi_j^e/q_j} = 0,
\end{aligned}
\tag{8.65}
$$

where we have substituted (8.63) and (8.62) in (8.65). Rewriting (8.65) in matrix form, shows that the stationary probabilities of a Markov process satisfy

$$
\pi Q = 0,
\tag{8.66}
$$

with the additional normalization requirement that $|\pi| = 1$.

Equations (8.66) are balance equations where probability flux into a state is equated to probability flux out of a state. The probability flux from state i to state j for continuous Markov processes is defined analogously to that of Markov chains (8.47):

$$
\mathbf{F}(i,j) \stackrel{\text{def}}{=} \pi_i q_{i,j}, \quad i, j \in \mathcal{S}.
\tag{8.67}
$$

(Note that we use the symbol $\mathbf{F}(i,j)$ to distinguish the probability flux for continuous processes from that of discrete chains, which is denoted by $\tilde{\mathbf{F}}(i,j)$.) The probability flux between two sets of states is defined similarly to that of (8.48). Written in these terms, the global balance equations (8.66) are given by

$$
\mathbf{F}(\mathcal{U}, \mathcal{U}^c) = \mathbf{F}(\mathcal{U}^c, \mathcal{U}), \quad \text{for all} \quad \mathcal{U} \in \mathcal{S},
\tag{8.68}
$$

which is identical to (8.49) except that in (8.68) it is expressed for continuous time processes.

Example 8.11 (The M/M/1 Queue) Consider the M/M/1 queue with arrival and service rates of λ and μ, respectively. We will show that the process is ergodic if $\lambda < \mu$. Let $X(t)$ be the number of customers in the system at time t. Since the interarrival times and service times are exponentially distributed it is clear that $X(t)$ is a Markov process. The generator matrix for this process has positive entries $q_{i,i+1} = \lambda$ and $q_{i+1,i} = \mu$ for $i = 0, 1, \ldots$ and hence is tridiagonal

$$Q = \begin{bmatrix} -\lambda & \lambda & 0 & 0 & 0 & 0 \\ \mu & -(\lambda+\mu) & \lambda & 0 & 0 & 0 \\ 0 & \mu & -(\lambda+\mu) & \lambda & 0 & 0 \\ 0 & 0 & \mu & -(\lambda+\mu) & \lambda & 0 \\ \vdots & \vdots & \vdots & \vdots & \vdots & \vdots \end{bmatrix}.$$

The ith equation for (8.66) is given by

$$\pi_{i-1}\lambda - (\lambda+\mu)\pi_i + \pi_{i+1}\mu = 0, \quad i = 1, 2, \ldots, \tag{8.69}$$

with the boundary equation

$$-\pi_0\lambda + \pi_1\mu = 0. \tag{8.70}$$

These equations can be easily solved in a fashion similar to that used in Example 8.8. We write all probabilities in terms of π_0 and determine this quantity using the normalization constraint. From (8.70) we obtain

$$\pi_1 = \rho\pi_0, \tag{8.71}$$

where we have defined $\rho \overset{\text{def}}{=} \lambda/\mu$. We rewrite (8.69) as

$$\pi_{i+1} = (1+\rho)\pi_i - \rho\pi_{i-1}, \quad i = 1, 2, \ldots,$$

and note that for $i = 1$ this implies that

$$\begin{aligned} \pi_2 &= (1+\rho)\pi_1 - \rho\pi_0 \\[6pt] &= \rho^2\pi_0, \end{aligned} \tag{8.72}$$

where we have used (8.71) in (8.72). Continuing in this fashion shows that

$$\pi_i = \rho^i\pi_0, \quad i = 1, 2, \ldots . \tag{8.73}$$

The normalization constraint implies that

$$\begin{aligned} \pi_0 &= \left\{\sum_{i=0}^{\infty} \pi_i/\pi_0\right\}^{-1} = \left\{\sum_{i=0}^{\infty} \rho^i\right\}^{-1} \\[6pt] &= 1 - \rho. \end{aligned}$$

Hence stationary probabilities exist only if $\rho < 1$, and this implies that the system is ergodic if $\lambda < \mu$. The solution is given by

$$\pi_i = (1-\rho)\rho^i, \quad i = 0, 1, \ldots, \tag{8.74}$$

as we previously derived using Laplace transform methods in (7.52). An important observation to make about this problem is that the repetitive nature of the state transitions leads to a geometric form for the solution of the stationary state probabilities. This key observation suggests that whenever a state space has a similar repetitive structure (for an infinite portion of its state space) a good guess for the state probabilities is that they are geometric. We will see another example of this in Example 8.12.

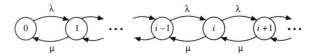

FIGURE 8.4. Rate Transition Diagram for the M/M/1 Queue

Analogous to the probability transition diagrams of Markov chains, a convenient way to represent the balance equations for Markov processes is with *rate transition diagrams*. In Figure 8.4 we show the rate transition diagram for the M/M/1 queue of Example 8.11. Balance equation (8.68) implies that for any collection of states in $\mathcal{U}$ the flow of probability flux out of the set equals the flow into the set. Setting $\mathcal{U} = \{0, 1, \ldots, i\}$, for example, implies that

$$\pi_i \lambda = \pi_{i+1} \mu, \quad i = 1, 2, \ldots,$$

and thus that

$$\pi_{i+1} = \rho \pi_i = \rho^i \pi_0, \quad i = 1, 2, \ldots, \tag{8.75}$$

as derived with a more lengthy calculation in (8.73).

Example 8.12 (A Variation of the M/M/1 Queue Revisited) It is interesting to note that with the M/M/1 queue of Example 8.11 we could have obtained the solution in a more straightforward manner by simply guessing that it had a geometric form. We do this for a variation of the M/M/1 queue that lends insight into how one establishes the validity of a guess. The procedure used here to derive the stationary solution will be used in Section 9.1 to solve for the stationary solution of a class of vector processes that have a repetitive structure. This solution technique is called the matrix geometric method and is the subject of Chapter 9.

Consider a model where customer arrivals to the system when the system is in state $i, i \geq 1$, have exponential interarrival times with a rate of λ and customer service times are exponential with rate μ. Interarrival times to an empty system are exponential with rate λ'. The system is ergodic if $\lambda < \mu$. The generator matrix of the process is given by

$$Q = \begin{bmatrix} -\lambda' & \lambda' & 0 & 0 & 0 \cdot \cdots \\ \mu & -(\lambda + \mu) & \lambda & 0 & 0 \cdots \\ 0 & \mu & -(\lambda + \mu) & \lambda & 0 \cdots \\ 0 & 0 & \mu & -(\lambda + \mu) & \lambda \cdots \\ \vdots & \vdots & \vdots & \vdots & \vdots \end{bmatrix}. \tag{8.76}$$

We call states 0 and 1 the *initial portion of the process* or, equivalently, the *boundary states*, and states $2, 3, \ldots$, the *repeating portion of the process* or, equivalently, the *repeating states*. It is clear from the generator matrix given in (8.76) that the transition rates for states in the repeating portion of the state space have a repetitive structure; the infinite portion of the state space from states $2, 3, \ldots$, has transitions that are congruent to the infinite portion of the state space from states $j, j+1, \ldots$, for any $j \geq 2$. This repetitive structure implies that the stationary state probabilities in the repeating portion of the process have a simple geometric form. This intuition follows from the similarly structured M/M/1 queue analyzed in Example 8.11. Suppose we guess that

$$\pi_i = \pi_1 \beta^{i-1}, \quad i = 1, 2, \ldots . \tag{8.77}$$

The ith global balance equation of this system is identical to (8.69) and is given by

$$\pi_{i-1}\lambda - (\lambda + \mu)\pi_i + \pi_{i+1}\mu = 0, \quad i = 2, 3, \ldots . \tag{8.78}$$

Substituting (8.77) into (8.78) and canceling common terms yields

$$\beta^2 - \beta(1 + \rho) + \rho = 0.$$

This quadratic equation has two solutions, $\beta = 1$ and $\beta = \rho$. Since $\beta = 1$ cannot be normalized, we conclude that (8.77) is a possible solution with $\beta = \rho$. To establish the validity of this solution, using Corollary 8.9, we need to determine values for π_0 and π_1 that satisfy the balance equations.

We can write equations for the initial portion of the matrix as

$$-\pi_0\lambda' + \pi_1\mu \qquad = \quad 0, \tag{8.79}$$

$$\pi_0\lambda' - \pi_1(\lambda + \mu) + \pi_2\mu \quad = \quad 0, \tag{8.80}$$

or in matrix form as

$$(\pi_0, \pi_1) \begin{bmatrix} -\lambda' & \lambda' \\ \mu & -\mu \end{bmatrix} = 0, \tag{8.81}$$

where we have used the equality $\pi_2 = \rho\pi_1$ in (8.80). It is clear that (8.81) does not have a unique solution since the second column is the negative of the first column (the rank of the matrix is one less than the number of unknowns). To determine the unique solution for these quantities we must use the normalization condition, which shows that

$$1 = \pi_0 + \pi_1 \sum_{j=1}^{\infty} \rho^{j-1} = \pi_0 + \pi_1(1 - \rho)^{-1}. \tag{8.82}$$

This, in combination with (8.81), yields a unique solution given by

$$\pi_0 \quad = \quad \frac{1}{1 + \rho'/(1 - \rho)}, \tag{8.83}$$

$$\pi_1 \quad = \quad \frac{\rho'}{1 + \rho'/(1 - \rho)}, \tag{8.84}$$

where $\rho' \overset{\text{def}}{=} \lambda'/\mu$. Equations (8.77), (8.83), and (8.84) constitute the complete solution, which can be checked by demonstrating that it solves global balance. Observe that if $\lambda' = \lambda$ then we obtain the solution for the M/M/1 queue as given in Example 8.11.

Suppose that we wish to calculate the expected number of customers waiting in the queue for this model. We can do this by assigning the weight $j - 1$ to state j and calculating

$$\sum_{j=1}^{\infty}(j-1)\pi_j \;=\; \pi_1\sum_{j=1}^{\infty}(j-1)\rho^{j-1}$$

$$=\; \pi_1\frac{\rho}{(1-\rho)^2} = \frac{\rho\rho'}{(1-\rho)(1-\rho+\rho')}.$$

More complex performance measures can be similarly calculated as a weighted sum of the stationary probabilities.

We used the following three steps to derive the solution for the modified M/M/1 queue considered in Example 8.12. This procedure will form the basis for subsequent derivations of more general forms of Markov processes discussed in Section 9.1.

1. We guessed a geometric solution form for the repeating portion of the Markov process. This form required us to calculate the value of an unknown constant.

2. To calculate the value of the constant, we substituted the geometric form for one of the balance equations in the repeating portion of the process and solved for the root of that equation.

3. The boundary portion of the process was then solved using the results of the repeating portion of the process and the normalization condition.

Often guessing solutions is the best initial approach. Consider the following comparison of a direct solution approach to that of an educated guess.

Example 8.13 (A Processor Model with Failures) Consider a single processor with an infinite waiting room. Customer arrivals are assumed to be Poisson with rate λ and service times are exponentially distributed with expectation $1/\mu$. Suppose that the processor fails at rate γ and, when failed, all of the customers in the system are assumed to be lost. We are interested in calculating the probability that there are i customers in the system. Observe that the model is similar to that of an M/M/1 queue but where there are

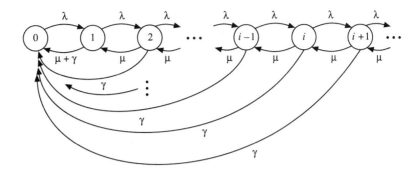

FIGURE 8.5. Rate Transition Diagram for Processor Failure Model

transitions from state i to state 0 at rate γ (these model processor failures). The rate transition diagram for the model is given in Figure 8.5.

To analyze this system, observe that the global balance equations imply that

$$\lambda\pi_0 = \mu\pi_1 + \gamma(\pi_1 + \pi_2 + \ldots)$$

$$= \mu\pi_1 + \gamma(1 - \pi_0).$$

Defining $\rho \overset{\text{def}}{=} \lambda/\mu$ and $\alpha \overset{\text{def}}{=} \gamma/\mu$ implies that the balance equation for the boundary can be written as

$$\pi_1 = \pi_0(\rho + \alpha) - \alpha. \tag{8.85}$$

The non-boundary balance equations are given by

$$(1 + \rho + \alpha)\pi_i = \pi_{i+1} + \rho\pi_{i-1}, \quad i = 1, 2, \ldots . \tag{8.86}$$

To solve these equations we can use generating functions and we define

$$\Pi(z) \overset{\text{def}}{=} \sum_{i=0}^{\infty} \pi_i z^i.$$

Multiplying (8.86) by z^{i+1} and summing over $i = 1, 2, \ldots$ yields

$$(1 + \rho + \alpha)z\left(\Pi(z) - \pi_0\right) = (\Pi(z) - \pi_1 z - \pi_0) + \rho z^2 \Pi(z).$$

Substituting (8.85) in this and simplifying implies that

$$\Pi(z) = \frac{\pi_0(1 - z) - \alpha z}{\rho z^2 - (1 + \rho + \alpha)z + 1}$$

$$= \frac{\pi_0(1 - z) - \alpha z}{(1 - z/r_1)(1 - z/r_2)}, \tag{8.87}$$

where $r_i, i = 1, 2$ are the roots of the quadratic polynomial in the denominator. These roots are given by

$$r_i = \frac{1 + \rho + \alpha + (-1)^i \sqrt{(1 + \rho + \alpha)^2 - 4\rho}}{2\rho}, \quad i = 1, 2.$$

To determine the value of π_0 we use the fact that $\Pi(z)$ must be bounded for $z < 1$ (see Appendix C). This implies that a root of the denominator that is less than 1 must also be a root of the numerator. The root r_1 is less than 1 and thus it must be the case that

$$\pi_0(1 - r_1) - \alpha r_1 = 0, \tag{8.88}$$

or that

$$\pi_0 = \frac{\alpha r_1}{1 - r_1}$$

$$= \frac{\alpha \left(1 + \alpha + \rho - \sqrt{(1 + \rho + \alpha)^2 - 4\rho}\right)}{\rho + \sqrt{(1 + \rho + \alpha)^2 - 4\rho} - (1 + \alpha)}.$$

A partial fraction expansion of (8.87) yields

$$\Pi(z) = \frac{\pi_0(1 - r_1) - \alpha r_1}{(1 - r_1/r_2)(1 - z/r_1)} + \frac{\pi_0(1 - r_2) - \alpha r_2}{(1 - r_2/r_1)(1 - z/r_2)}.$$

Equation (8.88) shows that the first summand equals 0 and thus

$$\Pi(z) = \frac{\pi_0(1 - r_2) - \alpha r_2}{(1 - r_2/r_1)(1 - z/r_2)}.$$

Expanding this in a power series (recall that $(1 - ax)^{-1} = 1 + ax + (ax)^2 + \ldots$) implies that

$$\Pi(z) = \pi_0(1 + \beta z + (\beta z)^2 + \ldots,$$

where $\beta = 1/r_2$, or equivalently

$$\beta = \frac{2\rho}{1 + \rho + \alpha + \sqrt{(1 + \rho + \alpha)^2 - 4\rho}}$$

$$= \frac{1 + \rho + \alpha - \sqrt{(1 + \rho + \alpha)^2 - 4\rho}}{2}. \tag{8.89}$$

This finally implies that

$$\pi_i = \pi_0 \beta^i, \quad i = 1, 2, \ldots,$$

and that $\pi_0 = 1 - \beta$. This solution can be easily checked algebraically. Hence, the final solution is given by

$$\pi_i = (1 - \beta)\beta^i, \quad i = 0, 1, \ldots . \tag{8.90}$$

The above example illustrates the algebra that is typically found in generating function solutions for the stationary distribution of a Markov process. Once one begins the algebra, it is easy to lose sight of the probabilistic nature of the problem. It is thus desirable to have a simpler probabilistic, approach to determine the stationary solution. In fact, for the processor failure model there is a direct method that avoids almost all calculation!

To anticipate the result, observe that the rate transition diagram for the process is identical to that of the M/M/1 queue except that arcs have been added from state i to 0. We can think of these arcs as "draining away" probability flux from the states. A critical observation is that the rate of this drainage is the same for each state. This suggests that the ratio of the stationary probabilities of state i to state j in this system should be the same as the corresponding ratio of the stationary probabilities in an M/M/1 queue. Hence, a natural guess is that the processor failure model has a geometric solution like that of the M/M/1 queue.

Suppose we guess the solution:

$$\pi_i = (1-a)a^i, \quad i = 0, 1, \ldots \tag{8.91}$$

Substituting (8.91) into (8.86) and canceling common terms yields

$$a^2 - (1 + \rho + \alpha)a + \rho = 0.$$

Substituting (8.91) into (8.85) yields

$$(1-a)a = (1-a)(\rho + \alpha) - \alpha.$$

There are two solutions to this

$$a = \frac{1 + \rho + \alpha \pm \sqrt{(1 + \rho + \alpha)^2 - 4\rho}}{2}.$$

The root with the positive radical is larger than 1, and thus

$$a = \frac{1 + \rho + \alpha - \sqrt{(1 + \rho + \alpha)^2 - 4\rho}}{2}.$$

Comparison of this to (8.89) shows that (8.91) matches the solution given in (8.90). A "good guess" is worth pages of algebra!

Example 8.14 (A Failure Model) Consider a computer system that has a total of n disk drives that fail at an exponential rate of λ. There is one repairman who fixes failed drives at an exponential rate of μ. What is the average number of failed disk drives?

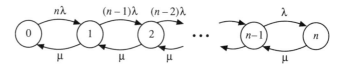

FIGURE 8.6. Transition Rate Diagram for Disk Failure Model

To answer the question, we model the system as a Markov process where the state of the system is the number of failed disks. The transition rate diagram of this system is shown in Figure 8.6 and shows that nonzero off-diagonal entries of the generator matrix are given by

$$q_{i,j} = \begin{cases} (n-i)\lambda, & j = i+1, \quad i = 0, 1, \ldots, n-1, \\ \\ \mu, & j = i-1, \quad i = 1, 2, \ldots, n. \end{cases}$$

To solve the system, observe that a cut corresponding to $\mathcal{U} = \{0, 1, \ldots, i\}, i = 0, 1, \ldots, n-1$, implies, from global balance (8.68), that

$$(n-i)\lambda \pi_i = \mu \pi_{i+1}, \quad i = 0, 1, \ldots, n-1.$$

Expressing probabilities in terms of π_0 implies that

$$\pi_i = n(n-1) \cdots (n-i)(\lambda/\mu)^i \pi_0 = (n)_i \beta^i \pi_i,$$

where $\beta \stackrel{\text{def}}{=} \lambda/\mu$. Normalizing this to determine π_0 yields

$$\pi_i = \frac{(n)_i \beta^i}{\sum_{j=0}^n (n)_j \beta^j}.$$

The expected number of failed disks is thus given by

$$\frac{\sum_{i=1}^n i(n)_i \beta^i}{\sum_{j=0}^n (n)_j \beta^j}. \tag{8.92}$$

Example 8.15 (Deriving the Pollaczek–Khinchin Equation) In this example we derive the Pollaczek–Khinchin transform equation for the number of customers in an M/G/1 queue (7.48). To do this, we consider an embedded Markov chain that is formed by examining an M/G/1 queue at customer departure epochs. Let the probability distribution of the number of customers in the system at these embedded points be denoted by $\pi_i, i = 0, 1, \ldots,$. From the level crossing law (7.17), π_i also equals the probability that an arriving customer sees i customers in the system. From the PASTA property of Poisson arrivals we know that this equals the probability that there are i customers in the system at a random point in time.

Let V be a random variable that equals the number of customers that arrive during a service interval and let $P[V = j] = v_j$, $j = 0, 1, \ldots$. We can write the probabilities v_j by conditioning on the length of a service epoch. This yields

$$v_j = \int_0^\infty \frac{(\lambda t)^j}{j!} e^{-\lambda t} f_S(t) dt, \quad j = 0, 1, \ldots,$$

where S is a random variable depicting a generic service time and λ is the customer arrival rate. The calculations of Section 7.3, see equation (7.47), show that

$$\mathcal{G}_V(z) = \mathcal{L}_S(\lambda(1 - z)). \tag{8.93}$$

The transition probabilities of the embedded chain satisfy

$$p_{i,j} = \begin{cases} 0, & j < i - 1, \\ v_j, & i = 0, \ j \geq 0, \\ v_{j-i+1}, & i = 1, 2, \ldots, \ j \geq i - 1, \end{cases}$$

and the global balance equations (8.30) are given by

$$\pi_j = \pi_0 v_j + \sum_{i=1}^{j+1} \pi_i v_{j-i+1}, \quad j = 0, 1, \ldots .$$

Let N be a random variable depicting the number of customers in the system and thus $P[N = i] = \pi_i$. This implies that

$$\begin{aligned} \mathcal{G}_N(z) &= (1 - \rho)\mathcal{G}_V(z) + \sum_{j=0}^\infty \sum_{i=0}^{j+1} \pi_i v_{j-i+1} z^j \\ &= (1 - \rho)\mathcal{G}_V(z) + \frac{1}{z} \sum_{i=1}^\infty \pi_i z^i \sum_{j=i-1}^\infty v_{j-i+1} z^{j-i+1} \\ &= (1 - \rho)\mathcal{G}_V(z) + \frac{(\mathcal{G}_N(z) - (1 - \rho))\mathcal{G}_V(z)}{z} \\ &= \frac{(1 - z)(1 - \rho)\mathcal{G}_V(z)}{\mathcal{G}_V(z) - z}. \end{aligned} \tag{8.94}$$

A simple calculation from (8.94) using the summation property, $\mathcal{G}_N(1) = 1$, shows that $\pi_0 = 1 - \rho$ where $\rho = \lambda E[S] < 1$. Substituting (8.93) into (8.94) shows that it is identical to the Pollaczek–Khinchin transform equation given in (7.48).

We can use the results from Chapter 7 to obtain more detailed stationary statistics for some of the M/G/1 systems considered in that chapter. One of the queues studied in Section 7.3.2 was the $M/H_2(k)/1$ queue where the service time was given by (7.33). Using transform analysis we obtained the stationary queue length statistics that were given by (7.62). In the next example we show how this analysis can be extended to obtain statistics not only of the number of customers in the system but also for the state of the server.

Example 8.16 (The $M/H_2(k)/1$ Queue-Continued) We can model the $M/H_2(k)/1$ queue defined in Chapter 7 as a bivariate Markov process. Let (N, I) be the state of the process where N is a random variable that denotes the number of customers in the system, and I denotes the state of the server. Random variable I equals 0 if no customer is in service, 1 if the rate of the server is $k\mu$, and 2 if the rate of the server is μ/k. By definition of the hyperexponential service, it is clear that (N, I) is a Markov process. The state transition diagram for the process is shown in Figure 8.7.

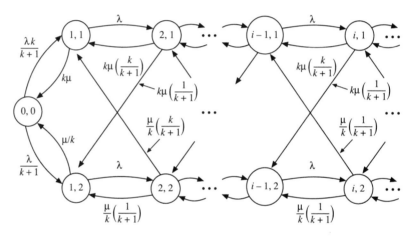

FIGURE 8.7. State Transition Diagram for the $M/H_2(k)/1$ Queue

To explain the diagram, consider state $(i, 1)$. The service rate in this state is $k\mu$. Upon service completion, the customer at the head of the queue begins service. This decreases the number of customers in the system by 1. With probability $k/(k+1)$ the new customer receives service at rate $k\mu$ and with probability $1/(k+1)$ it receives service at rate μ/k. This explains the transitions from state $(i, 1)$ to states $(i-1, 1)$ and $(i-1, 2)$, respectively. The transition from state $(i, 1)$ to state $(i+1, 1)$ at rate λ corresponds to an arrival of a new customer to the system.

Let $\pi_{i,j}, i = 0, 1, \ldots, j = 0, 1, 2$, be the stationary probability of state $\pi_{i,j}$ and let π_i be the stationary probability that there are i customers in the system. The values of π_i are given by

$$\pi_i = \begin{cases} \pi_{0,0}, & i = 0, \\ \pi_{i,1} + \pi_{i,2}, & i = 1, 2, \ldots . \end{cases} \tag{8.95}$$

The analysis of Section 7.3.2 shows that $\pi_i, i = 0, 1, \ldots$ is given by (7.62), that is,

$$\pi_i = q(1 - \alpha_1)\alpha_1^i + (1 - q)(1 - \alpha_2)\alpha_2^i, \quad i = 0, 1, \ldots, \tag{8.96}$$

where q and $r_i = 1/\alpha_i, i = 1, 2$ are given in (7.63) and (7.58), respectively. It is easy to check that

$$\pi_0 = 1 - \rho, \tag{8.97}$$

where $\rho = \lambda/\mu$. We can now determine the values of $\pi_{i,j}$ from (8.96).

To do this, let $\mathcal{U} = \{(n, j), n = 0, 1, \ldots, i - 1, \ j = 0, 1, 2\}$ be the set of states that have $i - 1$ or fewer customers. Forming the global balance equations for these states, (8.68) implies that

$$\lambda(\pi_{i-1,1} + \pi_{i-1,2}) = \pi_{i,1}\left[\frac{k^2\mu}{k+1} + \frac{k\mu}{k+1}\right] + \pi_{i,2}\left[\frac{\mu}{k+1} + \frac{\mu}{k(k+1)}\right],$$

for $i = 1, 2, \ldots$. Substituting (8.95) into this and simplifying yields

$$\lambda\pi_{i-1} = \pi_{i,1}k\mu + \pi_{i,2}\frac{\mu}{k}, \quad i = 1, 2, \ldots . \tag{8.98}$$

Equations (8.95) and (8.98) are linear and can be solved to yield

$$\pi_{i,1} = \frac{\rho k}{k^2 - 1}\pi_{i-1} - \frac{1}{k^2 - 1}\pi_i, \quad i = 1, 2, \ldots, \tag{8.99}$$

and

$$\pi_{i,2} = \frac{k^2}{k^2 - 1}\pi_i - \frac{\rho k}{k^2 - 1}\pi_{i-1}, \quad i = 1, 2, \ldots . \tag{8.100}$$

Thus the stationary distribution of the process is given by equations (8.95), (8.97), (8.99), and (8.100).

It is instructive to check this result by showing that it satisfies global balance. The balance equation for state $\pi_{i,1}$ is given by (the procedure is identical for $\pi_{i,2}$ and thus is not illustrated here)

$$\pi_{i,1}(\lambda + k\mu) = \frac{k^2\mu}{k+1}\pi_{i+1,1} + \frac{\mu}{k+1}\pi_{i+1,2} + \lambda\pi_{i-1,1}, \quad i = 1, 2, \ldots .$$

Substituting (8.99) and (8.100) into this yields, after some algebraic simplification, the following equation for the values of π_i,

$$\rho(k^2 + k\rho + 1)\pi_{i-1} - (\rho(k^2 - k + 1) + k)\pi_i - \rho^2 k\pi_{i-2} = 0, \quad i = 1, 2, \ldots .$$

Substituting (8.96) into this yields an equation of the form

$$q(1 - \alpha_1)\alpha_1^{i-1}f(r_1) + (1 - q)(1 - \alpha_2)\alpha_2^{i-1}f(r_2) = 0, \tag{8.101}$$

where f is a quadratic polynomial defined by

$$f(x) \stackrel{\text{def}}{=} \rho^2 kx^2 - \rho(k^2 + k\rho + 1)x + \rho(k^2 - k + 1) + k.$$

It thus suffices to show that $f(r_i) = 0, i = 1, 2$ for global balance to hold. This follows from the work already done in Chapter 7. Observe that the denominator of (7.57) equals $f(z)$. Equation (7.58) shows that $r_i, i = 1, 2$ are the roots of f and thus establishes (8.101).

The method used to determine the stationary probabilities of the $M/H_2(k)/1$ queue illustrated in Example 8.16 is quite involved, and it certainly is desirable to have a simpler approach. In Chapter 9 we will develop a technique, called the matrix geometric technique, that readily solves this and many other similar problems.

8.5 Variations of the M/M/1 Queue

The analysis presented in Section 8.4 can be used to derive the stationary distribution of a large class of birth–death processes. A *birth–death* process is a process that has transitions that can go up or down by at most one step. More precisely, the state transitions satisfy

$$q_{i,j} = \begin{cases} \lambda_i, & j = i+1, \quad i = 0, 1, \ldots, N-1, \\ \\ \mu_i, & j = i-1, \quad i = 1, 2, \ldots, N, \end{cases}$$

where N is the number of states in the system (N can be infinite).

The stationary probability distribution of birth–death processes have a particularly simple form. Similar to the derivation of (8.75), if we select $\mathcal{U} = \{0, 1, \ldots, i\}$ then (8.68) implies that the stationary probabilities satisfy

$$\pi_i \lambda_i = \pi_{i+1} \mu_{i+1}, \quad i = 0, 1, \ldots, N-1.$$

This can be written as

$$\pi_{i+1} = \rho_i \pi_i, \quad i = 0, 1, \ldots, N-1,$$

where we define

$$\rho_i \stackrel{\text{def}}{=} \frac{\lambda_{i-1}}{\mu_i}, \quad i = 1, 2, \ldots, N.$$

Iterating this equation and normalizing the solution implies that the stationary distribution of a birth–death process is given by

$$\pi_i = \frac{\psi_i}{\sum_{k=0}^{N} \psi_k}, \quad i = 0, 1, \ldots, N, \tag{8.102}$$

where

$$\psi_i \stackrel{\text{def}}{=} \begin{cases} 1, & i = 0, \\ \\ \prod_{k=1}^{i} \rho_k, & i \neq 0. \end{cases}$$

The process is ergodic if $N < \infty$ and if there are an infinite number of states then it is ergodic if

$$\sum_{k=0}^{\infty} \psi_k < \infty.$$

Birth–death processes have a wide range of applications, most notably in variations of the M/M/1 queue. These variations are obtained by changing the forms of λ_i, μ_i, and the value of N. For example, the M/M/1 queue has parameter values $\lambda_i = \lambda$, $\mu_i = \mu$ and $N = \infty$. We will consider several variations that frequently arise in applications (the queueing notation used in these examples was defined in the introduction to Chapter 7).

8.5.1 The M/M/1/k Queue

Parameter values $\lambda_i = \lambda$, $\mu_i = \mu$, and $N = k$: This queue can model a computer system that has a finite buffer. Customers arriving to the system when there are k customers already in the system are assumed to be lost. The corresponding Markov process for the queueing system corresponds to a state truncation of an M/M/1 queue. The stationary distribution given by (8.102) is

$$\pi_i = \frac{(1-\rho)\rho^i}{\sum_{j=0}^{k}(1-\rho)\rho^j} = \frac{(1-\rho)\rho^i}{1-\rho^{k+1}}, \quad i = 0, 1, \ldots, k. \quad (8.103)$$

8.5.2 The M/M/k Queue

Parameter values $\lambda_i = \lambda$, $\mu_i = \min\{i, k\}\mu$, and $N = \infty$: This models a processing system consisting of a single queue served by k processors. It is clear that the state space is identical to that of an M/M/1 queue except that transitions from i to $i - 1$ are given by $\min\{i, k\}\mu$ for $i = 1, 2, \ldots$ (see Figure 8.8).

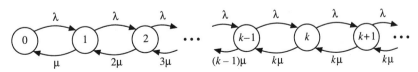

FIGURE 8.8. State Transition Diagram for the M/M/k Queue

The stationary distribution is given by

$$\pi_i = b \begin{cases} \rho^i/i!, & i = 0, 1, \ldots, k, \\ \rho^i/(k!k^{i-k}), & i = k+1, k+2, \ldots, \end{cases}$$

where

$$b = \left[\sum_{\ell=0}^{k-1} \frac{\rho^\ell}{\ell!} + \sum_{\ell=k}^{\infty} \frac{\rho^\ell}{k!m^{\ell-k}} \right]^{-1}$$

is a normalization constant.

The probability that an arriving customer must wait in the queue has special significance in switching systems that have buffer capacity. This probability is referred to as *Erlang's C formula* and is denoted by $C(k, \rho)$. It is given by

$$
C(k, \rho) = \sum_{i=k}^{\infty} \pi_i
$$

$$
= \frac{\rho^k / (k! (1 - \rho))}{\sum_{\ell=0}^{k-1} \rho^\ell / \ell! + \sum_{\ell=k}^{\infty} \rho^\ell / (k! m^{\ell-k})}.
$$

8.5.3 The M/M/k/k Queue

Parameter values $\lambda_i = \lambda$, $\mu_i = i\mu$, and $N = k$: This queue consists of k servers with no waiting room. Customer arrivals when all servers are busy are lost. Such a queue can be used to model a switching center that allows a maximum of k simultaneous calls. The stationary distribution is

$$
\pi_i = b \frac{\rho^i}{i!}, \quad i = 0, 1, \ldots, k,
$$

where

$$
b = \left[\sum_{\ell=0}^{k} \frac{\rho^\ell}{\ell!} \right]^{-1}
$$

is a normalization constant.

The probability π_k is the probability that all servers are busy, and consequently is the probability that an arriving customer is lost. This has special significance in telephone switching and is termed *Erlangs's B formula*. This is given by

$$
B(k, \rho) = \frac{\rho^k / k!}{\sum_{\ell=0}^{k} \rho^\ell / \ell!}.
$$

8.5.4 The M/M/1//p Queue

Parameter values $\lambda_i = \lambda(p-i)$, $\mu_i = \mu$, and $N = p$: This models a queue with a finite population of customers. If there are i customers already resident in the queue then the arrival rate is given by $\lambda(p-i)$. The arrival process thus "slows down" as the number of customers resident in the queue increases. The stationary distribution is

$$
\pi_i = bi! \begin{pmatrix} p \\ i \end{pmatrix} \rho^i, \quad i = 0, 1, \ldots, p,
$$

where

$$b = \left[\sum_{\ell=0}^{p} \ell! \begin{pmatrix} p \\ \ell \end{pmatrix} \rho^\ell \right]^{-1}$$

is a normalization constant.

8.6 The G/M/1 Queue

Example 8.15 showed that we could derive the stationary statistics of the M/G/1 queue by considering an embedded Markov chain. In this section we use the same method to derive the stationary statistics of the G/M/1 queue. In contrast to our derivation of the M/G/1 case, instead of embedding at departure instants to take advantage of the memoryless arrivals, for the G/M/1 case we embed at arrival instants to take advantage of the memoryless service intervals. Our first problem is to derive the state transition probabilities for the embedded chain. (By *state* we implicitly mean the total number of customers in the queueing system including the customer, if any, in the server.) It is clear that transitions from state i can increase to at most $i+1$ but can decrease by up to i, corresponding to emptying the queue during an arrival epoch. Let A be a random variable denoting a generic interarrival epoch and let its average arrival rate be given by $\lambda = 1/E[A]$. Since service times are exponential with parameter μ, the number of customers that leave the system in t seconds is distributed as a Poisson random variable with parameter μt. Hence we can write

$$p_{i,i+1-j} = \int_0^\infty \frac{e^{-\mu t}(\mu t)^j}{j!} f_A(t)dt, \quad j = 0, 1, \ldots, i, \quad (8.104)$$

and

$$\begin{aligned} p_{i,0} &= \int_0^\infty \left(1 - \sum_{\ell=0}^{i} \frac{e^{-\mu t}(\mu t)^\ell}{\ell!} \right) f_A(t)dt \\ &= \int_0^\infty \sum_{\ell=i+1}^{\infty} \frac{e^{-\mu t}(\mu t)^\ell}{\ell!} f_A(t)dt \quad i = 1, 2, \ldots . \quad (8.105) \end{aligned}$$

Equation (8.104) arises from a simple conditioning argument based on the supposition that the system does not go empty during an arrival epoch. Equation (8.105) handles the case where the system goes empty during an arrival epoch. The embedded stationary probabilities $\pi_i^e, i = 0, 1, \ldots$, satisfy global balance. This implies that

$$\pi_i^e = \sum_{j=0}^{\infty} \pi_j^e p_{j,i}$$

$$= \sum_{j=i-1}^{\infty} \pi_j^e \int_0^{\infty} \frac{e^{-\mu t}(\mu t)^{j+1-i}}{(j+1-i)!} f_A(t)dt, \qquad (8.106)$$

where the restriction in the summation of (8.106) arises from the fact that one can go up by at most one customer during an arrival epoch. We must also satisfy the normalization condition for the stationary embedded probabilities.

To solve (8.106) we use the now familiar method of guessing a solution. A simple guess is that the distribution is geometric:

$$\pi_i^e = (1 - \beta)\beta^i, \quad i = 0, 1, \ldots .$$

Substituting this guess into (8.106) yields

$$
\begin{aligned}
(1 - \beta)\beta^i &= (1 - \beta) \sum_{j=i-1}^{\infty} \beta^j \int_0^{\infty} \frac{e^{-\mu t}(\mu t)^{j+1-i}}{(j+1-i)!} f_A(t)dt, \\
&= (1 - \beta) \int_0^{\infty} e^{-\mu t}\beta^{i-1} \sum_{j=i-1}^{\infty} \frac{(\mu t\beta)^{j+1-i}}{(j+1-i)!} f_A(t)dt \\
&= (1 - \beta) \int_0^{\infty} e^{-\mu t(1-\beta)}\beta^{i-1} f_A(t)dt.
\end{aligned}
$$

Canceling common terms and recognizing that the last expression can be written in terms of Laplace transforms allows us to write

$$\beta = \mathcal{L}_A(\mu(1 - \beta)). \qquad (8.107)$$

If the solution to (8.107) is unique and yields $0 < \beta < 1$ then the stationary distribution of the embedded stationary state probabilities for the G/M/1 queue at the *time of arrival* are given by

$$\pi_i^e = (1 - \beta)\beta^i, \quad i = 0, 1, \ldots . \qquad (8.108)$$

If $\rho = \lambda/\mu$ is less than 1 then it can be shown that β, as determined by (8.107), is uniquely determined and is less than 1.

Because we embed at arrival instants we can easily determine waiting and response time statistics for the G/M/1 queue. To derive the density of the response time, we use the fact that the geometric sum of exponential random variables is exponentially distributed as established in Chapter 4. If an arriving customer J finds i customers in the system then J's response time equals the sum of $i + 1$ independent exponential random variables with parameter μ. Thus, as shown in (4.88), J's response time is Erlang with parameters $(i + 1, \mu)$. Unconditioning this implies that the response time density for the G/M/1 queue is given by

$$f_T(t) = \sum_{i=0}^{\infty}(1 - \beta)\beta^i \frac{\mu(\mu t)^i e^{-\mu t}}{i!}$$

$$= \mu(1 - \beta)e^{-\mu t(1-\beta)}, \tag{8.109}$$

and thus is exponentially distributed with parameter $\mu(1 - \beta)$. This implies that

$$E[T] = \frac{1}{\mu(1 - \beta)}, \tag{8.110}$$

and that

$$E[W] = \frac{\beta}{\mu(1 - \beta)}. \tag{8.111}$$

These results show that the main difficulty in determining the stationary statistics for a G/M/1 queue is to solve (8.107) to determine β. This is usually solved using numerical techniques.

Example 8.17 (Solution for the E_2/M/1 Queue) Consider a G/M/1 queue where arrivals are Erlang with parameter $(2, \lambda)$. To determine β we must solve (8.107) given by

$$\beta = \left(\frac{\lambda}{\mu(1 - \beta) + \lambda}\right)^2.$$

This yields a cubic equation

$$\beta^3 - 2(1 + \alpha)\beta^2 + (1 + \alpha)^2\beta - \alpha^2 = 0$$

where $\alpha \overset{\text{def}}{=} \lambda/\mu$. After some algebra we find that this cubic has roots of 1 and

$$\frac{1 + 2\alpha \pm \sqrt{1 + 4\alpha}}{2}.$$

The root with the positive radical is larger than 1 and thus cannot be a solution. Keeping the remaining root implies that

$$\beta = \frac{1 + 2\alpha - \sqrt{1 + 4\alpha}}{2}.$$

It is convenient to write this in terms of the utilization of the system, ρ. To do this, observe that the average arrival rate is given by $\lambda/2$ and hence $\rho = \alpha/2$. This yields

$$\beta = \frac{1 + 4\rho - \sqrt{1 + 8\rho}}{2}$$

and shows that $\lim_{\rho \to 1} \beta = 1$ (this must be the case since it corresponds to a saturated queue). We conclude that the expected response time is given by

$$E[T] = \frac{2}{\mu(1 - 4\rho + \sqrt{1 + 8\rho})}.$$

Observe that to obtain the response time of an $E_k/M/1$ queue we would have to solve for the roots of a polynomial of degree $k + 1$. For $k > 3$, a theorem by Galois shows that this must be done numerically (see footnote 7 in Chapter 3).

8.7 Uniformized Markov Processes

In this section we determine the conditions under which we can model a Markov process as a Markov chain. Such chains are termed *uniformized chains* and are often used in modeling. Let $X(t)$ be a Markov process with generator matrix Q with stationary distribution π, and let

$$q_{max} \stackrel{\text{def}}{=} \sup_{i \geq 0} q_i.$$

If $q_{max} < \infty$ then we can uniformize the Markov process so that it is equivalent to a Markov chain. We do this by defining a Markov chain, termed the *uniformized chain*, with state transition matrix P that has entries

$$p_{i,j} = \begin{cases} q_{i,j}/q_{max}, & i \neq j, \\ \\ 1 - q_i/q_{max}, & i = j. \end{cases} \tag{8.112}$$

Let π' be the stationary distribution of the uniformized chain. We claim that $\pi = \pi'$. This follows from the following calculation.

Using (8.112) and (8.58) we write the global balance equations for the uniformized chain as

$$\pi'_i = \frac{1}{q_i} \sum_{j \neq i} \pi'_j q_{j,i}, \quad i = 0, 1, \dots .$$

Using (8.62) and simplifying shows that

$$\sum_{j \neq i} \pi'_j q_{j,i} + \pi'_i q_{i,i} = 0, \quad i = 0, 1, \dots,$$

and thus, in matrix form, that

$$\pi' Q = 0.$$

The equality of π and π' thus follows from the fact that they satisfy the same set of the linear equations and both sum to unity.

Example 8.18 (Uniformized M/M/1 Queue) For the M/M/1 queue considered in Example 8.11 we have

$$q_{\max} = \lambda + \mu$$

and thus the equivalent uniformized Markov chain has state transitions given by

$$p_{i,j} = \begin{cases} \lambda/(\lambda + \mu), & j = i+1, \\[2mm] \mu/(\lambda + \mu), & j = i-1, \end{cases} \qquad i = 1, 2, \ldots,$$

and $p_{0,0} = \mu/(\lambda + \mu)$. The probability transition diagram for this is given in Figure 8.9 and shows that the uniformized chain has one self-loop. The solution for the stationary distribution for these transition probabilities is easily seen to be given by (8.74).

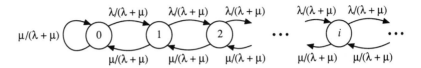

FIGURE 8.9. Probability Transition Diagram for Uniformized M/M/1 Queue

Example 8.19 (Probability Transition Matrix for a Markov Process) Let Q be the generator matrix for a Markov process that is uniformizable, that is, $q_{\max} < \infty$. Then it follows that the corresponding, equivalent, uniformized Markov chain has a state transition matrix given by

$$\boldsymbol{P} = \frac{1}{q_{\max}} \boldsymbol{Q} + \boldsymbol{I}.$$

This conversion between Markov processes and Markov chains will be used in Chapter 9 when we interpret the rate matrix for a matrix geometric solution.

8.8 The Chapman–Kolmogorov Differential Equations

In the discrete time case we developed the theory of Markov chains by deriving properties of the Chapman–Kolmogorov equations given in (8.9). In this section we show how one can, analogously, develop the theory of continuous time Markov processes by deriving properties of a continuous time form of these equations. In so doing we provide a mathematical explanation of the generator matrix $\boldsymbol{Q}$ used in determining the stationary probabilities of a Markov process as given in (8.66). Our results here establish the transient behavior for Markov processes. Since solving the differential equations derived here is typically difficult in practice, and since we do not later refer to the results of this section, our development here will be abbreviated. Rigorous proofs for statements made in this section can be found in [14].

We define the *transition function* of a continuous time Markov process as

$$P_{i,j}(t) \stackrel{\text{def}}{=} P\left[X(t+s) = j \mid X(s) = i\right].$$

The *Chapman–Kolmogorov* equations that are analogous to their discrete counterparts of (8.9), are given by

$$P_{i,j}(t+s) = \sum_{k=0}^{\infty} P_{i,k}(t) P_{k,j}(s), \quad t, s \geq 0.$$

These equations are established using the Markov property and the law of total probability. We assume that

$$\lim_{t \to 0} P_{i,i}(t) = 1, \quad i = 0, 1, \ldots,$$

and that

$$\lim_{t \to 0} P_{i,j}(t) = 0, \quad i \neq j.$$

Using the above notation, (8.60) and (8.61) can be written as

$$q_{i,j} = \lim_{t \to 0} \frac{P_{i,j}(t)}{t}, \tag{8.113}$$

and

$$q_i = \lim_{t \to 0} \frac{1 - P_{i,i}(t)}{t}. \tag{8.114}$$

These equations provide an interpretation for the entries of the generator matrix $\boldsymbol{Q}$; the probability that the Markov process jumps from state i to state j in the next dt seconds is given by $q_{i,j} dt$.

Rewriting the Chapman–Kolmogorov equations shows that

$$P_{i,j}(t + \Delta t) = \sum_{k=0}^{\infty} P_{i,k}(t) P_{k,j}(\Delta t)$$

$$= \sum_{k \neq j} P_{i,k}(t) P_{k,j}(\Delta t) + P_{j,j}(\Delta t) P_{i,j}(t). \qquad (8.115)$$

Rearranging this implies that

$$\frac{P_{i,j}(t + \Delta t) - P_{i,j}(t)}{\Delta t} = \sum_{k \neq j} P_{i,k}(t) \frac{P_{k,j}(\Delta t)}{\Delta t} - \frac{1 - P_{j,j}(\Delta t)}{\Delta t} P_{i,j}(t).$$

Letting $\Delta t \to 0$ and assuming that we can interchange limit and summation implies, after using (8.113) and (8.114), that

$$\frac{dP_{i,j}(t)}{dt} = \sum_{k \neq j} q_{k,j} P_{i,k}(t) - q_j P_{i,j}(t). \qquad (8.116)$$

This equation is termed the *Chapman–Kolmogorov forward equation* and permits an intuitive explanation in terms of probability flux. Note that given that the initial state at time 0 was state i, $P_{i,\ell}(t)$ is the probability flux of state ℓ at time t. The rate of the flow of probability flux into or out of state j at time t is given by $dP_{i,j}(t)/dt$. This equals the rate at which probability flux flows away from state j, a quantity given by $-q_j P_{i,j}(t)$, plus the rate at which probability flux flows into state j from states k, a quantity given by $q_{k,j} P_{i,k}(t)$ for each $k \neq j$.

We can rewrite (8.116) in matrix form to yield

$$\frac{d\boldsymbol{P}(t)}{dt} = \boldsymbol{P}(t) \boldsymbol{Q}$$

with initial conditions $\boldsymbol{P}(0) = \beta$. The solution to this differential equation is given by

$$\boldsymbol{P}(t) = \beta e^{\boldsymbol{Q} t}. \qquad (8.117)$$

Note that exponentiation of a matrix $\boldsymbol{M}$ can be represented in terms of a power series, that is,

$$\exp(\boldsymbol{M}) = \boldsymbol{I} + \boldsymbol{M} + \boldsymbol{M}^2/2! + \boldsymbol{M}^3/3! + \cdots .$$

Equation (8.116) is obtained by observing the flow of probability flux into state j over the interval of time $(t, t + \Delta t)$; hence to derive the equation we go "forward" in time to t and investigate what happens from this point on. This is why the equation is called the forward equation.

The implication inherent in the name is that there is a different version called the backward equation. To derive this we go "backward" in time to time Δt and then observe the probability flux into state j over the time interval $(\Delta t, t + \Delta t)$. To derive the backward equation we start from the Chapman–Kolmogorov equation to write[4]

$$P_{i,j}(t + \Delta t) = \sum_{k=0}^{\infty} P_{i,k}(\Delta t) P_{k,j}(t)$$

$$= \sum_{k \neq i} P_{i,k}(\Delta t) P_{k,j}(t) + P_{i,i}(\Delta t) P_{i,j}(t). \quad (8.118)$$

Rearranging this equation yields

$$\frac{P_{i,j}(t + \Delta t) - P_{i,j}(t)}{\Delta t} = \sum_{k \neq i} \frac{P_{i,k}(\Delta t)}{\Delta t} P_{k,j}(t) - \frac{1 - P_{i,i}(\Delta t)}{\Delta t} P_{i,j}(t).$$

Assuming that the limit and summation can be interchanged and taking the limit $\Delta t \to 0$, yields

$$\frac{dP_{i,j}(t)}{dt} = \sum_{k \neq i} q_{i,k} P_{k,j}(t) - q_i P_{i,j}(t). \quad (8.119)$$

This last equation is called the *Chapman–Kolmogorov backward equation*. Unlike the forward equation the backward equation does not lend itself to a simple intuitive interpretation.

Example 8.20 (Transient Probabilities for the M/M/1 Queue) Writing the forward equation for the M/M/1 queue yields

$$\frac{dP_{i,0}(t)}{dt} = \mu P_{i,1}(t) - \lambda P_{i,0}(t),$$

$$\frac{dP_{i,j}(t)}{dt} = \mu P_{i,j+1}(t) + \lambda P_{i,j-1}(t) - (\lambda + \mu) P_{i,j}(t).$$

The solution to these equations for this case is then given by (see [66])

$$P_{i,j}(t) = e^{-(\lambda + \mu)t} \left[\rho^{(j-i)/2} I_{j-i}(at) + \rho^{(j-i-1)/2} I_{j+i+1}(at) \right.$$

$$\left. + (1 - \rho) \rho^j \sum_{k=j+i+2}^{\infty} \rho^{-k/2} I_k(at) \right], \quad (8.120)$$

---

[4]The difference in the approach to deriving the forward and backward equations is readily manifest by seeing that we have swapped the evaluation of probabilities at times t and Δt in the two factors of equations (8.115) and (8.118).

where $\rho \overset{\text{def}}{=} \lambda/\mu$, $a \overset{\text{def}}{=} 2\mu\sqrt{\rho}$, and

$$I_k(x) \overset{\text{def}}{=} \sum_{m=0}^{\infty} \frac{(x/2)^{k+2m}}{(k+m)!m!}, \qquad k \geq -1,$$

is the series expansion for the modified Bessel function of the first kind.

The expression for the transient analysis of an M/M/1 queue given in (8.120) speaks more eloquently than words for our motivation to not dwell on transient analysis in this book! Even for the simple M/M/1 queue the equation for the transient state probabilities is hopelessly complicated. It is extremely difficult to have any intuition regarding the solution except for its limiting, and thus stationary, values. Observe that in the third term of expression (8.120) we see factors corresponding to the stationary distribution. As already established in previous sections, as $t \to \infty$ it must be the case that $\lim_{t\to\infty} P_{i,j}(t) = (1-\rho)\rho^j$ independent of i. Hence (8.120) suggests that the following limits must hold:

$$\lim_{t\to\infty} e^{-(\lambda+\mu)t} \rho^{(j-i)/2} I_{j-i}(at) = 0,$$

$$\lim_{t\to\infty} e^{-(\lambda+\mu)t} \rho^{(j-i-1)/2} I_{j+i+1}(at) = 0,$$

and

$$\lim_{t\to\infty} e^{-(\lambda+\mu)t} \sum_{k=j+i+2}^{\infty} \rho^{-k/2} I_k(at) = 1.$$

The interested student is more than welcome to try their hand at establishing these limits!

We can obtain equations for the stationary distributions from the forward or backward equations. If the Markov process admits a stationary distribution then it must be the case that

$$\lim_{t\to\infty} \frac{dP_{i,j}(t)}{dt} = 0.$$

Using the fact that

$$\lim_{t\to\infty} P_{i,j}(t) = \pi_j$$

implies that we can write the limiting forms of the forward and backward equations as

$$0 = \sum_{k\neq j} q_{k,j}\pi_k - q_j\pi_j = \sum_{k=0}^{\infty} q_{k,j}\pi_k, \qquad (8.121)$$

and

$$0 = \sum_{k \neq i} q_{i,k} \pi_j - q_i \pi_j = \pi_j \sum_{k=0}^{\infty} q_{i,k}, \qquad (8.122)$$

respectively, where we have used the fact that $q_{i,i} = -q_i$. We recognize (8.121) as being identical to (8.66), the global balance equations for the process, and (8.122) simply expresses the fact that the row sums of the Q matrix are zero.

Type of Process		Self-Loops	Holding Time
Semi-Markov Processes		No	Arbitrary
Markov chains	Model 1	Yes	$H_i = 1$
	Model 2	No	Geometric, $E[H_i] = (1 - p_{i,i})^{-1}$
Markov processes	Continuous time	No	Exponential, $E[H_i] = q_i^{-1}$
	Uniformized–Model 1	Yes	$H_i = 1$
	Uniformized–Model 2	No	Geometric, $E[H_i] = q_{\max}/q_i$

TABLE 8.2. Summary of Models of Markov Processes

8.9
Summary of
Chapter 8

We first studied Markov chains where state transitions occur at every step and self-loops are allowed. After classifying states of such processes we studied properties of chains where there is one set of ergodic states. We then showed that the stationary probabilities of such a chain are independent of its starting state and satisfies a set of linear equations called global balance equations. The study of Markov processes generally consists of methods to solve these sets of linear equations by making use of the special structure of the process. In the next chapter we will explore methods for solving such sets of equations. As mentioned in the chapter, since stationary probabilities are unique, one solution method is a "good guess." If this guess satisfies global balance then it is correct.

The most general form of a Markov process is a semi-Markov process. This is a generalization of a Markov chain where a time component is added to the state of the system. This destroys the Markov property if times are not exponentially distributed. The stationary probabilities of a semi-Markov process can be determined by first calculating the stationary distribution of the embedded Markov chain, and then using these values to weight average state holding times. In the special case where holding times are exponentially distributed and independent of the next state transition, we showed that the stationary probabilities satisfy a set of linear equations expressed in terms of the generator matrix of the process. If transition rates are bounded then one can uniformize the process and create an equivalent Markov chain that has the same stationary distributions. The different models of Markov processes that are studied in this chapter are summarized in Table 8.2. In the next chapter we will study special types of Markov processes that are frequently found in modeling applications and can be readily solved.

Bibliographic Notes

Most texts on stochastic processes have sections devoted to Markov processes. Additional reading can include sections from references mentioned in the bibliographic notes of Chapter 6 and texts by Kemeny and Snell [64, 63], Takács [111], and Wong and Hajek [127]. Continuous state Markov processes are called *diffusion processes*. Karlin and Taylor [58] have an extensive exposition of such processes.

8.10
Problems for
Chapter 8

Problems Related to Section 8.1

8.1 **[M3]** Suppose that $i \rightsquigarrow j$ and i is positive recurrent. Prove that j is also positive recurrent.

8.2 **[M3]** Consider state i of a Markov chain.

1. Suppose that $f_{i,i} = 1$. Prove that state i will be visited an infinite number of times if the chain is started in state i. Show that state i is recurrent if and only if

$$\sum_{n=0}^{\infty} p_{i,i}^n = \infty.$$

2. Suppose that $f_{i,i} < 1$. Prove that state i will be visited a finite number of times regardless of the initial state. Show that this implies that

$$\sum_{n=0}^{\infty} p_{i,i}^n < \infty.$$

8.3 **[M2]** Consider an irreducible Markov chain and suppose that a given state is periodic with period d. Show that this implies that all states are periodic with the same period.

8.4 **[M1]** Consider an irreducible Markov chain with transition matrix P. Suppose that $p_{i,i} > 0$ for some state i. Show that this implies that the chain is aperiodic.

8.5 **[E2]** Consider a finite state Markov chain. Show that there are no null-recurrent states. Show that all states cannot be transient.

8.6 **[E2]** Consider the following probability transition matrix for states 1, 2, 3, and 4:

$$P = \begin{bmatrix} 1-p & 0 & p & 0 \\ q & 0 & q & 0 \\ 0 & 0 & 1-p & p \\ 1 & 0 & 0 & 0 \end{bmatrix}.$$

1. What is the value of q?

2. Classify the states of the chain.

3. What is the probability that the chain is in state 4 at the fifth step given that it started in state 2?

4. Does the chain have a stationary distribution? If so, what are the stationary probabilities?

8.7 **[M3]** Consider a Markov chain with one ergodic set of states and let π_i be the stationary probability of being in state i. Define the generating function

$$P_{j,i}(z) = \sum_{n=0}^{\infty} p_{j,i}^n z^n$$

and the probability generating function

$$F_{j,i}(z) = \sum_{n=0}^{\infty} f_{j,i}^n z^n. \tag{8.123}$$

1. Show that

$$P_{j,i}(z) = F_{j,i}(z) P_{j,j}(z).$$

2. Let $\pi_{j,i}$ be the limiting probability of being in state i given that the chain starts in state j. Use (8.123) to show that $\pi_{j,i} = \pi_i$ for all j.

3. Show that

$$P_{i,i}(z) = \frac{1}{1 - F_{i,i}(z)}.$$

Use this to show that

$$\pi_i = \frac{1}{\sum_{n=0}^{\infty} n f_{i,i}^n}.$$

8.8 **[M2]** Let $\boldsymbol{P}$ be a nonnegative matrix with row and column sums both equal to 1 (such a matrix is termed *doubly stochastic*).

1. Derive a relationship between π_i and π_j for a Markov chain with $\boldsymbol{P}$ as its transition matrix.

2. Using this write the stationary distribution for the Markov chain if $\boldsymbol{P}$ is finite-dimensional.

3. If $\boldsymbol{P}$ is of infinite order does the relationship derived in part 1 still hold? If it does, why doesn't the same technique used in part 2 work for determining the stationary distribution? Does the matrix permit a stationary distribution?

8.9 **[M3](Stationary Probabilities are Independent of the Initial State)** In Section 8.1 we used results from renewal theory to show that the stationary probabilities of an irreducible ergodic Markov chain are independent of the initial state of the chain. In this problem we derive this result directly. Let $\pi_{i,j}$ be the stationary probability of being in state j given that $X_0 = i$,

$$\pi_{i,j} \overset{\text{def}}{=} \lim_{n \to \infty} p_{i,j}^n.$$

We will prove that $\pi_{i,j} - \pi_{j,j} = 0$ for all i.

1. Show that

$$p_{i,j}^n = \sum_{k=0}^{n} f_{i,j}^{n-k} p_{j,j}^k$$

and

$$\sum_{n=0}^{\infty} f_{i,j}^n = 1.$$

2. Show that

$$\pi_{i,j} - \pi_{j,j} = \lim_{n \to \infty} \left\{ \sum_{k=0}^{n} f_{i,j}^{n-k} \left(p_{j,j}^{k} - \pi_{j,j} \right) - \pi_{j,j} \sum_{k=n+1}^{\infty} f_{i,j}^{k} \right\}.$$

3. For an arbitrary value of $\epsilon > 0$, select $k(\epsilon)$ so that

$$\left| p_{j,j}^{k} - \pi_{j,j} \right| < \epsilon, \quad \text{for } k \geq k(\epsilon),$$

and let

$$m \overset{\text{def}}{=} \max_{0 \leq k \leq k(\epsilon)} \left\{ \left| p_{j,j}^{k} - \pi_{j,j} \right| \right\}.$$

Show that

$$\pi_{i,j} - \pi_{j,j} \leq \lim_{n \to \infty} \left\{ m \sum_{k=n-k(\epsilon)}^{n} f_{i,j}^{k} + \epsilon \sum_{k=0}^{n-k(\epsilon)-1} f_{i,j}^{k} - \pi_{j,j} \sum_{k=n+1}^{\infty} f_{i,j}^{k} \right\}.$$

4. Show that there exists a value $n(\epsilon)$ that simultaneously satisfies

$$\sum_{k=n-k(\epsilon)}^{n} f_{i,j}^{k} < \frac{\epsilon}{m} \quad \text{for } n > n(\epsilon)$$

and

$$\sum_{k=n+1}^{\infty} f_{i,j}^{k} < \epsilon \quad \text{for } n > n(\epsilon).$$

Use this to show that

$$\pi_{i,j} - \pi_{j,j} \leq \epsilon(2 + \pi_{j,j})$$

and hence establish the result.

8.10 **[R2]** We consider the problem of determining an optimal random walk strategy. Consider a random walk that takes place on the integers. An agent is placed at point 0 and can take unit steps in either the positive or negative direction. There are boundaries at $-B_\ell$ and B_r, where B_ℓ and B_r are positive integer random variables. We are interested in determining a strategy that minimizes the time needed to hit a boundary. The agent can implement a limited family of policies described as follows. The agent picks a probability p and integer values m_ℓ and m_r. At each step of the strategy the agent tosses a coin and with probability p walks m_ℓ steps to the left or with probability $1 - p$ walks m_r steps to the right. If a boundary has not been reached after these numbers of steps then the agent repeats this procedure until such time that a boundary is hit. The objective is to hit the boundary in a minimal expected number of steps. Assume that the agent knows the distribution of B_ℓ and B_r. As a trivial case suppose $B_\ell = b_\ell$ and $B_r = b_r$ deterministically. Then an optimal strategy is to pick $p = I_{b_\ell \leq b_r}$, $m_\ell = b_\ell$, and $m_r = b_r$. Can you determine the optimal strategy if:

1. $B_\ell = b_\ell$ deterministically and B_r is geometric with parameter α.

2. B_ℓ and B_r are independent geometric random variables with parameter α.

3. B_ℓ and B_r have distribution $F(x)$ and arise from the same sample.

4. Let X be a random variable with density

$$P[X = i] = \begin{cases} \alpha, & i = k, \quad k \geq 1, \\ 1 - \alpha, & i = ak, \quad a > 1. \end{cases}$$

Let B_ℓ and B_r be independent and equal in distribution to X.

Problem Related to Section 8.2

8.11 [M3] In this problem we derive another set of linear equations that the stationary probability vector π must solve. Let e be a vector of all 1s and let E be a matrix of all 1s. Observe that $\pi E = e$. Use this relationship and the global balance equations (8.32) to show that

$$\pi = e \left(P + E - I \right)^{-1}. \tag{8.124}$$

Explain why a normalization condition like (8.33) is not required in this formulation. Apply this result to determine the stationary distribution for the Markov chain with transition matrix

$$P = \begin{bmatrix} 1 - p & p \\ q & 1 - q \end{bmatrix}.$$

Problems Related to Section 8.3

8.12 [E2] Consider a Markov chain with transition matrix P and stationary distribution π. Suppose that state i has a holding time given by the random variable Z_i. Consider a semi-Markov process with embedded transition matrix P and holding times Z_i.

1. What is the stationary distribution of the semi-Markov process if $Z_i =_{st} Z_j$ for all states i, j?

2. Suppose that P is order n. What is the stationary distribution of the semi-Markov process if Z_i is exponentially distributed with parameter π_i^{-1}?

3. Suppose the stationary distribution of the semi-Markov process is π'. Can you determine the values of $E[Z_i]$.

8.13 **[M2]** Consider a semi-Markov process defined on points that lie on a set of states $0, 1, \ldots, n-1$, that lie in a ring of length n. Suppose that the transition matrix $\boldsymbol{P}$ of the embedded Markov chain satisfies $p_{i,j} = p_{i+k,j+k}$ where addition is taken modulo n. This implies that transition probabilities depend only on the number of nodes "skipped" over in the transition. In other words, there exists constants $q_1, q_2, \ldots, q_{n-1}$ that sum to 1 so that $p_{i,j} = q_k$ where $k = \max\{i, j\} - \min\{i, j\}$. Suppose the holding time of state i is distributed as random variable H_i.

1. Suppose that $H_i =_{st} H_j$ for all states i and j. What is the stationary distribution of the semi-Markov process?

2. Suppose that $E[H_i] = i + 1$. What is the stationary distribution of the semi-Markov process?

Problems Related to Section 8.4

8.14 **[M2]** For the $M/H_2(k)/1$ queue considered in Example 8.16 write expressions for the probability that the server is processing a customer at rate $k\mu$. (Hint: You can do this without knowing the stationary distribution of the process.)

8.15 **[E2]** Let $(i, j), i = 0, 1, \ldots, j = 1, 2$, be the state for the $M/E_2/1$ queue where i is the total number of customers in the queue and j is the number of stages left for the customer in service. Draw the state transition diagram. Derive the stationary state probabilities.

8.16 **[M3]** In this problem we extend the model considered in Example 8.14. As in Example 8.14 there are n disk drives that fail at exponential rates of λ and there is a repairman that fixes disks at rate μ. The system also has a single processor that fails at an exponential rate of λ_p and is repaired at a rate of μ_p. Observe that the system is a Markov process with state $(i, j), i = 0, 1, \ldots, n, j = 0, 1$, where i is the number of failed disks and j is the number of failed processors. Write expressions for:

1. the average number of failed disks.

2. the average number of failed processors.

3. the probability that the processor is failed when i disks are failed, for $i = 0, 1, \ldots, n$.

(Hint: Observe that $\pi_{i,0} + \pi_{i,1} = \pi_i$ where π_i is given by (8.92). Then use the same technique as that presented in Example 8.16.)

Problem Related to Section 8.8

8.17 [**M3**] Consider a pure birth process where the birth rate from state i is given by λ_i.

1. Show that the forward equation for the process is given by

$$\frac{dP_{i,i}(t)}{dt} = -\lambda_i P_{i,i}(t),$$

$$\frac{dP_{i,j}(t)}{dt} = \lambda_{j-1} P_{i,j-1}(t) - \lambda_j P_{i,j}(t), \quad j > i.$$

2. Using the initial condition $P_{i,i}(0) = 1$ show that

$$P_{i,i}(t) = e^{-\lambda_i t}.$$

3. Suppose $\lambda_j = \lambda$, $j = 0, 1, \ldots$. Write an expression for $P_{i,j}(t)$.

9

Matrix Geometric Solutions

In Example 8.12 we analyzed a scalar state process that was a modification of the M/M/1 queue. In the example, we classified two sets of states: boundary states and repeating states. Transitions between the repeating states had the property that rates from states $2, 3, \ldots$, were congruent to rates between states $j, j+1, \ldots$ for all $j \geq 2$. We noted in the example that this implied that the stationary distribution for the repeating portion of the process satisfied a geometric form. In this chapter we generalize this result to vector state processes that also have a repetitive structure. The technique we develop in this chapter to solve for the stationary state probabilities for such *vector state* Markov processes is called the *matrix geometric method*. (The theory of matrix geometric solutions was pioneered by Marcel Neuts; see [86] for a full development of the theory.) In much the same way that the repetition of the state transitions for this variation of the M/M/1 queue considered in Example 8.12 implied a geometric solution (with modifications made to account for boundary states), the repetition of the state transitions for vector processes implies a geometric form where scalars are replaced by matrices. We term such Markov processes *matrix geometric processes*.

Processes that have the repetitive structure required by matrix geometric techniques frequently arise in models of computer systems. The beginning portions of a state space for such models typically account for state configurations that arise before a queue begins to form in the system. As an example, consider a multiple heterogeneous server queue. The boundary states for such a system account for all possible server occupancies when the queue is empty. Once all servers are busy the addition of one more customer to the system merely "shifts" the state space up one level without changing the occupancies of the servers. This implies that the state space has a repetitive structure and is thus amenable to the solution technique developed in this chapter.

Matrix geometric techniques are applicable to both continuous and discrete time Markov processes. (Recall that we showed in the subsection on uniformization in Section 8.7 that we can create an equivalent discrete time Markov chain from a continuous Markov process, provided that the maximal transition rate is bounded by a known constant.) We will concentrate on continuous time matrix geometric processes in Sections 9.1 and 9.2. In Section 9.3 we consider discrete time chains. Various mathematical objects in the matrix geometric approach are interpreted in the discrete time case in Section 9.4. Section 9.5 shows how a large class of distributions, called *phase distributions*, can be incorporated into a matrix geometric framework.

A note on some of the notation used in this chapter: Previously we wrote the entries of matrix M as $m_{i,j}$. Many of the matrices in this chapter, however, are themselves subscripted, and this causes difficulties when we attempt to preserve our usual notation. In these cases we write entries as parenthesized pairs after the matrix name; thus we will write $M(i,j)$ to represent entry $m_{i,j}$.

9.1 Matrix Geometric Systems

Many of the Markov processes that we have considered until now have had a scalar representation. An exception to this is the $M/H_2(k)/1$ queue considered in Example 8.16 where there is a natural vector representation of the state space. In a vector state process, states are represented as pairs of the form (i, s) where $i = 0, 1, \ldots$ and $s = 0, 1, \ldots, m$. Typically i, which is unbounded, represents the number of customers in the system and s represents the state of the customer in service. Other interpretations of the states, of course, are possible. We shall say that states at level i are those states defined by $(i, 0), (i, 1), \ldots, (i, m)$ and we call the s component of the state pair the *inter-level* component of the state. If state transitions are repetitive then the stationary probabilities satisfy a "matrix geometric" form. A process has repetitive state transitions if the transition rate from state (i, j) to state $(i + k, j')$, for $k_d \leq k \leq k_u$, $0 \leq k_d, k_u < \infty$, is independent of the value of i for $i \geq i^\star$. This implies that the infinite portion of the state space consisting of levels $i^\star, i^\star + 1, \ldots$, is congruent to the infinite portion of the state space consisting of levels $i, i + 1, \ldots$, for all $i \geq i^\star$. In terms of the generator matrix, this repetitive structure implies that matrix entries eventually repeat diagonally. This repetition of state transitions is the essential feature that allows one to solve for the stationary distributions of such Markov processes.

9.1.1
A Matrix
Geometric
Example

It is best to start with an example to see how the repetitive structure helps to determine a solution. Our example is the vector equivalent of the modified M/M/1 queue scalar example considered in Example 8.12. Consider a queueing system with Poisson arrivals at rate λ' if the system is empty, and at rate λ otherwise. Customers require two exponential stages of service: the first at rate μ_1 and the second at rate μ_2. Let the state be given by $(i, s), i \geq 0, s = 0, 1, 2$, where i is the number of customers in the queue (not including any receiving service) and s is the current stage of service of the customer in service. By definition we set s equal to 0 if there are no customers in the system. The state transition diagram for this system is shown in Figure 9.1, where we have grouped states according to the number of customers in the queue.

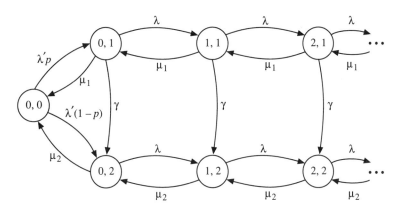

FIGURE 9.1. State Transition Diagram for Vector Process

We order states lexicographically, $(0, 0), (0, 1), (0, 2), (1, 1), (1, 2), \ldots$, and let $\pi_{i,s}$ be the stationary probability of state (i, s). The generator matrix is given by

$$
Q = \begin{bmatrix}
-\lambda' & \lambda' & 0 & 0 & 0 & 0 & 0 & 0 & 0 & \cdots \\
0 & -a_1 & \mu_1 & \lambda & 0 & 0 & 0 & 0 & \cdots \\
\mu_2 & 0 & -a_2 & 0 & \lambda & 0 & 0 & 0 & 0 & \cdots \\
0 & 0 & 0 & -a_1 & \mu_1 & \lambda & 0 & 0 & 0 & \cdots \\
0 & \mu_2 & 0 & 0 & -a_2 & 0 & \lambda & 0 & 0 & \cdots \\
0 & 0 & 0 & 0 & 0 & -a_1 & \mu_1 & \lambda & 0 & \cdots \\
0 & 0 & 0 & \mu_2 & 0 & 0 & -a_2 & 0 & \lambda & \cdots \\
\vdots & \vdots & \vdots & \vdots & \vdots & \vdots & \vdots & \vdots & \vdots & \ddots
\end{bmatrix}
\tag{9.1}
$$

where $a_i \overset{\text{def}}{=} \lambda + \mu_i, i = 1, 2$.

The structure of this generator matrix is similar to that found in the modified M/M/1 queue given in Example 8.12. This is easily seen by grouping the entries of (9.1) according to the number of customers in the system. Let $\boldsymbol{\pi}_i \stackrel{\text{def}}{=} (\pi_{i,1}, \pi_{i,2})$ for $i \geq 1$, $\boldsymbol{\pi}_0 \stackrel{\text{def}}{=} (\pi_{0,0}, \pi_{0,1}, \pi_{0,2})$, and $\boldsymbol{\pi} \stackrel{\text{def}}{=} (\boldsymbol{\pi}_0, \boldsymbol{\pi}_1, \boldsymbol{\pi}_2, \ldots)$. Define the following matrices

$$\boldsymbol{A}_0 = \begin{bmatrix} \lambda & 0 \\ 0 & \lambda \end{bmatrix}, \quad \boldsymbol{A}_1 = \begin{bmatrix} -a_1 & \mu_1 \\ 0 & -a_2 \end{bmatrix}, \quad \boldsymbol{A}_2 = \begin{bmatrix} 0 & 0 \\ \mu_2 & 0 \end{bmatrix},$$

$$\boldsymbol{B}_{1,0} = \begin{bmatrix} 0 & 0 & 0 \\ 0 & \mu_2 & 0 \end{bmatrix}, \quad \boldsymbol{B}_{0,1} = \begin{bmatrix} 0 & 0 \\ \lambda & 0 \\ 0 & \lambda \end{bmatrix},$$

and

$$\boldsymbol{B}_{0,0} = \begin{bmatrix} -\lambda' & \lambda' & 0 \\ 0 & -a_1 & \mu_1 \\ \mu_2 & 0 & -a_2 \end{bmatrix}.$$

With these definitions we can group the generator matrix into blocks as follows:

$$\boldsymbol{Q} = \begin{bmatrix} -\lambda' & \lambda' & 0 & 0 & 0 & 0 & 0 & 0 & 0 & \cdots \\ 0 & -a_1 & \mu_1 & \lambda & 0 & 0 & 0 & 0 & 0 & \cdots \\ \mu_2 & 0 & -a_2 & 0 & \lambda & 0 & 0 & 0 & 0 & \cdots \\ 0 & 0 & 0 & -a_1 & \mu_1 & \lambda & 0 & 0 & 0 & \cdots \\ 0 & \mu_2 & 0 & 0 & -a_2 & 0 & \lambda & 0 & 0 & \cdots \\ 0 & 0 & 0 & 0 & 0 & -a_1 & \mu_1 & \lambda & 0 & \cdots \\ 0 & 0 & 0 & \mu_2 & 0 & 0 & -a_2 & 0 & \lambda & \cdots \\ \vdots & \vdots & \vdots & \vdots & \vdots & \vdots & \vdots & \vdots & \vdots & \ddots \end{bmatrix}$$

$$= \begin{bmatrix} \boldsymbol{B}_{0,0} & \boldsymbol{B}_{0,1} & \boldsymbol{0} & \boldsymbol{0} & \boldsymbol{0} & \cdots \\ \boldsymbol{B}_{1,0} & \boldsymbol{A}_1 & \boldsymbol{A}_0 & \boldsymbol{0} & \boldsymbol{0} & \cdots \\ \boldsymbol{0} & \boldsymbol{A}_2 & \boldsymbol{A}_1 & \boldsymbol{A}_0 & \boldsymbol{0} & \cdots \\ \boldsymbol{0} & \boldsymbol{0} & \boldsymbol{A}_2 & \boldsymbol{A}_1 & \boldsymbol{A}_0 & \cdots \\ \vdots & \vdots & \vdots & \vdots & \vdots & \ddots \end{bmatrix}. \tag{9.2}$$

A $\boldsymbol{0}$ entry in (9.2) (and in other matrices that follow) is a matrix of all zeros of the appropriate dimension. This matrix is similar to that of the modified M/M/1 queue considered in Example 8.12 and given in (8.76); it is tridiagonal with *matrix* entries. Since $\boldsymbol{Q}$ is a generator matrix, its row sums equal 0.

Like Example 8.12 we call states $(0,0)$, $(0,1)$, and $(0,2)$ the *initial portion* of the process or the *boundary states*, and all other states are called the *repeating portion* of the process or the *repeating states*. We solve for the stationary probabilities in the same manner as in the scalar example of Example 8.12, except that here we deal with vectors instead of scalars. First we write down the equation for the repeating portion of the process, which, given in block matrix form, is

$$\pi_{j-1} A_0 + \pi_j A_1 + \pi_{j+1} A_2 = 0, \qquad j = 2, 3, \ldots . \tag{9.3}$$

Similar to the scalar case, since state transitions are between nearest blocks, it would not be surprising to find that the value of π_j is a function only of the transition rates between states with $j-1$ queued customers and states with j queued customers. Since these transition rates do not depend on j, this suggests that there is a *constant matrix* R such that

$$\pi_j = \pi_{j-1} R, \qquad j = 2, 3, \ldots, \tag{9.4}$$

and that the values of $\pi_j, j = 2, 3, \ldots$, have a *matrix geometric* form, that is,

$$\pi_j = \pi_1 R^{j-1}, \qquad j = 2, 3, \ldots . \tag{9.5}$$

Substituting this *guess* into (9.3) shows that

$$\pi_1 R^{j-2} A_0 + \pi_1 R^{j-1} A_1 + \pi_1 R^j A_2 = 0, \qquad j = 2, 3, \ldots . \tag{9.6}$$

This equation must be true for all j, for $j = 2, 3, \ldots$. Substituting $j = 2$ into (9.6) and simplifying yields

$$A_0 + R A_1 + R^2 A_2 = 0. \tag{9.7}$$

This is a quadratic equation in the matrix R, which is typically solved numerically (we discuss computational aspects of the matrix geometric method in Section 9.4.3). Recall that in the quadratic equation for the scalar case there were two possible solutions. One of these, $\rho = 1$, could not satisfy the normalization condition. It similarly follows that $R = 1$ (a matrix of all 1s) is a solution to (9.7) but, like its corresponding scalar solution, cannot be normalized. Similar to the scalar case, we pick the minimal matrix R that satisfies (9.7). If the normalization constant is satisfied for the vector state process, it must be the case that $\pi_1 \sum_{j=1}^{\infty} R^{j-1} e < \infty$, where e is defined to be a suitably dimensioned column vector of 1s. The analogous criteria to $\rho < 1$ in the scalar case is that the spectral radius of R must be less than unity.[1] This follows

---

[1] An eigenvector of matrix M is a vector x for which $Mx = \alpha x$. The value α is termed the eigenvalue corresponding to eigenvector x. The *spectral radius* of matrix M is the magnitude of the largest eigenvalue (this can, in general, be a complex number). Later we will show that the entries in matrix R are nonnegative. In this case the Perron–Frobenius theorem establishes that the eigenvalue of largest magnitude is real and positive. Hence, if the largest positive eigenvalue is less than unity, then the geometric sequence $I + R + R^2 + \cdots$ converges.

from the fact that all eigenvalues of R must be less than 1 for the sum to converge [52].

Assume that we have a solution for R, which is termed the *rate matrix*. To determine the stationary probabilities we continue as in the scalar state case. The equations for the initial portion of the matrix are given by

$$\pi_0 B_{0,0} + \pi_1 B_{1,0} \qquad = \quad 0,$$

$$\pi_0 B_{0,1} + \pi_1 A_1 + \pi_2 A_2 \quad = \quad 0, \qquad (9.8)$$

which can be written in matrix form as

$$(\pi_0, \pi_1) \begin{bmatrix} B_{0,0} & B_{0,1} \\ B_{1,0} & A_1 + R A_2 \end{bmatrix} = 0. \qquad (9.9)$$

We have substituted $\pi_2 = R \pi_1$ in (9.8) to obtain the submatrix in the lower right-hand corner of (9.9). As in the scalar case, equations (9.9) are not sufficient to determine the probabilities π_0 and π_1. To determine these probabilities requires that we use the normalization constraint, that is,

$$1 = \pi_0 e + \pi_1 \sum_{j=1}^{\infty} R^{j-1} e = \pi_0 e + \pi_1 (I - R)^{-1} e, \qquad (9.10)$$

which, together with (9.9), yields a unique solution.

We should digress momentarily to discuss how one closes an infinite summation like (9.10). This can sometimes be accomplished with the same technique that is used to close an infinite geometric series. Let S be the matrix obtained from the infinite sum. Then, provided that the sum converges, we can write

$$S = \sum_{j=1}^{\infty} R^{j-1} = I + R + R^2 + \cdots . \qquad (9.11)$$

Multiplying (9.11) by R yields

$$S R = R + R^2 + \cdots ,$$

which, upon subtracting from (9.11), yields

$$S(I - R) = (I + R + R^2 + \cdots) - (R + R^2 + \cdots) = I.$$

Multiplying both sides of this equation on the right by $(I - R)^{-1}$ shows that $S = (I - R)^{-1}$. This derivation depends on the assumption that the infinite geometric sequence converges. For a nonnegative scalar geometric sequence, $1 + x + x^2 + \cdots$, the criteria for convergence is that $x < 1$. For the analogous geometric sequence for the nonnegative matrix $\boldsymbol{R}$, the criteria is that the spectral radius of the matrix is less than unity (see footnote 1 of this chapter).

To write (9.10) in a form that is suitable for linear equation solvers, define $\boldsymbol{M}^\star$ to be matrix $\boldsymbol{M}$ with its first column eliminated and let $[1, \boldsymbol{0}]$ denote a row vector consisting of a 1 followed by 4 zeros (the "4" is explained by the fact that matrices $\boldsymbol{B}_{0,0}^\star$, $\boldsymbol{B}_{1,0}^\star$, $\boldsymbol{B}_{0,1}$, and $\boldsymbol{A}_1 + \boldsymbol{R}\boldsymbol{A}_2$ given in (9.12) each have two columns). We can then determine the solution for the boundary states by solving

$$(\pi_0, \pi_1) \begin{bmatrix} \boldsymbol{e} & \boldsymbol{B}_{0,0}^\star & \boldsymbol{B}_{0,1} \\ (\boldsymbol{I} - \boldsymbol{R})^{-1}\boldsymbol{e} & \boldsymbol{B}_{1,0}^\star & \boldsymbol{A}_1 + \boldsymbol{R}\boldsymbol{A}_2 \end{bmatrix} = [1, \boldsymbol{0}]. \quad (9.12)$$

The equation that results from multiplying (π_0, π_1) by the first column of the matrix corresponds to the normalization condition. The remaining equations are the non-redundant equations of (9.9). The linear equations given by (9.12) have a unique solution that also satisfies the normalization condition.

Suppose, as in the scalar case of Example 8.11, that we wish to calculate the expected number of queued customers, denoted by $\overline{N_q}$. We do this by assigning the weight of $j-1$ to states $(j, 1)$ and $(j, 2)$, for $j = 1, 2, \ldots$, and then calculate

$$\begin{aligned} \overline{N_q} &= \sum_{j=1}^{\infty}(j-1)\pi_j\boldsymbol{e} = \pi_1\sum_{j=1}^{\infty}(j-1)\boldsymbol{R}^{j-1}\boldsymbol{e} \\ &= \pi_1\boldsymbol{R}(\boldsymbol{I} - \boldsymbol{R})^{-2}\boldsymbol{e}. \end{aligned}$$

The summation in this equation is determined similarly to that obtained in (9.11). Table 9.1 summarizes the analogous derivations for the scalar and vector state processes.

9.1.2 Quasi Birth–Death Processes

The previous example of the queue where customers received two exponential stages of service is a case of a *quasi birth–death process*. Recall that a birth–death process allows only adjacent state transitions. Adding the name "quasi" reflects the fact that such nearest neighbor transitions are interpreted in terms of vectors of states; state transitions are only possible on the same level or between adjacent levels. Another example of a quasi birth–death process previously considered is that of the $M/H_2(k)/1$ queue analyzed in Example 8.16. Comparing the state transition diagram for this process as shown in Figure 8.7 with that of an

Step	Scalar Process	Vector Process
jth balance equation	$\pi_{j-1}\lambda - \pi_j(\lambda+\mu) + \pi_{j+1}\mu = 0$	$\pi_{j-1}\boldsymbol{A}_0 + \pi_j\boldsymbol{A}_1 + \pi_{j+1}\boldsymbol{A}_2 = 0$
Geometric solution	$\pi_j = \rho^{j-1}\pi_1$	$\pi_j = \boldsymbol{R}^{j-1}\pi_1$
Solution for the repeating portion	$\lambda - \rho(\lambda+\mu) + \rho^2\mu = 0$	$\boldsymbol{A}_0 + \boldsymbol{R}\boldsymbol{A}_1 + \boldsymbol{R}^2\boldsymbol{A}_2 = 0$
Solution for the initial portion	$(\pi_0,\pi_1)\begin{bmatrix} -\lambda' & \lambda' \\ \mu & -\mu \end{bmatrix} = 0,$	$(\pi_0,\pi_1)\begin{bmatrix} \boldsymbol{B}_{0,0} & \boldsymbol{B}_{0,1} \\ \boldsymbol{B}_{1,0} & \boldsymbol{A}_1+\boldsymbol{R}\boldsymbol{A}_2 \end{bmatrix} = 0$
Normalization condition	$1 = \pi_0 + \pi_1(1-\rho)^{-1}$	$1 = \pi_0\boldsymbol{e} + \pi_1(\boldsymbol{I}-\boldsymbol{R})^{-1}\boldsymbol{e}$
Performance measure	$\overline{N_q} = \pi_1\rho(1-\rho)^{-2}$	$\overline{N_q} = \pi_1\boldsymbol{R}(\boldsymbol{I}-\boldsymbol{R})^{-2}\boldsymbol{e}$

TABLE 9.1. Comparison of Scalar and Vector Geometric Processes

M/M/1 queue, Figure 8.4 reveals the quasi birth–death structure. The analysis for this queue, provided in Example 8.16, can be compared to the one obtained from a matrix geometric analysis. The matrices for this analysis are easily seen to be given by

$$\boldsymbol{A}_0 = \begin{bmatrix} \lambda & 0 \\ 0 & \lambda \end{bmatrix}, \quad \boldsymbol{A}_1 = \begin{bmatrix} -(\lambda+k\mu) & 0 \\ 0 & -(\lambda+\mu/k) \end{bmatrix},$$

$$\boldsymbol{A}_2 = \tfrac{\mu}{k+1}\begin{bmatrix} k^2 & k \\ 1 & 1/k \end{bmatrix}, \quad \boldsymbol{B}_{1,0} = \tfrac{\mu}{k+1}\begin{bmatrix} 0 & k^2 & k \\ 0 & 1 & 1/k \end{bmatrix},$$

and

$$\boldsymbol{B}_{0,1} = \begin{bmatrix} 0 & 0 \\ \lambda & 0 \\ 0 & \lambda \end{bmatrix}, \quad \boldsymbol{B}_{0,0} = \begin{bmatrix} -\lambda & \lambda k/(k+1) & \lambda/(k+1) \\ k\mu & -k\mu & 0 \\ \mu/k & 0 & -\mu/k \end{bmatrix}.$$

Note that $\boldsymbol{B}_{1,1} = \boldsymbol{A}_1$. We could proceed from here to numerically calculating the stationary state probabilities and performance measures using the matrix geometric technique outlined above. The ease of this solution approach should be contrasted to the solution in Example 8.16.

9.2 General Matrix Geometric Solutions

In this section we present the matrix geometric solution technique in a more general context than that previously considered. Since scalars are special cases of vectors, our results also hold for more general scalar state processes. The solution techniques used here are identical to those used in the simple cases considered in Section 9.1. Consider a Markov process that has a block generator matrix given by

$$
\boldsymbol{Q} = \begin{bmatrix}
\boldsymbol{B}_{0,0} & \boldsymbol{B}_{0,1} & \boldsymbol{0} & \boldsymbol{0} & \boldsymbol{0} & \cdots \\
\boldsymbol{B}_{1,0} & \boldsymbol{B}_{1,1} & \boldsymbol{A}_0 & \boldsymbol{0} & \boldsymbol{0} & \cdots \\
\boldsymbol{B}_{2,0} & \boldsymbol{B}_{2,1} & \boldsymbol{A}_1 & \boldsymbol{A}_0 & \boldsymbol{0} & \cdots \\
\boldsymbol{B}_{3,0} & \boldsymbol{B}_{3,1} & \boldsymbol{A}_2 & \boldsymbol{A}_1 & \boldsymbol{A}_0 & \cdots \\
\boldsymbol{B}_{4,0} & \boldsymbol{B}_{4,1} & \boldsymbol{A}_3 & \boldsymbol{A}_2 & \boldsymbol{A}_1 & \cdots \\
\vdots & \vdots & \vdots & \vdots & \vdots & \ddots
\end{bmatrix}, \tag{9.13}
$$

where $\boldsymbol{B}_{0,0}$ is a square matrix of dimension $(m_1 - m)$, $\boldsymbol{B}_{0,1}$ is of dimension $(m_1 - m) \times m$, matrices $\boldsymbol{B}_{k,0}$, $k \geq 1$, are of dimension $m \times (m_1 - m)$, and all other matrices are square with dimension m, where $m_1 \geq m$. The $\boldsymbol{B}$ matrices correspond to transitions from boundary states.

When referring to the states of the process it is convenient, as in the previous sections, to view the process in terms of levels of states. The 0th level of the process associated with $\boldsymbol{Q}$ of equation (9.13) are the $m_1 - m$ boundary states. The first level are the next m states, and generally the ith level, for $i \geq 1$, are those states indexed by $m_1 - m + (i-1)m + \ell$, $\ell = 0, 1, \ldots, m - 1$. We will index the repeating portion of the process by a pair (i, j), $i \geq 1, 0 \leq j \leq m - 1$, where i is the level of the state and j is the interlevel state. The state of the original process indexed by (i, j) is $m_1 - m + (i - 1)m + j$.

To write a general solution for the generator in (9.13) we proceed similarly to Section 9.1. We first write a balance equation for the repeating portion of the process

$$
\sum_{k=0}^{\infty} \pi_{j-1+k} \boldsymbol{A}_k = \boldsymbol{0}, \quad j = 2, 3, \ldots . \tag{9.14}
$$

We then guess a geometric solution

$$
\pi_j = \pi_{j-1} \boldsymbol{R}, \quad j = 2, 3, \ldots,
$$

or that

$$
\pi_j = \pi_1 \boldsymbol{R}^{j-1}, \quad j = 2, 3, \ldots .
$$

Substituting this guess into (9.14) and simplifying shows that $\boldsymbol{R}$ solves

$$
\sum_{k=0}^{\infty} \boldsymbol{R}^k \boldsymbol{A}_k = \boldsymbol{0}. \tag{9.15}
$$

To determine a solution for the boundary states we must solve the following linear equations

$$(\pi_0, \pi_1) \left[\begin{array}{cc} B_{0,0} & B_{0,1} \\ \sum_{k=1}^{\infty} R^{k-1} B_{k,0} & \sum_{k=1}^{\infty} R^{k-1} B_{k,1} \end{array} \right] = 0. \quad (9.16)$$

As in previous cases, these equations do not permit a unique solution and we must use the normalization condition to determine the unique solution. This is given by equation (9.10), which, together with (9.16), provides a unique solution. Similar to our previous discussion, we compute this unique solution using a linear equation solver by replacing one of the columns in the matrix in (9.16) with the column that represents the normalization condition and appropriately changing the right-hand side of (9.16). This yields

$$(\pi_0, \pi_1) \left[\begin{array}{ccc} e & B_{0,0}^{\star} & B_{0,1} \\ (I-R)^{-1}e & \left[\sum_{k=1}^{\infty} R^{k-1} B_{k,0}\right]^{\star} & \sum_{k=1}^{\infty} R^{k-1} B_{k,1} \end{array} \right] = [1, 0]$$

where $[1, 0]$ is a row vector consisting of a 1 followed by $m_1 - 1$ zeros. The form of this equation is suitable for most linear equation solvers and uniquely determines π_0 and π_1.

We end this section with some observations about how this general matrix form relates to problems often encountered in performance models. In practice one typically finds that matrices $B_{k,0}$ and $B_{k+1,1}$ are zero for values $k \geq K+1$, $K \geq 1$. Similarly one often finds that A_k are zero starting at $k > K+1$. The value of K is the maximum number of levels that the process can jump down in the repeating portion of the process, and for many performance models this value is small. For example, the Markov process considered in Section 9.1 with generator matrix given in (9.2) has $m_1 = 5$ and $m = 2$, $B_{1,1} = A_1$, $B_{2,1} = A_2$, and $K = 1$. Another example is that of a bulk service queueing system discussed later.

Example 9.1 (A Bulk Service Queue) A special case of a matrix geometric problem with a generator matrix given by (9.13) arises when one considers a bulk service queue. In such queues the server can process up to K customers at a time. Suppose that the bulk service time equals the sum of two exponential phases: the first at rate μ_1 followed by the second at rate μ_2. Arrivals to the system have exponentially distributed interarrival times with rates λ_i for $i = 0, 1, \ldots, K-1$ when there are i customers in the system and at rate λ when there are K or more customers in the system. When the server becomes available for service and there are i customers in the system it takes $\min(i, K)$ customers into service. The induced Markov process is of matrix geometric form and has a generator matrix given by (9.13). As a special case, suppose that $K = 3$ and thus that the state transition diagram is as shown in

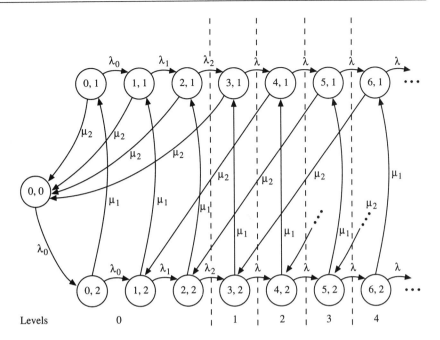

FIGURE 9.2. State Transition Diagram for Bulk Queue

Figure 9.2. In this figure (i, j) denotes a state consisting of i customers in the system and the customer in service has j remaining stages of service.

Typically the most difficult part of a matrix geometric problem lies in accurately delineating the matrices that comprise the generator matrix. From this point, computer software can produce the solution. It is instructive to write the corresponding matrices for the bulk queue depicted in Figure 9.2. The generator matrix for this problem is given by

$$
Q = \begin{bmatrix}
B_{0,0} & B_{0,1} & 0 & 0 & 0 & 0 & \cdots \\
B_{1,0} & B_{1,1} & A_0 & 0 & 0 & 0 & \cdots \\
B_{2,0} & 0 & A_1 & A_0 & 0 & 0 & \cdots \\
B_{3,0} & 0 & 0 & A_1 & A_0 & 0 & \cdots \\
0 & A_4 & 0 & 0 & A_1 & A_0 & \cdots \\
0 & 0 & A_4 & 0 & 0 & A_1 & \cdots \\
\vdots & \vdots & \vdots & \vdots & \vdots & \vdots & \ddots
\end{bmatrix},
$$

where the component submatrices are given by

$$
A_0 = \begin{bmatrix} \lambda & 0 \\ 0 & \lambda \end{bmatrix}, \quad
A_1 = \begin{bmatrix} -(\lambda + \mu_2) & 0 \\ \mu_1 & -(\lambda + \mu_1) \end{bmatrix}, \quad
A_4 = \begin{bmatrix} 0 & \mu_2 \\ 0 & 0 \end{bmatrix},
$$

and

$$
B_{0,0} = \begin{bmatrix}
-\lambda_0 & 0 & \lambda_0 & 0 & 0 & 0 & 0 \\
\mu_2 & -b_{1,2} & 0 & \lambda_1 & 0 & 0 & 0 \\
0 & \mu_1 & -b_{1,1} & 0 & \lambda_1 & 0 & 0 \\
\mu_2 & 0 & 0 & -b_{2,2} & 0 & \lambda_2 & 0 \\
0 & 0 & 0 & \mu_1 & -b_{2,1} & 0 & \lambda_2 \\
\mu_2 & 0 & 0 & 0 & 0 & -b_{0,2} & 0 \\
0 & 0 & 0 & 0 & 0 & \mu_1 & -b_{0,1}
\end{bmatrix},
$$

$$
B_{0,1} = \begin{bmatrix}
0 & 0 \\
0 & 0 \\
0 & 0 \\
0 & 0 \\
0 & 0 \\
\lambda & 0 \\
0 & \lambda
\end{bmatrix}, \qquad
B_{1,0} = \begin{bmatrix}
\mu_2 & 0 & 0 & 0 & 0 & 0 & 0 \\
0 & 0 & 0 & 0 & 0 & 0 & 0
\end{bmatrix},
$$

$$
B_{2,0} = \begin{bmatrix}
0 & 0 & 0 & 0 & \mu_2 & 0 & 0 \\
0 & 0 & 0 & 0 & 0 & 0 & 0
\end{bmatrix}, \qquad
B_{3,0} = \begin{bmatrix}
0 & 0 & 0 & 0 & 0 & 0 & \mu_2 \\
0 & 0 & 0 & 0 & 0 & 0 & 0
\end{bmatrix},
$$

and $B_{1,1} = A_1$. In these matrices we define $b_{i,j} \stackrel{\text{def}}{=} \lambda_i + \mu_j$ for $j = 1, 2$ and $i = 0, 1, 2$ where $\lambda_0 \stackrel{\text{def}}{=} \lambda$.

A special case of matrix Q that frequently occurs in practice is where $m_1 = m$. This leads to a generator of the form

$$
Q = \begin{bmatrix}
B_0 & A_0 & 0 & 0 & 0 & \cdots \\
B_1 & A_1 & A_0 & 0 & 0 & \cdots \\
B_2 & A_2 & A_1 & A_0 & 0 & \cdots \\
B_3 & A_3 & A_2 & A_1 & A_0 & \cdots \\
B_4 & A_4 & A_3 & A_2 & A_1 & \cdots \\
\vdots & \vdots & \vdots & \vdots & \vdots & \ddots
\end{bmatrix}, \tag{9.17}
$$

where all matrices are square with dimension m. For this case, the above analysis shows that the stationary probabilities satisfy

$$
\pi_j = \pi_0 R^j, \quad j = 1, 2, \ldots,
$$

where the vector π_0 is determined by simultaneously satisfying

$$
\pi_0 \sum_{k=0}^{\infty} R^k B_k = 0 \tag{9.18}
$$

and

$$
\pi_0 (I - R)^{-1} e = 1. \tag{9.19}
$$

Example 9.2 (Scalar Geometric Processes) Let us first revisit a scalar process that is of the form given in (9.17). Recall that Example 8.13 involved a variation of the M/M/1 queue that included processor failures. In the model, when the processor failed, all customers in the system were assumed to be lost. Using the notation from that example allows us to write the generator matrix in block form. Although all matrices considered here are of dimension 1×1 and are thus scalars, it is convenient to continue to refer to them as "matrices," keeping in line with the matrix geometric approach. Of course, scalar processes with repetitive structure are simply special cases of matrix geometric processes.

The matrices for the failure model are given by

$$\boldsymbol{B}_0 = \begin{bmatrix} -\lambda \end{bmatrix}, \quad \boldsymbol{B}_1 = \begin{bmatrix} \mu + \gamma \end{bmatrix}, \quad \boldsymbol{B}_k = \begin{bmatrix} \gamma \end{bmatrix}, \quad k = 2, 3, \dots,$$

and

$$\boldsymbol{A}_0 = \begin{bmatrix} \lambda \end{bmatrix}, \quad \boldsymbol{A}_1 = \begin{bmatrix} -(\lambda + \mu + \gamma) \end{bmatrix}, \quad \boldsymbol{A}_2 = \begin{bmatrix} \mu \end{bmatrix}.$$

Using the solution approach outlined above allows us to easily solve for the stationary state probabilities. The ease of our approach here contrasts with the rather more complex analysis presented in Example 8.13.

We first determine the matrix $\boldsymbol{R}$ (once again, this is a 1×1 matrix that is equivalent to the scalar β in Example 8.13). To do this we must solve (9.7), which, substituting the above matrices, yields

$$\begin{bmatrix} \lambda \end{bmatrix} - \boldsymbol{R} \begin{bmatrix} \lambda + \mu + \gamma \end{bmatrix} + \boldsymbol{R}^2 \begin{bmatrix} \mu \end{bmatrix} = 0.$$

This is a quadratic equation that can be solved to yield

$$\boldsymbol{R} = \begin{bmatrix} \frac{1}{2} \left(1 + \rho + \alpha - \sqrt{(1 + \rho + \alpha)^2 - 4\rho} \right) \end{bmatrix} = \begin{bmatrix} \beta \end{bmatrix},$$

where $\rho = \lambda/\mu$ and $\alpha = \gamma/\mu$, thus corroborating (8.89). Equation (9.18) implies that

$$\pi_0 = \boldsymbol{I} - \boldsymbol{R} = \begin{bmatrix} 1 - \beta \end{bmatrix}.$$

It is easy to check that (9.19) corroborates that π_0 is determined correctly. The stationary solution, written now in terms of scalars, is thus given by

$$\pi_i = (1 - \beta)\beta^i$$

as given in (8.90).

Example 9.3 (A General Matrix Geometric Problem) The matrix geometric technique allows one to generalize scalar models along many lines without changing the overall solution approach. Suppose that we consider a multiple-processor version of the failure problem considered in Example 9.2 (solved using transform techniques in Example 8.13). We use the same notation from these examples. Assume now that there are K homogeneous

processors that are subject to failure. When k, for $k = 1, \ldots, K$, processors
are functional the service rate of customers is given by $k\mu$. When no processors
are functional, $k = 0$, then we assume that all customers are lost. Processor
failures are assumed to be exponential with rate γ, and thus if there are k
functional processors then the failure rate is given by $k\gamma$. Repairs to failed
processors are assumed to occur at a single server repair facility (i.e., there is
a single repair person). The repair rate is assumed to be given by α. The state
of the system is (i, k), where i, for $i = 0, 1, \ldots$, is the number of customers in
the system and k, for $k = 0, 1, \ldots, K$, is the number of functional processors.
The state transition diagram for the induced Markov process is given in Figure
9.3.

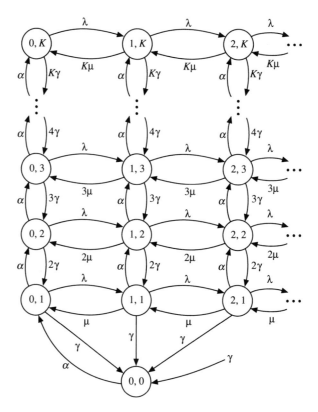

FIGURE 9.3.   State Transition Diagram for Multiple-Processor Failure
Model

It is clear that the process is matrix geometric and is of the form given by
(9.13). The component matrices for the boundary portion of the process are
given by

$$\boldsymbol{B}_{0,0} = \begin{bmatrix} -\alpha \end{bmatrix}, \quad \boldsymbol{B}_{0,1} = \begin{bmatrix} \alpha & 0 & \cdots & 0 \end{bmatrix},$$

and

$$
\boldsymbol{B}_{j,0} = \begin{bmatrix} \gamma \\ 0 \\ \vdots \\ 0 \end{bmatrix}, \quad j = 1, 2, \ldots, \qquad \boldsymbol{B}_{j,1} = 0, \quad j = 2, 3, \ldots,
$$

and

$$
\boldsymbol{B}_{1,1} = \begin{bmatrix}
-b_1 & \alpha & 0 & 0 & \cdots & 0 & 0 & 0 \\
2\gamma & -b_2 & \alpha & 0 & \cdots & 0 & 0 & 0 \\
0 & 3\gamma & -b_3 & \alpha & \cdots & 0 & 0 & 0 \\
\vdots & \vdots & \vdots & \vdots & & \vdots & \vdots & \vdots \\
0 & 0 & 0 & \cdots & (K-1)\gamma & -b_{K-1} & \alpha & 0 \\
0 & 0 & 0 & \cdots & 0 & K\gamma & -b_K & \alpha
\end{bmatrix},
$$

where we define $b_i \overset{\text{def}}{=} \lambda + i\gamma + \alpha$, for $i = 1, 2, \ldots, K$. The matrices of the repeating portion of the process are given by

$$
\boldsymbol{A}_0 = \lambda \boldsymbol{I}, \quad \boldsymbol{A}_1 = \boldsymbol{B}_{1,1}, \quad \boldsymbol{A}_2 = \begin{bmatrix}
\mu & 0 & \cdots & 0 & 0 \\
0 & 2\mu & \cdots & 0 & 0 \\
\vdots & \vdots & & \vdots & \vdots \\
0 & 0 & \cdots & (K-1)\mu & 0 \\
0 & 0 & \cdots & 0 & K\mu
\end{bmatrix}.
$$

We can solve for the stationary state probabilities using the matrix geometric technique outlined earlier. It would be difficult to solve for these probabilities using a generalization of the transform techniques given in Example 8.13.

9.3 Matrix Geometric Solutions for Markov Chains

The matrix geometric method can also be applied to discrete time Markov chains. In this section we delineate the basic equations for this case. Since the material here is analogous to that for the continuous case, our exposition will be abbreviated. Consider the general case of a matrix geometric chain (see (9.13)) with a state transition matrix given by (we put primes on the matrices to indicate that they are probabilities rather than transition rates)

$$
\boldsymbol{P} = \begin{bmatrix}
\boldsymbol{B}'_{0,0} & \boldsymbol{B}'_{0,1} & 0 & 0 & 0 & \cdots \\
\boldsymbol{B}'_{1,0} & \boldsymbol{B}'_{1,1} & \boldsymbol{A}'_0 & 0' & 0 & \cdots \\
\boldsymbol{B}'_{2,0} & \boldsymbol{B}'_{2,1} & \boldsymbol{A}'_1 & \boldsymbol{A}'_0 & 0 & \cdots \\
\boldsymbol{B}'_{3,0} & \boldsymbol{B}'_{3,1} & \boldsymbol{A}'_2 & \boldsymbol{A}'_1 & \boldsymbol{A}'_0 & \cdots \\
\boldsymbol{B}'_{4,0} & \boldsymbol{B}'_{4,1} & \boldsymbol{A}'_3 & \boldsymbol{A}'_2 & \boldsymbol{A}'_1 & \cdots \\
\vdots & \vdots & \vdots & \vdots & \vdots & \ddots
\end{bmatrix},
$$

where $B'_{0,0}$ is a square matrix of dimension $(m_1 - m)$, $B'_{0,1}$ is of dimension $(m_1 - m) \times m$, matrices $B'_{k,0}$, $k \geq 1$, are of dimension $m \times (m_1 - m)$, and all other matrices are square of dimension m, where $m_1 \geq m$. The B' matrices here correspond to state probability transitions from boundary states. Because P is a probability transition matrix it must have row sums equal 1.

The stationary solution for P is the unique nonnegative vector that sums to 1 and satisfies the global balance equations, $\pi^0 = \pi^0 P$. (We use the superscript 0 to distinguish the stationary state probabilities for discrete and continuous Markov processes. Similarly we will subscript matrix R to make the same distinction.) The associated balance equations for the repeating portion of the process are given by

$$\pi_j^0 = \sum_{k=0}^{\infty} \pi_{j-1+k}^0 A'_k, \quad j = 2, 3, \dots . \tag{9.20}$$

Guessing the geometric solution

$$\pi_j^0 = \pi_1^0 R_0^{j-1}, \quad j = 2, 3, \dots,$$

where R_0 is some unknown matrix, and substituting this guess into (9.20) shows that R_0 solves

$$R_0 = \sum_{k=0}^{\infty} R_0^k A_k. \tag{9.21}$$

To determine a solution for the boundary states we must solve the following linear equations

$$(\pi_0^0, \pi_1^0) = (\pi_0^0, \pi_1^0) \left[\begin{array}{cc} B'_{0,0} & B'_{0,1} \\ \sum_{k=1}^{\infty} R_0^{k-1} B'_{k,0} & \sum_{k=1}^{\infty} R_0^{k-1} B'_{k,1} \end{array} \right]. \tag{9.22}$$

As before, to obtain a unique solution we must satisfy the normalization condition,

$$\pi_0^0 e + \pi_1^0 (I - R_0)^{-1} e = 1.$$

This, with (9.22), provides a unique solution.

9.3.1 Uniformization for Matrix Geometric Markov Processes

Because of the special structure of a matrix geometric process, we can uniformize a continuous process by uniformizing transitions from each interlevel (for the repeating portion of the process). The results that we establish here hold generally but are most easily seen through an example. Consider a matrix geometric Markov process with generator (9.2). Concentrate on the repeating portion of the process and define

$q_{max}(j)$ to be the total transition rate for all transitions from interlevel j. It is clear that, by definition, we have

$$q_{max}(j) = -A_1(j,j), \quad j = 0, 1, \ldots, m-1.$$

Define the vector

$$\boldsymbol{q}_{max} \stackrel{\text{def}}{=} (q_{max}(0), q_{max}(1), \ldots, q_{max}(m-1)),$$

and let $\boldsymbol{A}_{max}$ be a diagonal matrix of these entries, that is,

$$\boldsymbol{A}_{max}(i,j) = \begin{cases} 0, & i \neq j, \\ q_{max}(j), & i = j, \quad j = 0, 1, \ldots, m-1. \end{cases}$$

Observe that the inverse, $\boldsymbol{A}_{max}^{-1}$, has off-diagonal entries equal to 0 and diagonal entries given by $1/q_{max}(j)$. Hence, premultiplying a transition rate matrix, such as $\boldsymbol{A}_k$, by $\boldsymbol{A}_{max}^{-1}$ has the same effect as dividing rates by $\boldsymbol{q}_{max}$ to create the state transition probabilities for a general uniformized process (see Section 8.7).

The matrices for the repeating portion of the uniformized process, denoted by $\boldsymbol{A}'_k$, are then given by

$$\boldsymbol{A}'_k = \begin{cases} \boldsymbol{A}_{max}^{-1}\boldsymbol{A}_k, & k \neq 1, \\ \boldsymbol{A}_{max}^{-1}\boldsymbol{A}_1 + \boldsymbol{I}, & k = 1. \end{cases} \tag{9.23}$$

In a similar manner we uniformize the boundary portion of the continuous time process. To do this, let $\boldsymbol{B}_{max}$ be the matrix of total transition rates from each boundary state. We can write the entries of this matrix as

$$\boldsymbol{B}_{max}(i,j) = \begin{cases} 0, & i \neq j, \\ -\boldsymbol{B}_{0,0}(j,j), & i = j. \end{cases}$$

We have the following probability transition matrices for the boundary portion

$$\boldsymbol{B}'_{0,0} = \boldsymbol{B}_{max}^{-1}\boldsymbol{B}_{0,0} + \boldsymbol{I}, \qquad \boldsymbol{B}'_{0,1} = \boldsymbol{B}_{max}^{-1}\boldsymbol{B}_{0,1},$$

and

$$\boldsymbol{B}'_{1,0} = \boldsymbol{A}_{max}^{-1}\boldsymbol{B}_{1,0}. \tag{9.24}$$

Equation (9.24) reflects the fact that the total rate of transitions from the first level of the repeating portion to the boundary states is given by matrix $\boldsymbol{A}_{max}$ rather than by $\boldsymbol{B}_{max}$.

The relationship between the continuous process rate matrix $\boldsymbol{R}$ and its uniformized discrete counterpart $\boldsymbol{R}_0$ is then given by

$$\boldsymbol{R} = \boldsymbol{A}_{\max} \boldsymbol{R}_0 \boldsymbol{A}_{\max}^{-1}. \tag{9.25}$$

To see this substitute (9.25) into (9.7) to obtain

$$
\begin{aligned}
0 &= \boldsymbol{A}_0 + \boldsymbol{R}\boldsymbol{A}_1 + \boldsymbol{R}^2 \boldsymbol{A}_2 \\[2mm]
&= \boldsymbol{A}_0 + \boldsymbol{A}_{\max} \boldsymbol{R}_0 \boldsymbol{A}_{\max}^{-1} \boldsymbol{A}_1 + \left(\boldsymbol{A}_{\max} \boldsymbol{R}_0 \boldsymbol{A}_{\max}^{-1}\right)^2 \boldsymbol{A}_2 \\[2mm]
&= \boldsymbol{A}_0 + \boldsymbol{A}_{\max} \boldsymbol{R}_0 (\boldsymbol{A}_1' - \boldsymbol{I}) + \boldsymbol{A}_{\max} \boldsymbol{R}_0^2 \boldsymbol{A}_2', \tag{9.26}
\end{aligned}
$$

where we have substituted (9.23) to obtain (9.26). Simplifying this equation implies that

$$\boldsymbol{A}_{\max} \boldsymbol{R}_0 = \boldsymbol{A}_0 + \boldsymbol{A}_{\max} \boldsymbol{R}_0 \boldsymbol{A}_1' + \boldsymbol{A}_{\max} \boldsymbol{R}_0^2 \boldsymbol{A}_2'.$$

Multiplying both sides of this equation on the left by $\boldsymbol{A}_{\max}^{-1}$ and using (9.23) shows that the form for $\boldsymbol{R}_0$ given in (9.25) satisfies (9.21).

We can relate the stationary probabilities for the continuous and uniformized case by defining

$$\nu \overset{\text{def}}{=} \left(\boldsymbol{\pi}_0 \boldsymbol{B}_{\max} \boldsymbol{e} + \boldsymbol{\pi}_0 \boldsymbol{R} (\boldsymbol{I} - \boldsymbol{R})^{-1} \boldsymbol{A}_{\max} \boldsymbol{e}\right)^{-1}.$$

It then follows that

$$
\boldsymbol{\pi}_i^0 = \begin{cases} \nu \boldsymbol{\pi}_0 \boldsymbol{B}_{\max}, & i = 0, \\[2mm] \nu \boldsymbol{\pi}_0 \boldsymbol{R}^i \boldsymbol{A}_{\max}, & i = 1, 2, \ldots . \end{cases} \tag{9.27}
$$

Problem 9.4 establishes this result.

To interpret (9.27), we view the uniformized process as a semi-Markov process. Recall that the stationary distribution of a semi-Markov process can be determined as a weighted sum of the stationary probabilities of the embedded process (see (8.52)) repeated here by

$$\pi_i = \frac{\pi_i^e E\left[H_i\right]}{\sum_{j=0}^{\infty} \pi_j^e E\left[H_j\right]}. \tag{9.28}$$

In this equation H_i is the holding time for state i. Rewriting (9.27) in terms of the probabilities of the original Markov process yields

$$
\pi_i = \begin{cases} \nu^{-1} \boldsymbol{\pi}_0^0 \boldsymbol{B}_{\max}^{-1}, & i = 0, \\[2mm] \nu^{-1} \boldsymbol{\pi}_i^0 \boldsymbol{A}_{\max}^{-1}, & i = 1, 2, \ldots . \end{cases} \tag{9.29}
$$

In the matrix geometric version of the semi-Markov process induced by uniformizing over levels, the expected holding times, $E[H_i]$ of (9.28) are given by a matrix equation. In particular, we recognize that the expected holding time of interlevel state i of the boundary portion of the process is given by the (i,i) entry of the matrix $\boldsymbol{B}_{\max}^{-1}$. Similarly the expected holding time of interlevel state i of the repeating portion of the process is given by the ith diagonal entry of $\boldsymbol{A}_{\max}^{-1}$. Equation (9.29) thus shows that the uniformized version of a matrix geometric process is simply a semi-Markov process view of a matrix geometric process where the embedding is by levels.

9.4 Properties of Solutions

In this section we establish a condition for the stability of the process in terms of its block matrices, provide an interpretation for the rate matrix $\boldsymbol{R}$, and discuss computational procedures associated with the method. We start with an equation for the stability of the process.

9.4.1 Stability

Loosely speaking, the stability of a process depends on the expected *drift* of the process for states in the repeating portion of the process. For example, in the M/M/1 queueing model, if we select a given state in the repeating portion then its expected drift toward higher states is given by λ (a rate of λ times a distance of plus 1 unit) and its expected drift toward lower states by $-\mu$ (a rate of μ times a distance of minus 1 unit). The drift of the process is given by $\lambda - \mu$ and the process is stable if this total drift is negative, leading to the well-known stability condition of $\lambda < \mu$. Intuitively, this means that the expected direction of the process is toward lower valued states. If the total drift is positive, then the process tends to move toward higher valued states and is unstable.

In the calculation of the total drift of a process we must include a notion of distance. For example, suppose in the scalar process that transitions can go up by 1 step and down by at most K steps. Also suppose that the transition rate for ℓ steps is given by $r(\ell), \ell = -K, -K-1, \ldots, -1, 1$. Then the drift upward is given by $r(1)$, the drift downward is given by $-\sum_{\ell=1}^{K} \ell r(-\ell)$, and the total drift is given by the sum of these two components. Stability of the process implies that

$$r(1) < \sum_{\ell=1}^{K} \ell r(-\ell). \tag{9.30}$$

We now investigate how we apply these concepts to processes with a matrix geometric structure.

Analogous to the scalar case, we will think of the drift of the process in terms of levels. Assume, as above, that transitions can go up one level or down by at most K levels. We wish to calculate the total drift from a level in the repeating portion of the process analogous to (9.30). To do this we consider the process for levels that are far from the boundary, that is, level i such that $i \gg 0$. It is clear that the stability of the system depends only on the expected drift from these levels. There is one complication that arises when the process is of matrix geometric form. To see this, consider a transition from level $i, i \gg 0$, to level $i - k, 1 \le k \le K$, that is, a distance of k downward. To calculate the drift of such a transition, we must know which rate from level i to apply. If, for example, we knew that the transition was from interlevel state j, $0 \le j \le m - 1$, then the component of the drift for this transition is given by

$$-k \sum_{\ell=0}^{m-1} A_{k+1}(j, \ell).\tag{9.31}$$

In (9.31) $A_{k+1}(j, \ell)$ is the transition rate from state j in level i to state ℓ in level $i - k$. To define the average drift, let f_j, $0 \le j \le m - 1$, be the probability that the process is in interlevel state j of the repeating portion of the process of level $i \gg 0$ (intuitively speaking the values of f_j, $0 \le j \le m - 1$, are independent of the level i since i is assumed to be extremely far from the boundary). The average drift from level i to level $i - k$ is given by

$$-k \sum_{j=0}^{m-1} f_j \sum_{\ell=0}^{m-1} A_{k+1}(j, \ell)\tag{9.32}$$

and to obtain the total rate we sum (9.32) over all $k, 0 \le k \le K + 1$.

The difficulty in performing this calculation is to determine the values of f_j. To determine these values we can consider a derived Markov process on state space $0, 1, \ldots, m-1$, which identifies transitions only in terms of their interlevel states. Specifically, if the original process has a transition from state (i, j) to state $(i-k, \ell)$, then in the derived process we consider that a transition occurs from interlevel state j to interlevel state ℓ. It can be easily seen that the generator for this derived process is given by the *interlevel transition matrix*, A, defined by

$$A \overset{\text{def}}{=} \sum_{\ell=0}^{K+1} A_\ell,\tag{9.33}$$

and that the probabilities f_j are given by its stationary measure, that is, they solve

$$fA = 0,\tag{9.34}$$

$$fe = 1,\tag{9.35}$$

where $\boldsymbol{f} \overset{\text{def}}{=} (f_0, f_1, \ldots, f_{m-1})$. Using this in the above drift analysis and simplifying yields a stability condition given by

$$\boldsymbol{f} \boldsymbol{A}_0 e < \sum_{k=2}^{K+1} (k-1) \boldsymbol{f} \boldsymbol{A}_k e. \qquad (9.36)$$

9.4.2 Interpretation of the Rate Matrix

We now discuss an interpretation of the rate matrix. Our discussion of this is simplified if we first consider the discrete time case. We can use the results of Section 9.3.1 to determine the analogous interpretation for the continuous time case. We show how one can interpret the entries in $\boldsymbol{R}_0$ (see [87] for the first derivation of these results). To do this we construct a transient Markov process. Consider the repeating portion of the process beginning with some given level and reindex levels from this point starting with the index 0. To start the process we select a *starting interlevel state*, say j, $0 \le j \le m-1$, in level 0 and let the process run according to state transitions given by $\boldsymbol{A}'_k, 0 \le k \le K+1$. We stop the process (i.e., it *absorbs*) the first time it makes a transition to a state below level 1. Thus if the first transition of the process is not to level 1 (i.e., the first transition is not one of the transitions $\boldsymbol{A}_0(j, j')$ where $0 \le j' \le m-1$) then the process absorbs at the first step. Clearly the induced transient process depends on the starting interlevel state j and we will denote the transient process that starts from level 0 in interlevel state j by Υ_j, $0 \le j \le m-1$.

Define

$$\boldsymbol{P}_i^{(n)} = \{\boldsymbol{P}_i^{(n)}(j, j'), \ i \ge 1, \ n \ge i, \ 0 \le j, j' \le m-1\}$$

to be a square matrix of dimension m where the value in position (j, j') is the probability that after n steps process Υ_j is in interlevel state j' of level i. We let $\boldsymbol{P}_0^{(n)} = \boldsymbol{I}$, for all n. Note that since the process can go up by at most one level at each transition, it is clear that $\boldsymbol{P}_i^{(n)} = \boldsymbol{0}$ for $i > n$. We also define

$$\boldsymbol{N}_i = \{\boldsymbol{N}_i(j, j'), \ i \ge 1, 0 \le j, j' \le m-1\}$$

to be a square matrix of dimension m where the value in position (j, j') is the expected number of visits made in Υ_j to interlevel state j' of level i before absorption. We set $\boldsymbol{N} = \boldsymbol{N}_1$. It follows that

$$\boldsymbol{N}_i = \sum_{n=i}^{\infty} \boldsymbol{P}_i^{(n)}, \quad i \ge 1. \qquad (9.37)$$

We now make the observation that if process Υ_j, $0 \le j \le m-1$, is in any interlevel state of level i, $i > 1$ at step n, $n \ge i$, it must have passed through level $i - 1$. Partition all possible paths according to their last visit to level $i - 1$. Suppose that at step $\ell, \ell < n$, Υ_j was in interlevel state $j', 0 \leq j' \leq m - 1$, and that this was its last visit to level $i - 1$ prior to step n. Then if we reindex levels so that level $i - 1$ becomes level 0 it is clear that the future evolution of the process is statistically identical to $\Upsilon_{j'}$. This follows from the fact that step ℓ is assumed to be the last visit to level $i - 1$ and thus the process can be thought of as stopping for any transition into levels less than i. Partitioning on all possible values of ℓ permits us to write

$$P_i^{(n)} = \sum_{\ell=i-1}^{n-1} P_{i-1}^{(\ell)} P_1^{(n-\ell)}, \quad i \geq 2. \tag{9.38}$$

The matrix $P_{i-1}^{(\ell)}$ in (9.38) is the probability that Υ_j visits level $i - 1$ in ℓ steps and the factor $P_1^{(n-\ell)}$ accounts for the fact that this is the last visit to level $i - 1$ since it is the probability that Υ_j starting from level $i - 1$ visits level i in $n - \ell$ steps without ever going below level i.

Summing this over n and using (9.37) yields

$$\begin{aligned}
N_i &= \sum_{n=i}^{\infty} \sum_{\ell=i-1}^{n} P_{i-1}^{(\ell)} P_1^{(n-\ell)} \\
&= \sum_{\ell=i-1}^{\infty} P_{i-1}^{(\ell)} \sum_{n=\ell+1}^{\infty} P_1^{(n-\ell)} \\
&= N_{i-1} N, \quad i \geq 2,
\end{aligned}$$

which implies that

$$N_i = N^i, i \geq 2. \tag{9.39}$$

Assume now that we observe process Υ_j, $0 \leq j \leq m - 1$, at step $n - 1$ and it makes a transition into level 1. Conditioning on all such possible transitions yields

$$P_1^{(n)} = \sum_{\ell=0}^{K+1} P_\ell^{(n-1)} A'_\ell.$$

Summing this over n and using (9.39) yields

$$\begin{aligned}
N &= \sum_{\ell=0}^{K+1} \sum_{n=1}^{\infty} P_\ell^{(n-1)} A'_\ell \\
&= \sum_{\ell=0}^{K+1} N^\ell A'_\ell, \tag{9.40}
\end{aligned}$$

where we select the minimal solution. Comparing (9.40) with (9.21) demonstrates that $\boldsymbol{R}_0 = \boldsymbol{N}$. (The fact that both $\boldsymbol{R}_0$ and $\boldsymbol{N}$ satisfy the same equation is not sufficient to imply that they are identical. Our derivation provides the intuition for why they are the same; see [86] for a rigorous argument that establishes the equality.) Thus an interpretation for $\boldsymbol{R}_0$ is that the entry in the (j, j') position is the expected number of visits to interlevel state j' of level 1 before absorption given that the process is started with an interlevel index of j of level 0. This interpretation implies that the entries of the rate matrix are nonnegative.

In the case of continuous time processes, the interpretation of the $\boldsymbol{R}$ matrix is complicated by the fact that we must include the time the process spends in each state in the calculation. It is convenient to consider the uniformized discrete chain and then to apply the above derivation for the interpretation of $\boldsymbol{R}_0$. Equation (9.25) forms the bridge between these two equivalent systems. Recall that we defined $q_{max}(j)$ to be the total transition rate out of interlevel j and set $\boldsymbol{A}_{max}(j, j) = q_{max}(j)$. Using (9.25) we can write the (j, k) entry of $\boldsymbol{R}$ as

$$\boldsymbol{R}(j, k) = \frac{q_{max}(j)}{q_{max}(k)} \boldsymbol{R}_0(j, k)$$

$$= \frac{\boldsymbol{R}_0(j, k)}{q_{max}(k)} \frac{1}{q_{max}(j)^{-1}}. \tag{9.41}$$

To interpret (9.41), first observe that $q_{max}(x)^{-1}$ is the expected length of time the process spends in a given state of interlevel x. Hence the first part of (9.41), given by $\boldsymbol{R}_0(j, k)/q_{max}(k)$, equals the total expected time (in units to be specified) spent in state $(i+1, k)$ before first hitting level i or below given that the process is started in state (i, j). The second part of (9.41) can be interpreted as specifying the units in which time is measured. The equations show that one unit of time equals $q_{max}(j)^{-1}$, that is, the expected length of time the process spends in a given state of interlevel j. Observe that if $q_{max}(j) = q_{max}(k)$ then we trivially obtain that $\boldsymbol{R} = \boldsymbol{R}_0$.

Example 9.4 (Interpretation of Rate Matrix for Birth–Death Processes) As previously noted, a scalar state Markov process that has a repetitive structure is a special case of a matrix geometric process and hence the interpretation of the rate matrix $\boldsymbol{R}$ should be applicable. Consider, as an example, the modified M/M/1 queue considered in Example 8.12. The "rate matrix" for this process equals $\rho = \lambda/\mu$. According to the above calculations, ρ should equal the expected time (measured in units of $1/\mu$) that the process spends in state $i + 1$ prior to its first visit to state i after starting in state i of the repeating portion of the process (i.e., states $2, 3, \ldots$). To derive this from first principles, let us first consider the embedded discrete time chain uniformized process obtained by using $q_{max} = \lambda + \mu$.

For i in the repeating portion, transitions to state $i+1$ occur with probability $p \overset{\text{def}}{=} \lambda/(\lambda+\mu)$ and to state $i-1$ with probability $1-p$. The probability that the chain visits state $i+1$ a total of k times before returning to state i after first starting in state i can be calculated as follows. The case of $k=0$ occurs if the first transition is downward, an event that occurs with probability $1-p$. Otherwise if $k>0$ then the first visit occurs at step 1 with probability p and the number of subsequent visits is geometrically distributed with parameter p starting from 0. To see this, consider the first transition made after hitting state $i+1$ on step 1 (i.e., consider the second transition). If that transition is upwards(with probability p) then clearly state $i+1$ will be visited again before the first visit to state i since all paths to i must first go through state $i+1$. Hence the process stops the first time there is a downward transition from state $i+1$. The probability that k visits occur is given by $p^k(1-p)$ for $k=1,2,\dots$. The expected number of visits to state $i+1$ before first hitting state i is thus given by

$$\sum_{k=1}^{\infty} kp^k(1-p) = \frac{p}{1-p} = \rho.$$

The expected holding time of state $i+1$ is the same as the expected holding time of state i, and thus the continuous process and uniformized process have the same value for their "rate matrix."

For some performance models the interpretation of the rate matrix can be used to determine explicit solutions for some of its elements. For example, if there is no path from interlevel state j at level i to interlevel state j' at level $i+1$ then $\boldsymbol{R}(j,j') = 0$. We can also derive explicit expressions if interlevel j is *isolated* from other interlevels. To illustrate this, consider a simple case where $K=1$ and there are transitions between interlevel states j and hence $\boldsymbol{A}_0(j,j) \neq 0$ and $\boldsymbol{A}_2(j,j) \neq 0$. Interlevel j is isolated if there are no transitions from it into other interlevels and thus $\boldsymbol{A}(j,j') = 0$ for $j \neq j'$ where $\boldsymbol{A}$ is given by (9.33). The entry in $\boldsymbol{R}(j,j)$ is what would be obtained from a scalar state birth–death process such as that considered in Example 9.4, that is, $\boldsymbol{R}(j,j) = \boldsymbol{A}_0(j,j)/\boldsymbol{A}_2(j,j)$. Other generalizations of isolation that lead to explicit entries in the rate matrix are clearly possible. A way to classify such cases is by the structure of the interlevel transition matrix $\boldsymbol{A}$ defined by (9.33). As discussed in the section on stability, we can interpret this matrix as the generator matrix for a Markov process defined on the interlevels. If the corresponding matrix is reducible then its states can be divided into irreducible components. If some of these components can be solved individually, then one can often use these solutions to help solve for an analytic representation of the rate matrix of the process. Consider the following example.

Example 9.5 (Reducible Matrix Geometric Problem)   Consider a computer system consisting of two processors, a main processor and a backup.

When the main processor fails the system immediately switches over to the backup processor. We assume that the backup processor operates at a slower rate than the main processor. Tasks executed on the main processor are assumed to be exponential with rate μ_1, whereas for the backup processor the rate reduces to μ_2, for $\mu_2 < \mu_1$. We assume that failures to the main processor occur at a rate of γ and that the main processor is immediately repaired when the system empties. This might be reflected in an actual system by the fact that the main processor cannot be repaired without losing customers in the queue. If the cost of losing customers is higher than the performance degradation due to running jobs on the backup processor, a system implementer might opt to wait until the system empties before repairing a failed processor. In this example we do not model failures to the system while the backup processor is in operation (see Problem 9.5). Customer arrivals are Poisson with rate λ. The state transition diagram for the induced Markov process is shown in Figure 9.4.

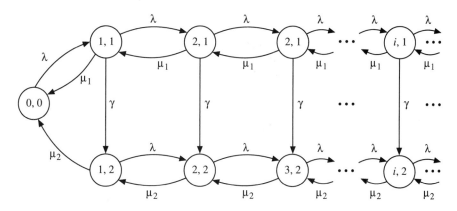

FIGURE 9.4. State Transition Diagram for a Reducible Matrix Geometric Problem

In the figure state (i, j), for $i = 0, 1, \ldots, \quad j = 1, 2$, corresponds to the situation where i customers are in the queue in the system and the main processor is not failed if $j = 1$ or is failed and the backup is processing customers if $j = 2$. The state $(0, 0)$ corresponds to an empty system where, by the definition of the model, the main processor is operational. The generator matrix for the process has the form given in (9.2) where the matrices are given by

$$\boldsymbol{A}_0 = \begin{bmatrix} \lambda & 0 \\ 0 & \lambda \end{bmatrix}, \quad \boldsymbol{A}_1 = \begin{bmatrix} -(\lambda + \mu_1 + \gamma) & \gamma \\ 0 & -(\lambda + \mu_2) \end{bmatrix}, \quad \boldsymbol{A}_2 = \begin{bmatrix} \mu_1 & 0 \\ \mu_2 & 0 \end{bmatrix},$$

and

$$\boldsymbol{B}_{1,0} = \begin{bmatrix} 0 & \mu_1 & 0 \\ 0 & 0 & \mu_2 \end{bmatrix}, \quad \boldsymbol{B}_{0,1} = \begin{bmatrix} 0 & 0 \\ \lambda & 0 \\ 0 & \lambda \end{bmatrix},$$

$$\boldsymbol{B}_{0,0} = \begin{bmatrix} -\lambda & \lambda & 0 \\ \mu_1 & -(\lambda + \mu_1 + \gamma) & \gamma \\ \mu_2 & 0 & -(\lambda + \mu_2) \end{bmatrix}.$$

Note that $\boldsymbol{B}_{1,1} = \boldsymbol{A}_1$.

It is clear that there is no path from a state on the second interlevel to a state on the first interlevel that does not pass through state $(0,0)$. Hence the interlevel transition matrix for the process $\boldsymbol{A}$ defined by (9.33) is reducible. In the model, this reducibility corresponds to the fact that a failed main processor is repaired only when the system empties. From the interpretation of the rate matrix this clearly implies that $\boldsymbol{R}(2,1) = 0$. To avoid cumbersome notation in what follows we revert back to our typical matrix entry notation and let $r_{i,j} = \boldsymbol{R}(i,j)$. Hence, we have just established that $r_{2,1} = 0$. The portion of the state space represented by the lower sequence of states in Figure 9.4 corresponds to an isolated birth–death process. This implies that the $(2,2)$ entry of $\boldsymbol{R}$ can be determined as with a birth–death scalar process. Example 9.4 implies that $r_{2,2} = \lambda/\mu_2$.

Using this information we can write a closed-form solution for the rate matrix. We proceed by observing that we can write the defining equation for $\boldsymbol{R}$ given in (9.7) using the above values for the entries of the rate matrix as

$$\begin{bmatrix} \lambda & 0 \\ 0 & \lambda \end{bmatrix} + \begin{bmatrix} r_{1,1} & r_{1,2} \\ 0 & \lambda/\mu_2 \end{bmatrix} \begin{bmatrix} -(\lambda + \mu_1 + \gamma) & \gamma \\ 0 & -(\lambda + \mu_2) \end{bmatrix}$$
$$+ \begin{bmatrix} r_{1,1}^2 & r_{1,2}(r_{1,1} + \lambda/\mu_2) \\ 0 & (\lambda/\mu_2)^2 \end{bmatrix} \begin{bmatrix} \mu_1 & 0 \\ 0 & \mu_2 \end{bmatrix} = 0. \quad (9.42)$$

We have to solve four simultaneous equations. It is easy to check that, because we know the values of $r_{2,1}$ and $r_{2,2}$, two of the equations are trivially satisfied. The remaining two equations are given by

$$\lambda - r_{1,1}(\lambda + \mu_1 + \gamma) + \mu_1 r_{1,1}^2 = 0, \quad (9.43)$$

and

$$\gamma r_{1,1} - r_{1,2}(\lambda + \mu_2) + \mu_2(r_{1,1}r_{1,2} + r_{1,2}\lambda/\mu_2) = 0. \quad (9.44)$$

Solving (9.44) for $r_{1,2}$ in terms of $r_{1,1}$ yields

$$r_{1,2} = \frac{\gamma r_{1,1}}{\mu_2 - \mu_2 r_{1,1}}.$$

Equation (9.43) can be solved using the quadratic formula (recall that the $\boldsymbol{R}$ matrix is the *minimal* solution and thus only the negative radical remains), yielding

$$r_{1,1} = \frac{\lambda + \mu_1 + \gamma - \sqrt{(\lambda + \mu_1 + \gamma)^2 - 4\mu_1\lambda}}{2\mu_1}.$$

**9.4.3
Computational
Properties**

One iterative procedure that is often used to solve for matrix $\boldsymbol{R}$ is given by

$$\boldsymbol{R}(0) \quad = \quad \boldsymbol{0}, \tag{9.45}$$

$$\boldsymbol{R}(n+1) \quad = \quad -\sum_{\ell=0,\,\ell\neq1}^{\infty} \boldsymbol{R}^{\ell}(n)\boldsymbol{A}_{\ell}\boldsymbol{A}_{1}^{-1}, \quad n=0,1,\ldots, \tag{9.46}$$

where the iteration halts when entries in $\boldsymbol{R}(n+1)$ and $\boldsymbol{R}(n)$ differ in absolute value by less than a given small constant. The equation in (9.46) is obtained by multiplying equation (9.15) from the right by $\boldsymbol{A}_{1}^{-1}$. The sequence $\{\boldsymbol{R}(n)\}$ are entry-wise nondecreasing and converge monotonically to a nonnegative matrix $\boldsymbol{R}$ that satisfies (9.15). This follows from the fact that $-\boldsymbol{A}_{1}^{-1}$ is a nonnegative matrix. To establish this, note that $\boldsymbol{A}_{1}$ has positive non-diagonal elements and negative elements along the diagonal. Let q_{max} be the absolute value of the smallest diagonal element. We can write $-\boldsymbol{A}_{1}^{-1}$ as

$$-\boldsymbol{A}_{1}^{-1} \quad = \quad [q_{max}\boldsymbol{I} - (q_{max}\boldsymbol{I} + \boldsymbol{A}_{1})]^{-1}$$

$$= \quad q_{max}^{-1}\left[\boldsymbol{I} - \left(\boldsymbol{I} + \frac{1}{q_{max}}\boldsymbol{A}_{1}\right)\right]^{-1}$$

$$= \quad q_{max}^{-1}\sum_{n=0}^{\infty}\left(\boldsymbol{I} + \frac{1}{q_{max}}\boldsymbol{A}_{1}\right)^{n}.$$

The nonnegativity of $-\boldsymbol{A}_{1}^{-1}$ thus follows from the fact that the matrix $\boldsymbol{I} + \boldsymbol{A}_{1}/q_{max}$ has nonnegative entries.

We note that the number of iterations needed for convergence increases as the spectral radius of $\boldsymbol{R}$ increases. Similar to the scalar case where ρ played the part of $\boldsymbol{R}$, the spectral radius of $\boldsymbol{R}$ in many performance models can be thought of as a measure of the utilization of the system. This implies that, for these cases, as the utilization of the system increases it becomes computationally more difficult to compute the entries in $\boldsymbol{R}$. Most of the computational effort associated with the matrix geometric method is, in fact, expended in computing $\boldsymbol{R}$.

For many performance models the $\boldsymbol{A}_{0}$ matrix is a diagonal matrix with entries $\boldsymbol{A}_{0}(j,j) = \lambda, 0 \leq j \leq m-1$, where λ is the arrival rate of customers to the system. This corresponds to a state-independent Poisson arrival stream of customers. As λ increases, the spectral radius of the $\boldsymbol{R}$ matrix also increases and performance measures become increasingly more compute-intensive because (9.46) takes more iterations to converge. Typically one wishes to evaluate some performance measure, say expected response time, for many different arrival rates $0 < \lambda_1 < \lambda_2 < \ldots < \lambda_L$. Using the probabilistic interpretation for the $\boldsymbol{R}$ matrix, it is easy to see that for some models $\boldsymbol{R}_\ell \leq \boldsymbol{R}_{\ell+1}$, $1 \leq \ell \leq L-1$, where $\boldsymbol{R}_\ell$ denotes the rate matrix corresponding to an arrival rate of λ_ℓ. Computational effort can be saved in these cases if one starts the iteration for $\ell > 1$ with the previously calculated rate matrix, that is, using $\boldsymbol{R}_\ell(0) = \boldsymbol{R}_{\ell-1}$ rather than $\boldsymbol{R}_\ell(0) = 0$ in (9.45).

Example 9.6 (Replicated Database)   Consider the performance of a database system where there are two replications of the data. Each replication is independently accessible and modeled by a server. Requests to access the database are assumed to queue in a central location and correspond to read or write operations. To preserve the integrity of both copies of the database, we assume that write requests must wait until both copies of the database are available before beginning execution. Both copies are assumed to be updated in parallel and released simultaneously. Read requests can be processed by any copy of the database. Both types of requests are assumed to wait in the queue in the order in which they arrive. We assume that requests arrive to the system from a Poisson point source with intensity of λ and that the probability a given request is a read (resp., write) is given by r (resp., $1-r$). Service times for both read and write requests are assumed to be exponential with an average value of μ^{-1}. Since we assume that writes are served in parallel, the total service time for write requests equals the maximum of two exponential random variables with parameter μ.

We let I_t be the number of requests at time t that are waiting in the queue and let $J_t, 0 \leq J_t \leq 2$, be the number of replications that are involved in a read or write operation at time t. Our assumptions above imply that (I_t, J_t) is a Markov process. The state transition diagram for the process is given in Figure 9.5.

We explain some of the transitions from the repeating portion of the process. In state $(2,2)$ both servers are busy serving customers and the customer at

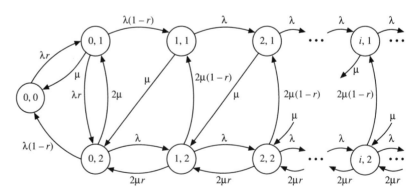

FIGURE 9.5. State Transition Diagram for Replicated Database

the head of the queue is equivalent to an unexamined arrival. Thus it is a read (resp., write) request with probability r (resp., $(1-r)$). Upon service completion, at rate 2μ, there are two possibilities depending on the type of customer at the head of the queue. If the customer is a read request, then the next state will be $(1,2)$ and hence the transition rate from $(2,2)$ to $(1,2)$ is given by $2\mu r$. On the other hand, if the head of the queue is a write request then the next state is $(2,1)$ since the write must wait until all servers are available before beginning execution. Hence the transition rate from $(2,2)$ to $(2,1)$ is given by $2\mu(1-r)$. The rest of the transitions can be explained in a similar manner.

If we order the state lexicographically, the generator matrix of the process is given by

$$
Q = \begin{bmatrix}
-\lambda & \lambda r & \lambda(1-r) & 0 & 0 & 0 & \cdots \\
\mu & -(\lambda+\mu) & \lambda r & \lambda(1-r) & 0 & 0 & \cdots \\
0 & 2\mu & -(\lambda+2\mu) & 0 & \lambda & 0 & \cdots \\
0 & 0 & \mu & -(\lambda+\mu) & 0 & \lambda & \cdots \\
0 & 0 & 2\mu r & 2\mu(1-r) & -(\lambda+2\mu) & 0 & \cdots \\
0 & 0 & 0 & 0 & \mu & -(\lambda+\mu) & \cdots \\
\vdots & \vdots & \vdots & \vdots & \vdots & \vdots & \ddots
\end{bmatrix}.
$$

This generator is of matrix geometric form. To easily see this, identify the matrices of (9.13) as follows:

$$
B_{0,0} = \begin{bmatrix}
-\lambda & \lambda r & \lambda(1-r) \\
\mu & -(\lambda+\mu) & \lambda r \\
0 & 2\mu & -(\lambda+2\mu)
\end{bmatrix}, \quad
B_{0,1} = \begin{bmatrix}
0 & 0 \\
\lambda(1-r) & 0 \\
0 & \lambda
\end{bmatrix},
$$

$$
B_{1,0} = \begin{bmatrix}
0 & 0 & \mu \\
0 & 0 & 2\mu r
\end{bmatrix}, \quad
B_{1,1} = A_1 = \begin{bmatrix}
-(\lambda+\mu) & 0 \\
2\mu(1-r) & -(\lambda+2\mu)
\end{bmatrix},
$$

$$
A_0 = \begin{bmatrix}
\lambda & 0 \\
0 & \lambda
\end{bmatrix}, \quad \text{and} \quad
A_2 = \begin{bmatrix}
0 & \mu \\
0 & 2\mu r
\end{bmatrix}.
$$

We can use the procedure outlined in Section 9.1 to solve for its stationary distribution and performance measures.

To determine the stability of the system we first form the matrix A of equation (9.33), which is given by

$$
A = \begin{bmatrix}
-\mu & \mu \\
2\mu(1-r) & -2\mu(1-r)
\end{bmatrix}.
$$

Solving equations (9.34) and (9.35) yields

$$
f_1 = \frac{2(1-r)}{3-2r} \quad \text{and} \quad f_2 = \frac{1}{3-2r},
$$

which, when used in (9.36), yields the stability condition

$$\lambda < \frac{2\mu}{3 - 2r}.$$

Observe that if $r = 1$ then the stability is identical to that obtained in an M/M/2 queue, that is, $\lambda < 2\mu$ and for $r = 0$ it is identical to that obtained for an M/G/1 queue where the expected service time is the maximum of two exponentials at rate μ, that is, $\lambda < 2\mu/3$.

9.4.4
Specifying
Matrices

It is relatively straightforward to write a computer program to solve matrix geometric problems. One requires a linear equation solver to solve for the stability condition of the process and for the boundary state probabilities. A simple matrix iteration can be used to determine the values of the rate matrix. To compute performance measures requires routines that calculate the inverse of a matrix. Routines that solve linear equations and calculate inverses can be found in standard linear algebra packages. The difficulties in using the method do not lie in the core programs for the technique but rather lie in specifying the matrices. Such a specification is detailed, error-prone, and sometimes difficult to code in a general fashion. These problems are more severe when the underlying matrices are large or must be specified in terms of unknown parameter values such as the size of the problem.

A general procedure can be used to aid the specification of the matrices. One writes a program for each of the underlying matrices of the problem. This program loops over all possible pairs of indices and, for each pair of indices, applies a set of disjoint Boolean conditions (i.e., at most one condition can be satisfied). Each condition corresponds to a non-zero value of the corresponding matrix. In this way, to specify the matrices one only has to determine the conditions that correspond to non-zero values. This can usually be determined from the semantics of the model being analyzed. As a simple example, consider matrix A_1 given above for the replicated database example (see Example 9.6). To avoid presenting too trivial a problem, suppose we consider the general case where, instead of two databases, there are $k \geq 2$ databases. Hence matrix A_1 is a $k \times k$ matrix. In the computer program we first initialize matrix $A_1 = 0$. The value corresponding to the entry in position (i, j) can be specified in a *programming table* as follows:

i-value	j-value	entry
$i, \quad 1 \leq i \leq k$	i	$-(\lambda + (i-1)\mu)$
$i, \quad 2 \leq i \leq k$	$i-1$	$i\mu(1-r)$

A computer program loops over i and j checking each of the conditions in the table. The second condition would read something like (this is not to represent any particular computer language):

if $(i \geq 2$ and $i \geq k$ and $j = i-1)$ then $a1(i,j) = i\mu(1-r)$

If no conditions are satisfied, then the entry in position (i,j) remains a 0. Similar tables can be constructed for each of the underlying matrices (see Problem 9.6). In this way a computer implementation only requires changing logical conditions specified in the program associated with the matrix.

9.5
Phase
Distributions

A continuous *phase distribution* is the distribution of time until absorption in an absorbing Markov process. Similarly a discrete phase distribution is the distribution of the number of steps prior to final absorption in an absorbing Markov chain. Phase distributions can be easily incorporated into matrix geometric models, and we will explore several examples of how such extensions increase the modeling capabilities of such models. We first concentrate on the continuous case.

9.5.1
Continuous
Phase
Distributions

To make the definition of a continuous phase distribution precise, assume that states $1, 2, \ldots, n$ are transient and that state $n+1$ is an absorbing state of a Markov process. Define $\beta_i, i = 1, 2, \ldots, n+1$ to be the probability that the process is started in state i and define $\beta \stackrel{\text{def}}{=} (\beta_1, \beta_2, \ldots, \beta_n)$. To avoid a trivial process assume that $\beta_{n+1} < 1$. The distribution of time from when the process is started until it eventually absorbs in state $n+1$ is a phase distribution. An Erlang random variable, which is simply a series of exponential stages, and a hyperexponential distribution are simple examples of phase distributions that have been previously analyzed.

Phase distributions can easily be incorporated into a matrix geometric framework and, in fact, several of the examples previously considered are phase extensions to simpler matrix geometric problems. To see this, recall that we motivated the matrix geometric approach by considering a generalization of the modified M/M/1 queue scalar example of Example 8.12. This extension changed the distribution of customer service time from a single exponential state to that of two exponential stages: the first being at rate μ_1 and the second at rate μ_2. To accommodate this extension, we added a component to the state description to account for the current stage of the customer in service. In other words, we went from a state description consisting of an integer j that represented the number of customers in the system (including any in the server) to a state description (i, s) where i represented the number of customers in the queue (not including any in the server) and s represented the current stage of service for the customer in service (the case $s = 0$ accounts for the fact that no customer is in service and explains the difference in the definitions of j and i). Generally speaking, we can modify any matrix geometric problem to accommodate a phase distribution in a similar manner, that is, augment the state description to include the "stage" of the phase variable and possibly change the semantic meaning of the already existing state descriptors. The usefulness of phase distributions lies in the fact that any distribution can be approximated by a phase distribution (these distributions are dense in the space of distribution functions defined on $[0, \infty)$). By this we mean that for any distribution function F, one can create a sequence of phase distributions F_n so that $F_n(x) \to F(x)$ as $n \to \infty$ for all points x where F is continuous. For modeling purposes this implies that all distributions of interest can be approximated by a phase distribution.

9.5.2 The Distribution of a Continuous Phase Distribution

To calculate an equation for the distribution function of a phase distribution, let the generator matrix be partitioned as

$$\left[\begin{array}{c|c} U & A \\ \hline 0 & 0 \end{array} \right], \tag{9.47}$$

where U is an $n \times n$ matrix that delineates the transition rates between the transient states $\{1, 2, \ldots, n\}$ and A is an $n \times 1$ column matrix that delineates the transition rates between the transient states and the absorbing state $n + 1$. The solution to forward Chapman–Kolmogorov equations, as given in (8.117), implies that we can write

$$P(t) = \beta e^{Ut},$$

where $P(0) = (\beta, \beta_{n+1})$. The probability that the process has not reached absorption by time t is thus given by $P(t)e$. We can write the distribution function of the absorption time for transient matrix U, a random variable denoted by X, as

$$F_X(t) = 1 - P(t)e = 1 - \beta e^{Ut}e. \tag{9.48}$$

From this it follows that the density function for the time until absorption can be determined by differentiating (9.48) yielding

$$f_X(t) = \beta_{n+1}\delta_0(0) + \beta e^{\boldsymbol{U}t}\boldsymbol{A}, \tag{9.49}$$

where we note that matrix $\boldsymbol{A}$ of (9.47) satisfies

$$\boldsymbol{A} = -\boldsymbol{U}\boldsymbol{e}. \tag{9.50}$$

The impulse in (9.49) arises from the fact that, from (9.48), we have that $F_X(0) = 1 - \beta\boldsymbol{e} = \beta_{n+1}$. The moments of X can be obtained by integration and yield

$$E[X^n] = (-1)^n n! \beta \boldsymbol{U}^{-n}\boldsymbol{e}. \tag{9.51}$$

Equation (9.50) shows that $\boldsymbol{A}$ can be determined from $\boldsymbol{U}$. We can thus represent a phase distribution with generator (9.47) by a pair $(\beta, \boldsymbol{U})$.

9.5.3 The PH-Markov Process and PH-Renewal Process

Consider modifying a phase distribution defined by (9.47) in the following manner. Upon absorption, the process immediately jumps to state i, $i = 1, 2, \ldots, n$, with probability $\beta_i/(1 - \beta_{n+1})$. This induces a Markov process on states $\{1, 2, \ldots, n\}$. The generator of the induced process is given by

$$\boldsymbol{Q}_{\text{ph}} = \boldsymbol{U} + (1 - \beta_{n+1})^{-1}\boldsymbol{A}\beta. \tag{9.52}$$

Note that $\boldsymbol{A}$ is an $n \times 1$ vector and that β is a $1 \times n$ vector and thus $\boldsymbol{A}\beta$ is an $n \times n$ matrix. The value of the term of (9.52) associated with position (i, j) of this matrix, given by $(1 - \beta_{n+1})^{-1}\boldsymbol{A}(i)\beta(j)$, is the transition rate from state i to state j that arises from transitions that would hit absorbing state $n + 1$ in the original phase process.

The stationary probability vector of this generator, $\boldsymbol{\xi} \stackrel{\text{def}}{=} (\xi_1, \xi_2, \ldots, \xi_n)$, satisfies $\xi_i \geq 0$, $i = 1, 2, \ldots, n$, and

$$\boldsymbol{\xi}\boldsymbol{Q}_{\text{ph}} = 0$$

and

$$\sum_{i=1}^{n} \xi_i = 1.$$

The value of ξ_i is the probability that the phase distribution is in state i for the induced Markov process. We call this process the *PH-Markov process* and represent this process by the triple $(\beta, \boldsymbol{U}, \boldsymbol{\xi})$. We shall refer to this induced process when discussing symmetric queues in Section 10.4.

The points in time that the process is absorbed and immediately jumps back to state $\{1, 2, \ldots, n\}$ correspond to renewal points for a renewal process termed the *PH-renewal process*.

9.5.4
Discrete Phase
Distributions

Because discrete phase distributions are analogous to their continuous counterparts, our discussion here will be abbreviated. The problems explore some of the properties of discrete phase distributions. To precisely define a discrete phase distribution, let a transient Markov chain have transient states $\{1, 2, \ldots, n\}$ and an absorbing state $n+1$. Partition the state transition matrix as

$$\left[\begin{array}{c|c} \boldsymbol{U}_0 & \boldsymbol{A}_0 \\ \hline \mathbf{0} & \mathbf{1} \end{array} \right], \tag{9.53}$$

where $\boldsymbol{U}_0$ is an $n \times n$ matrix that delineates the transition rates between the transient states $\{1, 2, \ldots, n\}$ and $\boldsymbol{A}_0$ is an $n \times 1$ column matrix that delineates the transition rates between the transient states and the absorbing state $n+1$. The entry in the lower right-hand corner of the transition matrix, $\mathbf{1}$, represents the fact that once the absorbing state is reached the probability of staying there equals 1. Again let β_0 be the initial state probabilities. Then the probability that k steps are taken prior to absorption, a quantity denoted by $p_k, k = 0, 1, \ldots$, is given by (see Problem 9.11)

$$p_k = \begin{cases} \beta_{n+1}, & k = 0, \\ \beta_0 \boldsymbol{U}^{k-1} \boldsymbol{A}_0, & k = 1, 2, \ldots. \end{cases} \tag{9.54}$$

Discrete phase distributions can be incorporated into discrete matrix geometric chains in a manner similar to that illustrated above for continuous phase distributions, that is, expanding the state description to include a phase variable and possibly modifying the semantics of the existing state variables.

9.6
Summary of
Chapter 9

In this chapter we studied matrix geometric processes that have an infinite state space possessing a repetitive structure. This repetitive structure allows one to formulate a simple algorithmic solution for the stationary distribution of such processes. In applications, such processes frequently arise in systems consisting of a single queue served by a perhaps complex service center. In such models the initial states of the process correspond to situations where not all of the service centers are occupied. The repetitive state transitions for models of this type arise when the queue length increases with arriving customers. Generally speaking, matrix geometric techniques are applicable when there is a vector process where only one component of the vector can take on an infinite number of values and, for sufficiently large values, transitions from this component satisfy repetitive constraints that are outlined in the chapter.

The matrix geometric technique is applicable to both discrete and continuous time Markov processes and in this chapter we discussed properties of both types of processes. We discussed a type of uniformization for continuous time matrix geometric processes that lead to an equivalent discrete time process. This allows us to interpret the rate matrix for the continuous process through its relationship with the discrete process. The same technique can be used to relate the stability conditions for both types of processes.

In essence, the matrix geometric technique is a method to solve the infinite number of linear equations required by global balance. It seems clear that to solve an infinite number of equations one must either use some special structure of the system of equations or resort to approximate techniques. In the next chapter we discuss two other type of processes possessing a special structure that allows us to easily determine their stationary distributions.

We motivated matrix geometric solutions by first discussing a scalar state birth–death process, and we will motivate our discussion of *reversible processes* through a similar comparison. Reversible processes, even those with an infinite state space, have exceedingly simple solutions. To find such solutions we use special properties of reversible processes. The second type of process where we can determine a simple solution form is for processes that are quasi-reversible. These processes frequently arise in models of computer systems and have a structure that leads to a product form solution. We discuss in detail the properties of such a characterization that permit such a simple solution.

Bibliographic Notes

A comprehensive treatment of matrix geometric techniques can be found in Neuts [86]. Our exposition is only applicable to queues of the G/M/1-type since the number of "up" steps is bounded. See Neuts [88] for an exposition of techniques useful for M/G/1 type queues where the number of up steps is unbounded. See Lipsky [79] and Tijms [114] for other texts that have a linear algebra or a computational approach to queueing theory. A new procedure for solving quasi birth–death processes that is computationally efficient can be found in [74].

9.7
Problems for
Chapter 9

Problems Related to Section 9.1

9.1 [**R2**] Given a generator matrix Q, is there an algorithmic way to determine if it has a matrix geometric structure?

9.2 [**M3**] Provide a specification for a matrix geometric formulation of the $M/E_k/1$ queue. Assume that arrivals occur at rate λ and that each stage of service is exponential with rate $k\mu$. Specify the matrices and derive expressions for:

    1. the stability condition.

    2. the expected number of customers in the queue.

    3. the probability that the server is processing a customer in the ith stage for $i = 1, 2, \ldots, k$.

    4. the expected customer response time.

9.3 [**E2**] Write expressions for the following - assume that the spectral radius of M is less than 1:

    1. $\sum_{i=0}^{m} M^i$.

    2. $\sum_{i=1}^{\infty} i M^{i-1}$.

    3. $\sum_{i=1}^{\infty} M^{2i+1}$.

Problem Related to Section 9.3

9.4 [**E2**] Show that (9.27) relates the stationary probabilities of a continuous matrix geometric process to that of its equivalent uniformized version.

Problems Related to Section 9.4

9.5 [**M3**] In this problem we extend the problem of Example 9.5 where we modeled failures to a processor that was backed up by a spare processor that was possibly slower. In that example we assumed that the backup processor could not fail and that repairs to the main processor occur only during system idle times. Modify the problem in the following way. Consider that repairs to a failed main processor occur during the operation of the system at an exponential rate of α_1 and that if the main processor is repaired while there are customers in the queue then the system immediately reverts to serving these customers on the main processor. Also assume that there is a second backup processor that is used if the first backup fails. Such failures are assumed to occur at an exponential rate of γ_1. Upon failure of the first backup processor a second backup processor immediately begins serving the customers in the system. Assume that the servicing rate of the second backup is given by μ_3 and that the second backup is not subject to failures. When both the main and first backup processors are failed we assume that they are repaired and brought back online simultaneously at a rate of α_2. Processing of customers then immediately switches over to the main processor.

1. Draw the state transition diagram for the model.

2. Write the corresponding matrices for the transition rate matrix.

3. Derive the stability condition for the system assuming that $\mu_3 \leq \mu_2 \leq \mu_1$ and $\gamma_1 \leq \gamma$.

4. Is the rate matrix $\boldsymbol{R}$ reducible in this case as with Example 9.5?

5. (Optional) Write a program to solve for the stationary distribution of the process.

6. (Optional) Plot the expected customer response time for such a system for $\mu_1 = 1$, $\mu_2 = \mu_3 = 0.5$ and $\gamma = 0.6$, $\gamma_1 = 0.3$ for values of λ in its appropriate range.

9.6 **[E3]** Specify the remaining programming tables for other matrices of the replicated database example given in Example 9.6.

9.7 **[E3]** Specify the programming tables for the general case of a bulk service queue described in Example 9.1.

9.8 **[M2]** In Section 9.4.1 we derived the stability condition for continuous time Markov processes that are of matrix geometric form. Derive the analogous stability condition for discrete time chains.

Problems Related to Section 9.5

9.9 **[E1]** Write down the generator matrices for the transient Markov processes for the phase distributions created by an Erlang and hyperexponential distribution. Assume that the Erlang random variable has n exponential stages at rate λ and that the hyperexponential random variable selects rate λ_i with probability p_i, for $i = 1, 2, \ldots, n$.

9.10 **[R1]** Which of the following distributions are phase distributions: exponential, hypergeometric, negative binomial, normal, Cauchy, Poisson.

9.11 **[M2]** Establish the validity of the distribution for the number of steps taken prior to absorption for a discrete phase distribution given by (9.54).

9.12 **[M2]** Recall that the kth factorial moment of a random variable X is the quantity

$$E\left[X(X-1)\cdots(X-k+1)\right].$$

Show that the kth factorial moment of a discrete phase distribution with representation (9.53) is given by

$$k!\boldsymbol{\beta}_0\boldsymbol{U}_0^{k-1}(\boldsymbol{I}-\boldsymbol{U}_0)^{-k}\boldsymbol{e}\,.$$

9.13 [**M3**] Consider an M/M/1 queue with arrival rate λ and service rate μ. Suppose that $\lambda < \mu$ and thus the process is ergodic. Suppose the process is started in state $i > 0$.

1. What is the expected number of state changes made before the process first hits state 0?

2. What is the expected number of times the process visits state $i+1$ before first hitting state $i-1$?

10

Queueing Networks

In this chapter we derive the mathematics of product form queueing networks. In such networks the stationary distribution of the network is the product of the distributions of each queue analyzed in isolation from the network (for closed networks this is subject to a normalization constant). When first encountered, such a solution is enigmatic since for open networks it implies independence of the stationary distributions of the individual queues, and for closed networks it implies that the dependence between the queues is captured by normalizing the independent solution over a truncated state space. The derivations presented in this chapter provide insight into why simple solutions of this type exist for such complex networks.

To derive properties of product form networks, we first investigate time reversed Markov processes. A reversed Markov process is obtained by "running" a Markov process backwards in time. This view of a Markov process may seem unnatural at first but such a viewpoint often proves to be surprisingly insightful. When time reversal arguments are applicable they typically lead to elegant arguments that avoid algebraic manipulation. Recall that in Chapter 3 we frequently used probabilistic arguments to derive combinatoric equalities. This approach provided insight, often lacking in an algebraic derivation, into the reasons why an equality is valid. In a similar fashion, time reversal arguments lead to insights that are difficult to obtain using algebraic techniques.

A notable application of time reversal techniques is to determine the input–output properties of a queueing system. The output process from a queue can be determined by viewing the process in reversed time; departures from the queue (resp., arrivals) correspond to arrivals to the queue (resp., departures) in the time reversed process. For some queues, time reversal arguments elegantly establish that the departure process from the queue is statistically identical to the arrival process. Results of this type can be established using standard algebraic techniques but are not as insightful.

There is one type of Markov process, called a reversible process, where time reversal arguments can be used to derive the structure of their stationary distributions. These distributions can be written as a product of the ratio of transition rates, like that of birth–death processes (see (8.102) of Section 8.5). The special structure of reversible Markov processes will be analyzed in detail in this chapter.

In Section 10.1 we define a reversed Markov process and derive properties for such a process. This leads to a derivation of detailed balance equations that must be satisfied by Markov processes that are reversible. Detailed balance is a conservation law of probability flux that is more restrictive than global balance. The main mathematical result of this chapter, Proposition 10.2, establishes a relationship that must hold between reversed transition rates and stationary probabilities. Other properties of reversible processes are derived in Section 10.2.

Section 10.3 derives important input–output properties of queues. We first discuss properties of reversible queues that yield a product form solution. These results lead us to define a key mathematical property called quasi-reversibility that broadens the scope of networks with product form. Input–output properties of quasi-reversible queues are also established in this section. Section 10.4 derives properties of a rich class of quasi-reversible queues called symmetric queues. These queues frequently appear in mathematical models.

In Section 10.5 we show that quasi-reversible queues satisfy balance equations called partial balance. Like detailed balance, partial balance is a conservation law of probability flux that is more restrictive than global balance. Mathematical properties of partial balance are derived in Section 10.6. These properties imply that one can create complex networks composed of quasi-reversible queues that have product form solutions.

A word on the notation in this chapter: To avoid notations that are sometimes difficult to read, we will write transition rates from state i to state j as $q(i,j)$ rather than as $q_{i,j}$, and similarly, we will write stationary probabilities in a parenthesized form. We also will call a network of queues a *product form network* if its stationary distribution is a product of the marginal distributions of each queue. To avoid repetitious language in this chapter, we will frequently omit the word "Markov" when speaking of processes. For example, we will use the short form, "forward process," in lieu of the longer form, "forward Markov process."

10.1
Reversibility

The time reversal of a stationary Markov process $X(t)$ on the same state space about time τ is defined to be the process $X^r(t) \stackrel{\text{def}}{=} X(\tau - t)$. We can think of the process X^r in terms of the original process run backwards in time about the time τ. The set of sample paths of the reversed process is the "reversal" of the sample paths of the original process. As an example, consider the sample path

$$\ldots (i_0, t_0), (i_1, t_1), (i_2, t_2), \ldots, (i_{k-1}, t_{k-1}), (i_k, t_k), \ldots,$$

which depicts a process that is in state i_k over the time period $[t_k, t_{k+1})$. If this sample path is contained in the set of sample paths for the original process, X, then the sample path

$$\ldots (i_{k+1}, \tau - t_{k+1}), (i_k, \tau - t_k), \ldots, (i_2, \tau - t_2), (i_1, \tau - t_1), (i_0, \tau - t_0), \ldots$$

is contained in the set of sample paths of the reversed process, X^r.

Throughout this section we will assume that the original non-reversed process is irreducible, stationary, and time homogeneous. This implies that X has a stationary distribution and has state transition rates that are not a function of time. Under these assumptions it can be shown that X^r is also a Markov process that is time homogeneous, irreducible, and stationary (see Problem 10.1). These assumptions imply that selection of τ is arbitrary since properties of the reversed process do not depend on absolute time. We henceforth set the pivot of time reversal to be $\tau = 0$ and we will call process $X(t)$ and $X^r(t) \stackrel{\text{def}}{=} X(-t)$ the forward and reversed processes, respectively. Our comments for reversed processes apply to both discrete and continuous time Markov processes, and we will focus our attention on the continuous time case.

Let $q^r(i, j)$, $i, j \in \mathcal{S}$, be the transition rates of the reversed process, let $q^r(i)$, $i \in \mathcal{S}$, be the total transition rate out of state i and let $\pi^r(i)$ denote the stationary distribution. In general the transition rates of the reversed process differ from those of the forward process. To see this, suppose that $q(i, j) > 0$ and $q(j, i) = 0$ for the forward process. This implies that the process can jump directly from state i to state j but cannot jump directly from state j to i. Viewing the process in reversed time, however, implies that the process can jump directly from state j to state i but cannot jump directly from state i to state j. Hence the reversed transition rates satisfy $q^r(j, i) > 0$ and $q(j, i) = 0$. It is thus clear that the forward and reversed transition rates are not equal.

Intuitively, reversing time does not affect the fraction of time a process spends in a state. This implies that the stationary distributions of both the forward and reversed processes are identical, that is, $\pi^r(i) = \pi(i), i \in \mathcal{S}$. The forward and reversed process, however, generally have different joint distributions. More precisely, for $t_1 < t_2 < \cdots < t_m$, $m \geq 1$, the joint distribution of X is the quantity

$$P[X(t_1) = i_1, X(t_2) = i_2, \ldots, X(t_m) = i_m]. \qquad (10.1)$$

In words, the joint distribution is the total probability of the set of paths where X is found in state i_j at time t_j, for $1 \leq j \leq m$. The joint distribution of the reversed process is defined similarly to that of the forward process,

$$P[X^r(t_1) = i_1, \ldots, X^r(t_m) = i_m] = P[X(-t_m) = i_m, \ldots, X(-t_1) = i_1].$$
$$(10.2)$$

We define the probability flux between two states for the reversed process in a similar way to that given by (8.67),

$$\mathbf{F}^r(i,j) \overset{\text{def}}{=} \pi^r(i)q^r(i,j) = \pi(i)q^r(i,j), \quad i,j \in \mathcal{S}. \qquad (10.3)$$

This definition has a similar form for sets of states,

$$\mathbf{F}^r(\mathcal{U}, \mathcal{V}) = \sum_{u \in \mathcal{U}, v \in \mathcal{V}} \mathbf{F}^r(u, v), \quad \mathcal{U}, \mathcal{V} \in \mathcal{S}.$$

By definition, the joint distributions of the forward and reversed process satisfy

$$P[X(t_1) = i_1, \ldots, X(t_m) = i_m] = P[X^r(-t_m) = i_m, \ldots, X^r(-t_1) = i_1].$$
$$(10.4)$$

It is important to note that equality of the stationary probabilities of the forward and reversed processes does not imply equality of their joint distributions. The joint distribution is a more detailed characterization of the process since it specifies the probability that the process *evolves along a certain set of paths* rather than simply specifying the fraction of time the process spends in a given set of states. Consider the following example.

Example 10.1 (Difference in the Forward and Reversed Joint Distributions) Let X be a Markov process with N states, numbered from $0, 1, \ldots, N-1$, and with transition rates given by $q(i, (i+1, \bmod N)) = \lambda$. This process is a clockwise circulation around the states $0, 1, \ldots, N-1$ and clearly has a stationary distribution given by $\pi(i) = 1/N$. The reversed process is obtained by reversing these transition rates, that is, a counterclockwise circulation around the states, and hence satisfies $q^r(i, (i-1, \bmod N)) = \lambda$ and $\pi^r(i) = 1/N$.

To see that the joint distributions of the forward and reversed processes differ, consider the joint probability $P[X(0) = 0, X(t) = 1]$. If $X(t) = 1$, conditioned on the event $\{X(0) = 0\}$, then there is a first transition prior to time t from state 0 to state 1. This is followed until time t by k, for $k = 0, 1, \ldots$, clockwise rotations starting and ending at state 1. Let the last entrance into state 1 prior to time t be denoted by Υ where $0 \leq \Upsilon < t$. The probability that $X(t) = 1$ conditioned on the event $\{\Upsilon = \tau\}$ equals the survivor of an exponential at rate λ evaluated at $t - \tau$,

$$P[X(t) = 1 \mid X(0) = 0, \Upsilon = \tau] = e^{-\lambda(t-\tau)}. \tag{10.5}$$

Observe that the distribution of Υ, given that k rotations have taken place, is Erlang with parameters $(kN + 1, \lambda)$. The parameter $kN + 1$ counts the first transition into state 1 and the kN transitions that lead to the last entrance into state 1 prior to time t.

To evaluate the joint distribution of the reversed process, we calculate

$$P[X^r(0) = 0, X^r(t) = 1].$$

It is more convenient to view this in terms of the forward process. Using (10.2) we can equivalently calculate the joint probability $P[X(0) = 1, X(t) = 0]$. For $X(t) = 0$ conditioned on the event $\{X(0) = 1\}$, there must be a first time that X enters state 0, followed by k, for $k = 0, 1, \ldots$, rotations starting and ending at 0. If the last entrance into state 0 occurs at time Υ then it is clear that the probability that X is in state 0 at time t is given by (10.5). The distribution of Υ, given that k rotations have taken place, is that of an Erlang random variable with parameters $(kN + N - 1, \lambda)$. From this it is immediately clear that the joint distributions of the forward and reverse processes differ.

10.1.1 Reversible Processes

Example 10.1 shows that, generally, the joint distributions of the forward and reversed processes are not identical. Processes that have the property that the joint distribution of the forward and reversed processes are equal are said to be *reversible*. More precisely, a process is reversible if

$$P[X(t_1) = i_1, \ldots, X(t_m) = i_m] \quad = \quad P[X^r(t_1) = i_1, \ldots, X^r(t_m) = i_m]$$

$$= \quad P[X(-t_m) = i_m, \ldots, X(-t_1) = i_1]. \tag{10.6}$$

The definition of a reversible process implies that the process is statistically identical in forward or reversed time. (Intuitively, one cannot determine the direction of time by observing the sample paths of the process.) This implies that $\pi(i)q(i,j) = \pi^r(i)q^r(i,j)$ and, since $\pi(i) = \pi^r(i)$, that

$$q(i,j) = q^r(i,j). \tag{10.7}$$

A consequence of this equality is that the probability flux of the forward and reversed processes must be equal for reversible processes,

$$\mathbf{F}(\mathcal{U}, \mathcal{V}) - \mathbf{F}^r(\mathcal{U}, \mathcal{V}) = 0, \quad \textbf{Reversibility Balance Equations.} \tag{10.8}$$

It is important to make the distinction between a *reversed process* and a *reversible process*. Running a Markov process backwards in time yields a reversed process. In general, the reversed Markov process X^r has a different joint distribution than that of the forward process X, as illustrated in Example 10.1. A reversible process is a special case of a reversed process where both the forward and reverse processes have the same joint distribution.

10.1.2
The Reversed
Process

We now derive equations satisfied by the reversed transition rates (processes discussed here are not necessarily reversible processes). Use (10.4) to write

$$P[X(t+dt) = k, X(t) = i] \quad = \quad P[X^r(-t) = i, X^r(-t-dt) = k]$$

$$= \quad P[X^r(t+dt) = i, X^r(t) = k], \quad (10.9)$$

where (10.9) arises by shifting time by $2t + dt$ units. Since the process is time homogeneous and stationary, this shift in time does not change the joint probability. We rewrite (10.9) as

$$\pi(i)P[X(t+dt) = k \,|\, X(t) = i] = \pi^r(k)P[X^r(t+dt) = i \,|\, X^r(t) = k].$$

Dividing both sides of this equation by dt and letting $dt \to 0$ implies that

$$\pi(i)q(i,k) = \pi(k)q^r(k,i). \quad (10.10)$$

Summing this equation over k and using global balance for the reversed process shows that

$$q(i) = q^r(i). \quad (10.11)$$

(Observe that (10.7) for reversible processes is a more detailed equality than that given in (10.11).) It is convenient to record (10.10) in terms of probability flux; for any two disjoint subsets $\mathcal{U}, \mathcal{V} \in \mathcal{S}$, we have

$$\mathbf{F}(\mathcal{U},\mathcal{V}) - \mathbf{F}^r(\mathcal{V},\mathcal{U}) = 0, \quad \textbf{Reversed Balance Equations.} \quad (10.12)$$

Equations (10.10-10.12) must be satisfied by the reversed transition rates. They show that one can deduce the reversed transition rates after one knows the stationary distribution or, alternatively, that one can deduce the stationary distribution after one knows the reversed transition rates. Suppose we can guess values for *both* the reversed transition rates and the stationary probabilities that simultaneously satisfy (10.10) and

(10.11). Then this guess corresponds to the actual values of the stationary probability distribution and the reversed transition rates. To see this, sum (10.10) over i to obtain

$$\sum_{i \in S} \pi(i) q(i, k) = \sum_{i \in S} \pi(k) q^r(k, i)$$

$$= \pi(k) \sum_{i \in S} q(k, i).$$

This shows that the guessed values of $\pi(i)$ and $q^r(i, j)$ satisfy global balance and thus, by Corollary 8.9, are the unique stationary probabilities. This is an important result, and we record it in the following proposition.

Proposition 10.2 *Consider a Markov process X with state space S and transition rates $q(i, k)$, $i, k \in S$. Suppose we are given nonnegative numbers $q'(i, k)$ that satisfy*

$$\sum_{k \in S} q(i, k) = \sum_{k \in S} q'(i, k), \quad i \in S, \tag{10.13}$$

and a set of nonnegative numbers $\alpha(i)$, $i \in S$, that sum to unity and satisfy

$$\alpha(i) q(i, k) = \alpha(k) q'(k, i), \quad i, k \in S. \tag{10.14}$$

Then $q'(i, k)$, $i, k \in S$, are the transition rates for the reversed process X^r and $\alpha(i), i \in S$, are the stationary probabilities for both X and X^r.

Without doubt, Proposition 10.2 is the *most important result* of this chapter. This is a surprising statement since, initially, it would appear that the proposition has limited applicability because it requires guessing *both* the reversed transition rates and the stationary probabilities. Throughout this chapter, we will use intuition regarding the reversed process to guess reversed transition rates. This, combined with a tractable guess for the stationary distribution (for queueing networks we guess a *product form* solution) leads to results that would be difficult to obtain without knowing Proposition 10.2. The main queueing network result of this chapter that shows that quasi-reversible queues with Markov routing have product form, established in Section 10.6, will be directly derivable from Proposition 10.2.

If the stationary state probabilities are known, then determining the reversed transition matrix is an easy exercise. Suppose that π is the stationary probability vector for a Markov process with generator matrix $\boldsymbol{Q}$. Then the generator matrix for the reversed process, $\boldsymbol{Q}^r$, is given by

$$\boldsymbol{Q}^r = \boldsymbol{\Pi}^{-1} \boldsymbol{Q}^T \boldsymbol{\Pi}, \tag{10.15}$$

where $\mathbf{\Pi} \overset{\text{def}}{=} \text{diag}(\pi)$ is a diagonal matrix composed of the entries of π and $\boldsymbol{Q}^T$ is the transpose of $\boldsymbol{Q}$. We can use this in the following example to define a reversed phase distribution.

Example 10.3 (A Reversed Phase Distribution)   Consider a phase distribution on $n+1$ states with representation $(\beta, \boldsymbol{U})$ and expected value $E[X] = \mu^{-1}$. Assume that $\beta_{n+1} = 1 - \beta e = 0$ and thus that the random variable X associated with the phase distribution satisfies $P[X > 0] = 1$. The generator for such a phase distribution has the form

$$\left[\begin{array}{c|c} U & A \\ \hline 0 & 0 \end{array} \right],$$

and our object is to define $\boldsymbol{U}^r$, $\boldsymbol{A}^r$, and β^r for a "reversed phase distribution." The transient Markov process that defines a phase distribution is not a stationary process and thus the analysis for reversed processes just discussed is not applicable. The PH-Markov process associated with the phase distribution (see Section 9.5) has a representation given by $(\beta, \boldsymbol{U}, \xi)$ and is a stationary process. This suggests that we use its properties to define a reversed phase distribution.

The generator of the PH-Markov process is given by (see (9.52))

$$\boldsymbol{Q}_{\text{ph}} = \boldsymbol{U} + \boldsymbol{A}\beta. \tag{10.16}$$

The (i, j) entry of $\boldsymbol{Q}_{\text{ph}}$ is given by

$$Q_{\text{ph}}(i, j) = U(i, j) + A(i)\beta(j). \tag{10.17}$$

Using (10.15) we can write the generator for the time reversed PH-Markov process as

$$\boldsymbol{Q}_{\text{ph}}^r = \boldsymbol{E}^{-1} \boldsymbol{Q}_{\text{ph}}^T \boldsymbol{E},$$

where $\boldsymbol{E} \overset{\text{def}}{=} \text{diag}(\xi)$. Using (10.16) to calculate the transpose, writing this in terms of the (i, j) entry and some rearrangement of terms yields

$$Q_{\text{ph}}^r(i, j) = \underbrace{\xi_j U(j, i)/\xi_i}_{U^r(i,j)} + \underbrace{\beta(i)\mu/\xi_i}_{A^r(i)} \underbrace{\xi_j A(j)/\mu}_{\beta^r(j)}. \tag{10.18}$$

The underbraces in (10.18) point out similarities to (10.17) and suggest candidates for the components of the reversed phase distribution. In particular, the form of (10.18) suggests that we define

$$U^r(i, j) \overset{\text{def}}{=} \frac{\xi_j U(j, i)}{\xi_i}, \tag{10.19}$$

$$A^r(i) \overset{\text{def}}{=} \frac{\beta(i)\mu}{\xi_i} \tag{10.20}$$

and

$$\beta^r(j) \stackrel{\text{def}}{=} \frac{\xi_j A(j)}{\mu}. \tag{10.21}$$

To explain these definitions, note that (10.19) can be written in matrix terms, which yields

$$\boldsymbol{U}^r = \boldsymbol{E}^{-1} \boldsymbol{U}^T \boldsymbol{E}.$$

This form is consistent with (10.15). The reason we multiply and divide by μ in the second summand of (10.18) is to make the terms associated with $\boldsymbol{A}^r$ into rates and to make the terms associated with β^r into probabilities that sum to unity (see Problem 10.3). This implies that $\beta^r_{n+1} = 0$ and thus that the random variable X^r also has the property that $P[X^r > 0] = 1$.

Equations (10.19), (10.20), and (10.21) are the components of the reversed phase distribution with a representation given by $(\beta^r, \boldsymbol{U}^r)$. It is a simple exercise to show that the reversal of this phase distribution $(\beta^r, \boldsymbol{U}^r)$ is the original phase distribution $(\beta, \boldsymbol{U})$ (see Problem 10.4).

Transition rates for reversed processes are often not as apparent as in Example 10.3. Consider the following example:

Example 10.4 (An Iterative Algorithm)   Consider an algorithm that processes a request to a computer system by alternating execution on two different processors. Upon receiving a request, assume that processor 1 executes for an exponential period of time until it obtains an intermediate result that is passed to processor 2. Processor 2 then executes for an exponential period of time and either finishes processing the request or returns control to processor 1. This alternation of executions by processors 1 and 2 continues until processor 2 determines that the request is satisfied. Such an algorithm might be used in an iterative calculation or in a simulation.

A Markov model of such a process is shown in Figure 10.1. State 0 corresponds to the system waiting for a request, and state 1 (resp., state 2) corresponds to processor 1 (resp., processor 2) in execution. For the transition rates shown in that figure, it is easy to determine that the stationary state probabilities are given by

$$\pi(0) = \frac{\alpha\lambda}{\alpha(\gamma + \lambda) + 2\lambda\gamma}, \quad \pi(1) = \frac{\gamma(\alpha + \lambda)}{\alpha(\gamma + \lambda) + 2\lambda\gamma}, \quad \pi(2) = \frac{\lambda\gamma}{\alpha(\gamma + \lambda) + 2\lambda\gamma}.$$

Using these values and (10.10) yields the reversed transition rates that are also shown in Figure 10.1. It is not at all obvious that the reversed transition rates would have this form nor that both processes depicted in the figure have the same stationary probability distributions.

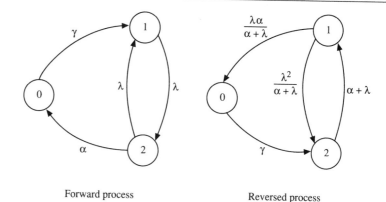

Forward process   Reversed process

FIGURE 10.1. Forward and Reversed Transition Rates for the Iterative Algorithm

10.1.3
Detailed
Balance

We can easily deduce the reversed transition rates for one type of Markov process. Comparing the reversibility balance equations (10.8) to the reversed balance equations (10.12) shows that if a process is reversible then

$$\mathbf{F}(\mathcal{U}, \mathcal{V}) = \mathbf{F}(\mathcal{V}, \mathcal{U}), \quad \textbf{Detailed Balance Equations,}$$

and thus

$$\pi(i)q(i,j) = \pi(j)q(j,i), \quad i, j \in \mathcal{S}. \tag{10.22}$$

These equations are termed *detailed balance equations*, and our arguments have just shown that detailed balance is a sufficient condition for reversibility.

We can also establish that detailed balance is a necessary condition for reversibility. To prove this, assume that (10.22) holds and consider the process over the time interval $[-t, t]$. Suppose that the process X makes m jumps within this interval at times $-t < t_1 < t_2 < \ldots < t_m < t$. Let i_k be the state of the system over the time period $[t_k, t_{k+1})$, $k = 1, 2 \ldots, m-1$ and let $X(-t) = i_0$. We can write the density function for such a path by analyzing each jump. We start with the process in state i_0 at time $-t$.

The probability density that the process is in state i_0 at time $-t$ and stays in this state for $(t_1 - (-t)) = t_1 + t$ time units and then jumps to state i_1 equals

$$\pi(i_0)\frac{q(i_0, i_1)}{q(i_0)} \, q(i_0)e^{-q(i_0)(t_1+t)} = \pi(i_0)q(i_0, i_1)e^{-q(i_0)(t_1+t)}.$$

Factors in the left-hand side of this equation include the state transition probability and the density function for the exponential holding time of state i_0. In a similar fashion the probability density that the process stays in state i_k, $k = 1, 2 \ldots, m - 1$, for $(t_{k+1} - t_k)$ time units and then jumps to state i_{k+1} is given by

$$\frac{q(i_k, i_{k+1})}{q(i_k)} \, q(i_k) e^{-q(i_k)(t_{k+1} - t_k)} = q(i_k, i_{k+1}) e^{-q(i_k)(t_{k+1} - t_k)}.$$

The last component of the path described above is the probability density that the holding time for state i_m is greater than $(t - t_m)$. This is given by

$$e^{-q(i_m)(t - t_m)}.$$

Define $h_0 \overset{\text{def}}{=} t + t_1$, $h_m \overset{\text{def}}{=} t - t_m$, and $h_k \overset{\text{def}}{=} t_{k+1} - t_k$, $k = 1, 2, \ldots, m - 1$. With these values we can write the probability density for the path as

$$\pi(i_0) \prod_{j=0}^{m-1} q(i_j, i_{j+1}) e^{-q(i_j)h_j} \, e^{-q(i_m)h_m}. \tag{10.23}$$

By assumption, the detailed balance equations hold (10.22), and thus we can write

$$\pi(i_0) \prod_{j=0}^{m-1} q(i_j, i_{j+1}) = \pi(i_m) \prod_{j=0}^{m-1} q(i_{j+1}, i_j)$$

$$= \pi(i_m) \prod_{j=1}^{m} q(i_j, i_{j-1}).$$

Substituting this into (10.23) and rearranging implies that (10.23) can be rewritten as

$$\pi(i_m) \prod_{j=1}^{m} q(i_j, i_{j-1}) e^{-q(i_j)h_j} e^{-q(i_0)h_0}. \tag{10.24}$$

Equation (10.24) is the probability density of the path for the reversed process, X^r. To see this, observe that (10.24) is the probability density for a process that starts in state i_m at time t and jumps to state i_{m-1} after h_m time units. After staying in state i_{m-1} for h_{m-1} time units the process jumps to state i_{m-2}. The evolution of the process continues in this fashion until it jumps to state i_0 at time t_1. The holding time of state i_0 is longer than h_0, and thus the process is in state i_0 at time $-t$. Since the probability density of this path is identical to that of the reversed process the joint distributions of the forward and reversed processes are identical. Hence the process is reversible, and we conclude that detailed balance (10.22) is a necessary and sufficient condition for reversibility.

10.1.4 Stationary Distribution of a Reversible Process

The stationary probabilities of a reversible process can be obtained in a straightforward manner. Pick a starting state, $s \in \mathcal{S}$, and, for each state $i \in \mathcal{S}$, find a path

$$s = j(i, 1) \mapsto j(i, 2) \mapsto \cdots \mapsto j(i, m_i) = i, \quad m_i \geq 1.$$

The existence of such a path is guaranteed since X is assumed to be irreducible. Guess a solution of the form

$$\psi(i) = \begin{cases} 1, & i = s, \\ \prod_{k=1}^{m_i} \alpha(i, k), & i \neq s, \quad i \in \mathcal{S}, \end{cases} \tag{10.25}$$

where

$$\alpha(i, k) \stackrel{\text{def}}{=} \frac{q(j(i, k), j(i, k+1))}{q(j(i, k+1), j(i, k))}, \quad k = 1, 2, \ldots, m_i - 1.$$

Normalize (10.25) to obtain

$$\pi(i) = \frac{\psi(i)}{\sum_{k \in \mathcal{S}} \psi(k)}, \quad i \in \mathcal{S}. \tag{10.26}$$

It is easy to check that (10.26) satisfies global balance (Problem 10.6) and thus is the form of the solution for the stationary distribution for a reversible process. This is an extremely important result, and it implies that all reversible processes have tractable solutions. Notice that the solution form of (10.26) is similar to that obtained for birth–death Markov processes considered in Section 8.5 (see (8.102)). This follows from the fact (soon to be established) that all birth–death processes are reversible.

10.1.5 Kolmogorov's Criterion

Recall that in Example 10.1 we showed that the joint probability of the forward and reversed processes are different. Even for the simple process described in that example calculating the joint distribution for the process is a nontrivial exercise (Problem 10.2). For more complicated processes the joint distribution is extremely difficult to calculate. Thus the definition of a reversible process given in (10.6) does not offer a practical method to determine if a process is reversible.

One can determine if a Markov process is reversible by inspecting its transition rates. To show this, let $j_1, j_2, \ldots j_m, j_1$, $m \geq 0$, be any cycle of states. If the process is reversible then detailed balance implies that

$$\pi(j_k)q(j_k, j_{k+1}) = \pi(j_{k+1})q(j_{k+1}, j_k), \quad k = 1, 2, \ldots, m - 1, \tag{10.27}$$

and that

$$\pi(j_m)q(j_m, j_1) = \pi(j_1)q(j_1, j_m). \tag{10.28}$$

Multiplying the left-hand and right-hand side of (10.28) and (10.27) for $k = 1, 2, \ldots, m - 1$, and canceling common factors implies that

$$q(j_1, j_2)q(j_2, j_3) \cdots q(j_m, j_1) = q(j_1, j_m)q(j_m, j_{m-1}) \cdots q(j_2, j_1) \tag{10.29}$$

is satisfied if the process is reversible. Another argument can be used to establish that if (10.29) is satisfied for all sequences of states, then the process is reversible (Problem 10.7). Hence we conclude that if (10.29) is satisfied for all possible cycles in the state space of a Markov process then the process is reversible. This criterion, called *Kolmogorov's criterion*, is often used to establish the reversibility of a process. Each side of (10.29) can be thought of as a flow of transition rates along one direction of the cycle and the equality thus implies that there is no net circulation of this flow in the state space.

10.2 Properties of Reversible Processes

Several properties of reversible processes follow immediately from Kolmogorov's criterion and from the form of the stationary distribution given by (10.26). These properties elucidate the special characteristics that are found in reversible processes. Let X be a reversible Markov process on state space $\mathcal{S}$ and define $\mathcal{C}$ to be the set of all possible cycles of states.

10.2.1 Scaling of Transition Rates

The first consequence of Kolmogorov's criterion concerns scaling of the transition rates of the process. Consider two states, i and j, and suppose that $q(i, j) > 0$. Kolmogorov's criterion trivially implies that $q(j, i) > 0$ (see Problem 10.8). Consider a new process, X', that is obtained from X by modifying the transition rates between states i and j by a multiplicative factor,

$$q'(i, j) = a(i, j)q(i, j), \quad i, j \in \mathcal{S}, \tag{10.30}$$

where $a(i, j) = a(j, i)$ are positive symmetric constants. Kolmogorov's criterion implies that X' is reversible and the form of the stationary probabilities (10.26) implies that the stationary distribution of X' is identical to that of X, and thus $\pi'(i) = \pi(i)$. Not all Markov processes have stationary distributions that are invariant to scaling of their transition rates. For example, consider the processor failure model of Example 8.13 and assume that we scale interarrival rates and service rates by a factor a, that is, $a\lambda$ and $a\mu$, respectively. Then it is a simple exercise to see that the stationary distribution of the process is modified by this scaling (see Problem 10.9).

10.2.2 Spanning Tree Representation of a Reversible Process

Kolmogorov's criterion is trivially satisfied if there are no cycles in the state space. To derive a consequence of this observation, recall that a *tree* is a set of nodes that are connected by edges that have no cycles. A *spanning tree* is a subgraph containing all of the nodes of a graph that forms a tree. For our purposes, an edge exists between two nodes i and j if $i \mapsto j$ and $j \mapsto i$. These definitions imply that an immediate consequence of Kolmogorov's criterion is that a Markov process that has a state transition diagram that forms a tree with bidirectional arcs is reversible *regardless of its transition rates*. For state spaces that form a tree, equation (10.29) is trivially satisfied. Hence all birth–death processes are reversible. It also follows from (10.26) that one can determine the stationary distribution of a reversible process from transition rates found along *any spanning tree* of its state space. This has a rather surprising consequence, as illustrated in the following example:

Example 10.5 (Changing Rates in a Reversible Process) Let X be a reversible process on state space $\mathcal{S}$ and consider modifying the transition rates of the process to produce a new reversible process X'. Suppose that in creating X' we do not change any of the rates along a given spanning tree of the states. The form of (10.26) implies that the stationary distribution for the modified process is identical to that of the original process, and thus that $\pi'(i) = \pi(i)$.

10.2.3 Joining of Two Independent Reversible Processes

Suppose that X_1 and X_2 are independent processes that are each reversible. It can be shown that the joint process (X_1, X_2) is also reversible (see Problem 10.10). This implies that we can join independent reversible queues and preserve reversibility. Consider the following example.

Example 10.6 (Independent Queues) Consider two independent queues each with its own Poisson arrival process at rate λ. Queue 1 has exponential service with rate μ_1 and queue 2 has exponential service with rate μ_2. Let $N_i, i = 1, 2$ be a random variable denoting the number of customers in the ith queue. The state transition diagram for the state vector (N_1, N_2) is shown in Figure 10.2.

Kolmogorov's criteria is clearly satisfied. The joint distribution is thus given by

$$P[N_1 = n_1, \ N_2 = n_2] = (1 - \rho_1)\rho_1^{n_1}(1 - \rho_2)\rho_2^{n_2}, \quad 0 \leq n_1, n_2 < \infty, \quad (10.31)$$

where $\rho_i \stackrel{\text{def}}{=} \lambda_i/\mu_i, \ i = 1, 2$.

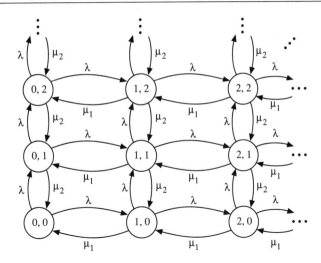

FIGURE 10.2. State Transition Diagram for Independent Queues

10.2.4 State Truncation in Reversible Processes

Consider a reversible Markov process, X, on state space $\mathcal{S}$ and let $\mathcal{S}' \in \mathcal{S}$ be a subset of the states. Suppose we consider the *truncated process*, X', on this set defined by the transition rates

$$q'(i,j) = \begin{cases} q(i,j), & i,j \in \mathcal{S}', \\ \\ 0, & \text{otherwise.} \end{cases}$$

Suppose that X' is irreducible. Then it follows that X' is reversible and has a stationary distribution given by

$$\pi'(i) = \frac{\pi(i)}{\sum_{k \in \mathcal{S}'} \pi'(k)}. \tag{10.32}$$

In other words, the stationary probabilities of X' are the same as that of X up to a normalization constant (they have the same relative values for both processes). This result easily follows from Kolmogorov's criterion. Consider the set of all cycles of states $\mathcal{C}$ and let $\mathcal{C}'$ be those cycles that contain states only from $\mathcal{S}'$. Then it is clear that if $\mathcal{C}$ satisfies Kolmogorov's criterion then $\mathcal{C}'$ must also satisfy the criterion and thus X' is reversible. The equation for the stationary distribution of the process given in (10.32) follows directly from the form of the stationary distribution (10.26). Reversibility is thus a sufficient condition for (10.32) to hold. In Section 10.5.2 we establish necessary and sufficient conditions for (10.32) to hold for truncated state spaces.

The state truncation property of reversible processes has important ramifications in performance models of computer systems. It implies, for example, that if we know the stationary distribution of a reversible process that models a queueing system with an infinite waiting room, then we also know the distribution when waiting rooms are finite. Consider the following examples:

Example 10.7 (The M/M/1/k Queue) Consider an M/M/1 queue with arrival and service rates of λ and μ, respectively. Recall that the stationary distribution of an M/M/1 queue is given by $\pi(i) = (1 - \rho)\rho^i, i = 0, 1, \ldots$, where $\rho \overset{\text{def}}{=} \lambda/\mu$ (see (7.52)). Suppose that arrivals to the system when it contains k customers are lost. This corresponds to a truncated M/M/1 queue, the M/M/1/k queue discussed in Section 8.5. Since the Markov process associated with the M/M/1 queue is reversible, the stationary distribution for the M/M/1/k queue is simply a normalized version of the M/M/1 unconditional distribution, that is,

$$\pi'(i) = (1 - \rho)\rho^i/(1 - \rho^{k+1}), \quad i = 0, 1, \ldots, k.$$

This corroborates our previous derivation of this result in (8.103).

Example 10.8 (Processors Sharing a Common Buffer Space) Consider p processors that share a common buffer space. Customer arrivals for processor i, for $i = 1, 2, \ldots, p$, enter the common buffer and wait, in a first-come-first-served manner, to be processed by that processor. The buffer has room for at most n customers, and customer arrivals, destined for any processor, are lost when the buffer is full. Assume that arrivals for processor i, for $i = 1, 2, \ldots, p$, are Poisson with rate λ_i and that customer service times for the processor are exponential with rate μ_i. Let $N_i(t)$ be the number of processor i's customers in the system at time t and let N_i be a random variable corresponding to the stationary distribution of the number of processor i's customers in the system.

We could model the operation of the system as a set of independent M/M/1 queues if it were not for the common buffer. The existence of the common buffer, however, implies that $N_i(t)$, for $i = 1, 2, \ldots, p$, are *dependent random variables*; for example, if $N_i(t) = n$ then the capacity constraints of the buffer imply that $N_j(t) = 0$, $j \neq i$. The Markov process that models this system, however, corresponds to a truncated version of a p queue version of Example 10.6. Hence, the state truncation property for reversible Markov processes implies that the stationary distribution of the system is given by a renormalized version of (10.31),

$$P[N_1 = n_1, \ldots, N_p = n_p] = b \prod_{i=1}^{p} (1 - \rho_i)\rho_i^{n_i}, \ 0 \le n_i, i = 1, \ldots, p, \ \sum_{i=1}^{p} n_i \le n,$$
$$(10.33)$$

where b is a normalization constant.

It is important to note that the "product form" distribution of Example 10.8 as given by (10.33) *does not imply independence* of the queues in the system. As mentioned in the example, the random variables N_i, for $i = 1, 2, \ldots, p$, are *dependent*. This dependency is reflected in the fact that the normalization constant b of (10.33) is not equal to 1. If $b = 1$ then each term of the product is a distribution function and the product implies independence of the random variables N_i, for $i = 1, 2, \ldots, p$. In (10.33) the value of b differs from 1 and one cannot decompose (10.33) into a product of distribution functions. In other words, there is no factorization of the normalization constant, $b = b_1 b_2 \cdots b_p$, so that $b_i(1 - \rho_i)\rho_i^{n_i}$ is a distribution function for all i.

10.3 Input–Output Properties

To model complex computer systems one joins queues into a network where the output from one queue forms the input to another. This motivates our interest in determining the properties of the output or *departure process* of a queue. The tractability that we have seen that arises from Poisson assumptions suggests that queues that have Poisson departure processes can be joined into networks that have mathematically tractable solutions. One of our goals is to investigate departure properties of reversible queues, the most notable of which is the M/M/1 queue. A word on nomenclature: we say a queue is a *reversible queue* if it has a corresponding Markov process that is reversible. (Simple changes to state descriptors can destroy reversibility without altering the "intrinsic" nature of the Markov process, see Problem 10.11.)

10.3.1 Input–Output Properties of Reversible Queues

We now show that the departure process from a reversible queue has the same joint distribution as its arrival process. Let $X(t)$ be the number of customers in the queue at time t. The points in time where $X(t)$ increases by 1 correspond to arrival times. Similarly the points in time where $X^r(t)$ increases by 1 correspond to arrival times for the reversed process. The points in time where $X^r(t)$ increase by 1, however, correspond to points in time where $X(t)$ decreases by 1. These points correspond to departures from the queue, and thus we conclude that the departure process from the queue has the same joint statistics as the arrival process into the queue. This argument is summarized in Figure 10.3.

To continue this line of reasoning, assume that a reversible queue has arrival rates that are Poisson and independent of the state. It then follows that the departure process is also Poisson with rate λ. This seems to violate intuition since it is invariant to the way customers are served in the queue. Although it is clear that the average customer departure rate must be λ if the queue is stationary, it is not clear that the departure process must have independent interdeparture intervals. Problem 10.12 explores a paradoxical implication of this result.

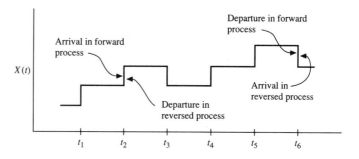

FIGURE 10.3. Forward Arrivals (Departures) = Reversed Departures (Arrivals)

10.3.2 Tandem Reversible Queues

Assume that a reversible queue has a state-independent arrival process and consider the state of the system at a particular point in time, say t. It is clear that the arrival process after time t is independent of the state of the system at time t. The correspondence of arrivals in the forward process and departures in the reversed process, however, implies that the *departure process prior to time t is independent of the state of the system at time t*. This is a striking conclusion since it would initially seem that the departure process prior to time t *determines* the state at time t and thus cannot be independent of $X(t)$ (see Problem 10.20 for some interesting consequences of these conclusions). This observation has important ramifications when reversible queues are joined together into a network as illustrated in the following example.

Example 10.9 (Tandem Reversible Queues) Consider a tandem queueing network consisting of two M/M/1 queues. Customers arrive to the first queue, are served there, and upon departure are immediately fed into the second queue. After being served in the second queue, customers leave the system. Suppose that the arrival rate of customers to the first queue equals λ and the servicing rates of the first and second queue equal μ_1 and μ_2, respectively, where we assume that $\lambda < \max\{\mu_1, \mu_2\}$. Such a system could be a crude model for a computer system where customers require processing at a CPU followed by processing at an I/O device, such as a hard disk. In Figure 10.4 we show such a system. (Clearly, such a simple system has a limited modeling capability and in the remainder of this chapter we progressively develop the theory of product form networks that is required to analyze more complex networks.)

To analyze this simple tandem queueing network model, first note that the above arguments show that the departure process from the first queue is Poisson with rate λ. Since these departures form the arrivals into the second queue, it follows that the departure process from the second queue is Poisson. Viewed in isolation, each queue acts like an M/M/1 queue and has a marginal queue

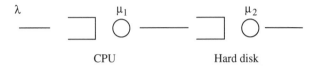

FIGURE 10.4. A Simple Computer Model

length distribution given by

$$P[N_i = n_i] = (1 - \rho_i)\rho_i^{n_i}, \quad i = 1, 2, \quad n_i = 0, 1, \ldots,$$

where N_i, for $i = 1, 2$, is a random variable that denotes the stationary number of customers in queue i and $\rho_i \stackrel{\text{def}}{=} \lambda/\mu_i$. Our objective is to determine the joint distribution of the joint process (N_1, N_2). In general, if we join two queues together in tandem their queue lengths are not independent, and hence the joint distribution is not a product form. For tandem M/M/1 queues, however, we can show that the queue lengths are independent.

To establish this, observe that the length of queue i at time t, $N_i(t)$, $i = 1, 2$, is determined by the service requirements and arrival times of customers who enter queue i prior to time t. The arrival process prior to time t for the second queue is the departure process from the first queue prior to time t. Since $N_1(t)$ is independent of this departure process, it follows that $N_2(t)$ is independent of $N_1(t)$. This implies that the joint queue length distribution is given by the product of the two marginals. Hence the stationary distribution of the tandem queue has a product form solution given by

$$P[N_1 = n_1, \ N_2 = n_2] = (1 - \rho_1)\rho_1^{n_1}(1 - \rho_2)\rho_2^{n_2}, \quad 0 \leq n_1, n_2 < \infty. \quad (10.34)$$

The argument given in Example 10.9 can be generalized to any number of queues arranged in tandem. The routing mechanism in the network can also be generalized and still preserve product form. Consider the following example of a feed forward network.

Example 10.10 (Random Feed Forward Networks) We say that a network is a *random feed forward network* if customers are randomly routed in the network and cannot cycle. In such a network departures from queue j, for $j = 1, 2, \ldots, J - 1$, are routed to queue k for $k = j + 1, j + 2, \ldots, J$, with probability $p_{j,k}$. We require that $p_{j,i} = 0$, for $i = 1, 2, \ldots, j$, and that $\sum_{k=j+1}^{m} p_{j,k} \leq 1$. Note that $1 - \sum_{k=j+1}^{m} p_{j,k}$ is the probability that a customer departs the system from queue j. Let $\boldsymbol{P}$ be a lower diagonal matrix with entries $\{p_{j,k}\}$, termed the *routing matrix*. Departures from queue J are assumed to leave the network. An example of a feed forward network is shown in Figure 10.5. This is a generalization of the model depicted in Figure 10.4, where two I/O devices are included in the model: a hard disk and a floppy disk.

Assume that external arrivals (those that are not routed from another queue) to queue j, for $i = 1, 2, \ldots, J$, are independent Poisson processes with rate λ_j

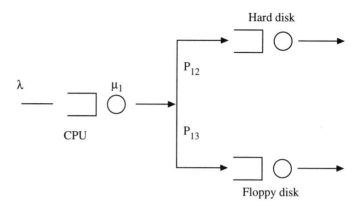

FIGURE 10.5. A Feed Forward Network

and let $\boldsymbol{\lambda} \stackrel{\text{def}}{=} (\lambda_1, \lambda_2, \ldots, \lambda_m)$. Suppose that queue j is an M/M/1 queue with service rate μ_j. Let $\boldsymbol{n} \stackrel{\text{def}}{=} (n_1, n_2, \ldots, n_J)$ be a state vector where queue j has n_j customers and let $\pi(\boldsymbol{n})$ be its stationary distribution. We claim that the stationary distribution is a product form given by

$$\pi(\boldsymbol{n}) = \prod_{j=1}^{m}(1 - \rho_j)\rho_j^{n_j}, \tag{10.35}$$

where $\rho_j \stackrel{\text{def}}{=} \gamma_j/\mu_j$ and γ_j is the jth component of the vector $\boldsymbol{\gamma}$ given by

$$\boldsymbol{\gamma} = \boldsymbol{\lambda}(\boldsymbol{I} - \boldsymbol{P})^{-1}. \tag{10.36}$$

We call γ_j the *throughput* of queue j since it equals the total arrival rate of customers into and out of the queue. (We assume that the Markov process is ergodic and thus that the arrival and departure rates are equal and $\rho_j < 1$, $j = 1, 2, \ldots, J$.)

To establish this result, note that the routing of customers departing from a queue in the network corresponds to splitting a Poisson process. Proposition 6.7 shows that the departure process from queue j is split into independent Poisson processes in this fashion. The arrival process into queue j is the superposition of Poisson processes generated by the external arrival process and departures from queues $1, 2, \ldots, j-1$ that are routed to queue j. Since these are independent Poisson processes, it follows from Section 6.2.7 that the arrival process into each queue is a Poisson process.

An argument similar to that given in Example 10.9 can be used to show that the queues are independent. In particular, arrivals to queue j prior to time t consist of external arrivals and routed departures from queues $1, 2, \ldots, j-1$, that occur prior to time t. These processes are independent of $N_i(t)$ for $i = 1, 2, \ldots, j-1$, and thus $N_j(t)$ is independent of $(N_1(t), N_2(t), \ldots, N_{j-1}(t))$. A continuation of this argument implies that the stationary distribution is a product form.

We have yet only to explain the definition of ρ_j used in (10.35). Let γ_j be the arrival rate of customers into queue j. We can write this as

$$\gamma_j = \lambda_j + \sum_{i=1}^{j-1} \gamma_i p_{i,j}$$

or, in matrix form, as

$$\boldsymbol{\gamma} = \boldsymbol{\lambda} + \boldsymbol{\gamma} \boldsymbol{P}.$$

It is clear that (10.36) solves this equation and yields the utilization given for each queue.

The remainder of this section will explore subtleties associated with having a product form solution. Our objective is to extract the mathematical features of the tandem queue of Example 10.9 and of the feed forward queue of Example 10.10 that imply product form. We will do this by developing variations of these examples that lead to defining a key property (quasi-reversibility) that implies product form. Before delving into these examples, it is important to first clear up a misconception that might arise from the tandem queue example.

**10.3.3
The Tandem
Queue Is Not
Reversible**

It is interesting to note that the product form solution given in (10.34) shows that the two queues in the tandem are independent. It is tempting to believe, since each queue in isolation is reversible, that the joint process (N_1, N_2) for the tandem queue is also reversible. This plausible hypothesis can easily be shown to be false. (This seems to contradict a result we discussed in Section 10.2 that showed that joining two independent reversible processes yields a reversible process (see Problem 10.21)).

The state space for the tandem queues depicted in Figure 10.6 shows that Kolmogorov's criterion is not satisfied. For example, consider the cycle $(1,1)$, $(2,1)$, $(2,0)$, $(1,1)$. The flow of rates in one direction of this cycle is the product $\lambda\mu_1\mu_2$ whereas it is 0 in the other direction. Just using the state space diagram it is difficult to see that the stationary distribution is given by the product form of (10.34). It is an easy exercise, however, to show that (10.34) satisfies the global balance equations (see Problem 10.22). Without using reversibility to show independence it would be difficult to guess such a solution form.

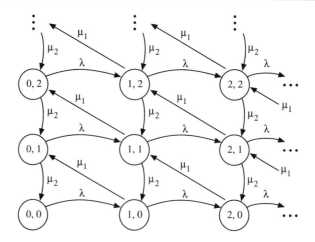

FIGURE 10.6. State Transition Diagram for Tandem M/M/1 Queues

10.3.4
Poisson Arrivals and Departures $\nRightarrow$ Product Form

A salient feature of the tandem queue example of Example 10.9 and the feed forward example of Example 10.10 is that each queue has Poisson arrival and departure processes. This could lead one to believe that such input–output properties are necessary and/or sufficient to obtain product form. The following two examples show that having Poisson arrival and departure processes from each queue is neither sufficient (Example 10.11) nor necessary (Example 10.12) for product form to hold.

Example 10.11 (Tandem Queues with Poisson Departure Processes) Consider joining queues in a tandem. Arrivals to the first queue are Poisson with rate λ and service times are exponential with rate μ_1. This queue also has state transitions identical to an M/M/1 queue except that state 0 is split into two states, states $0a$ and $0b$. Transitions from state 1 to state $0a$ occur at rate $\mu_1 I_X$ where I_X is an indicator random variable that equals 1 if the second queue in the tandem is idle. Transitions from state 1 to state $0b$ occur at rate $\mu_1(1 - I_X)$. Transitions for queue 1 in isolation are shown in Figure 10.7. It is important to observe that the process depicted in this figure is not Markovian since the value of X is not contained in the description of the state.

Assume that the second queue has an exponential service at rate μ_2. The state transition diagram for the tandem is depicted in Figure 10.8. It is clear that states $(0a, j)$ for $j = 2, 3, \ldots$, have zero probability and therefore are not depicted in the figure. In contrast to Figure 10.7 the process depicted in Figure 10.8 is Markovian and it is interesting to see that there is no need to refer to random variable X in the state transition diagram.

It is clear that the departure process from the first queue is identical to that of an M/M/1 queue and hence is Poisson with rate λ. It thus follows that both

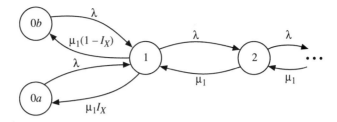

FIGURE 10.7. State Transition Diagram for First Queue in Isolation

the arrival and departure processes for each queue are Poisson. It is tempting to conclude that the stationary state probabilities satisfy a product form. If product form holds, then

$$P[N_1 = 0a, \ N_2 = 0] = P[N_1 = 0a] \, P[N_2 = 0]. \tag{10.37}$$

In this exercise we show that (10.37) does not hold, and hence the tandem queue is not product form.

To establish this we calculate the joint distribution of (N_1, N_2). Since only the boundary states of N_1 are influenced by our construction, a natural guess for non-boundary probabilities is that they are identical to those derived for the tandem queue considered in Example 10.9 given by (10.34),

$$P[N_1 = n_1, \ N_2 = n_2] = (1 - \rho_1)\rho_1^{n_1}(1 - \rho_2)\rho_2^{n_2}, \quad n_1 = 1, 2, \ldots, \quad n_2 = 0, 1, \ldots,$$

where $\rho_i \overset{\text{def}}{=} \lambda/\mu_i$ for $i = 1, 2$. The left-hand column in Figure 10.8 depicting transitions between states $(0b, j)$, for $j = 1, 2, \ldots$, appears to have a geometric form. Hence a natural guess is that these states satisfy

$$P[N_1 = 0b, \ N_2 = n_2] = (1 - \rho_1)(1 - \rho_2)\rho_2^{n_2}, \quad n_2 = 2, 3, \ldots .$$

With these guesses, if global balance is satisfied then

$$P[N_1 = 0b, \ N_2 = 0] = (1 - \rho_1)(1 - \rho_2)\frac{\lambda}{\lambda + \mu_2},$$

$$P[N_1 = 0b, \ N_2 = 1] = (1 - \rho_1)(1 - \rho_2)\rho_2\frac{\lambda}{\lambda + \mu_2},$$

$$P[N_1 = 0a, \ N_2 = 0] = (1 - \rho_1)(1 - \rho_2)\frac{\mu_2}{\lambda + \mu_2},$$

and

$$P[N_1 = 0a, \ N_2 = 1] = (1 - \rho_1)(1 - \rho_2)\frac{\lambda}{\lambda + \mu_2}.$$

It is easy to check that these equations satisfy global balance and thus are the stationary distribution for (N_1, N_2).

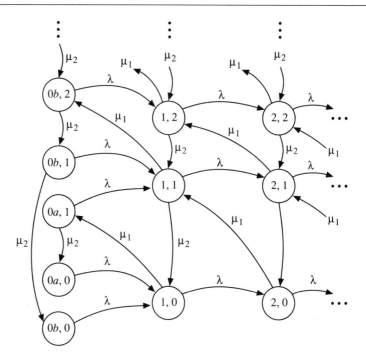

FIGURE 10.8. State Transitions Diagram for the Modified Tandem Queue

To see that (10.37) is not satisfied note that

$$P\left[N_1 = 0a\right] \;=\; P\left[N_1 = 0a,\ N_2 = 0\right] + P\left[N_1 = 0a,\ N_2 = 1\right]$$

$$=\; (1 - \rho_1)(1 - \rho_2)$$

and that

$$P\left[N_2 = 0\right] = 1 - \rho_2.$$

It is thus clear that (10.37) doesn't hold since

$$(1 - \rho_2)(1 - \rho_2)\frac{\mu_2}{\lambda + \mu_2} \neq (1 - \rho_1)(1 - \rho_2)^2.$$

Example 10.11 shows that, even though the arrival and departure process for each queue is Poisson, the tandem queue does not have a product form solution and thus the queues are not independent. This shows that *Poisson arrival and departure processes do not suffice for product form.* The next example shows that such an input–output condition is also not necessary for product form.

Example 10.12 (**Networks of M/M/1 Queues**) Consider a generalized version of Example 10.10 where the routing matrix is not constrained to be feed forward. Let the routing probabilities, $p_{j,k}$, be unconstrained except that $\sum_{k=1}^{m} p_{j,k} \leq 1$, for $j = 1, 2, \ldots, m$. The probability that a customer departs the system from queue j is given by $1 - \sum_{k=1}^{J} p_{j,k}$, and we require that customers can leave the system from at least one queue in the network. Such a network clearly has a richer modeling capability than that of a feed forward network considered in Example 10.10. In Figure 10.9 we show how the network of Figure 10.5 can be generalized to include multiple passes through the CPU and I/O devices.

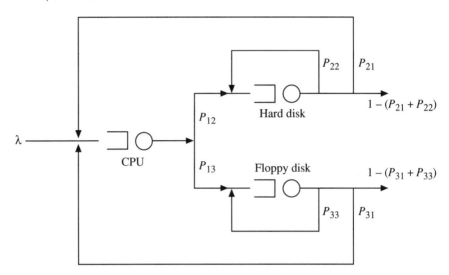

FIGURE 10.9. A Computer Model with Feedback Loops

We claim that the stationary state probabilities have the same product form given by (10.35) where the throughput of queue j satisfies

$$\gamma_j = \lambda_j + \sum_{k=1}^{J} \gamma_k p_{k,j}. \tag{10.38}$$

We cannot establish this result in the same manner as in Example 10.10 since the arrival process into the queues of the network are not necessarily Poisson since we allow loops in the routing of customers (Problem 10.24). We show that the product form of (10.35) holds by using Proposition 10.2. A natural guess for the reversed network is that it is similar to the original network but with a *reversed routing matrix*. To determine the form of this reversed routing matrix, denoted by $\boldsymbol{P}^r$, observe that the total customer flow from queue j to queue k in the forward process equals the total customer flow from queue k to queue j in the reversed process. This argues that

$$\gamma_j p_{j,k} = \gamma_k p_{k,j}^r \tag{10.39}$$

or, in matrix form, that

$$\boldsymbol{P}^r = \Gamma^{-1}\boldsymbol{P}\Gamma,$$

where $\Gamma = \mathrm{diag}(\boldsymbol{\gamma})$. This form is consistent with (10.15).

To use Proposition 10.2 we must establish that the total transition rate from any state for the forward and reverse processes are equal (10.11), and establish that the reversed balance equations, (10.14), hold for all possible transitions. From any state, the possible state transitions are arrivals to queue j, departures from queue j or routed customers from j to k, where $1 \leq j, k \leq J$. We can thus write the total transition rate from any state, $\boldsymbol{n}$, as

$$q(\boldsymbol{n}) = \sum_{\boldsymbol{n}'} q(\boldsymbol{n}, \boldsymbol{n}')$$

$$= \underbrace{\sum_{j=1}^{J} \lambda_j}_{\text{arrivals}} + \underbrace{\sum_{\{j \,|\, n_j > 0\}} \mu_j \left(1 - \sum_{k=1}^{J} p_{j,k}\right)}_{\text{departures}} + \underbrace{\sum_{\{j \,|\, n_j > 0\}} \mu_j \sum_{k=1}^{J} p_{j,k}}_{\text{routed customers}}$$

$$= \sum_{j=1}^{J} \lambda_j + \sum_{\{j \,|\, n_j > 0\}} \mu_k. \tag{10.40}$$

Notice that (10.40) does not depend on the routing probabilities of the forward process. It is thus clear that the total transition rate $q^r(\boldsymbol{n}, \boldsymbol{n}')$ for the reversed process also satisfies (10.40) and that (10.11) is satisfied.

To show that the reversed balance equations are satisfied we must show that

$$\pi(\boldsymbol{n})q(\boldsymbol{n}, \boldsymbol{n}') = \pi(\boldsymbol{n}')q^r(\boldsymbol{n}', \boldsymbol{n}) \tag{10.41}$$

holds for all states $\boldsymbol{n}$ and $\boldsymbol{n}'$. We consider all possible types of transitions.

Arrival Transition. Suppose that $\boldsymbol{n} \mapsto \boldsymbol{n}'$ is an arrival of an external customer to queue j. The forward transition rate is given by $q(\boldsymbol{n}, \boldsymbol{n}') = \lambda_j$. The transition $\boldsymbol{n}' \mapsto \boldsymbol{n}$ in the reversed process corresponds to a departure of a customer from queue j. This transition rate equals

$$q^r(\boldsymbol{n}', \boldsymbol{n}) = \mu_j \left(1 - \sum_{k=1}^{J} p_{j,k}^r\right)$$

$$= \mu_j \frac{\left(\gamma_j - \sum_{k=1}^{J} \gamma_k p_{k,j}\right)}{\gamma_j}$$

$$= \frac{\mu_j \lambda_j}{\gamma_j},$$

where we use (10.39) in the second equality and (10.38) in the last equality. Observe that $\prod_{j=1}^{J}(1 - \rho_j)\rho_j^{n_j}$ is a common factor of both $\pi(\boldsymbol{n})$ and $\pi(\boldsymbol{n}')$ and thus the reversed balance equations (10.41) are satisfied if

$$\lambda_j = \frac{\mu_j \lambda_j}{\gamma_j} \rho_j.$$

This follows from the definition of ρ_j. Departure transitions are shown to satisfy (10.41) in a similar fashion.

Routing Transitions. Suppose that $n \mapsto n'$ is a departure of a customer from queue j that is routed to queue k. Transitions for the forward and reversed processes are easily seen to be given by

$$q(n, n') = \mu_j p_{j,k}$$

and

$$q^r(n', n) = \mu_k p_{k,j}^r.$$

Canceling common terms between $\pi(n)$ and $\pi(n')$ implies that (10.41) is satisfied if

$$\mu_i p_{j,k} \rho_j = \mu_k p_{k,j}^r \rho_k.$$

This follows from the definition of the reversed transition rates given in (10.39).

We established the product form result of Example 10.12 using M/M/1 queues. The result actually holds more generally and will be discussed in Section 10.6. As with the tandem queue example, the Markov process created by joining reversible queues into a network is not, in general, reversible.

10.3.5 Features of Product Form Networks

Comparison of Examples 10.11 and 10.12 shows that having Poisson arrivals and departures from a queue is not a necessary nor a sufficient condition for a network to have a product form solution. A review of the derivations presented in the tandem network of Example 10.9 and the feed forward network of Example 10.10 shows that the property used to establish product form concerned a type of independence of the arrival and departure processes associated with a queue and its state. Specifically, for each queue in these networks, customer arrival times after time t and customer departure times prior to time t are *independent* of the state of a queue at time t. This is a key property of product form networks and motivates us to define quasi-reversibility.

Definition 10.13 *Suppose there are C classes of customers that arrive and are serviced by a queueing system. The Markov process associated with this system is said to be quasi-reversible if the state of the process for all classes c, $1 \leq c \leq C$, at time t is independent of the arrival process of class c customers after time t and is independent of the departure process of class c customers prior to time t.*

The identification of arrivals (resp., departures) in the forward process with departures (resp., arrivals) in the reversed process, implies that the reversed process of a quasi-reversible queue is also quasi-reversible.

10.3.6 Input–Output Properties of Quasi-Reversible Queues

We will now show that if a quasi-reversible queue has class-dependent Poisson arrival processes then the departure process of class c customers is also Poisson. Let t be an arbitrary point in time and let $\mathcal{S}_c(i), 1 \leq c \leq C, i \in \mathcal{S}$, be the set of states that contain one more class c customer than that of state i with the same number of customers of other classes. The arrival rate of class c customers, given that $X(t) = i$, is given by

$$\lambda_{c,i} = \sum_{k \in \mathcal{S}_c(i)} q(i,k), \quad i \in \mathcal{S}. \tag{10.42}$$

Thus the average arrival rate of class c customers equals

$$\lambda_c = \sum_{i \in \mathcal{S}} \pi(i)\lambda_{c,i}. \tag{10.43}$$

By the definition of quasi-reversibility, the arrival process of class c customers subsequent to t is independent of the state at time t, and thus $\lambda_{c,i}$ is independent of i. Using (10.42) and (10.43) we write

$$\lambda_c = \sum_{k \in \mathcal{S}_c(i)} q(i,k), \tag{10.44}$$

where i is any state in $\mathcal{S}$. The probability that a class c customer arrives in the time interval $(t, t+dt)$ is independent of any event prior to time t and equals $\lambda_c dt$. This shows that the arrival process of class c customers has independent increments and has constant rate. Thus, according to the results of Section 6.3.7, it is a Poisson process.

Assume that all arriving customers leave the system and that the queue is in equilibrium. The identification of arrivals (resp., departures) of the forward process with departures (resp., arrivals) of the reverse process and the fact that the queue viewed in reversed time is also quasi-reversible, implies that the departure process from the queue is Poisson with rate λ_c. This argument also shows that

$$\lambda_c = \sum_{k \in \mathcal{S}_c(j)} q^r(j,k), \quad j \in \mathcal{S}. \tag{10.45}$$

The above analysis provides a more direct operational definition of quasi-reversibility than that given in Definition 10.13. Specifically, a multiple-class queue is quasi-reversible if, for every class c and every state i, there is a constant λ_c that satisfies

$$\lambda_c = \sum_{k \in \mathcal{S}_c(i)} q(i,k) = \sum_{k \in \mathcal{S}_c(i)} q^r(i,k), \quad i \in \mathcal{S}. \tag{10.46}$$

This definition can be used to establish that a queue within a network is quasi-reversible (see Problem 10.27).

**10.3.7
Quasi-
Reversibility and
Reversibility**

The above arguments imply that an M/M/1 queue with a fixed arrival rate λ is quasi-reversible whereas an M/M/1 queue with state-dependent arrival rates λ_i, $i = 0, 1, \ldots$, is not. Both systems, it is important to note, are reversible. On the other hand, one can construct examples of Markov processes that are quasi-reversible but not reversible. Consider the following example.

Example 10.14 (A Quasi-Reversible Process that is Not Reversible). Consider the M/M/1 queue where we split state 1 into two states, $1a$ and $1b$. Set the transition from state 0 to state $1a$ (resp., state $1b$) equal to $p\lambda$ (resp., $(1-p)\lambda$) and the transition from state 2 to state $1a$ (resp., state $1b$) equal to $(1-p)\mu$ (resp., $p\mu$). It is simple to see that the sum of the stationary probabilities of state $1a$ and $1b$ in this modified process equals the stationary probability of state 1 in the original birth–death process. It is also clear that splitting state 1 does not influence the departure process from the system. The arrival process after a given time t, and departure process prior to time t are independent of the state of the system and thus the system is quasi-reversible. The modified process, however, is not reversible. This is easily seen from Kolmogorov's criteria by comparing the product of transition rates in both directions of the cycle $0 \rightarrow 1a \rightarrow 2 \rightarrow 1b \rightarrow 0$. Equality of these two products is only achieved if $p = 1/2$.

The above discussion shows that reversibility and quasi-reversibility are *entirely separate notions*. Quasi-reversibility (resp., reversibility) does not imply reversibility (resp., quasi-reversibility). Example 10.12 showed that product form resulted when one joins *reversible queues* into a network using random routing. Since reversible queues and quasi-reversible queues are entirely different notions, the example sheds no light on whether quasi-reversible queues can be joined in a network in this fashion and have a product form solution.

10.3.8 Multiple-Class Quasi-Reversible M/M/1 Queue

An example of a quasi-reversible queue is an M/M/1 queue with C classes of customers. Assume that class c customers arrive at rate λ_c and that *all classes* have exponential service times with mean μ^{-1}. It is important to notice that all customer classes have the same service rate. The queue is not quasi-reversible if this condition is relaxed. The state of the system when there are n customers is given by $\boldsymbol{c} = (c_1, c_2, \ldots, c_n)$ where c_i is the class of the customer in the ith position of the queue. (The first position is the server and the last is the tail of the queue.) Let the total arrival rate of customers of all classes be defined by $\lambda \stackrel{\text{def}}{=} \sum_{c=1}^{C} \lambda_c$ and let $\rho \stackrel{\text{def}}{=} \lambda/\mu$ be the utilization of the system and $\rho_c \stackrel{\text{def}}{=} \lambda_c/\mu$ be the utilization of class c customers. The probability that a particular customer in the queue is of class c, denoted by φ_c, is given by

$$\varphi_c \stackrel{\text{def}}{=} \frac{\rho_c}{\rho} = \frac{\lambda_c}{\lambda}. \tag{10.47}$$

We claim that the stationary probability distribution equals

$$\pi(\boldsymbol{c}) = (1-\rho)\rho^n \prod_{i=1}^{n} \varphi_{c_i} \tag{10.48}$$

and that the queue is quasi-reversible.

To establish these assertions we investigate the reversed process. A natural candidate for this process is an M/M/1 queue with class-dependent Poisson arrivals at rate λ_c, service times that are exponential with rate μ, and where the positions of the queue are *reversed*. To describe this last criteria define an operator ϑ that reverses the components of a vector,

$$\vartheta(\boldsymbol{c}) \stackrel{\text{def}}{=} (c_n, c_{n-1}, \ldots, c_2, c_1).$$

We claim that the reversed transition rates satisfy

$$q^r(\boldsymbol{c}, \boldsymbol{c}') = q(\vartheta(\boldsymbol{c}), \vartheta(\boldsymbol{c}')). \tag{10.49}$$

Proposition 10.2 shows that to establish (10.48) with these proposed values of q^r we must satisfy the reversed balance equations given by

$$\pi(\boldsymbol{c})q(\boldsymbol{c}, \boldsymbol{c}') = \pi(\boldsymbol{c}')q^r(\boldsymbol{c}', \boldsymbol{c}), \tag{10.50}$$

for all possible transitions.

There are two possible transitions: arrivals and departures. Consider an arrival to state $\boldsymbol{c}$ of a class d customer leading to $\boldsymbol{c}' = (\boldsymbol{c}, d)$. It is clear that $q(\boldsymbol{c}, \boldsymbol{c}') = \lambda_d$ and, from (10.49) that $q^r(\boldsymbol{c}', \boldsymbol{c}) = \mu$. Substituting this into (10.50) shows that (10.48) is satisfied. To see this, observe that $(1-\rho)\rho^n \prod_{i=1}^{n} \varphi_{c_i}$ is a common factor of both $\pi(\boldsymbol{c})$ and $\pi(\boldsymbol{c}')$ and thus (10.48) is satisfied if $\lambda_d = \mu\rho\varphi_d$, which follows from (10.47). Showing that (10.50) holds for departure processes is entirely similar, thus we conclude that (10.48) is the stationary distribution.

The quasi-reversibility of the queue follows from a straightforward argument. Arrivals to the M/M/1 queue are state-independent Poisson processes and hence the arrival process after a given time t is independent of the state at time t. The reversed process is also an M/M/1 queue with Poisson arrivals. Hence, the identification of arrivals (resp., departures) of the forward process with departures (resp., arrivals) of the reversed process shows that the departure process of class c customers from an M/M/1 queue is Poisson with rate λ_c. This shows that the queue is quasi-reversible.

Sometimes we do not require a detailed specification of the class of customers in given positions of the queue but rather are only concerned with the number of class c customers resident in the queue. Let

$$\boldsymbol{N} \stackrel{\text{def}}{=} (N_1, N_2, \ldots, N_C)$$

be a vector of random variables where N_i denotes the number of class c customers in the queue. Let $\boldsymbol{n} \stackrel{\text{def}}{=} (n_1, n_2, \ldots, n_C)$ be a possible state vector where there are n_c class c customers in the queue. The stationary distribution given in (10.48) implies that we can write the probability of state $\boldsymbol{n}$ as

$$P\left[\boldsymbol{N} = \boldsymbol{n}\right] = (1 - \rho)\binom{n}{n_1, n_2, \ldots, n_C}\prod_{c=1}^{C}\rho_c^{n_c}. \qquad (10.51)$$

The multinomial term in (10.51) counts all possible combinations of states that have the same number of class c customers. Let $N \stackrel{\text{def}}{=} N_1 + N_2 + \cdots + N_C$ be a random variable that denotes the total number of customers in the queue. The distribution (10.51) implies that we can write

$$P\left[N = n\right] = (1 - \rho)\rho^n, \quad n = 0, 1, \ldots, \qquad (10.52)$$

which is the same as that of an M/M/1 queue. The expected number of class c customers when there are n customers in the queue can be written as

$$E\left[N_c \mid N = n\right] = n\varphi_c, \quad c = 1, 2, \ldots, C,$$

which implies that

$$E\left[N_c\right] = \frac{\varphi_c \rho}{1 - \rho}, \quad c = 1, 2, \ldots, C.$$

Little's law implies that the class c response time, denoted by T_c, has an expectation given by

$$E\left[T_c\right] = \frac{1}{\lambda \varphi_c} \frac{\varphi_c \rho}{1 - \rho} = \frac{1}{\mu(1 - \rho)}, \qquad (10.53)$$

which is independent of the class of the customer and identical to that of an M/M/1 queue. We next discuss a large class of quasi-reversible queues called symmetric queues.

10.4 Symmetric Queues

In this section we study a rich class of quasi-reversible queues termed *symmetric queues*. Consider a queue containing n customers that are ordered in positions $1, 2, \dots, n$, in the queue. Customers are assumed to be one of C classes and the distribution of service for customers depends on their class. The operation of the queue is governed by a set of discrete probability distributions. In particular, we assume that vectors

$$\kappa_n \overset{\text{def}}{=} (\kappa_n(1), \kappa_n(2), \dots, \kappa_n(n)), \quad n = 1, 2, \dots, \qquad (10.54)$$

are given so that $\kappa_n(i) \geq 0$, and $\sum_{i=1}^{n} \kappa_n(i) = 1$, $n = 1, 2, \dots$. The queue is a symmetric queue if it operates in the following manner:

1. The total service rate provided to the n customers in the queue is given by $r(n) > 0$, $n = 1, 2, \dots$.

2. The customer in position i of the queue receives a fraction $\kappa_n(i)$ of the service rate, for $i = 1, 2, \dots, n$.

3. When customer i completes service, for $i = 1, 2, \dots, n-1$, customers in positions $i+1, i+2, \dots, n$, move down to positions $i, i+1, \dots, n-1$. If customer n completes service, the positions of the other customers in the queue are not changed.

4. An arriving customer is placed in position i of the queue, for $i = 1, 2, \dots, n$, with probability $\kappa_{n+1}(i)$. Customers that were in positions $i, i+1, \dots, n$, move up to positions $i+1, i+2, \dots, n+1$. An arriving customer is placed in position $n+1$ with probability $\kappa_{n+1}(n+1)$. In this case the positions of the other customers in the queue are not changed.

A rich set of queues and corresponding scheduling disciplines arise from variations of the probability distributions κ_n and the rate functions $r(n)$ of symmetric queues. Some queues frequently used in modeling include:

Last-Come-First-Served with Preemptive Resume. In this discipline customers are served in the reverse order of their time of arrival. An arrival causes the server to stop working on the customer in service and immediately begin serving the arriving customer. After a service completion, the server returns to servicing the last preempted customer, if there was one. We assume that the server resumes servicing the customer where it left off, and thus the remaining service time of a customer is identical to when its service was last interrupted. The parameter values are given by $r(n) = 1, n = 1, 2, \ldots$, and

$$
\kappa_n(i) = \begin{cases} 0, & i = 1, 2, \ldots, n-1, \\ 1, & i = n, \end{cases} \quad n = 1, 2, \ldots .
$$

Infinite Server. In an infinite server queue, each arriving customer is assigned to an available server. The parameter values are given by $r(n) = n, \ n = 1, 2, \ldots$, and $\kappa_n(i) = 1/n, \ i = 1, 2, \ldots, n$.

Processor Sharing. In a processor sharing queue, all jobs receive an equal allocation of the processor and thus $\kappa_n(i) = 1/n, \ i = 1, 2, \ldots, n$. It is distinguished from the infinite server case by the fact that typically $r(n) \neq n$. If $r(n) = 1$ then a single server is shared by all customers in the system and thus the remaining server time of a particular customer, when n customers are in the queue, decreases at a rate of $1/n$. A processor sharing system with K servers can be specified by setting

$$
r(n) = \begin{cases} n, & n = 1, 2, \ldots, K, \\ K, & n = K+1, K+2, \ldots . \end{cases}
$$

In such a system, each customer has its own server if there are K or fewer total customers in the system. Otherwise, all customers equally share the processing capacity of the K queues.

Finite Capacity Queues. We can restrict a given symmetric queue so that it approximates a finite capacity queue in the following manner. Let a given symmetric queue have parameter values $r(n)$ and κ_n for $n = 1, 2, \ldots$ and let M be a large constant. To approximate the operation of the queue under the constraint that it has at most N customers we modify the parameters as follows:

$$
q_n'(i) = \begin{cases} \kappa_n(i), & i = 1, 2, \ldots, n, & n = 1, 2, \ldots, N, \\ 0, & i = 1, 2, \ldots, n-1, & n = N+1, N+2, \ldots, \\ 1, & i = n, & n = N+1, N+2, \ldots, \end{cases}
$$

and

$$
r'(n) = \begin{cases} r(n), & n = 1, 2, \ldots, N, \\ M, & n = N+1, N+2, \ldots . \end{cases}
$$

These modified parameter values approximate a finite capacity queue by providing extremely fast service rates to customers that arrive to positions $N+1, N+2, \ldots$. For large M, these customers arrive to the system and then leave almost immediately without disrupting customers in positions $1, 2, \ldots, N$.

We will refer to symmetric queues that have $r(n) = r(1)$ for $n = 1, 2, \ldots$, as *single server symmetric queues* and, without loss of generality, for such queues we set $r(1) = 1$.

10.4.1 Quasi-Reversible Symmetric Queues

We now show that symmetric queues with class-dependent Poisson arrival processes are quasi-reversible and establish the form of their stationary distributions. We will do this for the case where customers have class-dependent phase service times. This is not a restriction since, as stated in Section 9.5, any distribution can be approximated to any degree of accuracy by a phase distribution.

Let class c, for $c = 1, 2, \ldots, C$, have a service time given by a phase distribution with $n_c + 1$ states and representation $(\beta_c, \boldsymbol{U}_c)$, where $\beta_c \overset{\text{def}}{=} (\beta_c(1), \beta_c(2), \ldots, \beta_c(n_c))$ (see Section 9.5). To avoid trivial cases we assume that $\beta_c(n_c + 1) = 0$ and thus service times are non-zero with probability 1. Recall that $\boldsymbol{A}_c = -\boldsymbol{U}\boldsymbol{e}$ is a column vector that contains the transition rates into the absorbing state. The expected service time of a class c customer is given by (see (9.51)), $\mu_c^{-1} = -\beta_c (\boldsymbol{U}_c)^{-1} \boldsymbol{e}$. Assume that class c arrivals are Poisson with rate λ_c and define the utilization of a class c customer by

$$\rho_c \overset{\text{def}}{=} \lambda_c/\mu_c.$$

The PH-Markov process associated with the phase distribution $(\beta_c, \boldsymbol{U}_c)$ has a representation given by $(\beta_c, \boldsymbol{U}_c, \xi_c)$, where $\xi_c \overset{\text{def}}{=} (\xi_c(1), \ldots, \xi_c(n))$, is the stationary probability vector of the PH-Markov process.

The state of the symmetric queue, $\boldsymbol{x}$, can be represented by

$$\boldsymbol{x} \overset{\text{def}}{=} ((c_1, s_1), (c_2, s_2), \ldots, (c_n, s_n)), \tag{10.55}$$

where c_i, is the class of the customer in position i of the queue and s_i, for $s_i \in \{1, 2, \ldots, n_{c_i}\}$, is the state of service of the customer. We will equivalently say that the customer is in *stage* s_i of service. We claim that the stationary probability of state $\boldsymbol{x}$, denoted by $\pi(\boldsymbol{x})$, is given by

$$\pi(\boldsymbol{x}) = \frac{b}{r(1)r(2)\cdots r(n)} \prod_{i=1}^{n} \rho_{c_i} \, \xi_{c_i}(s_i), \tag{10.56}$$

where b is a normalization constant. We will use Proposition 10.2 to prove this statement. Before doing so, it is interesting to note that the stationary distribution of the symmetric queue does not depend on the probability distributions κ_n, $n = 1, 2, \ldots$, that define its operation!

To establish (10.56) we must determine the transition rates of the reversed process. Notice that, by definition, arrivals and departures from a symmetric queue are governed by the same mechanism - the probability distributions κ_n, $n = 1, 2, \ldots$. It is this *symmetry* (hence the name "symmetric queue") that suggests a natural candidate for the reversed process. The proposed reversed process is a symmetric queue that is managed according to the same probability distributions κ_n. Arrival of class c customers to this queue are Poisson with rate λ_c and the service times have a *time reversed* phase distribution (see Example 10.3). The representation for this distribution is given by (β_c^r, U_c^r), as specified by (10.19), (10.20), and (10.21). The column matrix containing the absorbing transition rates is given by $A_c^r = -U_c^r e$.

We now apply Proposition 10.2 to establish the validity of (10.56). Consider a transition from state x to state x' where we assume that state x has n customers. There are three possibilities: the transition arises from an arrival, a departure or a completion of a stage of service for one of the customers. For each transition type we must check that the reversed balance equations are satisfied. This implies that equations

$$\pi(x)q(x, x') = \pi(x')q^r(x', x) \tag{10.57}$$

must hold for all possible transition types.

Arrival Transition. Suppose that $x \mapsto x'$ corresponds to an arrival of a class c customer to position i of the queue. Assume that the customer starts in stage s of service. The operation of the symmetric queue implies that we can write

$$q(x, x') = \lambda_c \beta_c(s)\kappa_{n+1}(i). \tag{10.58}$$

The reversed transition rate corresponds to the departure of a class c customer from position i of the queue after completion of service in stage s. This transition rate can be written as

$$q^r(x', x) = A_c^r(s)\kappa_{n+1}(i)r(n+1)$$

$$= \frac{\beta_c(s)\mu_c}{\xi_c(s)}\,\kappa_{n+1}(i)r(n+1), \tag{10.59}$$

where we have used (10.20) to obtain (10.59). Simple algebra shows that (10.57) holds. To see this, note that $\pi(x)$ and $\pi(x')$ have common factors of $b \prod_{k=1}^{n} \rho_{c_k} \xi_{c_k}(s_k) / \prod_{k=1}^{n} r(k)$. Canceling these common factors and using (10.58) and (10.59) shows that (10.57) is satisfied if

$$\lambda_c \beta_c(s) \kappa_{n+1}(i) = \frac{\rho_c \xi_c(s)}{r(n+1)} \frac{\beta_c(s) \mu_c \kappa_{n+1}(i) r(n+1)}{\xi_c(s)}. \qquad (10.60)$$

It is clear that (10.60) holds as an equality.

Departure Transition. Suppose that $x \mapsto x'$ corresponds to a departure of a class c customer from position i of the queue after completion of stage s of service. The transition rate for such a transition can be written as

$$q(x, x') = A_c(s) \kappa_n(i) r(n). \qquad (10.61)$$

The reversed transition rate corresponds to an arrival of a class c customer to the queue when there are $n - 1$ customers. This customer starts service in stage s and is placed in position i of the queue. The reversed transition rate is thus given by

$$q^r(x', x) = \lambda_c \beta_c^r(s) \kappa_n(i)$$

$$= \frac{\lambda_c \xi_c(s) A_c(s) \kappa_n(i)}{\mu_c}, \qquad (10.62)$$

where we have used (10.21) to obtain (10.62). The common factors of $\pi(x)$ and $\pi(x')$ are $b \prod_{k=1}^{n-1} \rho_{c_k} \xi_{c_k}(s_k) / \prod_{k=1}^{n-1} r(k)$. Canceling these common factors and using (10.61) and (10.62) shows that (10.57) is satisfied if

$$\frac{\rho_c \xi_c(s)}{r(n)} A_c(s) \kappa_n(i) r(n) = \frac{\lambda_c \xi_c(s) A_c(s) \kappa_n(i)}{\mu_c}. \qquad (10.63)$$

It is clear that (10.63) holds as an equality.

Completion of a Stage of Service. Suppose that $x \mapsto x'$ corresponds to the completion of stage s of service for a customer of type c in position i of the queue. This completion results in a new stage s' of service for this customer without changing the states of the other customers in the queue. It is clear that

$$q(x, x') = \kappa_n(i) r(n) U_c(s, s'). \qquad (10.64)$$

The reversed transition corresponds to completion of service state s' by the reversed process and continuing service in stage s. This transition rate is given by

$$q^r(x', x) = \kappa_n(i) r(n) U_c^r(s', s)$$

$$= \kappa_n(i) r(n) \frac{\xi_c(s) U_c(s, s')}{\xi_c(s')}, \qquad (10.65)$$

where we have used (10.19) to obtain (10.65). Common factors of $\pi(\boldsymbol{x})$ and $\pi(\boldsymbol{x}')$ include $b \prod_{k=1, k \neq i}^{n} \rho_{c_k} \xi_{c_k}(s_k) / \prod_{k=1}^{n} r(k)$. Canceling these common factors and using (10.64) and (10.65) shows that (10.57) is satisfied if

$$\rho_c \xi_c(s) \kappa_n(i) r(n) U_c(s, s') = \rho_c \xi_c(s') \kappa_n(i) r(n) \, \frac{\xi_c(s) U_c(s, s')}{\xi_c(s')}. \tag{10.66}$$

It is clear that (10.66) holds as an equality.

The transition flow conditions (10.13) of Proposition 10.2 are easily established and are thus left to the reader. We have established the validity of the stationary distribution for symmetric queues given in (10.56). Symmetric queues have a complex operation, and it is somewhat surprising to find that they have such simple stationary distributions. The method we used to establish (10.56) attests to the power of viewing Markov processes in reverse time. This view shows that the time reversed process of a symmetric queue that has Poisson arrivals is a symmetric queue with Poisson arrivals. It thus follows that symmetric queues with Poisson arrival processes are quasi-reversible. (This results from the same argument used in Section 10.3 to establish that the M/M/1 queue with class-dependent arrivals is quasi-reversible.)

10.4.2 Distribution of Number of Customers in a Symmetric Queue

The state $\boldsymbol{x}$ of the symmetric queue as discussed above specifies the class and stage of service for each customer in the queue. Often, we do not require this level of detail and, for example, might only be interested in the number of class c customers in the queue regardless of their stage of service. Let the total arrival rate of work be denoted by

$$\rho \stackrel{\text{def}}{=} \sum_{c=1}^{C} \rho_c,$$

and define

$$\varphi_c \stackrel{\text{def}}{=} \frac{\rho_c}{\rho}. \tag{10.67}$$

The value of φ_c is the probability that a given position in the symmetric queue is a class c customer, and it clearly follows that $\sum_{c=1}^{C} \varphi_c = 1$.

Using the same notation as for the multiple-class M/M/1 queue analyzed at the end of Section 10.3.1 we can sum (10.56) to obtain (see Problem 10.30)

$$P\left[\boldsymbol{N} = \boldsymbol{n}\right] = b \rho^n \left(\begin{array}{c} n \\ n_1, n_2, \ldots, n_C \end{array} \right) \frac{\varphi_1^{n_1} \varphi_2^{n_2} \cdots \varphi_C^{n_C}}{r(1) r(2) \cdots r(n)}, \tag{10.68}$$

where b is the same normalization constant as in (10.56) and the multinomial term arises from the fact that different combinations of states correspond to the same number of class c customers. Notice that for single server queues where $r(n) = 1, n = 1, 2, \ldots$, the stationary vector for the number of customers in a symmetric queue as given in (10.68) is the same as that for a multiple-class quasi-reversible M/M/1 queue given in (10.51).

The probability that there are n customers in the system is given by

$$
\begin{aligned}
P[N = n] &= b \frac{\rho^n}{r(1) \cdots r(n)} \sum_{n_1 + \cdots + n_C = n} \binom{n}{n_1, \ldots, n_C} \varphi_1^{n_1} \cdots \varphi_C^{n_C} \\
&= b \frac{\rho^n}{r(1)r(2) \cdots r(n)}, \qquad n = 0, 1, \ldots \; . \qquad (10.69)
\end{aligned}
$$

We now focus on the probability that there are n_c class c customers in a symmetric queue. It can be shown that, given n customers in the queue, the classes of these customers as specified by $(c_1, c_2, \ldots, c_n)$ are independent (see Problem 10.29). This fact, and the definition of φ_c given in (10.67), allows us to write

$$
P[N_c = n_c \,|\, N = n] = \binom{n}{n_c} (1 - \varphi_c)^{n - n_c} \varphi_c^{n_c}, \qquad n_c = 0, 1, \ldots, n.
$$

We can use the expectation of a binomial, as given in (5.4), to write

$$
E[N_c \,|\, N = n] = n \varphi_c, \qquad c = 1, 2, \ldots, C. \qquad (10.70)
$$

There are two special cases of the above equations that are of particular interest: an infinite server queue with phase service times and a single server symmetric queue. In the infinite server case, (10.68) is given by (see Problem 10.31)

$$
P[\boldsymbol{N} = \boldsymbol{n}] = e^{-\rho} \prod_{c=1}^{C} \frac{\rho_c^{n_c}}{n_c!}, \qquad (10.71)
$$

and (10.69) simplifies to

$$
P[N = n] = e^{-\rho} \frac{\rho^n}{n!}, \qquad n = 0, 1, \ldots \; . \qquad (10.72)
$$

Using (10.70) implies that

$$
E[N_c] = \rho_c
$$

and, using Little's law, that

$$
E[T_c] = \frac{1}{\mu_c}, \qquad c = 1, 2, \ldots, C.
$$

For the single server case, $r(n) = 1$, and (10.68) is given by

$$P\left[\boldsymbol{N} = \boldsymbol{n}\right] = (1 - \rho) \binom{n}{n_1, n_2, \ldots, n_C} \prod_{c=1}^{C} \rho_c^{n_c}, \quad (10.73)$$

and (10.69) simplifies to

$$P\left[N = n\right] = (1 - \rho)\rho^n, \quad n = 0, 1, \ldots. \quad (10.74)$$

Using (10.70) this implies that

$$E\left[N_c\right] = \frac{\varphi_c \rho}{1 - \rho}, \quad c = 1, 2, \ldots, C,$$

and, using Little's law, that

$$E\left[T_c\right] = \frac{1}{\mu_c(1 - \rho)}, \quad c = 1, 2, \ldots, C. \quad (10.75)$$

There are several interesting properties of these equations. Equation (10.75) shows that for single sever symmetric queues, expected class response times are differentiated by their expected service times. Higher moments of the phase service time do not enter into the expected response time. The response time for such systems is said to be *insensitive* to higher moments. If the queue were empty then the expected response time of a class c customer is its expected service time, μ_c^{-1}. We can interpret the $(1 - \rho)^{-1}$ factor in (10.75) as scaling this service time to account for the presence of other customers sharing the server. This embodies a "fairness" principle since response time grows linearly with the expected service time required by a customer. It is not obvious that this principle would hold for all single server symmetric queues that include such diverse service policies as processor sharing and last-come-first-served preemptive.

It is also interesting to compare the results for the single server symmetric queue with the multiple-class M/M/1 quasi-reversible queue considered in Section 10.3.6. Comparison of (10.51) and (10.73) shows that the distribution of the number of customers in the queue is the same for both queues provided that the utilizations ρ_c of both queues is the same. Notice the similar form of the expected response time for both queues as given in (10.53) and (10.75). A principal conclusion drawn from comparing these equations is summarized in the following proposition:

Proposition 10.15 *Consider a multiple-class quasi-reversible M/M/1 queue as described in Section 10.3.6 where all classes have expected service times of μ^{-1}. Let the response time of class c customers, for*

$c = 1, 2, \ldots, C$, *be denoted by* $T_c^{m/m/1}$. *Consider an independent symmetric queue where class c customers have an expected service time of* μ_c^{-1}. *Denote its class c response time by* T_c^{sym}. *For each queue the arrival of class c customers forms a Poisson point process with rate* λ_c. *Set the service rate of the M/M/1 queue according to*

$$\mu = \frac{\sum_{c=1}^{C} \lambda_c/\mu_c}{\sum_{c=1}^{C} \lambda_c}.$$

Then the expected class c response time of the symmetric queue can be written in terms of the M/M/1 queue as follows:

$$E\left[T_c^{\text{sym}}\right] = \frac{\mu}{\mu_c} E\left[T_c^{m/m/1}\right].$$

We will use this proposition later when analyzing closed queue networks using the mean value analysis algorithm where it is not clear how to write the expected response time of a symmetric queue but where the form of the expected response time of an M/M/1 queue is obvious (see Section 10.5.2).

10.5 Partial Balance and Quasi-Reversible Queues

This section shows that quasi-reversible queues satisfy partial balance. Partial balance, like detailed balance for reversible queues, equates probability flux into and out of a set of states. Partial balance is more restrictive than global balance and less restrictive than detailed balance.

Consider the reversed balance equations (10.12). These equations show that

$$\pi(i) \sum_{k \in \mathcal{S}_c(i)} q^r(i,k) = \sum_{k \in \mathcal{S}_c(i)} \pi(k)q(k,i). \tag{10.76}$$

Using (10.44) and (10.45) in (10.76) shows that

$$\pi(i) \sum_{k \in \mathcal{S}_c(i)} q(i,k) = \sum_{k \in \mathcal{S}_c(i)} \pi(k)q(k,i)$$

and (this is obtained by subtracting this from global balance) that

$$\pi(i) \sum_{k \in \mathcal{S}_c{}^c(i)} q(i,k) = \sum_{k \in \mathcal{S}_c{}^c(i)} \pi(k)q(k,i).$$

Rewriting in terms of probability flux we have for all c, $1 \le c \le C$,

$$\mathbf{F}(i, \mathcal{S}_c(i)) - \mathbf{F}(\mathcal{S}_c(i), i) \quad\quad = \quad 0, \; i \in \mathcal{S}, \tag{10.77}$$

$$\mathbf{F}(i, \; \mathcal{S}-\mathcal{S}_c(i)) - \mathbf{F}(\mathcal{S}-\mathcal{S}_c(i), i) \quad = \quad 0, \; \textbf{Partial Balance Equations.}$$

The first of these equations implies that the probability flux due to arrivals of class c jobs from a state i equals the probability flux due to departures of class c jobs that result in state i. In contrast to detailed balance, which was necessary and sufficient for the reversibility of a process, partial balance is only a necessary condition for quasi-reversibility. There are processes that satisfy partial balance that are not quasi-reversible (e.g., processes that do not have Poisson arrival and departure processes).

10.5.1 Local Balance

To discuss other versions of partial balance, let $\mathcal{V}_c(i)$ be the set of states that have one less class c customer than state i and let $\mathcal{Y}(i)$ be the set of states that have the same number of class c customers as state i for $c = 1, 2, \ldots, C$. Transitions between state i and set $\mathcal{Y}(i)$ will be termed *internal transitions* since they can be viewed as transitions within a queue that do not change the number of its customers. Such transitions could correspond to completions of a service stage in a symmetric queue with phase service times.

We will also term *external transitions* those that correspond to external arrival or departure events. If all transitions cause class changes for some class, that is, if all transitions are external transitions, then set $\mathcal{Y}(i)$ is empty. State transitions from i are contained in the set $\{\mathcal{Y}(i), \mathcal{S}_c(i), \mathcal{V}_c(i)\}_{c=1}^C$. Global balance implies that

$$\mathbf{F}(i, \{\mathcal{Y}(i), \mathcal{S}_c(i), \mathcal{V}_c(i)\}_{c=1}^C) - \mathbf{F}(\{\mathcal{Y}(i), \mathcal{S}_c(i), \mathcal{V}_c(i)\}_{c=1}^C, i) = 0, \quad i \in \mathcal{S}. \tag{10.78}$$

Summing (10.77) over all c implies that partial balance holds for the set $\{\mathcal{S}_c(i)\}_{c=1}^C$ and that $\mathbf{F}(i, \{\mathcal{S}_c(i)\}_{c=1}^C) - \mathbf{F}(\{\mathcal{S}_c(i)\}_{c=1}^C, i) = 0$. Using this in (10.78) shows that for systems that satisfy partial balance, the following balance equation also holds:

$$\mathbf{F}(i, \{\mathcal{Y}(i), \mathcal{V}_c(i)\}_{c=1}^C) - \mathbf{F}(\{\mathcal{Y}(i), \mathcal{V}_c(i)\}_{c=1}^C, i) = 0, \quad i \in \mathcal{S}. \tag{10.79}$$

The partial balance equations (10.77), along with (10.79), are sometimes collectively termed *local balance* equations. Equation(10.79) implies that the probability flux due to internal transitions and departures of customers from state i equals the probability flux due to internal transitions and arrivals that result in state i.

10.5.2 General Definition of Partial Balance

The notion of customer classes is a useful paradigm within which to couch partial balance and when discussing quasi-reversible queues we find it convenient to express it in terms of the sets $\mathcal{S}_c(i)$ as in (10.77). More generally we say that partial balance holds on set $\mathcal{U}$ if

$$\mathbf{F}(i, \mathcal{U}) - \mathbf{F}(\mathcal{U}, i) \quad = \quad 0, \tag{10.80}$$

$$\mathbf{F}(i, \mathcal{U}^c) - \mathbf{F}(\mathcal{U}^c, i) \quad = \quad 0, \quad i \in \mathcal{U}.$$

(In this last equation $\mathcal{U}^c$ is the complement of set $\mathcal{U}$; the superscript "c" should not be confused as a class index.) Observe that global balance for state i results by summing these two equations. It is important to note that if partial balance holds over a particular set $\mathcal{U}$ then it does not necessarily hold over all possible sets.

10.5.3 State Truncation Property

Recall that we previously showed that the stationary solution of a reversible process truncated to a subset of irreducible states is simply a renormalization of the original stationary distribution. This result followed directly from detailed balance. We show that this result holds more generally if partial balance is satisfied for the truncated states. Let $\mathcal{U}$ be a set of states and consider the process Y that restricts X to $\mathcal{U}$ by setting

$$q_Y(\boldsymbol{y},\boldsymbol{y}') = \begin{cases} q(\boldsymbol{y},\boldsymbol{y}'), & \boldsymbol{y},\boldsymbol{y}' \in \mathcal{U}, \\ \\ 0, & \boldsymbol{y} \in \mathcal{U}^c \ \text{ or } \ \boldsymbol{y}' \in \mathcal{U}^c. \end{cases} \tag{10.81}$$

If partial balance holds over $\mathcal{U}$ and if Y is irreducible then

$$\pi(\boldsymbol{y}) \sum_{\boldsymbol{y}' \in \mathcal{U}} q(\boldsymbol{y},\boldsymbol{y}') = \sum_{\boldsymbol{y}' \in \mathcal{U}} \pi(\boldsymbol{y}')q(\boldsymbol{y}',\boldsymbol{y}) \qquad \boldsymbol{y} \in \mathcal{U}, \tag{10.82}$$

and it is easy to see that (10.82) is unchanged if $\pi(\boldsymbol{y})$ is replaced by

$$\pi_Y(\boldsymbol{y}) = b\pi(\boldsymbol{y}), \quad \boldsymbol{y} \in \mathcal{U}, \tag{10.83}$$

where $b = 1/\sum_{\boldsymbol{y} \in \mathcal{U}} \pi(\boldsymbol{y})$ is a normalizing constant. Since Y is restricted to $\mathcal{U}$, this shows that $\pi_Y(\boldsymbol{y})$ is its stationary distribution. Conversely, suppose that (10.83) is Y's stationary distribution and thus satisfies the global balance equations,

$$\pi_Y(\boldsymbol{y}) \sum_{\boldsymbol{y}' \in \mathcal{U}} q_Y(\boldsymbol{y},\boldsymbol{y}') = \sum_{\boldsymbol{y}' \in \mathcal{U}} \pi_Y(\boldsymbol{y}')q_Y(\boldsymbol{y}',\boldsymbol{y}), \quad \boldsymbol{y} \in \mathcal{U}. \tag{10.84}$$

Global balance for X can be written as

$$\pi(\boldsymbol{y})\{ \sum_{\boldsymbol{y}' \in \mathcal{U}} q(\boldsymbol{y},\boldsymbol{y}') \ + \sum_{\boldsymbol{y}' \in \mathcal{U}^c} q(\boldsymbol{y},\boldsymbol{y}')\}$$

$$= \sum_{\boldsymbol{y}' \in \mathcal{U}} \pi(\boldsymbol{y}')q(\boldsymbol{y}',\boldsymbol{y}) + \sum_{\boldsymbol{y}' \in \mathcal{U}^c} \pi(\boldsymbol{y}')q(\boldsymbol{y}',\boldsymbol{y}) \ \ \boldsymbol{y} \in \mathcal{U}.$$

Substituting (10.81 and 10.83) into (10.84) and subtracting it from this equation shows that partial balance is satisfied. Thus stationary probabilities are identical (up to a normalization) if a process is truncated to a set $\mathcal{U}$ *if and only if* partial balance holds over $\mathcal{U}$. We call this property the state truncation property of processes that satisfy partial balance. If the detailed balance equations are satisfied then partial balance is also satisfied and thus the state truncation property of partial balance is a generalization of the truncation property of reversible systems as discussed in Section 10.2.

10.5.4 Summary of Reversibility and Quasi-Reversibility

We close this section by reviewing the differences between reversible and quasi-reversible queueing systems in the context of a queueing system with C customer classes. Reversible systems satisfy detailed balance and have arrival and departure processes that are statistically identical. The arrival and departure processes of class c customers are generally state-dependent. A reversible process is quasi-reversible only if arrivals of customers to the queue are Poisson with state-independent rates. Quasi-reversible queues satisfy partial balance, a less restrictive condition than detailed balance, and always have Poisson arrival and departure processes. A queueing system can be reversible without being quasi-reversible (as in the birth–death queue with state-dependent arrival rates), and a quasi-reversible queue is not necessarily reversible since partial balance does not imply detailed balance. We summarize these statements in Table 10.1.

Property	Quasi-Reversible	Reversible
Arrival rates	Poisson arrivals	
	State-independent (stronger condition)	State-dependent (weaker condition)
Balance equations	Partial balance (weaker condition)	Detailed balance (stronger condition)

TABLE 10.1. Summary of Reversible and Quasi-Reversible Processes

10.6
Properties of Partial Balance and Product Form Networks

In this section we derive basic properties of quasi-reversibility and partial balance that lead to product form networks. We will derive properties that encapsulate an *algebra* for how quasi-reversible queues can be joined together so as to preserve quasi-reversibility and thus also have product form.

Clearly we cannot join queues in an arbitrary fashion and still preserve quasi-reversibility but the algebra for how such queues can be joined is surprisingly flexible. To derive elements of this algebra we will use a multiple-class network of quasi-reversible queues that can be thought of as an extension of the network of M/M/1 queues considered in Example 10.12. We show that if a network of quasi-reversible queues is joined together into a network that uses a type of random routing, termed *Markov routing*, then the stationary state probabilities are product form and the resultant network is quasi-reversible. This contrasts to the networks of reversible queues that do not generally remain reversible when joined together using Markov routing.

A word on notation: In this section we will denote the Markov process describing the queueing network as X and states of quasi-reversible queues by vectors $\boldsymbol{x}_j$. We will not be concerned with the nature of these states but note that their descriptors can be complex. Since a network of quasi-reversible queues is also a quasi-reversible queue (we will prove this statement in this section), the state of a queue can include such features as the number of customers in each queue, the class of these customers, the stages of service for each customer, and the relative positions in each of the queues. When deriving general properties of quasi-reversible queues and partial balance we will denote states by $\boldsymbol{y}$ and $\boldsymbol{y}'$. This avoids a possible confusion that might arise, leading one to think that the derived property holds only for the quasi-reversible network example.

To preserve a consistent notation in this section we will write the arrival rate of class c customers as $\lambda(c)$. Also to avoid a cluttered presentation we will dispense with putting bounds on variables in this section. Throughout this section, the variables j and k corresponds to queue indices and are contained in set $\{1, 2, \ldots, J, \}$. Similarly, the variables c and c' are class indices and are contained in $\{1, 2, \ldots, C\}$. Values of ℓ in set $\{1, 2, \ldots, L\}$ denote chains that are defined in Section 10.6.4.

10.6.1
Description of the Network

Consider a multiple-class network composed of J quasi-reversible queues. The state of the system is denoted by

$$\boldsymbol{x} \stackrel{\text{def}}{=} (\boldsymbol{x}_1, \boldsymbol{x}_2, \ldots, \boldsymbol{x}_J), \tag{10.85}$$

where $\boldsymbol{x}_j$ is the *state of queue j*. To avoid confusion when talking about states, we term $\boldsymbol{x}$ the *global state vector*. We let $\pi_j(\boldsymbol{x}_j)$ be the stationary distribution of queue j when analyzed in isolation with class-dependent Poisson arrival processes (the rates of these processes are specified later). We also let $q_j()$ denote the state transition rates for queue j. We assume that the reversed transition rates $q_j^r()$ are known and that the reversed balance equations are satisfied,

$$\pi_j(\boldsymbol{x}_j)q_j(\boldsymbol{x}_j, \boldsymbol{x}'_j) = \pi_j(\boldsymbol{x}'_j)q_j^r(\boldsymbol{x}'_j, \boldsymbol{x}_j). \tag{10.86}$$

We let $\pi(\boldsymbol{x})$ be the stationary distribution for the global state and let the forward and reversed transition rates be denoted by $q()$ and $q^r()$, respectively. We assume that there are C classes of customers.

Assume that the arrival rate of class c customers to the system is Poisson with rate $\lambda(c)$. The probability that an arriving class c customer is routed to queue j is denoted by $p_j^{\mathrm{arr}}(c)$. We assume that all arriving customers are assigned to a queue and thus that

$$\sum_{j=1}^{J} p_j^{\mathrm{arr}}(c) = 1.$$

10.6.2 Markov Routing Policy

Customers that are routed from one queue to another can change their class and we let $p_{j,k}^{\mathrm{rou}}(c, c')$ denote the probability that a class c customer that finishes service at center j is routed to queue k as a class c' customer. The probability that a class c customer departs from the network after service completion at queue j is given by

$$p_j^{\mathrm{dep}}(c) = 1 - \sum_{c'=1}^{C} \sum_{k=1}^{J} p_{j,k}^{\mathrm{rou}}(c, c'). \tag{10.87}$$

Since the network is assumed to be open, it follows that for every class there is at least one value of j so that $p_j^{\mathrm{dep}}(c) > 0$. Thus all customers eventually leave the system (we make the usual ergodic assumptions). The total arrival rate of class c customers to queue j is denoted by $\gamma_j(c)$. Similar to (10.38) we can write

$$\gamma_j(c) = \lambda(c)p_j^{\mathrm{arr}}(c) + \sum_{k=1}^{J} \sum_{c'=1}^{C} \gamma_k(c')p_{k,j}^{\mathrm{rou}}(c', c). \tag{10.88}$$

The values of $\gamma_j(c)$ are the *throughput of class c customers through queue j* and are uniquely determined by (10.88).

Since routing probabilities depend only on a customer's class and queue index, a rich set of possible routes can be specified and still be within the scope of our analysis. For example, we can define a definite path that a customer must take through the network. We might specify that a particular type of customer arrives to queue j_1 as a class c_1 customer. After finishing execution at j_1 the customer changes class to c_2 and is routed to j_2. This generates a *queue-class path*, which is generally described by specifying that, after finishing service at j_s as a class c_s customer, the customer is routed to queue j_{s+1} as class c_{s+1} customer. It is clear that we can assign unique class indexes and arrange routing probabilities to establish a desired queue-class path for this customer. Such paths can easily have random components. For example, a customer could cycle through a set of queues a random number of times before proceeding to the next queue in the queue-class path.

10.6.3
Types of
Transitions

State transitions $x \mapsto x'$ can be one of four possible types: arrival transitions, departure transitions, routing transitions and internal state transitions. If $x \mapsto x'$ corresponds to an arrival transition then one component of the global state vector x is changed and the remaining $J - 1$ components are unaltered. Such a transition can be written as

$$(x_1, \ldots, x_{j-1}, x_j, x_{j+1}, \ldots, x_J) \longmapsto (x_1, \ldots, x_{j-1}, x'_j, x_{j+1}, \ldots, x_J).$$
$$(10.89)$$

The set of all transitions corresponding to a class c customer arriving at queue j is denoted by $\mathcal{S}_j^{\mathrm{arr}}(c)$. We denote an element of this set by (x, x'), where

$$(x, x') \in \mathcal{S}_j^{\mathrm{arr}}(c)$$

implies that equation (10.89) holds and that state x'_j has one more class c customer than that of state x_j with the same number of class c' customers for $c \neq c'$.

In a similar fashion, departures of a customer from the system only change one component of the global state vector. If $x \mapsto x'$ corresponds to a departure of a class c customer from queue j then (10.89) holds and the state x'_j has one less class c customer than that of state x_j with the same number of class c' customers for $c \neq c'$. We denote the set of all transitions corresponding to a class c departure from queue j by $\mathcal{S}_j^{\mathrm{dep}}(c)$.

Internal transitions correspond to changes in only one component of the global state vector and hence also satisfy (10.89). These transitions correspond to a different state for service center j but where the number of class c customers in state x_j is the same as in state x'_j. We denote the set of all possible internal transitions for queue j by $\mathcal{S}_j^{\mathrm{int}}$.

Routing transitions change exactly two components of the global state vector. We can write such a transition by

$$(x_1, \ldots, x_j, \ldots x_k \ldots, x_J) \longmapsto (x_1, \ldots, x'_j, \ldots x'_k \ldots, x_J).$$

We let $\mathcal{S}^{\mathrm{rou}}_{j,k}(c, c')$ denote the set of all possible routing transitions corresponding to a class c customer completing service at queue j and being routed as a class c' customer to queue k. The network model just described increases the expressive capabilities of the previous computer networks depicted in Figures 10.4, 10.5 and 10.9.

Example 10.16 (A General Quasi-Reversible Network) We can generalize the M/M/1 queue network model of Figure 10.9 to include more general queues and allow for different classes of customers. An example of such a network is shown in Figure 10.10.

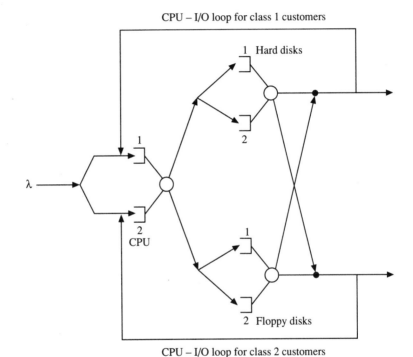

FIGURE 10.10. A General Quasi-Reversible Network

The network consists of a CPU and two I/O devices - a hard disk and a floppy disk. Assume that the CPU is a processor sharing queue and that the I/O devices are single server symmetric queues. There are two classes of customers, labeled "1" and "2." The two waiting rooms that are depicted for each queue in the network are used to visually distinguish between the different classes of customers. Each class can have different routing probabilities. For a given class, a customer enters the CPU and, after being executed, is served by one of the I/O devices before either leaving the system or returning to the CPU. The statistics for each of these *CPU - I/O loops* can be class-dependent.

In this section we show that the stationary distribution for networks of this type have a product form. This tractability allows one to easily calculate performance measures such as the stationary number of customers in the system, the expected number of customers of each class, or, using Little's result, the expected response time for customers of each class.

To continue the specification of the quasi-reversible network discussed above, we must now specify transition rates for the process. This is relatively straightforward for arrival, departure, and internal transitions since they affect the state of only one queue in the system. The specification for routed customers, however, is more delicate and leads us to an important property of quasi-reversible queues - *the distribution property*.

10.6.4 The Distribution Property

Suppose we observe an arrival of a class c customer to a quasi-reversible queue that is in state $\boldsymbol{y}$. We know that this arrival causes a state transition to some state $\boldsymbol{y}' \in \mathcal{S}_c(\boldsymbol{y})$, and here we wish to derive its distribution.

This probability is written as

$$P[X(t+dt) = \boldsymbol{y}' \mid X(t) = \boldsymbol{y}, \text{ Arrival of class } c \text{ in } dt]$$

$$= \frac{P[X(t+dt) = \boldsymbol{y}', X(t) = \boldsymbol{y}, \text{ Arrival of class } c \text{ in } dt]}{P[X(t) = \boldsymbol{y}, \text{ Arrival of class } c \text{ in } dt]}$$

$$= \frac{\pi(\boldsymbol{y})q(\boldsymbol{y}, \boldsymbol{y}')dt}{\pi(\boldsymbol{y})\lambda(c)dt} = \frac{q(\boldsymbol{y}, \boldsymbol{y}')}{\lambda(c)}. \tag{10.90}$$

We have used the fact that the arrival rate of class c customers is independent of the state of the system (10.44). We call this the distribution property of quasi-reversible queues. Observe that since $\lambda(c) = \sum_{\boldsymbol{y}' \in \mathcal{S}_c(\boldsymbol{y})} q(\boldsymbol{y}, \boldsymbol{y}')$ equation (10.90) embodies an "independence property" since the probability that an arrival "sees state" $\boldsymbol{y}'$ depends only on its frequency of occurrence. This explains why the stationary distribution of a network of quasi-reversible queues has a product form solution (we will prove this momentarily).

The distribution property holds only for quasi-reversible queues and does not generally hold for processes that satisfy partial balance. It is the mechanism that allows us to specify how state changes occur for routed customers in the quasi-reversible network example.

10.6.5
Open Product
Form Networks

We can now specify the transition rates of the quasi-reversible network. The transition rates for the global state vector can be written in terms of transition rates of the individual queues. Specifically, the global transition rates are given by

$$
q(\boldsymbol{x},\boldsymbol{x}') = \begin{cases}
q_j(\boldsymbol{x}_j,\boldsymbol{x}'_j), & (\boldsymbol{x},\boldsymbol{x}') \in \mathcal{S}_j^{\mathrm{int}}, \\[2ex]
q_j(\boldsymbol{x}_j,\boldsymbol{x}'_j)\left\{\frac{\lambda(c)p_j^{\mathrm{arr}}(c)}{\gamma_j(c)}\right\}, & (\boldsymbol{x},\boldsymbol{x}') \in \mathcal{S}_j^{\mathrm{arr}}(c), \\[2ex]
q_j(\boldsymbol{x}_j,\boldsymbol{x}'_j)p_j^{\mathrm{dep}}(c), & (\boldsymbol{x},\boldsymbol{x}') \in \mathcal{S}_j^{\mathrm{dep}}(c), \\[2ex]
q_k(\boldsymbol{x}_k,\boldsymbol{x}'_k)\left\{\frac{q_j(\boldsymbol{x}_j,\boldsymbol{x}'_j)p_{j,k}^{\mathrm{rou}}(c,c')}{\gamma_k(c')}\right\}, & (\boldsymbol{x},\boldsymbol{x}') \in \mathcal{S}_{j,k}^{\mathrm{rou}}(c,c').
\end{cases}
$$
$$(10.91)$$

Transition rates for internal and departure transitions are self-explanatory. Class c customers that enter queue j can be either external arrivals or customers that have been routed from another queue. The probability that any given arrival is an external arrival is given by $\lambda(c)p_j^{\mathrm{arr}}(c)/\gamma_j(c)$. This explains our definition of arrival transitions given in (10.91). Routing transitions are defined to satisfy the distribution property (10.90) and thus explains the form given in (10.91). Notice that defining routing transitions in this way mimics how a queue would behave in isolation with the modification that arrivals to the queue come from departures from another network queue rather than from an external Poisson source.

We are interested in the stationary distribution $\pi(\boldsymbol{x})$ and guess the only tractable solution; the distribution of the network is the product of the distributions of the individual queues analyzed in isolation,

$$\pi(\boldsymbol{x}) = \pi_1(\boldsymbol{x}_1)\pi_2(\boldsymbol{x}_2)\cdots\pi_J(\boldsymbol{x}_J), \qquad (10.92)$$

where $\pi_j(\boldsymbol{x}_j)$ is the stationary distribution of queue j analyzed as if it had Poisson arrival rates of class c customers with rates $\gamma_j(c)$. (As discussed in Example 10.12 the total arrival rate of class c customers to a given queue is not, generally, a Poisson process.) To establish the validity of this guess we use Proposition 10.2 and thus must determine the reversed transition rates.

10.6.6
Transition Rates
for the Reversed
Process

A possibility for the reversed process is that it also has class-dependent Poisson arrivals at rate $\lambda(c)$ and, as in Example 10.12, it follows reversed routing probabilities. The definition of these probabilities is similar to that of (10.39) for each class. Specifically, let $p_{k,j}^{\mathrm{r-rou}}(c',c)$ denote the probability that a class c' customer leaving queue k is routed to queue j as a class c customer in the reversed process. Then, following (10.39), we guess that these probabilities satisfy

$$\gamma_j(c)p_{j,k}^{\mathrm{rou}}(c,c') = \gamma_k(c')p_{k,j}^{\mathrm{r-rou}}(c',c)$$

and thus that

$$p_{k,j}^{\mathrm{r-rou}}(c',c) = \frac{\gamma_j(c)p_{j,k}^{\mathrm{rou}}(c,c')}{\gamma_k(c')}. \tag{10.93}$$

Let $p_j^{\mathrm{r-arr}}(c)$ be the probability that, in the reversed process, a class c customer arrives to queue j. An arrival of a class c customer to queue j in the reversed process corresponds to a departure of a class c customer from queue j in the forward process. The total departure rate of class c customers from service center j in the forward process is given by $\gamma_j(c)p_j^{\mathrm{dep}}(c)$. Setting this equal to the arrival rate of class c customers to queue j in the reversed process implies that

$$\lambda(c)p_j^{\mathrm{r-arr}}(c) = \gamma_j(c)p_j^{\mathrm{dep}}(c)$$

or

$$p_j^{\mathrm{r-arr}}(c) = \frac{\gamma_j(c)p_j^{\mathrm{dep}}(c)}{\lambda(c)}. \tag{10.94}$$

Similarly, let $p_j^{\mathrm{r-dep}}(c)$ be the probability that a class c customer departs the system from queue j in the reversed process. Using an argument similar to that of reversed arrivals leading to equation (10.94) implies that

$$p_j^{\mathrm{r-dep}}(c) = \frac{\lambda(c)p_j^{\mathrm{arr}}(c)}{\gamma_j(c)}. \tag{10.95}$$

The guessed reversed transition rates can now be specified by

$$q^r(\boldsymbol{x}',\boldsymbol{x}) = \begin{cases} q_j^r(\boldsymbol{x}'_j,\boldsymbol{x}_j), & (\boldsymbol{x},\boldsymbol{x}') \in \mathcal{S}_j^{\mathrm{int}}, \\[2ex] q_j^r(\boldsymbol{x}'_j,\boldsymbol{x}_j)p_j^{\mathrm{r-dep}}(c), & (\boldsymbol{x},\boldsymbol{x}') \in \mathcal{S}_j^{\mathrm{arr}}(c), \\[2ex] q_j^r(\boldsymbol{x}'_j,\boldsymbol{x}_j)\left\{\dfrac{\lambda(c)p_j^{\mathrm{r-arr}}(c)}{\gamma_j(c)}\right\}, & (\boldsymbol{x},\boldsymbol{x}') \in \mathcal{S}_j^{\mathrm{dep}}(c), \\[2ex] q_j^r(\boldsymbol{x}'_j,\boldsymbol{x}_j)\left\{\dfrac{q_k^r(\boldsymbol{x}'_k,\boldsymbol{x}_k)p_{k,j}^{\mathrm{r-rou}}(c',c)}{\gamma_j(c)}\right\}, & (\boldsymbol{x},\boldsymbol{x}') \in \mathcal{S}_{j,k}^{\mathrm{rou}}(c,c'). \end{cases} \tag{10.96}$$

To avoid a possible source of confusion in these equations, observe that if $(\boldsymbol{x},\boldsymbol{x}') \in \mathcal{S}_j^{\mathrm{dep}}(c)$ is a transition of the forward process then the transition for the reversed process satisfies $(\boldsymbol{x}',\boldsymbol{x}) \in \mathcal{S}_j^{\mathrm{arr}}(c)$. This explains why we have a "departure" term like $p_j^{\mathrm{r-dep}}(c)$ associated with an "arrival" transition set, $\mathcal{S}_j^{\mathrm{arr}}(c)$, in the second case of (10.96). Our purpose in writing equations (10.96) in the given sequence is to preserve the correspondence between forward and reversed transition rates for each type of transition. This facilitates application of Proposition 10.2 since to establish that (10.96) are the reversed transition rates we simply need to equate probability fluxes for the forward and reversed process by progressing line by line through equations (10.91) and (10.96). We now establish these equalities.

Internal Transitions. For internal transitions we must check that

$$\pi_j(\boldsymbol{x}_j)q_j(\boldsymbol{x}_j) = \pi_j(\boldsymbol{x}'_j)q_j^r(\boldsymbol{x}'_j, \boldsymbol{x}_j), \qquad (\boldsymbol{x}, \boldsymbol{x}') \in \mathcal{S}_j^{\mathrm{int}}.$$

This follows directly from the reversed balance equations (10.86).

Arrival Transitions. Arrival transitions, $(\boldsymbol{x}, \boldsymbol{x}') \in \mathcal{S}_j^{\mathrm{arr}}(c)$, must satisfy

$$\frac{\pi_j(\boldsymbol{x}_j)q_j(\boldsymbol{x}_j, \boldsymbol{x}'_j)\lambda(c)p_j^{\mathrm{arr}}(c)}{\gamma_j(c)} = \pi_j(\boldsymbol{x}'_j)q_j^r(\boldsymbol{x}'_j, \boldsymbol{x}_j)p_j^{\mathrm{r-dep}}(c).$$

This, however, follows from the reversed balance equations (10.86) and the definition of the reversed departure probabilities given in (10.95). External departures are just as easily checked and follow from (10.86) and (10.94).

Routing Transitions. Routing transitions, $(\boldsymbol{x}, \boldsymbol{x}') \in \mathcal{S}_{j,k}^{\mathrm{rou}}(c, c')$, must satisfy

$$\frac{\pi_j(\boldsymbol{x}_j)\pi_k(\boldsymbol{x}_k)q_j(\boldsymbol{x}_j, \boldsymbol{x}'_j)q_k(\boldsymbol{x}_k, \boldsymbol{x}'_k)p_{j,k}^{\mathrm{rou}}(c, c')}{\gamma_k(c')}$$

$$= \frac{\pi_j(\boldsymbol{x}'_j)\pi_k(\boldsymbol{x}'_k)q_j^r(\boldsymbol{x}'_j, \boldsymbol{x}_j)q_k^r(\boldsymbol{x}'_k, \boldsymbol{x}_k)p_{k,j}^{\mathrm{r-rou}}(c', c)}{\gamma_j(c)}.$$

This follows from the definition of the reversed routing probabilities given in (10.93) and the reversed balance equations for queue j and k as given in (10.86).

It is straightforward to show that the total transition rates for the forward and reversed processes satisfy (10.13) of Proposition 10.2. We have thus shown that the quasi-reversible network generated by the Markov routing policy delineated above has a product form solution! Again, it is remarkable that such a simple solution form exists for such a complex network of queues. Although it is possible to establish such a result algebraically, the intuition for why such a solution exists is found by viewing the process in reversed time. Proposition 10.2, which initially appears to be of little value since it requires guessing stationary probabilities *and* reversed transition rates, proves to be the critical mathematical tool that allows one to establish such results. Proposition 10.2 represents the essence of mathematics: a simple relationship that elegantly leads to complex results that embody profound beauty.

It is important to make the following observations about the product form solution given by (10.92). First, observe that the routing probabilities do not explicitly enter into the equations for the stationary distribution. All assignments of routing probabilities with the same throughput, $\gamma_j(c)$, lead to the same stationary distribution (see Problem 10.33). This is similar to the result obtained for symmetric queues where the stationary distribution does not depend on the distributions κ_n defining its operation (see comments after equation (10.56)).

Also observe that the only property that we require in defining the transition rates (10.91) is the distribution property. By the nature of the reversed transition rates (10.96), it is clear that the entire network is quasi-reversible and thus could be considered to be a "single queue." Such a queue could be joined with other quasi-reversible queues and, as long as the distribution property is satisfied, the resultant network would have product form. We thus have an "algebra" for forming quasi-reversible networks of queues that preserve product form solutions. It naturally follows that product form implies that the states of the queues of the network are independent random variables. It is not essential to know the nature of these states nor to specify the internal operation of each of the queues in the network. Product form is assured as long as each queue is quasi-reversible. We can conclude that quasi-reversibility is a sufficient condition for product form to hold (it is not a necessary condition).

Recall that in Section 10.3.1 we presented four examples of networks consisting of reversible queues. Three of these examples have product form solutions (the tandem queue of Example 10.9, the feed forward network of Example 10.10, and the network of M/M/1 queues of Example 10.12) and one does not (the tandem queue of Example 10.11). We are now in a position to provide the reason for these results - all of the queues with product form are both reversible and quasi-reversible.

The first queue of Example 10.11, which does not have product form, is a simple modification of an M/M/1 queue where the service rate depends on the number of customers in the queue. This makes the departure process prior to time t dependent on the state of the queue at time t and thus does not satisfy the conditions for quasi-reversibility. Since quasi-reversibility is only a sufficient condition, we cannot immediately conclude that this modified tandem does not have product form. A simple calculation, however, showed that this change results in a network with dependent queues.

10.6.7 Distribution of Number in Queue

The product form result of (10.92) just established uses a detailed description, $\boldsymbol{x}_j$, for the states of the queues of the network. As previously mentioned, such detail is often not required in a model, and all that is required is to specify stationary state probabilities for the number of customers of each class at each of the queues. It is clear that such a distribution can be readily calculated from the detailed probabilities. Let $\mathcal{N}_{c,j}(i)$ be the set of states of queue j that contain i class c customers, and let

$$\boldsymbol{N}_j \overset{\text{def}}{=} (N_{1,j}, N_{2,j}, \ldots, N_{C,j}) \tag{10.97}$$

be a random vector where $N_{c,j}$ is the number of class c customers at queue j. Let $\boldsymbol{n}_j \overset{\text{def}}{=} (n_{1,j}, n_{2,j}, \ldots, n_{C,j})$ denote a feasible state for queue j. The value $n_{c,j}$ denotes the number of class c customers at queue j. It is clear that the stationary probability of $\boldsymbol{n}_j$ is given by

$$P[\boldsymbol{N}_j = \boldsymbol{n}_j] = \sum_{\boldsymbol{x}_j \in \mathcal{N}_j} \pi_j(\boldsymbol{x}_j), \tag{10.98}$$

where

$$\mathcal{N}_j \overset{\text{def}}{=} \bigcap_{c=1}^{C} \mathcal{N}_{c,j}(n_{c,j}). \tag{10.99}$$

The stationary probability for the global state $\boldsymbol{n} \overset{\text{def}}{=} (\boldsymbol{n}_1, \boldsymbol{n}_2, \ldots, \boldsymbol{n}_J)$ is thus given by

$$P[\boldsymbol{N} = \boldsymbol{n}] = P[\boldsymbol{N}_1 = \boldsymbol{n}_1] P[\boldsymbol{N}_2 = \boldsymbol{n}_2] \cdots P[\boldsymbol{N}_J = \boldsymbol{n}_J],$$

where $\boldsymbol{N} \overset{\text{def}}{=} (\boldsymbol{N}_1, \boldsymbol{N}_2, \ldots, \boldsymbol{N}_J)$. It is interesting to note that, although the process associated with the detailed state $\boldsymbol{x}$ is Markovian this is not generally true for the aggregated state $\boldsymbol{n}$ (see Problem 10.34).

We can write (10.98) in a common form for the cases where queue j is a quasi-reversible multiple-class M/M/1 queue or a symmetric queue. For these cases (10.51) and (10.68) can be simplified to yield

$$P[\boldsymbol{N}_j = \boldsymbol{n}_j] = b_j \begin{pmatrix} n_j \\ n_{1,j}, n_{2,j}, \ldots, n_{C,j} \end{pmatrix} \frac{\rho_{1,j}^{n_{1,j}} \rho_{2,j}^{n_{2,j}} \cdots \rho_{C,j}^{n_{C,j}}}{r_j(1) r_j(2) \cdots r_j(n)}. \tag{10.100}$$

In this equation

$$\rho_{c,j} \overset{\text{def}}{=} \frac{\gamma_j(c)}{\mu_j(c)},$$

is the class c utilization at queue j where $\mu_j^{-1}(c)$ is the expected service time for class c customers at queue j. The value of b_j, the normalization constant of the queue, is given by

$$
b_j = \left\{ \sum_{n_j=0}^{\infty} \sum_{n_{1,j}+\cdots+n_{C,j}=n_j} \binom{n_j}{n_{1,j},\ldots,n_{C,j}} \frac{\rho_{1,j}^{n_{1,j}} \cdots \rho_{C,j}^{n_{C,j}}}{r_j(1) \cdots r_j(n_j)} \right\}^{-1}.
$$

For the special case of symmetric queues equations (10.71) and (10.73) show that (10.100) can be simplified to yield

$$
P\left[\boldsymbol{N}_j = \boldsymbol{n}_j\right] = \begin{cases} e^{-\rho_j} \prod_{c=1}^{C} \rho_{c,j}^{n_{c,j}} / n_{c,j}!, & \text{infinite server} \\[2em] (1-\rho_j) \binom{n_j}{n_{1,j},\ldots,n_{C,j}} \prod_{c=1}^{C} \rho_{c,j}^{n_{c,j}}, & \text{single server.} \end{cases}
$$

10.6.8 Distribution of Chains

We can provide a different aggregation of states by defining a *chain*. To precisely define a chain, write $c \overset{1}{\leadsto} c'$ if there exists a j and k so that $p_{j,k}^{\mathrm{rou}}(c,c') > 0$. We write $c \overset{n}{\leadsto} c'$ if there is a sequence of classes $d_1, d_2, \ldots, d_n$ so that $c = d_1 \overset{1}{\leadsto} d_2 \overset{1}{\leadsto} \cdots \overset{1}{\leadsto} d_n = c'$ and write $c \leadsto c'$ if there exists an n so that $c \overset{n}{\leadsto} c'$ (we allow an infinite sequence). We say that c' is *reachable* from c if $c \overset{n}{\leadsto} c'$ for some n and, by convention, assume that $c \leadsto c$.

A chain consists of a set of classes that are mutually reachable. That is, if two classes c and c' are in a particular chain then necessarily $c \leadsto c'$ and $c' \leadsto c$. Assume that there are L chains denoted by $\mathcal{K}_\ell$, for $\ell = 1, 2, \ldots, L$. Then it is clear that the chains are disjoint and that every class is contained in exactly one chain. In other words, chains form a partition of the set of classes (see Problem 10.35). A customer with a class contained in $\mathcal{K}_\ell$ is termed a *chain ℓ customer*. A chain $\mathcal{K}_\ell$ is said to be *closed* if none of the classes in the chain have external arrivals or departures, that is, if

$$
p_j^{\mathrm{arr}}(c) = p_j^{\mathrm{dep}}(c) = 0, \quad j = 1, 2, \ldots, J, \quad c \in \mathcal{K}_\ell.
$$

Otherwise, the chain has external arrivals and departures and is said to be *open*.

We can specify the stationary distribution of the number of chain ℓ customers in a similar way to that for classes. Let $m_{\ell,j}$ be the number of chain ℓ customers in service center j and define $\boldsymbol{m}_j \overset{\text{def}}{=} (m_{1,j}, \ldots, m_{L,j})$.

Let $\mathcal{M}_{\ell,j}(i)$ be the set of states for queue j that contain i chain ℓ customers and let

$$\boldsymbol{M}_j \stackrel{\text{def}}{=} (M_{1,j}, M_{2,j}, \ldots, M_{L,j}) \tag{10.101}$$

be a vector of random variables where $M_{\ell,j}$ is the number of chain ℓ customers at queue j. It is clear that the stationary probability of $\boldsymbol{m}_j$ is given by

$$P[\boldsymbol{M}_j = \boldsymbol{m}_j] = \sum_{\boldsymbol{x}_j \in \mathcal{M}_j} \pi_j(\boldsymbol{x}_j), \tag{10.102}$$

where

$$\mathcal{M}_j \stackrel{\text{def}}{=} \bigcap_{\ell=1}^{L} \mathcal{M}_{\ell,j}(m_{\ell,j}). \tag{10.103}$$

The stationary probability for state $\boldsymbol{m} \stackrel{\text{def}}{=} (\boldsymbol{m}_1, \boldsymbol{m}_2, \ldots, \boldsymbol{m}_J)$ is thus given by

$$P[\boldsymbol{M} = \boldsymbol{m}] = bP[\boldsymbol{M}_1 = \boldsymbol{m}_1]P[\boldsymbol{M}_2 = \boldsymbol{m}_2] \cdots P[\boldsymbol{M}_J = \boldsymbol{m}_J],$$

where $\boldsymbol{M} \stackrel{\text{def}}{=} (\boldsymbol{M}_1, \boldsymbol{M}_2, \ldots, \boldsymbol{M}_J)$.

We now investigate properties of closed quasi-reversible networks. The stationary solution for such networks is a normalization of the corresponding solution for open networks.

10.6.9 Closed Product Form Networks

Consider a closed queueing network of quasi-reversible queues as described earlier where the class c population is given by P_c and call

$$\boldsymbol{P} \stackrel{\text{def}}{=} (P_1, P_2, \ldots, P_C) \tag{10.104}$$

the *population vector*. It is important to note that P_c is not a random variable but is a constant that indicates the fixed population of class c customers in the network. Let $\mathcal{P}_c(P_c)$ be a set that contains all states with P_c class c customers, and let

$$\mathcal{P} \stackrel{\text{def}}{=} \cap_{c=1}^{C} \mathcal{P}_c(P_c) \tag{10.105}$$

be the set of states that have a population vector of $\boldsymbol{P}$. In the closed network the only transitions that are allowed are internal and routing transitions. In particular, we require that all customers that leave a queue are routed to another queue and thus that

$$\sum_{k=1}^{J} \sum_{c'=1}^{C} p_{j,k}^{\text{rou}}(c, c') = 1.$$

Equation (10.87) implies that $p_j^{\mathrm{dep}}(c) = 0$. The network does not allow external customers to arrive to the system, and thus we set $\lambda(c) = 0$, which implies that the flow equations (10.88) are given by

$$\gamma_{c,j}(\boldsymbol{P}) = \sum_{k=1}^{J} \sum_{c'=1}^{C} \gamma_{c',k}(\boldsymbol{P}) p_{k,j}^{\mathrm{rou}}(c',c). \qquad (10.106)$$

These equations do not have a unique solution and the normalization constant for the network depends on the particular solution selected.

The stationary distribution of the closed network is given by

$$\pi(\boldsymbol{x}) = b\pi_1(\boldsymbol{x}_1)\pi_2(\boldsymbol{x}_2)\cdots\pi_J(\boldsymbol{x}_J) \qquad (10.107)$$

where b is a normalization constant that is determined by

$$b^{-1} = \sum_{\boldsymbol{x} \in \mathcal{P}} \pi(\boldsymbol{x}).$$

We have essentially done all of the work necessary to derive (10.107). The forward and reversed process have rates given by (10.91) and (10.96), respectively, where only internal and routing transitions are non-zero. The solution (10.107) then follows from Proposition 10.2 where one needs to check that probability fluxes for the forward and reversed processes satisfy the proposition for internal and routing transitions. These were already checked in the open case, and the existence of a normalization constant does not alter the comparisons. As in the open case, it is easily shown that the total transition rates for the forward and reversed processes satisfy (10.13) of Proposition 10.2.

Closed networks are used to model computer systems where a fixed set of users shares a computing facility. One typically assumes that each customer in the system has a terminal through which they access the system. Since each customer has its own terminal, a convenient model is that the terminals are an infinite server symmetric queue. Customers are said to be "thinking" when they are receiving service at a terminal. A closed version of the quasi-reversible network model of Example 10.16 is depicted in Figure 10.11.

It is clear that we can obtain the stationary distributions of the number of customers of each class at each queue and similarly for the number of chain ℓ customers at each queue. Using the same notation as that defined in Section 10.6.4 we can write the stationary probability for state $\boldsymbol{n}$ as

$$P[\boldsymbol{N} = \boldsymbol{n}] = bP[\boldsymbol{N}_1 = \boldsymbol{n}_1]P[\boldsymbol{N}_2 = \boldsymbol{n}_2]\cdots P[\boldsymbol{N}_J = \boldsymbol{n}_J].$$

Again, the expressions for $P[\boldsymbol{N}_j = \boldsymbol{n}_j]$ are given by (10.51) for quasi-reversible M/M/1 queues and (10.68) for symmetric queues. Similarly

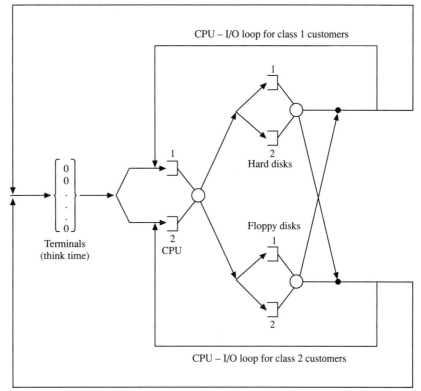

FIGURE 10.11. A Closed Network of Quasi-Reversible Queues

we can write the distribution for chains for a closed network as

$$P\left[\boldsymbol{M} = \boldsymbol{m}\right] = bP\left[\boldsymbol{M}_1 = \boldsymbol{m}_1\right] P\left[\boldsymbol{M}_2 = \boldsymbol{m}_2\right] \cdots P\left[\boldsymbol{M}_J = \boldsymbol{m}_J\right],$$
(10.108)

where $P\left[\boldsymbol{M}_j = \boldsymbol{m}_j\right]$ is given by (10.102).

A network is said to be *mixed* if it contains both open and closed chains. The stationary solution for chains in such networks is given by (10.108) but where the normalization constant b is determined by summing over all closed chains.

10.6.10 Calculating the Normalization Constant

Up to this point we have not specified a procedure for how one actually calculates normalization constants. We can provide such a procedure if we restrict our attention to quasi-reversible M/M/1 queues and single server and infinite server symmetric queues. We furthermore restrict our derivation to aggregated states corresponding to the distribution of the number in the queue. The common form for the stationary distributions for these queues analyzed in isolation is given by (10.100).

We first consider single-class networks. Let P be the total population of the network and let $b(P)$ be the normalization constant for this population. (For reasons that will become clear, we will explicitly write the normalization constant as a function of the population in this section.) We can write the stationary distribution as

$$P\left[\boldsymbol{N} = \boldsymbol{n}\right] = b(P) \prod_{j=1}^{J} \frac{\rho_j^{n_j}}{r_j(1)r_j(2)\cdots r_j(n_j)},$$

where $n_1 + n_2 + \cdots + n_J = P$ and

$$b^{-1}(P) = \sum_{n_1+n_2+\cdots+n_J=P} \prod_{j=1}^{J} \frac{\rho_j^{n_j}}{r_j(1)r_j(2)\cdots r_j(n_j)}. \qquad (10.109)$$

Assume that there are i infinite server queues and $J - i$ single server queues. We label the infinite server queues $1, 2, \ldots, i$ and the single server queues $i + 1, i + 2, \ldots, J$. For this case we can write (10.109) as

$$b^{-1}(P) = \sum_{n_1+n_2+\cdots+n_J=P} \prod_{k=1}^{i} \frac{\rho_k^{n_k}}{n_k!} \prod_{\ell=i+1}^{J} \rho_\ell^{n_\ell}. \qquad (10.110)$$

We use generating functions to obtain a procedure for calculating (10.110). For the jth queue, define the generating function

$$\mathcal{H}_j(z) \stackrel{\text{def}}{=} \sum_{n=0}^{\infty} \frac{\rho_j^n z^n}{r_j(1)r_j(2)\cdots r_j(n)}$$

which for infinite and single server queues can be written as

$$\mathcal{H}_j(z) = \begin{cases} e^{\rho_j z}, & j = 1, 2, \ldots, i, \\ 1/(1 - \rho_j z), & j = i+1, i+2, \ldots, J. \end{cases} \qquad (10.111)$$

Let $\mathcal{G}_j(z)$ be a generating function obtained from the first j queues, that is,

$$\mathcal{G}_j(z) \stackrel{\text{def}}{=} \mathcal{H}_1(z)\mathcal{H}_2(z)\cdots\mathcal{H}_j(z),$$

where the coefficients of $\mathcal{G}_j()$ are denoted by $g_j(i)$, that is,

$$\mathcal{G}_j(z) = g_j(0) + g_j(1)z + g_j(2)z^2 + \cdots .$$

The coefficient $g_j(n)$ of z_n in $\mathcal{G}_j(z)$, is the sum of the product of all factors corresponding to the first j queues where $n_1 + n_2 + \cdots + n_j = n$. More precisely,

$$g_j(n) = \begin{cases} \sum_{n_1 + \cdots + n_j = n} \prod_{k=1}^{j} \rho_k^{n_k}/n_k!, & j = 1, 2, \ldots, i, \\[2em] \sum_{n_1 + \cdots + n_j = n} \prod_{k=1}^{i} \rho_k^{n_k}/n_k! \prod_{\ell=i+1}^{j} \rho_\ell^{n_\ell}, & j = i+1, \ldots, J. \end{cases}$$

This implies that the desired normalization constant is given by

$$b(P) = \frac{1}{g_J(P)}.$$

Equation (10.111) implies that

$$\mathcal{G}_i(z) = e^{(\rho_1 + \rho_2 + \cdots + \rho_i)z}$$

and thus that

$$g_j(n) = \frac{(\rho_1 + \rho_2 + \cdots + \rho_j)^n}{n!}, \quad j = 1, 2, \ldots, i, \quad n \neq 0. \tag{10.112}$$

For $j = i+1, i+2, \ldots, J$, equation (10.111) implies that

$$\mathcal{G}_j(z) = \mathcal{G}_{j-1}(z)\frac{1}{1 - \rho_j z}, \quad j = i+1, i+2, \ldots, J,$$

and thus that

$$\mathcal{G}_j(z) = \mathcal{G}_{j-1}(z) + \rho_j z \mathcal{G}_j(z), \quad j = i+1, i+2, \ldots, J. \tag{10.113}$$

The "$z\mathcal{G}_j(z)$' term on the right-hand side of this equation corresponds to a shift operation (see Table 3.6). Equating coefficients of (10.113) thus yields

$$g_j(n) = g_{j-1}(n) + \rho_j g_j(n-1), \quad j = i+1, i+2, \ldots, J, \quad n \neq 0. \tag{10.114}$$

Equations (10.112) and (10.114) specify a computational procedure for computing the normalization constant $b(P)$. The recurrence can be started with an initial condition of $g_j(0) = 1$, $j = 1, 2, \ldots, J$, and is summarized by

$$g_j(n) = \begin{cases} 1, & j = 1, 2, \ldots, J, & n = 0, \\[1em] (\rho_1 + \rho_2 + \cdots + \rho_j)^n/n!, & j = 1, 2, \ldots, i, & n \neq 0, \\[1em] g_{j-1}(n) + \rho_j g_j(n-1), & j = i+1, i+2, \ldots, J, & n \neq 0. \end{cases} \tag{10.115}$$

We can easily generalize this procedure to include multiple classes (see Problem 10.36). The multiple-class recurrence corresponding to (10.115) is given by

$$
g_j(\boldsymbol{n}) = \begin{cases} 1, & j = 1, \ldots, J, & \boldsymbol{n} = \boldsymbol{0} \\[2mm] \prod_{c=1}^{C} (\rho_{c,1} + \cdots + \rho_{c,i})^{n_c}/n_c!, & j = 1, \ldots, i, & \boldsymbol{n} \neq \boldsymbol{0}, \\[2mm] g_{j-1}(\boldsymbol{n}) + \displaystyle\sum_{\{c \,|\, n_c \neq 0\}} \rho_{c,j} g_j(\boldsymbol{n} - \boldsymbol{e}_c), & j = i+1, \ldots, J, & \boldsymbol{n} \neq \boldsymbol{0}. \end{cases}
$$

$$(10.116)$$

In these equations $\boldsymbol{0}$ is a vector of all 0s. The normalization constant for a population of $\boldsymbol{P}$ is given by

$$
b(\boldsymbol{P}) = \frac{1}{g_J(\boldsymbol{P})}.
$$

We now proceed to the *the arrival–departure property* of partial balance.

10.6.11 The Arrival–Departure Property

The equation given in (10.80) permits an interesting probabilistic interpretation. Suppose one defines the point process Y_d obtained by observing X just before making a transition that leaves a set $\mathcal{U}$. Let $\psi_{Y_d}(\boldsymbol{y}), \boldsymbol{y} \in \mathcal{U}$, be the probability that the system is in state $\boldsymbol{y}$ just prior to the transition. Similarly let Y_a be the point process formed by observing X just after a transition into $\mathcal{U}$ and let $\psi_{Y_a}(\boldsymbol{y}), \boldsymbol{y} \in \mathcal{U}$, be its distribution. We will speak of the distribution of Y_d and Y_a as being the distribution of states as *seen by a transition out of and into* set $\mathcal{U}$, respectively. What is the relationship between these two distributions?

We can write the following equations for $\boldsymbol{y} \in \mathcal{U}$:

$$
\psi_{Y_d}(\boldsymbol{y}) = \frac{\pi(\boldsymbol{y}) \sum_{\boldsymbol{y}' \in \mathcal{U}^c} q(\boldsymbol{y}, \boldsymbol{y}')}{\sum_{\boldsymbol{y} \in \mathcal{U}} \pi(\boldsymbol{y}) \sum_{\boldsymbol{y}' \in \mathcal{U}^c} q(\boldsymbol{y}, \boldsymbol{y}')} = \frac{F(\boldsymbol{y}, \mathcal{U}^c)}{F(\mathcal{U}, \mathcal{U}^c)} \quad (10.117)
$$

and

$$
\psi_{Y_a}(\boldsymbol{y}) = \frac{\sum_{\boldsymbol{y}' \in \mathcal{U}^c} \pi(\boldsymbol{y}') q(\boldsymbol{y}', \boldsymbol{y})}{\sum_{\boldsymbol{y}' \in \mathcal{U}^c} \pi(\boldsymbol{y}') \sum_{\boldsymbol{y} \in \mathcal{U}} q(\boldsymbol{y}', \boldsymbol{y})} = \frac{F(\mathcal{U}^c, \boldsymbol{y})}{F(\mathcal{U}^c, \mathcal{U})}. \quad (10.118)
$$

Global balance (8.68) shows that the denominators of (10.117) and (10.118) are equal. Partial balance shows that the numerators are also equal, and thus $\psi_{Y_d}(\boldsymbol{y}) = \psi_{Y_a}(\boldsymbol{y}), \boldsymbol{y} \in \mathcal{U}$. It is easy to see, conversely, that if these two distributions are equal then partial balance must hold. Thus the distribution as seen by transitions out of set $\mathcal{U}$ is identical to that seen by transitions into set $\mathcal{U}$ if and only if partial balance holds on set $\mathcal{U}$.

We call this property the *partial balance arrival–departure property*. Note that, in general, the distributions seen by transitions out of and into a set $\mathcal{U}$ that satisfies partial balance are not equal to the stationary distribution of the process. As a counterexample, assume that only one state in $\mathcal{U}$, say $\boldsymbol{u} \in \mathcal{U}$, permits transitions out of $\mathcal{U}$. Then it must be the case that $\psi_{Y_d}(\boldsymbol{u}) = 1$, which is clearly not equal to $\pi(\boldsymbol{u})$.

Quasi-reversible queues inherit the partial balance arrival–departure property since they satisfy partial balance. They also satisfy the additional property that the distributions seen by transitions out of and into set $\mathcal{U}$ *are equal* to the stationary distribution. We show this by recasting the above argument in terms of customer classes, and we will talk of distributions seen by transitions out of (resp., into) a set $\mathcal{U}$ in terms of distributions seen by customer departures (resp., arrivals) that leave (resp., enter) set $\mathcal{U}$. Let $Z_a(\boldsymbol{y}), \boldsymbol{y} \in \mathcal{S}$ be the point process seen by an arriving class c customer which, similar to above, has a distribution given by

$$\psi_{Z_a}(\boldsymbol{y}) = \frac{\pi(\boldsymbol{y}) \sum_{\boldsymbol{y}' \in \mathcal{S}_c(\boldsymbol{y})} q(\boldsymbol{y}, \boldsymbol{y}')}{\sum_{\boldsymbol{y} \in \mathcal{S}} \pi(\boldsymbol{y}) \sum_{\boldsymbol{y}' \in \mathcal{S}_c(\boldsymbol{y})} q(\boldsymbol{y}, \boldsymbol{y}')}. \tag{10.119}$$

Equation (10.44) allows us to write

$$\lambda(c) = \sum_{\boldsymbol{y}' \in \mathcal{S}_c(\boldsymbol{y})} q(\boldsymbol{y}, \boldsymbol{y}') = \sum_{\boldsymbol{y} \in \mathcal{S}} \pi(\boldsymbol{y}) \sum_{\boldsymbol{y}' \in \mathcal{S}_c(\boldsymbol{y})} q(\boldsymbol{y}, \boldsymbol{y}')$$

and substituting this into (10.119) shows that $\psi_{Z_a}(\boldsymbol{y}) = \pi(\boldsymbol{y})$ as claimed. A similar argument shows that departing class c customers also see the system in equilibrium. Analogous to (10.119) we write

$$\begin{aligned}
\psi_{Z_d}(\boldsymbol{y}) &= \frac{\sum_{\boldsymbol{y}' \in \mathcal{S}_c(\boldsymbol{y})} \pi(\boldsymbol{y}') q(\boldsymbol{y}', \boldsymbol{y})}{\sum_{\boldsymbol{y}' \in \mathcal{S}_c(\boldsymbol{y})} \pi(\boldsymbol{y}') \sum_{\boldsymbol{y} \in \mathcal{S}} q(\boldsymbol{y}', \boldsymbol{y})} \\
&= \frac{\pi(\boldsymbol{y}) \sum_{\boldsymbol{y}' \in \mathcal{S}_c(\boldsymbol{y})} q^{r}(\boldsymbol{y}, \boldsymbol{y}')}{\sum_{\boldsymbol{y} \in \mathcal{S}} \pi(\boldsymbol{y}) \sum_{\boldsymbol{y}' \in \mathcal{S}_c(\boldsymbol{y})} q^{r}(\boldsymbol{y}, \boldsymbol{y}')},
\end{aligned}$$

which, using (10.45), shows that $\psi_{Z_d}(\boldsymbol{y}) = \pi(\boldsymbol{y})$ from the partial balance arrival–departure property. Thus for quasi-reversible queues, arrivals, and departures of class c queues see the system in equilibrium. We call this property, the *arrival–departure property of open quasi-reversible queues*. We will sometimes refer to this as the *arrival theorem* for open networks (see [75, 107] for theorems of this type).

10.6.12 Arrival–Departure Property for Closed Quasi-Reversible Queues

Consider the process Y obtained from X that corresponds to a quasi-reversible queue restricted to $\mathcal{P}(\boldsymbol{P})$, the set of states with population vector $\boldsymbol{P}$, and thus has stationary distribution equal to

$$\pi_Y(\boldsymbol{y}) = b\pi(\boldsymbol{y}), \quad \boldsymbol{y} \in \mathcal{P}(\boldsymbol{P}),$$

where $b^{-1} = \sum_{\boldsymbol{y}' \in \mathcal{P}(\boldsymbol{P})} \pi(\boldsymbol{y}')$. This corresponds to a closed system where the population vector is given by $\boldsymbol{P} = (P_1, P_2, \ldots, P_C)$. What are the distributions seen by a class c arrival or departure from this closed system? (When we talk about such a customer, we are assuming the customer is in transit between queues, that is, is not resident at any queue.)

Consider a departing class c customer and denote the distribution it sees at time of departure by $\psi_d(\boldsymbol{y})$. Note that there must be a class c customer for one to depart and that there is one less class c customer in the system after departure. Thus the states that can be seen by the departing customer are in set $\mathcal{P}(\boldsymbol{P}')$, which consists of all states that have a population vector of $\boldsymbol{P}' \stackrel{\text{def}}{=} (P_1, P_2, \ldots, P_c - 1, \ldots, P_C)$. We can write the distribution as

$$\psi_d(\boldsymbol{y}) = b' \sum_{\boldsymbol{y}' \in \mathcal{S}_c(\boldsymbol{y})} \pi(\boldsymbol{y}')q(\boldsymbol{y}', \boldsymbol{y}), \quad \boldsymbol{y} \in \mathcal{P}(\boldsymbol{P}')$$

$$= b'\pi(\boldsymbol{y}) \sum_{\boldsymbol{y}' \in \mathcal{S}_c(\boldsymbol{y})} q^r(\boldsymbol{y}, \boldsymbol{y}') = \pi(\boldsymbol{y})\lambda(c), \quad (10.120)$$

where we have used (10.45) and (10.76). This has to be normalized over $\mathcal{P}(\boldsymbol{P}')$ (the normalization constant b') and thus the class c departure sees the system in equilibrium with one less class c customer. Considering the process in reversed time shows that this is also true for an arriving class c customer and thus arrivals or departures of class c customers see the system in equilibrium with one less class c customer. We call this property the *arrival–departure property of closed quasi-reversible queues*.

An important application of this property is a procedure called mean value analysis which allows one to calculate expected values without calculating a normalization constant.

10.6.13 Mean Value Analysis

Consider a closed quasi-reversible queueing network with population vector $\boldsymbol{P}$. Suppose that queue j, for $j = 1, 2, \ldots, J$, is a quasi-reversible M/M/1 queue with exponential service times with expectation $1/\mu_j$ for all classes. Let $T_{c,j}(\boldsymbol{P})$, $L_j(\boldsymbol{P})$, and $\gamma_{c,j}(\boldsymbol{P})$ be the response time, queue length and throughput, respectively, for queue j when the population of the network is $\boldsymbol{P}$. We define

$$\theta_{c,j} \stackrel{\text{def}}{=} \frac{\gamma_{c,j}(\boldsymbol{P})}{\gamma_{c,1}(\boldsymbol{P})}$$

to be the frequency with which a class c customer visits queue j relative to the frequency with which it visits queue 1. The values of $\theta_{c,j}$ can be determined by finding any solution to the set of linear equations that arise from the routing matrix (10.106).

Consider a class c arrival to queue j. From the arrival–departure property of closed quasi-reversible queues it follows that the expected number of customers in queue j found by this arrival (while in transit and not in any queue) equals the expected number of customers in queue j with population vector $\boldsymbol{P} - \boldsymbol{e}_c$ where $\boldsymbol{e}_c$ is a vector of length C consisting of 0s except for a 1 in position c. Using the arrival–departure property of closed quasi-reversible queues allows us to write the expected response time for the newly arrived customer as

$$E\left[T_{c,j}(\boldsymbol{P})\right] = \frac{1}{\mu_j}\left(1 + E\left[L_j(\boldsymbol{P} - \boldsymbol{e}_c)\right]\right). \tag{10.121}$$

Applying Little's result [80] to the individual queues implies that

$$E\left[L_{c,j}(\boldsymbol{P})\right] = \theta_{c,j}\gamma_{c,1}(\boldsymbol{P})E\left[T_{c,j}(\boldsymbol{P})\right], \tag{10.122}$$

and summing (10.122) over all queues yields

$$\gamma_{c,1}(\boldsymbol{P}) = \frac{P_c}{\sum_{j=1}^{J}\theta_{c,j}E\left[T_{c,j}(\boldsymbol{P})\right]}. \tag{10.123}$$

Equations (10.121), (10.122), and (10.123) are a special case of the celebrated *Mean Value Analysis* (MVA) equations [94]. These equations, with the obvious boundary condition of $E\left[L_{c,j}(0)\right] = 0$, can be used to recursively calculate the expected response time and queue length for increasing values of $\boldsymbol{P}$ without needing to calculate a normalizing constant.

We can generalize the above argument to relax the assumption that queue j is an M/M/1 queue. Suppose that center j is a symmetric queue with phase service times having mean $\mu_{c,j}^{-1}$. The only equation that must be altered in the mean value analysis formulation is the expected response time for an arriving class c customer at queue j as given by (10.121). We will consider two possible extensions: infinite server queues and single server symmetric queues. (The analysis can be extended to symmetric queues with arbitrary rate functions but will not be considered here.)

If queue j has an infinite number of servers then the response time equation is given by

$$E\left[T_{c,j}(\boldsymbol{P})\right] = \frac{1}{\mu_{c,j}}.$$

Suppose that queue j is a single server symmetric queue. To calculate an equation for its response time we consider an independent multiple-class quasi-reversible queue as described in Section 10.3.6 where class c has Poisson arrivals with rate λ_c and all classes have expected service times given by μ_j^{-1}. Denote class c's response time by $T_{c,j}^{\mathrm{m/m/1}}$ and set μ_j so that the utilizations of the M/M/1 queue are the same as that of the symmetric queue,

$$
\mu_j = \frac{\sum_{c=1}^{C} \lambda_{c,j}/\mu_{c,j}}{\sum_{c=1}^{C} \lambda_{c,j}}.
$$

The expected class c response time for this queue is given by

$$
E\left[T_{c,j}^{\mathrm{m/m/1}}(\boldsymbol{P})\right] = \frac{1}{\mu_j}\left(1 + E\left[L_j(\boldsymbol{P} - \boldsymbol{e}_c)\right]\right). \qquad (10.124)
$$

Proposition 10.15 implies that we can write

$$
E\left[T_{c,j}(\boldsymbol{P})\right] = \frac{\mu_j}{\mu_{c,j}} E\left[T_{c,j}^{\mathrm{m/m/1}}(\boldsymbol{P})\right]
$$

$$
= \frac{1}{\mu_{c,j}}\left(1 + E\left[L_j(\boldsymbol{P} - \boldsymbol{e}_c)\right]\right), \qquad (10.125)
$$

where we have used (10.124) to obtain (10.125).

We can now recast (10.121) in a uniform manner:

$$
E\left[T_{c,j}(\boldsymbol{P})\right] = \begin{cases} \mu_{c,j}^{-1}, & \text{infinite server queues,} \\[2mm] \dfrac{1+E\left[L_j(\boldsymbol{P}-\boldsymbol{e}_c)\right]}{\mu_j}, & \text{quasi-reversible M/M/1 queues,} \\[2mm] \dfrac{1+E\left[L_j(\boldsymbol{P}-\boldsymbol{e}_c)\right]}{\mu_{c,j}}, & \text{single server symmetric queues.} \end{cases}
$$

The remaining equations (10.122) and (10.123) of the mean value analysis algorithm remain the same.

We next derive our final property of partial balance - *the state aggregation property.*

10.6.14 The State Aggregation Property

The last property of partial balance that we discuss is related to state truncation. Suppose we partition S into sets $\mathcal{U}^m, m = 0, 1, \ldots, M, M \geq 1$ and let X^m denote X truncated to $\mathcal{U}^m$. We assume that X^m is irreducible and let $\pi^m(\boldsymbol{y}), \boldsymbol{y} \in \mathcal{U}^m$, denote the stationary distribution of

X^m analyzed in isolation. Also assume that the sets satisfy nearest neighbor transitions, that is, $q(\boldsymbol{y}, \boldsymbol{y}') = 0, \boldsymbol{y} \in \mathcal{U}^m, \boldsymbol{y}' \in \mathcal{U}^{m'}$, if $|m - m'| > 1$. We call each set an *aggregated state* and define a birth–death process with transition rates, $Q(m, m')$ given by

$$Q(m, m') \stackrel{\text{def}}{=} \sum_{\boldsymbol{y} \in \mathcal{U}^m} \sum_{\boldsymbol{y}' \in \mathcal{U}^{m'}} \pi^m(\boldsymbol{y})q(\boldsymbol{y}, \boldsymbol{y}'), \quad m, m' \geq 0, \quad |m - m'| = 1.$$
(10.126)

Let $\Pi(m), m \geq 0$, be the stationary distribution of this birth–death process. These values satisfy the following detailed balance equations

$$\Pi(m)Q(m, m + 1) - \Pi(m + 1)Q(m + 1, m) = 0, \quad m = 0, 1, \ldots, M - 1,$$
(10.127)

and thus can be easily solved using $Q(m, m')$.

What relationship does the distribution of this aggregated birth–death process have to the original process? We claim that if partial balance holds on sets $\mathcal{U}^m$ then $\Pi(m) = \sum_{\boldsymbol{y} \in \mathcal{U}^m} \pi(\boldsymbol{y})$. To show this, observe that if partial balance is satisfied then the state truncation property implies that

$$\pi^m(\boldsymbol{y}) = \frac{\pi(\boldsymbol{y})}{\sum_{\boldsymbol{y} \in \mathcal{U}^m} \pi(\boldsymbol{y})}, \quad \boldsymbol{y} \in \mathcal{U}^m,$$

and thus substituting this into (10.126) shows that (10.127) is satisfied with $\Pi(m) = \sum_{\boldsymbol{y} \in \mathcal{U}^m} \pi(\boldsymbol{y})$. Note that if partial balance holds then $Q(m, m')$ equals the average transition rate from states in set $\mathcal{U}^m$ to states in set $\mathcal{U}^{m'}$ in the original process. Summarizing this result, we say that if partial balance holds on sets $\mathcal{U}^m$ then the distribution of the aggregated process is identical to what would be obtained in the original process by summing the stationary distribution of the aggregated states. We call this property the *state aggregation* property of processes that satisfy partial balance.

We can view this in terms of an open quasi-reversible network with customer classes. Partial balance among classes (10.77) implies that

$$\mathbf{F}(\mathcal{N}_c(m), \mathcal{N}_c(m + 1)) - \mathbf{F}(\mathcal{N}_c(m + 1), \mathcal{N}_c(m)) = 0, \quad m = 0, \ldots, M - 1.$$

Let states in the aggregated system correspond to the number of class c customers and let $\Pi_c(m) \stackrel{\text{def}}{=} \sum_{\boldsymbol{y} \in \mathcal{N}_c(m)} \pi(\boldsymbol{y})$ be the probability that there are m class c customers in the original system. Since we assume the process is quasi-reversible, the arrival rate of class c customers, $\lambda(c)$, is independent of the system state. Let $\mu_c(m)$ be the average departure rate of class c customers conditioned on m class c customers being in the queue. This is given by

$$\mu_c(m) = \sum_{\boldsymbol{y} \in \mathcal{N}_c(m)} \sum_{\boldsymbol{y}' \in \mathcal{N}_c(m-1)} \frac{\pi(\boldsymbol{y})}{\Pi_c(m)} q(\boldsymbol{y}, \boldsymbol{y}'), \quad m = 0, 1, \ldots, M.$$

Thus the detailed balance equations satisfied by the aggregated process (analogous to (10.127)) are

$$\Pi_c(m)\lambda(c) - \Pi_c(m+1)\mu_c(m+1) = 0, \qquad m \geq 0. \quad (10.128)$$

Notice that (10.26) implies that we can write the solution to (10.128) as

$$\Pi_c(m) = b \prod_{j=1}^{m} \frac{\lambda(c)}{\mu_c(j)},$$

where b is a normalization constant; thus, the arrival rate and departure rates of the aggregated process determine its stationary distribution. Thus if a property of a given quasi-reversible queue depends only on the distribution of its aggregated process, then the stationary statistics for that property are identical to that of a system where we replace the given queue by a birth–death queue that has the appropriate state-dependent service rates. This result has been termed Norton's theorem [13].

10.6.15
Flow Equivalent
Servers

The main application of state aggregation property is to create a *flow equivalent* server for a complex set of queues in a network. The flow equivalent server is equivalent to the birth–death queue with state-dependent service rates mentioned above. For example, suppose we consider the model of a computer system given in Example 10.16. This model consists of a CPU and I/O system and might form a component of a larger model of a computer system. For example, it might be one of several workstations in a local area network. If our main interest lies in analyzing the performance of the local area network, then we can create a flow equivalent server for the workstation model. This effectively hides the details of the workstation model. If parameter values for the workstation model do not change (e.g., routing probabilities and service times) then one can do a parametric study of the local area network using flow equivalent servers for the workstations without needing to repeatedly solve the workstation model.

10.7
Summary of
Chapter 10

This chapter derived the mathematics of product form networks within a general framework. The main result is that a network of quasi-reversible queues with Markov routing has a product form solution. Our approach depended on viewing Markov processes in reversed time, and the key result of the chapter is Proposition 10.2. The results of the chapter are summarized in Figure 10.12. In this figure, $A \to C$ means that A *implies* C. If $A \to C$ and $B \to C$, then if both A and B hold they imply that C holds.

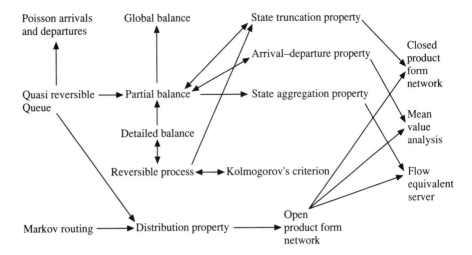

FIGURE 10.12. Summary of Implications

The concepts presented in this chapter are mathematically subtle and to avoid possible misconceptions we take this opportunity to briefly review the main results. We started with the time reversal of a Markov process. Reversing time does not change the stationary distribution of a Markov process but does, generally, alter the joint distribution of the process. A special case arises when the joint distribution is not changed under time reversal. These are *reversible* processes and they have special properties: most notably they have simple stationary solutions as given by (10.26) and transition rates that satisfy Kolmogorov's criterion (10.29). Queues that have corresponding reversible Markov processes have arrival and departure processes with the same joint distributions. This implies that a reversible queue with a Poisson arrival process has a Poisson departure process.

In this chapter we developed several models of queueing networks using reversible queues to motivate the definition of quasi-reversibility. It is important to recall that reversibility and quasi-reversibility are entirely different notions. We established that Poisson arrivals and departures from the queues in a network were not necessary or sufficient conditions for the network to have a product form solution. Knowledge of the arrival and departure process from a queue are not sufficient to establish independence of queues within a network. The key property that can be used to establish such independence is quasi-reversibility; the arrival process after a given time t and the departure process prior to t are independent of the state of the system at time t. As a consequence of this definition, we showed that quasi-reversible queues have Poisson arrival and departure processes.

We showed that quasi-reversible queues satisfy the distribution property, which implies that when such queues are joined in an open network using Markov routing, the stationary distribution has a product form. An important property of quasi-reversible queues is that they satisfy partial balance. For such queues this is defined in terms of classes of customers but partial balance is more generally defined. Three properties follow from a general definition of partial balance: the state truncation property, the arrival–departure property, and the state aggregation property. These properties, when applied to quasi-reversible networks queues, imply that product form solutions have special properties. In particular, the arrival–departure property implies that there is a simple recurrence relationship, called mean value analysis, that relates expected response times for closed networks as a function of the number of customers. The state aggregation property implies that one can create a simple representation of a complex quasi-reversible network, the flow equivalent server, which can facilitate parametric studies of computer models.

We review the properties that were used above to derive properties of queueing networks:

Distribution Property.   In an open quasi-reversible network the probability that an arriving class c customer causes a state transition to a given state equals the ratio of the state transition rate to the arrival rate of class c customers.

Arrival–Departure Property.   The distribution seen by transitions out of a set $\mathcal{U}$ is identical to that seen by transitions into set $\mathcal{U}$ if and only if partial balance on set $\mathcal{U}$ holds. If partial balance holds on set $\mathcal{U}$ and the queue is quasi-reversible, then: for open systems, arriving (resp., departing) class c customers that enter (resp., leave) set $\mathcal{U}$ see the stationary distribution, and for closed systems, arriving (resp., departing) class c customers that enter (resp., leave) set $\mathcal{U}$ see the stationary distribution, calculated as if they were not in the system.

State Aggregation Property.   The stationary distribution of an aggregated system $\mathcal{U}^m$, for $m = 0, 1, \ldots, M$, with constant arrival and departure rates is the same as that found by summing stationary probabilities of the aggregated states in the original process if partial balance holds on sets $\mathcal{U}^m$.

It is important to observe that we have now come full circle in the development of the mathematics required to analyze complex computer systems. A simple comparison of the computer network of the first chapter that was used to motivate performance modeling, Figure 1.4, with

Figure 10.11 shows that they are the same models. The mathematics required to analyze such networks can be directly traced back to the axioms of probability first stated in Chapter 2!

Bibliographic Notes

Kolmogorov [72] appears to have been the first to view Markov processes in reversed time. The first product form solution was obtained by Jackson [53]. The important paper by Muntz [83] established the relationship between queues that have Poisson departures when fed Poisson arrival streams and product form solutions. Whittle [123] first defined partial balance and Kelly [62] first defined quasi-reversibility. The important paper by Baskett, Chandy, Muntz, and Palacios [4] established a general class of networks with product form. This class is a subset of the networks considered in this chapter. The routing policies considered in this chapter can be extended and still preserve product form. See Towsley [117] for results along these lines.

Our exposition of product form networks is indebted to the treatments found in Kelly [61] and Walrand [121]. The development of the three properties of partial balance is based on work found in [85]. Other texts that contain material related to product form networks include Disney and Kiessler [28], Gelenbe and Pujolle [38], Kant [56], Robertazzi [97], Trivedi [118], and Wolff [125].

10.8

Problems for Chapter 10

Problems Related to Section 10.1

10.1 [M2] Show that if X is an irreducible, stationary and time homogeneous Markov process then so is its reversed process X^r.

10.2 [E2] Write expressions for the joint probabilities $P[X(0) = 0, X(t) = 1]$ and $P[X^r(0) = 0, X^r(t) = 1]$ of Example 10.1.

10.3 [M2] Show that β^r, as given by (10.21), is a probability distribution, that is, that its components are nonnegative and sum to unity.

10.4 [E2] In Example 10.3 we established parameter values for a reversed phase distribution. In particular, (10.19), (10.20) and (10.21) give forms for the phase distribution (β^r, U^r) which is the reversal of phase distribution (β, U). Show that the reversal of the phase distribution, (β^r, U^r), is the original phase distribution (β, U).

10.5 [R1] The analysis presented in Example 10.3 deriving the reversed phase distribution for random variable X assumed that $\beta_{n+1} = 0$ and thus $P[X > 0] = 1$. This implies that $\beta_{n+1}^r = 0$ and therefore that $P[X^r > 0] = 1$. How would you generalize the definition of a reversed phase distribution for the case where $\beta_{n+1} \neq 0$?

10.6 [E1] Show that (10.26) satisfies the global balance equations.

10.7 [M3] Show that Kolmogorov's criterion is a necessary and sufficient condition for reversibility.

10.8 [E1] Show that if $q_{i,j} > 0$ (resp., $q_{i,j} = 0$) for a reversible process then $q_{j,i} > 0$ (resp., $q_{j,i} = 0$).

Problems Related to Section 10.2

10.9 [E2] Show that the scaling of transition rates given in (10.30) does not change the stationary distribution of a reversible process.

10.10 [M2] Suppose that $X_i, i = 1, 2, \ldots, n$ are independent reversible processes. Show that the joint process $(X_1, X_2, \ldots, X_n)$ is also reversible.

10.11 [E2] Consider an M/M/1 queue that has the following modified state description. In the original process the transition $i \mapsto i+1$, for $i = 0, 1, \ldots$, occurs at rate λ and the transition $(i+1) \mapsto i$ occurs at rate μ. Consider splitting state 1 into two states $1a$ and $1b$ and modifying state transitions as follows (only non-zero transitions are given)

Transition	Rate
$0 \mapsto 1a$	λ
$1a \mapsto 0$	μ
$1a \mapsto 2$	λ
$1b \mapsto 0$	μ
$1b \mapsto 2$	λ
$2 \mapsto 1b$	μ

Is the modified process reversible?

10.12 **[R1]** Consider setting the service rates of a constant arrival birth–death process with Poisson arrival rate λ as follows:

$$
\mu_i = \begin{cases} \epsilon, & 1 \le i \le M, \\[2mm] \mu, & M < i, \end{cases}
$$

for a given value of $M, 0 \le M < \infty$, and $\epsilon > 0$. The system is stationary provided that $\lambda/\mu < 1$. For any value of M and $\delta, 0 < \delta < 1$, one can select ϵ so that the probability of having at least M customers in the queue is greater than δ. As an example, suppose one selects $M = 10^{134}$ and $\delta = 1 - 10^{-431}$. Thus the process has fewer than 10^{134} customers less than $10^{-431} \times 100$ percent of the time. This seems to imply that the departure process would frequently consist of a series of exponential interdeparture intervals at rate μ when the queue length is greater than M and exponential interdepartures at rate ϵ during the infrequent times that the queue length was below M. This conclusion is clearly false since we know that the departure process is Poisson with rate λ and has independent inter-departure intervals. How do you explain such a paradox?

10.13 **[M1]** Consider an irreducible Markov process with n states that has symmetric transition rates, that is, $q_{i,j} = q_{j,i}$. What is π_i/π_j?

10.14 **[R1]** Consider a Markov process X and its time reversal X^r. Suppose you know the state transition rates for both processes. Suppose you are allowed to view the sequence of state holding times and state transitions for a process Y that is either X or X^r. How could you determine which process Y corresponded to if X is not reversible?

10.15 **[R1]** In the text we show that a birth–death queue with constant birth rates has a Poisson departure process independent of the death rates as long as the process is ergodic. Suppose we consider the dual situation where the death rates are constant but there are possibly different birth rates. Since this is simply the "reverse" of the constant birth rate process it is plausible that it should also have a Poisson departure process. Does it?

10.16 **[M3]** Let $\boldsymbol{Q}$ be the rate transition matrix of a Markov process, X. Show that $\boldsymbol{Q}$ can be written as

$$
\boldsymbol{Q} = \boldsymbol{S}\boldsymbol{D},
$$

where $\boldsymbol{S}$ is a symmetric matrix and $\boldsymbol{D}$ is a diagonal matrix, if and only if X is reversible.

10.17 **[R2]** Let X be a Markov process defined on the nonnegative integers. Let $\mu(x)$ be a bounded positive function defined on nonnegative integers that has a minimum value of $\mu > \lambda$. Consider a birth–death single server queue with Poisson arrivals at rate λ and service rates given by $\mu(X)$. The assumptions on $\mu()$ imply that the process is ergodic. Is the departure process Poisson with rate λ?

10.18 **[M2]** Consider a stationary renewal process. Use time reversal arguments to show that $A =_{st} R$ where A and R are the age and residual life of a randomly selected point, respectively.

10.19 [R1] Prove or disprove the following statement: A stationary renewal process is time reversible.

Problems Related to Section 10.3

10.20 [R2] Consider a reversible queue $X(t)$ that has Poisson arrivals at rate λ. Answer the following questions (assume the general case):

1. Is there a method to determine the service rate μ of the queue by observing its sample paths?

2. Is $X(t)$ independent of the departure process after time t?

3. Is $X(t)$ independent of the arrival process prior to time t?

4. Is the departure process independent of the queue at time $t > 0$ given that $X(0) = 0$?

5. Does the departure process have independent increments given that there were n arrivals in $(0, t)$?

10.21 [R1] Recall that in Section 10.2 we showed that the vector process (X_1, X_2) is reversible if X_1 and X_2 are reversible and independent. Also recall that in Example 10.9 we showed that the stationary distribution of the tandem queue has a product form. This shows that the queue lengths N_1 and N_2 are independent random variables. In the same section we showed that the process for the tandem queue, (N_1, N_2), is not reversible. How do you explain this seeming contradiction?

10.22 [E1] Write out the balance equations for the tandem queue considered in Example 10.9 (the state transition diagram is depicted in Figure 10.6) and show that they satisfy the product form given in (10.34).

10.23 [E3] Comparison of Examples 10.9 and 10.6 show that a network of two M/M/1 queues and two independent M/M/1 queues have the same stationary distribution given by (10.34) and (10.31), respectively. Do the systems have the same joint distributions?

10.24 [M3] Consider an M/M/1 queue in which departing customers return to the tail of the queue with probability p and depart from the system with probability $1 - p$. Show that the total arrival rate into the queue is not a Poisson process.

10.25 [M3] For Example 10.11 show that the departure process of the first queue prior to time t is not independent of the state of the system at time t.

10.26 [M2] Consider the tandem queue of Example 10.11 with the modification that exogenous arrivals are state-dependent. Specifically, suppose that the arrival rate when the first queue is empty is given by λ_0 and equals λ otherwise. Show that the system does not satisfy product form.

10.27 [E3] Use the definition of quasi-reversibility given in (10.46) to show that the queues in the reversible queue network of Example 10.12 are quasi-reversible.

10.28 [**R3**] Determine conditions on the generator matrix of a Markov process that imply that the process is quasi-reversible.

Problems Related to Section 10.4

10.29 [**M3**] Show that the probability that a given position in a symmetric queue is class c is given by φ_c as given by (10.67). Show that if there are n customers in the queue then the classes of these customers $(c_1, c_2, \ldots, c_n)$ are independent.

10.30 [**E2**] Show that (10.68) follows from (10.56) by summing over the appropriate region.

10.31 [**E2**] In (10.71) we established the stationary distribution for an infinite server symmetric queue. The difficulty in establishing this result is to calculate the normalization constant b of (10.68). This involves showing that

$$e^\rho = \sum_{n=0}^\infty \rho^n \begin{pmatrix} n \\ n_1, n_2, \ldots, n_C \end{pmatrix} \frac{\varphi_1^{n_1} \varphi_2^{n_2} \cdots \varphi_C^{n_C}}{n!}.$$

Establish this result.

10.32 [**E1**] Show that the number of class c customers at an infinite server symmetric queue is Poisson with rate $\rho\varphi_c$ where φ_c is defined by (10.67).

Problems Related to Section 10.6

10.33 [**R1**] The product form solution given by (10.92) shows that the form of the stationary distribution does not explicitly depend on the routing probabilities of the network as long as they result in the same throughput values $\gamma_j(c)$. Does the transient solution of the network depend on the routing probabilities?

10.34 [**E2**] Provide an example of a product form queueing network of quasi-reversible queues where the detailed representation $\boldsymbol{x}$ is Markovian but the aggregated state representing the number of customers of each class at each queue $\boldsymbol{n}$ is not Markovian.

10.35 [**E2**] Show that chains form a partition of the set of classes.

10.36 [**E3**] Show that equation (10.116) is the general version of the convolution recurrence (10.115) for multiple-class networks.

11

Epilogue and Special Topics

The last ten chapters and the five appendices have encompassed a wide range of mathematics that is derivable from the axioms of probability. Lest the reader think that we have come to the end of the subject of probability, we allow ourselves in this final parting the indulgence of mentioning some extensions that naturally follow from the development in the text. Each of the three results that are derived in this epilogue can be considered to be the start of a field of mathematics; it is hoped that the brief descriptions that follow will whet the students' appetite to continue their study. It is more appropriate to view this closing chapter as a beginning rather than an end. This highlights the most intriguing aspect of mathematics - that it appears to be unending.

The three areas that are discussed in this section are selected to represent the wide diversity of problems that can be addressed by probability and include natural extensions of results from Chapters 4, 5, and 7. These extensions include stochastic comparisons, large deviation theory and bounds for the G/G/1 queue. Each section briefly derives a main result and provides an example of how the result is used in practice. We could have easily provided extensions for the other chapters but did not wish to tax the student's attention at this last juncture.

11.1 Stochastic Comparisons

Recall that in Chapter 4 we introduced random variables and noted that they were simply generalizations of algebraic variables. In Table 4.3 we noticed the differences between random variables and variables of classical algebra. Most notably, we showed that random variables can have different forms of equality: equality for all random experiments and equality in distribution. Regarding this last form of equality, recall that we write

$$X =_{st} Y$$

if random variables X and Y have the same distribution function. It is natural to extend this notion to relative comparisons. We say that X is *stochastically larger* than Y, written as

$$X \geq_{st} Y$$

if

$$P[X > z] \geq P[Y > z], \quad \text{for all } z. \tag{11.1}$$

Writing this in terms of distribution functions yields

$$F_X(z) \leq F_Y(z), \quad \text{for all } z.$$

The generalization to a strict comparison is straightforward and we write $X >_{st} Y$ if $P[X > z] > P[Y > z]$ for all z.

There is an alternative method of defining stochastic inequalities that can be readily generalized. It is not difficult to show that $X \geq_{st} Y$ if and only if

$$E[h(X)] \geq E[h(Y)], \tag{11.2}$$

for all increasing functions h. Choosing $h(x) = x$ thus implies that if $X \geq_{st} Y$ then $E[X] \geq E[Y]$. Furthermore, if X and Y are nonnegative random variables, then selecting $h(x) = x^n$ implies that if $X \geq_{st} Y$ then $E[X^n] \geq E[Y^n]$. This is one sense in which X is "larger" than Y, that is, all of its corresponding moments are larger.

There is a stronger sense in which X is larger than Y. If $X \geq_{st} Y$ then we can construct random variables $\hat{X}$ and $\hat{Y}$ on the same probability space that satisfy $\hat{X} =_{st} X$ and $\hat{Y} =_{st} Y$ and $\hat{X}(\omega) \geq \hat{Y}(\omega)$ for all random experiments ω. To construct these random variables, assume that the distribution functions F_X and F_Y are continuous. Since distribution functions are nondecreasing, this implies that they have inverses. If $X \geq_{st} Y$ then these inverses satisfy

$$F_X^{-1}(z) \geq F_Y^{-1}(z), \tag{11.3}$$

for all z. Let $\hat{X}$ and $\hat{Y}$ be random variables defined on the same probability space and define $\hat{X}$ so it has the same distribution as X, that is, $\hat{X} =_{st} X$. Define $\hat{Y}$ in terms of $\hat{X}$ as follows:

$$\hat{Y}(\omega) \stackrel{\text{def}}{=} F_Y^{-1}(F_X(\hat{X}(\omega))). \tag{11.4}$$

A simple calculation shows that Y and $\hat{Y}$ have the same distribution,

$$P\left[\hat{Y} \le z\right] = P\left[F_Y^{-1}\left(F_X(\hat{X})\right) \le z\right]$$

$$= P\left[F_X(\hat{X}) \le F_Y(z)\right]$$

$$= P\left[\hat{X} \le F_X^{-1}\left(F_Y(z)\right)\right]$$

$$= F_X\left(F_X^{-1}\left(F_Y(z)\right)\right) = F_Y(z).$$

It now follows from (11.3) that

$$\hat{X}(\omega) = F_X^{-1}\left(F_X(\hat{X}(\omega))\right) \ge F_Y^{-1}\left(F_X(\hat{X}(\omega))\right) = \hat{Y}(\omega)$$

and thus that random variable $\hat{X}$ is larger than random variable $\hat{Y}$ *for all random experiments* ω. This is a strong statement of what we mean by a random variable X being stochastically larger than Y.

Example 11.1 (Comparing Poisson Processes) Consider two Poisson processes with arrival rates λ_1 and λ_2, respectively, where we assume that $\lambda_1 < \lambda_2$. We assume that each process starts at time 0 and we denote the time of the nth arrival epoch by Υ_1^n and Υ_2^n, respectively. Since the first process has a "slower rate" one would expect that $E[\Upsilon_1^n] > E[\Upsilon_2^n]$. This can be easily established. Stochastic comparisons allow us to strengthen this comparison, however, and here we show that $\Upsilon_1^n >_{st} \Upsilon_2^n$.

Recall that the time of the nth arrival in a Poisson process stochastically equals the sum of n independent exponential random variables with the same parameter. Let X_n, $n = 1, 2, \ldots$, be a set of independent and identically distributed random variables with parameter λ_1 and let Y_n be a similarly defined set of exponential random variables but with parameter λ_2. It is clear that we can write

$$\Upsilon_1^n =_{st} X_1 + X_2 + \cdots + X_n, \tag{11.5}$$

and

$$\Upsilon_2^n =_{st} Y_1 + Y_2 + \cdots + Y_n. \tag{11.6}$$

A simple calculation shows that $X_k >_{st} Y_k$, $k = 1, 2, \ldots$. This follows directly from the definition (11.1),

$$P[X_k > z] = e^{-\lambda_1 z} > e^{-\lambda_2 z} = P[Y_k > z].$$

From (11.5) and (11.6), the desired result, $\Upsilon_1^n >_{st} \Upsilon_2^n$, follows if $X_k >_{st} Y_k$ for all $k = 1, 2, \ldots, n$ implies that

$$X_1 + X_2 + \cdots + X_n >_{st} Y_1 + Y_2 + \cdots + Y_n. \tag{11.7}$$

To establish (11.7) we use the construction defined in (11.4). Let $\hat{X}_k$ be a random variable defined on the same probability space as X_k and set $\hat{X}_k(\omega) = X_k(\omega)$ $k = 1, 2, \ldots$. Define $\hat{Y}_k$ using (11.4). It follows that $\hat{Y}_k =_{st} Y_k$ and thus that

$$\hat{Y}_1 + \hat{Y}_2 + \cdots + \hat{Y}_n =_{st} Y_1 + Y_2 + \cdots + Y_n. \tag{11.8}$$

By construction of $\hat{Y}_k$, it also follows that

$$X_k = \hat{X}_k > \hat{Y}_k, \quad k = 1, 2, \ldots. \tag{11.9}$$

Equation (11.7) thus follows from (11.8) and (11.9).

The definition in (11.2) invites a natural extension to define relative comparisons between random variables in terms of the expectation of different classes of functions. We say that X is *stochastically more variable* than Y, written $X \geq_{icx} Y$ if

$$E[h(X)] \geq E[h(Y)], \tag{11.10}$$

for all functions h that are increasing and convex. The comparison $\geq_{icx}$ is more refined than $>_{st}$ since if $X \geq_{st} Y$ then $X \geq_{icx} Y$. Other types of orderings between random variables have been defined and the interested reader is advised to consult references [101] and [110].

11.2 Large Deviation Theory

Recall that in Chapter 5 we derived Chernoff's bound from Markov's inequality. For a random variable X, Chernoff's bound implies that

$$P[X \geq t] \leq \inf_{\theta \geq 0} e^{-\theta t} \mathcal{M}_X(\theta), \tag{11.11}$$

where $\mathcal{M}_X(\theta)$ is the moment generating function of X. It is convenient if we write (11.11) in a slightly different form. Taking the log of both sides of this equation yields

$$\ln P[X \geq t] \leq \inf_{\theta \geq 0} (-\theta t + \ln \mathcal{M}_X(\theta))$$

$$= -\sup_{\theta \geq 0} (\theta t - \ln \mathcal{M}_X(\theta)). \tag{11.12}$$

Equation (11.12) motivates defining

$$I(t) \stackrel{\text{def}}{=} \sup_{\theta \geq 0}(\theta t - \ln \mathcal{M}_X(\theta)), \tag{11.13}$$

which is termed the *large deviation rate function*. It can be shown that $I(t)$ is convex and reaches its minimal value at $E[X]$ where $I(E[X]) = 0$. This function plays a key role in large deviation theory.

To derive a fundamental result of this theory, consider the statistical average of n independent random variables with the same distribution as X given by

$$S_n \stackrel{\text{def}}{=} \frac{X_1 + X_2 + \cdots + X_n}{n}. \tag{11.14}$$

The strong law of large numbers shows that

$$S_n \xrightarrow{a.s.} E[X], \quad \text{as } n \to \infty,$$

but provides no information about the rate of convergence. We are interested in the probability that S_n is larger than some value t, where $t \geq E[X]$. For large n, large deviation theory shows that

$$P[S_n \geq t] = e^{-nI(t)+o(n)}, \quad t \geq E[X]. \tag{11.15}$$

Equation (11.15) shows that deviations away from the mean decrease exponentially with n at a rate of $-I(t)$.

To derive (11.15), we first observe that

$$\mathcal{M}_{S_n}(\theta) = \mathcal{M}_X^n(\theta/n)$$

and thus Chernoff's bound, as given by (11.12), implies that

$$\ln P[S_n \geq t] \leq -\sup_{\theta \geq 0}(\theta t - n \ln \mathcal{M}_X(\theta/n))$$

$$= -n \sup_{\theta \geq 0}((\theta/n)t - \ln \mathcal{M}_X(\theta/n)). \tag{11.16}$$

In (11.16) we can replace the "dummy" variable θ with $n\theta$. Doing this, dividing by n, and using (11.13) implies that we can rewrite (11.16) in the form

$$\frac{1}{n}\ln P[S_n \geq t] \leq -I(t).$$

Since this holds for all n it must also hold for the limit supremum and thus we can write

$$\limsup_{n\to\infty} \frac{1}{n}\ln P[S_n \geq t] \leq -I(t). \tag{11.17}$$

So far all we have done in (11.17) is to rewrite Chernoff's bound for S_n in a different form. This result provides an upper bound on the value $(1/n) \ln P[S_n \geq t]$. Large deviation theory takes this a step further and shows that $-I(t)$ is also a lower bound on this value. The derivation of a lower bound requires that we define a new random variable. Suppose that $\theta^\star$ is the value obtained in the supremum of the definition of the rate function, that is,

$$I(t) = \theta^\star t - \mathcal{M}_X(\theta^\star). \tag{11.18}$$

(We are avoiding technicalities associated with proving that $\theta^\star$ exists and will suppress its dependency on variable t in our notation.) Define a new random variable Y with a density function given by

$$f_Y(z) = \frac{e^{\theta^\star z} f_X(z)}{\mathcal{M}_X(\theta^\star)}. \tag{11.19}$$

The density defined in (11.19) is said to be a *twisted distribution* or the *exponential change of measure*. A key feature of this distribution is that the random variable Y has an expectation given by

$$E[Y] = t.$$

To see this, observe that we can write the expected value of Y as

$$
\begin{aligned}
E[Y] &= \int_{-\infty}^{\infty} \frac{z e^{\theta^\star z} f_X(z) dz}{\mathcal{M}_X(\theta^\star)} \\
&= \frac{1}{\mathcal{M}_X(\theta^\star)} \frac{d}{d\theta} \int_{-\infty}^{\infty} e^{\theta z} f_X(z) dz \bigg|_{\theta=\theta^\star} \\
&= \frac{\mathcal{M}'_X(\theta^\star)}{\mathcal{M}_X(\theta^\star)}.
\end{aligned}
$$

We can rewrite this as

$$\frac{\mathcal{M}'_X(\theta^\star)}{\mathcal{M}_X(\theta^\star)} = \frac{d}{d\theta} \ln \mathcal{M}_X(\theta) \bigg|_{\theta=\theta^\star},$$

which, from (11.18), implies that

$$\frac{d}{d\theta} \ln \mathcal{M}_X(\theta) \bigg|_{\theta=\theta^\star} = t,$$

as claimed.

To obtain a lower bound on $(1/n) \ln P[S_n \geq t]$ we first note that we can write

$$P[S_n \geq t] = \int_{nt \leq z_1 + \cdots + z_n} f_X(z_1) \cdots f_X(z_n) dz_1 \cdots dz_n.$$

Rewriting this in terms of the density of Y using (11.19) yields

$$P[S_n \geq t] = \mathcal{M}_X^n(\theta^\star) \int_{nt \leq z_1 + \cdots + z_n} e^{-\theta^\star (z_1 + \cdots + z_n)} f_Y(z_1) \cdots f_Y(z_n) dz_1 \cdots dz_n.$$

(11.20)

Let ϵ be a positive constant that is used to restrict the range of the integral in (11.20) as follows:

$$P[S_n \geq t] \geq \mathcal{M}_X^n(\theta^\star) \int_{nt \leq z_1 \cdots + z_n \leq n(t+\epsilon)} e^{-\theta^\star (z_1 + \cdots + z_n)} f_Y(z_1) \cdots f_Y(z_n) dz_1 \cdots dz_n$$

$$\geq \mathcal{M}_X^n(\theta^\star) e^{-\theta^\star n(t+\epsilon)} \int_{nt \leq z_1 + \cdots + z_n \leq n(t+\epsilon)} f_Y(z_1) \cdots f_Y(z_n) dz_1 \cdots dz_n. \quad (11.21)$$

Since the expected value of Y is t, the strong law of large numbers implies that the integral in (11.21) converges to 1 as $n \to \infty$. This can be easily seen by rewriting the range as follows

$$\lim_{n \to \infty} \int_{t \leq \frac{z_1 + \cdots + z_n}{n} \leq t+\epsilon} f_Y(z_1) \cdots f_Y(z_n) dz_1 \cdots dz_n = 1.$$

Taking the log of both sides of (11.21) and dividing by n implies that

$$\liminf_{n \to \infty} \frac{1}{n} \ln P[S_n \geq t] \geq -I(t).$$

This, combined with (11.17), yields

$$\lim_{n \to \infty} \frac{1}{n} \ln P[S_n \geq t] = -I(t),$$

and thus it yields (11.15) for large values of n.

Example 11.2 (Large Deviation Bound for Exponential Random Variables) Let $X_i, i = 1, 2, \ldots$, be independent and identically distributed exponential random variables with expectation 1 and let S_n and $\theta^\star$ be defined as in (11.14) and (11.18). Recall that the moment generating function of X is given by

$$\mathcal{M}_X(\theta) = \int_0^\infty e^{\theta z} e^{-z} dz = \frac{1}{1 - \theta}. \quad (11.22)$$

The supremum in the definition of $I(t)$ in (11.13) can be determined using calculus and yields

$$0 = \frac{d}{d\theta}(\theta t - \ln \mathcal{M}_X(\theta)) = t - \frac{\mathcal{M}_X'(\theta)}{\mathcal{M}_X(\theta)}.$$

Substituting (11.22) into this and solving yields

$$\theta^\star = \frac{t - 1}{t},$$

and thus

$$I(t) = t - 1 - \ln t.$$

Substituting this into (11.15) yields

$$P[S_n \geq t] = t^n e^{-n(t-1)+o(n)}, \quad t \geq 1.$$

Readers interested in pursuing large deviations theory further are advised to consult references [10] and [108].

11.3 Bounds on the Response Time for a G/G/1 Queue

In Chapters 7 and 8 we derived response time equations for a variety of queueing systems. A common feature of these systems is that at least one of the arrival or service processes is Markovian, that is, they could be classified as an M/G/1 or a G/M/1 queue. There is a good reason why we did not address response times for G/G/1 queues: exact results require advanced techniques and yield tractable expressions only for special cases. In this section we derive a simple upper bound on the response time of a G/G/1 queue that can be easily used in performance models.

Recall that in Example 4.7 we derived an equation for the waiting time of a G/G/1 queue (4.10) given by

$$W_n = (W_{n-1} + S_{n-1} - A_n)^+,$$

where W_n is the waiting time, S_n is the service time of the n customer, and A_n is the time between the nth and $n - 1$st arrival to the system, for $n = 1, 2, \ldots$. We can write this equation as

$$W_{n+1} = (W_n + U_n)^+, \quad n = 1, 2, \ldots, \tag{11.23}$$

where we set $W_0 = 0$ and define

$$U_n \overset{\text{def}}{=} S_{n-1} - A_n. \tag{11.24}$$

Let W and U be limiting random variables corresponding to the sequences W_n and U_n, respectively. We claim that an upper bound on the tail probabilities is given by

$$P[W \geq t] \leq e^{-\theta^\star t}, \tag{11.25}$$

where

$$\theta^\star = \sup_{\theta \geq 0}\{\mathcal{M}_U(\theta) \leq 1\}. \tag{11.26}$$

We can establish this result by induction.

Assume that $P[W_i \geq t] \leq e^{-\theta^\star t}$ for $i = 1, 2, \ldots, n$. We will show that this implies that $P[W_{n+1} \geq t] \leq e^{-\theta^\star t}$. For $t \geq 0$ it is clear that

$$P\left[(W_n + U_n)^+ \geq t\right] = P[W_n + U_n \geq t],$$

and thus (11.23) implies that

$$P[W_{n+1} \geq t] = P[W_n + U_n \geq t] = \int_{-\infty}^{\infty} P[W_n \geq t - z] f_U(z)dz.$$

Applying the induction hypothesis and the fact that $\mathcal{M}_U(\theta^\star) \leq 1$ yields

$$P[W_{n+1} \geq t] \leq \int_{-\infty}^{\infty} e^{-\theta^\star(t-z)} f_U(z)dz$$

$$= e^{-\theta^\star t} \mathcal{M}_U(\theta^\star) \leq e^{-\theta^\star t}, \tag{11.27}$$

and thus the induction carries through to $n + 1$. This establishes the bound given in (11.25). Note that the definition of $\theta^\star$ given in (11.26) assures that the induction step of (11.27) can proceed.

Example 11.3 (Bounding the Tail of an $E_2/E_2/1$ Queue) Consider a G/G/1 queue with two-stage Erlang arrivals at rate 2λ and with two-stage Erlang service times at rate 2μ, where $\mu > \lambda$. We can write the generic random variable U defined by (11.24) as

$$U = S + (-A).$$

Make the observation that

$$\mathcal{M}_{-A}(\theta) = E\left[e^{\theta(-A)}\right] = E\left[e^{-\theta A}\right] = \mathcal{L}_A(\theta).$$

We can thus write the moment generating function of U as

$$\mathcal{M}_U(\theta) = \mathcal{M}_S(\theta)\mathcal{M}_{-A}(\theta) = \mathcal{M}_S(\theta)\mathcal{L}_A(\theta).$$

For the Erlang arrival and service processes described earlier, we have

$$\mathcal{M}_U(\theta) = \left(\frac{2\mu}{2\mu - \theta}\right)^2 \left(\frac{2\lambda}{2\lambda + \theta}\right)^2$$

$$= \left(\frac{4\lambda\mu}{4\lambda\mu + 2\theta(\mu - \lambda) - \theta^2}\right)^2. \tag{11.28}$$

To solve for $\theta^\star$ we set (11.28) equal to 1,

$$4\lambda\mu = 4\lambda\mu + 2\theta^\star(\mu - \lambda) - \theta^{\star 2},$$

which yields

$$\theta^\star = 2(\mu - \lambda).$$

We thus have the following bound on the waiting time

$$P[W \geq t] \leq e^{-2(\mu-\lambda)t}.$$

It is interesting to compare this result to what we would obtain from an M/M/1 queue. To make this comparison, observe that the expected interarrival and service time for the $E_2/E_2/1$ queue are given by λ^{-1} and μ^{-1}, respectively. For an M/M/1 queue with these expected interarrival and service times, we have (see (7.74))

$$P[W \geq t] = e^{-(\mu-\lambda)t}.$$

Thus the exponential decay of the tail of the waiting time distribution is slower than for an $E_2/E_2/1$ queue. This follows from the fact that Erlang 2 random variables are "less variable" than exponential random variables with the same expectation.

A

Types of Randomness

In this appendix we address the following questions: What do we mean by saying that an event is random? Are there different types of randomness? To answer these questions, we describe how randomness is manifest in physical and mathematical systems governed by deterministic rules. A simple case of such randomness is a toss of a fair coin. Experience tells us that the outcome of the toss is "random" and that there is an equal probability of the coin landing heads or tails up. Experience also tells us that if we could precisely control the experiment, then using simple laws of physics and the initial conditions of the velocity and angular velocity of the toss, we could predict the outcome with certitude. Thus we seem to be faced with the contradictory fact that something can be simultaneously random and deterministic.

To analyze coin tossing, in Section A.1 we define a phase space model of the experiment and derive some startling conclusions concerning invariant outcomes of its perturbations. Specifically, we show that the outcomes of a deterministic coin toss are random. This suggests a probabilistic model, called *intrinsic probability*, that is based on properties of the phase space of physical experiments. Such a model, although intuitively appealing, has operational handicaps in that desirable properties of probability are not satisfied within this framework. In particular, one cannot combine events with intrinsic probability and guarantee that the combined event also has intrinsic probability (see Problem A.4).

The development of intrinsic probability shows how a deterministic experiment can result in random behavior. We investigate this phenomena further in Section A.2 through the mechanism of an abstract dynamical system. In one such system we create an equivalence between the evolution of a dynamical system with that of the outcomes of a coin tossing experiment. Since coin tossing experiments are random, and dynamical systems are deterministic this suggests the surprising result that both systems can be viewed within the same framework. The dynamical system we consider does not have a physical analog because it is simply a defined mathematical object. This brings up the question of whether such behavior can be found in systems that obey physical laws. In answering this question, we form a bridge between the physical approach and the dynamical system approach and complement our understanding of randomness with results from isomorphism theory which establishes the equivalence of a dynamical system governed by the physics of Newton's laws with that of a random process based on coin tossing. The example we discuss reveals a different type of randomness from that previously considered.

A.1 Randomness: Physical Systems

What do we mean by saying that an event is random? To answer this consider the following experiment:

Example A.1 (Coin Tossing) Assume we have a symmetric coin of negligible width whose two sides can be distinguished as "heads" and "tails." The coin is oriented heads up and is tossed by a reliable mechanical machine and always lands with one side up. What is the chance that the coin lands heads up?

It seems fairly clear that if one were given sufficient information about how the machine tosses the coin, that is, the force applied to the coin and the angular momentum, then one could calculate the final outcome using the laws of physics. Intuitively, each toss would yield the same outcome. Kechen [59] and Keller [60] have derived equations that determine the final outcome for two different, simplified, versions of coin tossing experiments. The exact model of coin tossing developed in these references is not the issue here and will not be discussed. We will, however, discuss how each model establishes conditions under which the outcome of the toss is "random". These conditions imply, since the coin is symmetrical and thus unbiased, that the answer to the question posed in Example A.1 is that the coin comes up heads one half of the time even though the coin is tossed each time under identical initial conditions!

To explain how such randomness arises from a deterministic tossing machine, we need to consider the phase space of the experiment. The phase space for the model considered by Keller [60] is shown in Figure A.1. The x (horizontal) and y (vertical) axes correspond to the vertical and angular velocities, respectively, that were used to toss the coin. The equation

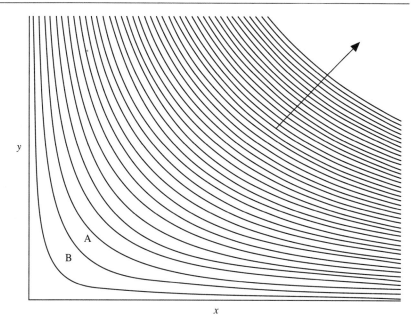

FIGURE A.1. Phase Space for Coin Toss

that determines when the outcome is heads or tails is given by

$$y = \frac{c}{x}(4n \pm 1), \quad n = 1, 2, \ldots, \tag{A.1}$$

where c is a constant and n indexes a region of phase space. Thus the nth region that results in heads is the portion of phase space between the two curves $xy = c(4n-1)$ and $xy = c(4n+1)$ (the nth region that results in tails is bounded by $xy = c(4n + 1)$ and $xy = c(4(n + 1) - 1)$). The regions corresponding to the outcomes that heads and tails alternate in the figure. Thus the band of parameter values that contains point A in the figure has heads landing up and the region that contains B results in tails. A key observation from the figure is that the width of the areas thus delineated decreases as one moves up in the figure. (We will define the "width" of a region as its greatest distance along a 45° line). In terms of the parameters, this corresponds to tossing the coin higher and/or with more rotational speed. It is easy to see that the width of the intervals decreases to 0 in the limit as we move up on the curves, that is, in the direction of the arrow in Figure A.1.

The conditions for each coin toss are assumed to be identical up to some limit of accuracy. Thermal fluctuations, for example, might slightly change the force applied by the machine so that a particular toss differs infinitesimally from previous tosses. For any size of perturbation, even one that changes the force in the trillionth decimal place, there exist parameter values for which the width of the curves are much smaller than the perturbation. If these perturbations are not known in advance, for example, if they cannot be measured, and if their size is large with respect to the width of the alternating bands in phase space, then the outcome of the toss is not predictable. We next show that under these conditions the coin will come up heads on half the tosses.

A.1.1 Intrinsic Probability

Consider that we have a square with sides α that we will call an "α-window." Imagine passing an α-window over different portions of the phase space of the coin toss. For each coordinate (x, y), let $H^\alpha_{(x,y)}$ be the fraction of phase space viewed through the α-window centered on point (x, y) that results in heads. It is clear that $0 \leq H^\alpha_{(x,y)} \leq 1$ and that for points A and B we have $H^\alpha_A = 1$ and $H^\alpha_B = 0$ provided that α is small with respect to the width of the regions containing these points. We are interested in what happens as we move the α-window in the direction of higher parameter values. Clearly, as the values of (x, y) increase, the α-window views an increasing number of alternating bands of decreasing width. It is not surprising then that, for any fixed α, one can show that (see Problem A.1)

$$\lim_{(x,y) \to \infty} H^\alpha_{(x,y)} = \frac{1}{2} \tag{A.2}$$

and thus the fractions of phase space under the α-window for heads and tails are equal. When the condition implied by (A.2) holds for all (x, y) in a portion of phase space we say that portion is *uniformly* distributed.

Now suppose the parameter values for the coin toss are given by $P = (x_0, y_0)$ and assume that perturbations are less than size Δ. We can thus imagine that we have a square of size Δ that is centered in phase space at point P. This implies that the possible range of parameter values for the coin toss are in this square, that is, the ranges $x_0 - \Delta/2 \leq x \leq x_0 + \Delta/2$, and $y_0 - \Delta/2 \leq y \leq y_0 + \Delta/2$. Assume that the width of the regions around the point P are small with respect to α and furthermore that $\alpha \ll \Delta$. We can now imagine the Δ square being covered (with no overlap and no uncovered space) by α-windows of size α and, letting $n = \lfloor \Delta/\alpha \rfloor$, this implies that there are approximately n^2 such α-windows. Suppose that n is large and for the ith α-window, for $i = 1, 2, \ldots, n^2$, let f_i be the fraction of perturbations that result in actual parameter values found in α-window i and let H^α_i be the fraction of phase space within the α-window that lands heads up. Then the chance that the outcome will be heads, denoted by $P(x, y)$, is approximated by

$$P(x, y) \approx \sum_{i=1}^{n^2} f_i H_i^{\alpha}$$

$$\approx \frac{1}{2} \sum_{i=1}^{n^2} f_i = \frac{1}{2}.$$

Thus, independently of the form of f_i, the chances of having heads land up equals $1/2$. (Some technical assumptions are needed here to avoid cases where the distribution of perturbations is sharply peaked and thus only a "few" of the n^2 α-windows have positive frequency.) This is a surprising result that explains how a seemingly deterministic system can produce random results.

We can use the preceding arguments to define a system of probability based on physical systems. To do this we introduce the notion of *intrinsic probability* [59]. We call the outcome of an experiment an *event* and a set of events corresponds to a set of possible outcomes. In the coin tossing experiment there were two events depending on whether the coin lands heads or tails up.

Definition A.2 *A portion of phase space U is uniformly distributed for events $E_i, 1 \leq i \leq I$, if, for small α, the fraction of phase space viewed by an α-window corresponding to event E_i is invariant of the location of the α-window in U. Furthermore, we call this fraction the intrinsic probability of event E_i.*

For the coin tossing experiment above, the phase space is uniformly distributed for large parameter values and the intrinsic probability of the event heads equals $1/2$. One property that holds for intrinsic probabilities is that long term averages converge to intrinsic probabilities. Specifically, suppose we repeat a physical experiment n times and let the outcome of the experiment be given by $X_j, 1 \leq j \leq n$. Let $1\{X_j \in E_i\}$ be an indicator that outcome X_j is in event E_i (i.e., $1\{X_j \in E_i\} = 1$ if outcome X_j is in event E_i and equal to 0 otherwise). Then it is the case, if event E_i has intrinsic probability p_i, that

$$\lim_{n \to \infty} \frac{1}{n} \sum_{j=1}^{n} 1\{X_j \in E_i\} = p_i, \quad 1 \leq i \leq I.$$

It is clear that many other physical systems that produce random outcomes are examples of intrinsic probabilities. Examples include roulette wheels, the ping pong ball selection technique used in state lotteries, steel balls falling down an array of equally spaced nails, and the outcome of throwing a pair of dice. In each case there are parameter values where phase space is uniformly distributed and the outcome produces random events. Intrinsic probabilities thus explain how some random phenomena are manifest in physical systems but operationally they are difficult to use. For most practical situations it is difficult to actually create a phase space model of the physical systems that simulates reality. Even in the coin tossing experiment the models were highly idealized. Even so, the insights provided by such simplified models are useful in understanding more complex systems. There is a problem, however, with the intrinsic probability framework that reduces its usefulness. Intuitively, it is desirable to have probabilities satisfy algebraic properties. Events with intrinsic probability, however, cannot be added together and always result in an event with intrinsic probability. In Problem A.4 the reader is asked to prove a result of this type. In this next section we expand on the notion of deterministic randomness.

A.2 Randomness: Deterministic Systems

In Section A.1 we developed a physical basis for randomness and intrinsic probability. In the coin toss experiment, randomness in the outcome arose from parameter values that were contained in a portion of phase space that was uniformly distributed. Small perturbations of the initial parameter values for this portion of phase space, result in large differences in the outcomes of the experiment and we concluded that if the perturbations were not sharply peaked then the outcome of the experiment would be random. The randomness of the outcome essentially results from the fact that it is impossible to obtain measurements of arbitrary accuracy. Thus we can never have sufficient information about the initial conditions of the toss to make a precise prediction of the outcome. For that example, this loss of information can be very slight, that is, not knowing the precision of the parameters within the trillionth decimal place is sufficient to make the outcome random for certain initial parameter values. The main point is that denied this information, a deterministic process can, for all purposes, appear to be random. In this section we elaborate on this fact by investigating deterministic sequences that appear to be random when denied certain information.

A.2.1 The Baker's Transformation

Consider the following transformation of the *unit square,* that is, all points (x, y), where $0 \le x, y \le 1$, into itself called the *baker's transformation.* We denote the transformation of the point (x, y) by $(x', y') =$

$B(x, y)$, where

$$(x', y') = \begin{cases} (2x, y/2), & x < 1/2, \\ (2x - 1, (y + 1)/2), & 1/2 \le x \le 1, \end{cases} \quad 0 \le x, y \le 1.$$

It is clear that this transformation is deterministic, maps the unit square into itself and has fixed points $(0, 0)$ and $(1, 1)$. Let $B^n(x, y)$ be the transformation applied n times to point (x, y). Thus $B(x, y) \overset{\text{def}}{=} B^1(x, y)$ and $B^n(x, y) = B^{n-1}(B^1(x, y))$.

FIGURE A.2. The Baker's Transformation

The reason the transformation is called the baker's transformation is that the mathematical operation is similar to what a baker does when dough is kneaded. A picturesque way to envision the transformation is shown in Figure A.2. As illustrated, one first flattens the unit square making it twice as wide and half as high. This is then split into left and right halves and the right half is placed on top of the left half. Repeating the operation many times "mixes" the unit square into itself. Let the collection of points $\{B^j(x, y)\}_{j=1}^{\infty}$ be called the *orbit* of point (x, y). It is interesting to note that the orbits of two points that are very close can be quite different.

Example A.3 (Orbits of Baker's Transformation) Let

$$(x, y) = (0.444444, 0.444444)$$

and suppose that $\epsilon = 0.000001$. Then straightforward calculation shows that

$$B^{20}(x, y) = (0.655872, 0.111111)$$

whereas

$$B^{20}(x + \epsilon, y) = (0.180160, 0.611111).$$

In this instance a small change in the sixth decimal place of the initial x value causes a large difference in the values at the twentieth iteration. Also observe that some initial points cause orbits that cycle.

Example A.4 (A Cyclic Orbit) The orbit of the point $(0.2, 0.8)$ is

$$(0.2, 0.8) \longrightarrow (0.4, 0.4) \longrightarrow (0.8, 0.2) \longrightarrow (0.6, 0.6) \longrightarrow (0.2, 0.8) \longrightarrow \ldots .$$

Suppose that one selects an initial point and successively applies the baker's transformation. Example A.3 implies that if there is a limit to the precision with which we can measure the initial conditions, then for some starting points the value of the nth transformation for sufficiently large n cannot be accurately predicted. Thus, as in Section A.1, small perturbations in the initial conditions can lead to very different outcomes. We should note that the sensitivity of the baker's transformation to initial conditions makes it very difficult to analyze orbits on a computer. Since there is only finite precision in the machine, orbits are inaccurately calculated due to roundoff and truncation errors. For example, iterating the transformation with initial point $(0.2, 0.8)$ eventually leads to the point $(0, 0)$ on some computers, whereas Example A.4 shows that the orbit is actually cyclic.

Now consider the following variation of the baker's transformation. Suppose we have a viewer through which we observe an orbit. This viewer clearly shows the value of the y coordinate for each point in the orbit but does not show the value of the x coordinate. We can imagine that through the viewer we watch the moving shadow of the orbit on the right vertical wall of the unit square. Suppose that the orbit extends infinitely in the past and that at step 0 we know the values of y_i for $i = 0, -1, -2, \ldots$. We will call these values the *past values of y*. Do the past values of y allow us to predict the value of the next iterate, y_1?

If the past values of y are cyclic, as in Example A.4, then it is plausible that the value of y_1 is predictable since it seems intuitive that the y values must continue to cycle. For the non-cyclic case it also seems clear that the longer we go in the past, the more information we obtain about the values of $x_i, i = 0, -1, -2, \ldots$ (which we call the *past values of x*). Since to predict the value of y_1 it suffices to determine if $x_0 \leq 1/2$ or $x_0 > 1/2$ it is easy to conjecture that the past values of y contain sufficient information to predict the value of y_1. Although it might be harder to establish this value in the non-cyclic case, intuitively speaking it should be possible. One could be thus tempted to prove both of the above conjectures.

In fact, both conjectures are false and there is no difference in predictability between the cyclic and non-cyclic cases. In both cases the value of y_1 is *totally unpredictable* and furthermore is equivalent to predicting the outcome of heads in the coin tossing experiment of Example A.1! To show these startling claims we next introduce the notion of a *dynamical system*. This will prove to be useful when we discuss randomness and Newton's laws later in the section.

**A.2.2
Dynamical
Systems**

A dynamical system is simply a specification of a set of rules for how a system evolves in time. A precise definition is given here:

Definition A.5 *A dynamical system consists of a set of possible system states S and a deterministic transition rule*

$$\mathbf{R} : S \longrightarrow S : s \longrightarrow \mathbf{R}(s)$$

that specifies how the system changes from state to state. We can think of the system starting at some initial state and then evolving as a sequence increasing in time according to the transition rules. The timed sequence of states obtained from the orbit is called the flow of the dynamical system.

We have already seen an example of a dynamical system.

Example A.6 (A Dynamical System) Let S be the set of all points of the unit square and let $\mathbf{R}(x, y)$ be the baker's transformation, that is, $\mathbf{R}(x, y) = B(x, y)$. Then the system is a dynamical system and the orbit of a given initial point is the flow of the dynamical system.

Other systems include physical systems that obey Newton's laws where the rule is the set of differential equations that describe the equations of motion. Consider the following example of an abstract dynamical system.

Example A.7 Let the set of states of a dynamical system be points in the unit square and let $z_i, i = 0, \pm1, \pm2, \dots$ be a given bidirectional infinite sequence of 0s and 1s. From this sequence *form a coordinate pair* (x, y), where

$$x = 0.z_0 z_1 z_2 \dots \tag{A.3}$$

and

$$y = 0.z_{-1} z_{-2} \dots . \tag{A.4}$$

The point (x, y) thus lies in the unit square and (A.3) and (A.4) are the binary expansions of x and y, respectively. The rule for going from one state to the other is obtained by shifting the values z_i and forming another point. Specifically, let $z_i' = z_{i-1}$ and using these shifted values form a new coordinate pair. This yields the new values (x', y') given by

$$x' = 0.z_0' z_1' z_2' \dots = \quad 0.z_1 z_2 \dots$$

and

$$y' = 0.z_{-1}' z_{-2}' \dots = \quad 0.z_0 z_{-1} z_{-2} \dots .$$

Thus

$$\mathbf{R}(0.z_0 z_1 z_2 \dots, \ 0.z_{-1} z_{-2} \dots) = (0.z_1 z_2 \dots, \ 0.z_0 z_{-1} z_{-2} \dots).$$

It is clear that the rule for the dynamical systems of Examples A.6 and A.7 are both mappings of the unit square to itself. What perhaps is not so clear is that the transition rule for going from state to state in both systems is identical to a baker's transformation and thus both examples are, in some sense, equivalent. (The proof of this is left to the reader in Problem A.3.) This observation allows us to answer the question about predicting the value of y_1 posed earlier. To do this we create the sequence z_i as follows. For $i < 0$ we let the values of z_i be given by the binary expansion of y_0, that is,

$$y_0 = 0.z_{-1}z_{-2}z_{-3}\ldots .$$

We leave the values of z_i, for $i \geq 0$, unspecified for the moment and call the values of z_i for $i \leq -1$ (resp., $i \geq 0$) the past (resp., future) values of z. From the transition rule in Example A.7, we have that

$$y_{-i} = 0.z_{-(i+1)}z_{-(i+2)}z_{-(i+3)}\cdots$$

and thus the past values of z determine the past values of y. But notice that we have not yet specified the future values of z. This implies that we obtain identical values for the past values of y regardless of the future values of z. But the future values of z are needed to determine the value of x_0 that is needed if we are to predict y_1. Thus y_1 is not predictable!

We can obtain much more from the above argument. Since we haven't specified the future values of z they can be obtained in an arbitrary manner. For example, they could be a set of all 1s or could arise as outcomes of a coin tossing experiment where heads (resp., tails) landing up is mapped to the value 1 (resp., 0). Thus the future sequence of y can depend on the outcomes of a random experiment. In fact, nothing precludes the past values of z also being obtained from random coin tosses. This implies that there is a correspondence between the dynamical system defined in Example A.7 and a random coin tossing experiment. In the one case, the sequence of state transitions are completely deterministic whereas in the other they arise from random experiments. Clearly, here our notions of determinism and randomness are intertwined.

A.3 Deterministic Randomness**

Recent results in ergodic and isomorphism theory take this a step further (see [89, 91] for a description of the results that follow). Consider a billiard table with a ball that travels at a constant speed. The phase space for such a system is the set of points (p, q) where p is a point on the billiard table and q is its velocity vector. The flow is simply the trajectory of the ball for all time. Clearly the flow of the ball satisfies

---

**This section can be skipped on first reading.

Newton's laws and the system is an example of a dynamical system. For the above system, it can be shown that the flow is completely predictable in that, given initial conditions of some accuracy, one can look sufficiently far into the past so that the future trajectory of the ball can be predicted. This information does not always allow one to make a prediction of this type, as we have just shown in Example A.7. But for the billiard example under consideration we *can* make predictions of this type. Now place a convex obstacle on the table. This simple alteration to the experiment changes things in quite an unexpected way. To describe this change we need to go into some details of isomorphism theory. We will keep this within the scope of our presentation by appealing to the reader's intuition.

A.3.1 Isomorphism Between Systems

We say that two dynamical systems are *isomorphic* if there is an invertible mapping of the phase space of one system into the other that takes the flows of one into the other in a time-preserving manner (up to a constant scaling of time). Two dynamical systems that are isomorphic can appear to be quite different, and it is often convenient to restrict the study to systems that have the same phase space. An isomorphism of this type is found between the systems of Examples A.6 and A.7 where the phase space for each system is the unit square. We say that two dynamical systems are α-*congruent* if they both have the same phase space, are isomorphic, and if the mapping of the two phase spaces does not move points by more than α (we assume that α is small). This is a way to directly compare the flows on both systems, that is, they are on the same phase space and are close (within α of each other). We need one more definition that corresponds to the fact that measurements have only finite precision. Suppose we view the phase space of the billiard table through a *viewer* that *deterministically* distorts the values of phase space. When the viewer is placed on a point (p, q) in phase space it sees point (p', q'). If all but ϵ of the viewed points p' are within ϵ of the actual points p in phase space then we say the viewer is ϵ-*reliable*. The viewer models what happens when we can obtain only a finite amount of information each time we view the system. As an example of a viewer, suppose we can only "see" within m digits of accuracy. Then in Example A.7 this means that the points (x, y) formed by a coordinate pair are α-reliable for $\alpha = 2^{-m}$. Recall in the physical model of Section A.1 the fact that we could only obtain a finite amount of information about the initial parameters implied that the outcome of the toss was random. Thus a loss of information was sufficient to cause randomness. In the case we consider here we will see that loss of information is not a necessary condition for randomness.

Before continuing with this development let us make the following observation about Example A.7 for the case where we can only view m decimal digits of each coordinate pair. If we are interested in determining the state at a particular time, say time 0, within a precision of m' digits for $m' > m$, then we can do so by evaluating the flow for a sufficiently long time into the future and the past. This implies that, although we only obtain a finite amount of information each time we view the system, we can eventually obtain an arbitrary amount of information about the state of the system at any given time by viewing a larger portion of the flow. In other words, we lose no information by only being able to view m digits at a time. Now we can show that loss of information is not necessary for randomness.

A.3.2
Random
Newtonian
Systems

Ornstein and Weiss [90] first established conditions where dynamical systems obeying Newton's laws, namely, geodesic flows on a surface of negative curvature, were isomorphic to a "random process." They have also shown that the dynamical system of the billiard table with a convex obstacle is α-congruent to a random process that is *semi-Markov*. A semi-Markov process, for our purposes here, is a random process that stays at point p_i for a time t_i and then jumps to one of possibly several points using a random selection process such as coin tosses. The theorem [91] that establishes the relationship between the billiard experiment and a coin tossing experiment states that, for any α, there exists an α-reliable viewer such that the system seen through this viewer is a semi-Markov process. In other words, if we can view the process with only finite precision then it is possible to construct a semi-Markov process that produces flows that are close to those of a billiard ball rolling around on a table with a convex obstacle. One further result that arises from this relationship is that we can construct the position of the ball at time 0 to any degree of accuracy by looking at the flow through the viewer far enough into the past and future. In other words, even though the viewer has a finite accuracy we can use it to gather sufficient information so that the position at any time can be reconstructed as in the case for the system of Example A.7. The randomness of the billiard ball thus does not arise from any loss of information as it did in the physical coin toss experiment of Section A.1.

The fact that the billiard table with convex obstacle is α-congruent to some semi-Markov process is rather shocking in that it says that there is no distinction between some systems governed by Newton's laws and other systems that are governed by random processes. Furthermore, it represents a type of randomness different from than that discussed in Section A.1 for physical systems. In the physical system loss of information arose from the fact that we could not discern between small changes in initial conditions. This causes the flows of two processes with slightly different initial conditions to diverge and explains the cause of random behavior. In the case of the billiard table with a convex obstacle, the flows of this and the semi-Markov process are always close to each other, and random behavior arises from factors unrelated to information loss. The fact that we see the process through a viewer that deterministically distorts points is sufficient to make the flow of the billiard ball be α-congruent to a semi-Markov process. We can, however, use observations of the process in the past and future to reconstruct the actual flow.

A.4 Summary of Appendix A

In this appendix we first established how randomness can arise from seemingly deterministic systems. In physical experiments randomness arises from the fact that initial conditions cannot be specified to within arbitrary accuracy. If the phase space of a physical experiment is uniformly distributed, then the outcome of an experiment will be random and its occurrence has a probability that depends on the fraction of phase space that corresponds to that outcome. This outcome is independent of the distribution of initial parameter values for non-peaked distributions. For the abstract dynamical system, randomness arose from the fact that the orbit of the system was very sensitive to initial conditions. Using the baker's transformation we showed that there is a deterministic way to view a random coin tossing experiment and thus random systems and deterministic systems have a common framework. We then stated some results that showed that some systems that obey Newton's laws are equivalent to a random process. We summarize the types of deterministic randomness discussed in this appendix in Table A.1.

The table shows that random behavior can arise in many diverse ways from deterministic dynamical systems and, if hard pressed to make a conclusion, such richness would force us to conclude that "God likes to play with dice."

Type of Randomness	Description
Physical	(Section A.1) Arises in physical experiments with uniformly distributed phase spaces where parameter values can be specified within a finite accuracy. Small perturbations in initial conditions cause the outcome of the experiment to be unpredictable.
Dynamical systems	(Example A.3) Arises in deterministic systems where specification of initial conditions with finite accuracy yield unpredictable orbits. The orbits of two systems with small differences in initial conditions diverge and knowledge of any finite portion of a system's past and future trajectory is not sufficient to determine the complete orbit.
	(Discussion following Example A.7) Arises in deterministic systems when denied knowledge of certain parameter values used in the transition rule. Even though the transition rule of the system is known, the orbit is unpredictable and cannot be determined from any finite portion of the system's past or future trajectory.
	(Discussion under the heading "Random Newtonian Systems") Arises in deterministic systems (possibly satisfying Newton's laws) where states can be viewed with only finite accuracy. Orbits of two systems with small differences in initial conditions stay close (i.e., they *do not diverge*). Knowledge of a sufficiently large portion of the past and future trajectory of a system is sufficient to determine the system's orbit within a given finite accuracy.

TABLE A.1. Summary of Types of Randomness

A.5
Problems for
Appendix A

A.1 [**M3**] Show that (A.1) implies that for large (x, y) values (A.2) holds and thus that the phase space for the coin tossing experiment is uniformly distributed.

A.2 [**R1**] Let circle C_0 have radius 1 and circle C_1 have radius r where r is an irrational number less than 1. Mark any set of m disjoint points on the circumference of C_1 with unique labels and roll circle C_1 around on the inside of circle C_0. Each time a marked point of C_1 makes contact with a point on C_0 mark the point of C_0 with the same label as that of C_1 (we allow points on C_0 to be relabeled). Roll circle C_1, n times and consider the marked points on the circumference of C_0 as a phase space for a physical experiment. In this experiment, a point on the circumference of C_0 is randomly selected and the outcome of the experiment is considered to be the label of the first marked point encountered in going clockwise around C_0. Prove that the phase space is uniformly distributed as $n \to \infty$ and that the intrinsic probability of any particular label is given by $1/n$. Are any points of C_0 labeled more than once? Suppose now that several of the marked points of C_1 receive the same label. Does that change any of the above conclusions?

A.3 [**M2**] Prove that the transformation in Example A.7 is a baker's transformation.

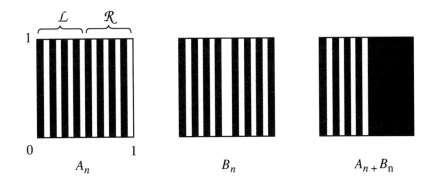

FIGURE A.3. Representations of $\mathcal{A}_n$, $\mathcal{B}_n$, and $\mathcal{A}_n + \mathcal{B}_n$.

A.4 [**M3**] In this problem we show that the sum of two events, each with intrinsic probability, do not necessarily correspond to an event with intrinsic probability. To do this let $\mathcal{L} = [0, 1/2) \times [0, 1)$ be the left half of the unit square, $\mathcal{R} = [1/2, 1) \times [0, 1)$ be the right half of the unit square, and let $B^{-1}(\mathcal{A})$ be the set of points $\mathcal{B}$ that map to $\mathcal{A}$ under a baker's transformation. Define

$$\mathcal{A}_n \stackrel{\text{def}}{=} B^{-n}(\mathcal{L})$$

and

$$\mathcal{B}_n \stackrel{\text{def}}{=} (\mathcal{L}B^{-n}(\mathcal{L})) + (\mathcal{R}B^{-n}(\mathcal{R})).$$

Suppose one selects an arbitrary point in the unit square; then $\mathcal{A}_n$ is the event that the marked point finally appears in $\mathcal{L}$ after n transformations and event $\mathcal{R}$ is the event where the marked point is initially selected in $\mathcal{L}$ (resp., $\mathcal{R}$) and after n transformations finally appears in $\mathcal{L}$ (resp., $\mathcal{R}$). Prove that if n is sufficiently large than $\mathcal{A}_n$ and $\mathcal{B}_n$ have intrinsic probabilities while $\mathcal{A}_n + \mathcal{B}_n$ does not. Show that the probability of event $\mathcal{A}_n + \mathcal{B}_n$ is necessarily between $1/2$ and 1. (Hint: Use Figure A.3)

B

Combinatorial Equalities and Inequalities

In this appendix we list some combinatorial relationships that are used in the text.

For our purposes we are not typically constrained when combinatorial definitions are restricted to integer values. They can be expressed, however, more generally and we will have recourse on occasion to use noninteger values for the upper index in binomial coefficients. A case frequently encountered is that of $\begin{pmatrix} 1/2 \\ k \end{pmatrix}$ which, following the definition implicitly given by (3.21), is defined as

$$\begin{pmatrix} 1/2 \\ k \end{pmatrix} = \frac{(1/2) \cdot (1/2 - 1) \cdots (1/2 - (k-1)) \cdot (1/2 - k)!}{k!(1/2 - k)!}$$

$$= (1/2)^k (-1)^{k-1} \frac{1 \cdot 3 \cdots 2k - 3}{k!}, \quad k = 1, 2, \ldots, \quad \text{(B.1)}$$

where $\begin{pmatrix} 1/2 \\ 0 \end{pmatrix} = 1$ by convention. Equation (B.1) can be simplified as

$$\begin{pmatrix} 1/2 \\ k \end{pmatrix} = (1/2)^k (-1)^{k-1} \frac{1 \cdot 3 \cdot 5 \cdots 2k - 3}{k!} \frac{2 \cdots 2(k-1)}{2 \cdots 2(k-1)}$$

$$= \frac{1}{k} \frac{(-1)^{k-1}}{2^{2k-1}} \begin{pmatrix} 2(k-1) \\ k-1 \end{pmatrix}, \quad k \geq 1. \quad \text{(B.2)}$$

In a similar fashion we have

$$\begin{pmatrix} -1/2 \\ k \end{pmatrix} = (-1)^k \begin{pmatrix} 2k \\ k \end{pmatrix} 2^{-2k}, \quad k = 0, 1, \ldots . \quad \text{(B.3)}$$

B.2 Binomial Formula

The binomial formula for nonnegative integer n is

$$(x+y)^n = \sum_{i=0}^{n} \binom{n}{i} x^i y^{n-i}; \qquad \text{(B.4)}$$

note that it is not restricted to integers. For our purposes we only require generalization to noninteger powers as found in *Newton's binomial formula* which, for any value a, is given as

$$(1+t)^a = 1 + \binom{a}{1} t + \binom{a}{2} t^2 + \cdots, \quad -1 < t < 1,$$

$$= \sum_{k=0}^{\infty} \frac{(a)_k}{k!} t^k. \qquad \text{(B.5)}$$

Power expansions for square roots are conveniently expressed in this manner and we record two such expressions that will be used later:

$$\sqrt{1+t} = \sum_{i=0}^{\infty} \binom{1/2}{i} t^i = 1 + \sum_{i=1}^{\infty} \frac{1}{i} \binom{2(i-1)}{i-1} \frac{(-1)^{i-1} t^i}{2^{2i-1}}, \qquad \text{(B.6)}$$

where we have used (B.2) and

$$\frac{1}{\sqrt{1+t}} = \sum_{i=0}^{\infty} \binom{-1/2}{i} t^i = \sum_{i=0}^{\infty} (-1)^i \binom{2i}{i} \frac{t^i}{2^{2i}},$$

where we have used (B.3).

B.3 Stirling's (de Moivre's) Formula

Factorial expressions are sometimes difficult to work with for large integer values, and it is often convenient in these cases to express factors of $n!$ in terms of an approximation first discovered by Abraham de Moivre (1667–1754)[1] which is misnamed *Stirling's formula* after the mathematician James Stirling (1692–1770) [32]. This formula is given by

$$n! \approx \sqrt{2n\pi} \left(\frac{n}{e}\right)^n. \qquad \text{(B.7)}$$

We refer the reader to Feller ([34] page 52 and [33]) for a derivation of this formula as well as for bounds on its accuracy that are given by

$$\sqrt{2n\pi} \left(\frac{n}{e}\right)^n B_L(n) < n! < \sqrt{2n\pi} \left(\frac{n}{e}\right)^n B_U(n), \qquad \text{(B.8)}$$

---

[1]De Moivre was an intimate friend of Isaac Newton who made many important discoveries in probability, the most important of which is a form of the central limit theorem, proved in Section 4.6.

where

$$B_L(n) \overset{\text{def}}{=} \exp\left(\frac{1}{12n+1}\right), \quad B_U(n) \overset{\text{def}}{=} \exp\left(\frac{1}{12n}\right).$$

These bounds show that even for small n Stirling's formula yields close approximations to $n!$. We note that using Stirling's formula we can approximate (3.21) by

$$\binom{n}{k} \approx \sqrt{\frac{n}{2\pi k(n-k)}} \; \frac{n^n}{(n-k)^{n-k}\, k^k} \qquad (B.9)$$

with bounds given by

$$\sqrt{\frac{n}{2\pi k(n-k)}} \frac{n^n \theta_L}{(n-k)^{n-k}\, k^k} < \binom{n}{k} < \sqrt{\frac{n}{2\pi k(n-k)}} \frac{n^n \theta_U}{(n-k)^{n-k}\, k^k},$$

where

$$\theta_L \overset{\text{def}}{=} \frac{B_L(n)}{B_U(k)B_U(n-k)} \quad \text{and} \quad \theta_U \overset{\text{def}}{=} \frac{B_U(n)}{B_L(k)B_L(n-k)}.$$

A useful formula derivable from (B.9) is

$$\binom{n}{\alpha n} \approx \frac{(1-\alpha)^{-n(1-\alpha)}\alpha^{-\alpha n}}{\sqrt{2\pi n\alpha(1-\alpha)}}, \qquad \alpha n \text{ integer},$$

from which one obtains

$$\binom{n}{n/2} \approx \sqrt{\frac{2}{\pi n}} \; 2^n, \qquad n \text{ even}.$$

B.4
Bounds on
Factorial
Expressions

We first establish that

$$1 - x < e^{-x}, \quad x > 0. \qquad (B.10)$$

The Taylor series expansion of e^{-x} is given by

$$e^{-x} = 1 - x + \frac{x^2}{2!} - \frac{x^3}{3!} + \cdots,$$

and thus (B.10) follows if

$$g(x) \overset{\text{def}}{=} \frac{x^2}{2!} - \frac{x^3}{3!} + \frac{x^4}{4!} - \cdots > 0.$$

This, however, follows from the fact that $g(0) = 0$ and g is an increasing function,

$$\frac{d}{dx} g(x) = x - \frac{x^2}{2!} + \frac{x^3}{3!} - \cdots$$

$$= 1 - e^{-x} > 0, \quad x > 0.$$

Using this we can obtain the following bound on $(n)_k$ using (B.10),

$$(n)_k < n^k \exp\left[-k(k-1)/(2n)\right].$$

B.5 Noninteger Factorials**

We close this appendix by addressing a question raised by integer constraints of factorials. In these definitions we have assumed that factorials are defined only for positive integers, that is, $5\frac{1}{2}!$ or $(-\frac{1}{2})!$ have not been defined (note that this does not preclude the definition of binomial coefficients such as (B.2) or (B.3), which can be defined without defining noninteger factorials). We make this restriction because this is all the generality we need in the book, but it is perfectly clear that mathematics is not satisfied with such specialty. In fact, there is an interesting history concerning the quest to generalize the factorials to noninteger values (see Davis [26] for a stimulating review of this history), which leads to the definition of the *gamma* function, something we will need later in the text. If we could follow a letter in 1729 from the 22-year-old Swiss Leonard Euler[2] (1707–1783) to the eminent German mathematician Christian Goldbach[3] (1690–1764) we would find the "solution" to the noninteger factorial problem. Evidently, while investigating some properties of infinite products, Euler discovered that $n!$ could be written as

$$n! = \left[\left(\frac{2}{1}\right)^n \frac{1}{n+1}\right]\left[\left(\frac{3}{2}\right)^n \frac{2}{n+2}\right]\left[\left(\frac{4}{3}\right)^n \frac{3}{n+3}\right]\cdots,$$

where n is not restricted to integers. It is easy to check that, provided the infinite product converges, all terms cancel for integer n leaving

---

**Of historical interest only.

[2]Leonard Euler was a prolific mathematician who published many books and 886 papers in his career. It was said that nothing, not even loud disturbances, could shake his concentration, and his output did not lessen when he became completely blind during the last 17 years of his life. He evidently was prolific in other areas as well, fathering 13 children.

[3]If you have plenty of time on your hands you might try a problem that Goldbach left behind. He conjectured that every even integer can be written as the sum of two prime numbers, such as $98 = 19 + 79$ or $21000 = 17 + 20983$. The conjecture still haunts us, and no counterexample or proof has yet been found. It might indeed be the case that the conjecture is true but not derivable from the axioms of arithmetic [84].

$n(n-1)(n-2)\cdots 2\cdot 1$ in agreement with our previous definition. This equation can also be written, perhaps more conveniently, as

$$n! = \lim_{m\to\infty}\frac{m!(m+1)^n}{(n+1)(n+2)\cdots(n+m)},$$

which is the basis for the derivation of properties of the gamma function. It is interesting to note that substituting $n=1/2$ in this expression and simplifying yields the curious expansion

$$\frac{2\cdot 2\cdot 4\cdot 4\cdot 6\cdot 6\cdot 8\cdots}{1\cdot 3\cdot 3\cdot 5\cdot 5\cdot 7\cdot 7\cdots} \tag{B.11}$$

that John Wallis[4] (1616–1703) had already shown to be equal to $\pi/2$. Hence was born the first noninteger factorial, $(1/2)! = \pi/2$. This excited Euler to look further and eventually led to the expression $n! = \int_0^1(-\ln x)^n dx$ which, by a change of variables, is usually expressed as (this last form is attributed to Adrien Marie Legendre (1752–1833))

$$\Gamma(x) = \int_0^\infty e^{-t}t^{x-1}dt, \tag{B.12}$$

the gamma function for which $\Gamma(n+1) = n!$ for integer n. Like integer factorials, this function satisfies the recurrence relationship (see (3.7))

$$x\Gamma(x) = \Gamma(x+1), \tag{B.13}$$

but this time we allow all real values for x ((B.13) follows from a simple application of integration by parts). This recurrence allows one to establish properties of factorials for noninteger values and, in fact, for the above mentioned values of $(5\frac{1}{2})!$ and $(-\frac{1}{2})!$, equations (B.12) and (B.13) yield

$$\left(5\frac{1}{2}\right)! = \frac{9}{2}\cdot\frac{7}{2}\cdot\frac{5}{2}\cdot\frac{3}{2}\cdot\frac{\pi}{2}$$

and $(-\frac{1}{2})! = \sqrt{\pi}$. It thus follows from this last equation and from (B.13) that

$$\Gamma\left(\frac{n}{2}\right) = \frac{\sqrt{\pi}(n-1)!}{2^{n-1}\left(\frac{n-1}{2}\right)!}.$$

Of course we still have bothersome constraints on the definition of factorials since $(\sqrt{-1})!$ is left undefined but we leave this pursuit to the undaunted reader!

---

[4]John Wallis, a British mathematician, was an immediate predecessor of Isaac Newton who spent considerable effort looking for an expression of π by determining the area of a circle in the first quadrant. Equation (B.11) is one of the fruits of that labor. Perhaps Wallis' most enduring accomplishment is his introduction of the symbol "∞" for infinity.

C

Tables of Laplace Transforms and Generating Functions

In this appendix we provide lists of Laplace transforms and generating functions, and we list some of their useful properties. More comprehensive lists of Laplace transforms can be found in [1, 30, 42]. We start with the Laplace transform.

C.0.1 Laplace Transforms

Recall from Section 5.4 (see (5.54)) that the definition of the Laplace transform for a nonnegative function $f(t)$ defined over $t \in [0, \infty)$ is given by

$$f^*(s) \overset{\text{def}}{=} \int_0^\infty e^{-sx} f(x) dx$$

provided the integral exists. Transform pairs are indicated by

$$f(t) \iff f^*(s).$$

If X is a random variable with density f_X then the Laplace transform is denoted by

$$\mathcal{L}_X(s) \overset{\text{def}}{=} \int_0^\infty e^{-sx} f_X(x) dx = E\left[e^{-sX}\right].$$

Let h be a nonnegative function defined over $[0, \infty)$ and let

$$c = \int_0^\infty h(x) dx.$$

Then if $c < \infty$ the normalized function, $h(t)/c$, is a density function. Letting X be a random variable with this density implies that

$$h^*(s) = c\mathcal{L}_X(s).$$

In this way, the functions in Table C.2 can be converted into Laplace transforms of density functions. As an example, the transform pair

$$e^{-at} \iff \frac{1}{s+a}$$

combined with the fact that

$$\frac{1}{a} = \int_0^\infty e^{-at} dt$$

yields the transform pair

$$ae^{-at} \iff \frac{a}{s+a}.$$

The properties of Laplace transforms given in Table C.1 (these are repeated in Table 5.6) can be used to expand on the entries of Table C.2. As an example, using the transform pair from Table C.2

$$\ln(t) \iff -\frac{\ln(\mathbf{C}s)}{s}, \quad \mathbf{C} \approx .57721 \qquad \text{Euler's Constant}$$

and the linear shift property of Table C.1

$$tf(t) \iff -\frac{d}{ds} f^*(s)$$

implies that

$$t\ln(t) \iff \frac{1 - \ln(\mathbf{C}s)}{s^2}.$$

Transforms are almost always inverted by using the above properties to decompose the transform into components that match entries in a table. One standard method for inverting transforms that are expressed as the ratio of two polynomials is to use a partial fraction expansion as illustrated in Section 5.4. In general, inversions using table lookup require ad hoc techniques that are pertinent to the particular problem at hand. For example, consider the transform

$$f^*(s) = e^{-as} \left[\frac{(1-s)(1+s)}{s^2} \right].$$

To invert this we first use some simple algebra to rewrite it as

$$f^*(s) = \frac{1}{s} \frac{e^{-as}}{s} - e^{-as}.$$

Property		Function	Laplace Transform
	Addition	$af(t) + bg(t)$	$af^*(s) + bg^*(s)$
	Convolution	$f \circledast g(t)dx$	$f^*(s)g^*(s)$
Shifting	Linear	$f(t-a)$	$e^{-as}f^*(s)$
	Exponential	$e^{-at}f(t)$	$f^*(s+a)$
Scaling	Power	$t^n f(t)$	$(-1)^n \frac{d^n f^*(s)}{ds^n}$
	Linear	$tf(t)$	$-\frac{d}{ds}f^*(s)$
	Inverse power	$f(t)/t^n$	$\int_{s_1=s}^{\infty} ds_1 \int_{s_2=s_1}^{\infty} ds_2 \cdots \int_{s_n=s_{n-1}}^{\infty} ds_n f^*(s_n)$
	Harmonic	$f(t)/t$	$\int_s^{\infty} f^*(s_1)ds_1$
Integrals	Single	$\int_0^t f(x)dx$	$\frac{f^*(s)}{s}$
	Multiple	$\underbrace{\int_0^{\infty} \cdots \int_0^{\infty}}_{n \text{ times}} f(x)dx^n$	$\frac{f^*(s)}{s^n}$
Derivatives	Single	$\frac{d}{dt}f(t)$	$sf^*(s) - f(0^-)$
	Multiple	$\frac{d^n f^*(s)}{dt^n}$	$s^n f^*(s) - \sum_{i=0}^{n-1} s^{n-1-i} f^{(i)}(0^-)$
Values	Initial	$\lim_{t\to 0} f(t)$	$\lim_{s\to\infty} sf^*(s)$
	Integral	$\int_0^{\infty} f(x)dx$	$f^*(0)$
	Final	$\lim_{t\to\infty} f(t)$	$\lim_{s\to 0} sf^*(s)$

TABLE C.1. Algebraic Properties of Laplace Transforms

$f(t)$	$f^*(s)$	$f(t)$	$f^*(s)$
$\delta_0(t-a)$	e^{-as}	$(t-a)$	$\frac{e^{-as}}{s}$
$\frac{t^{n-1}}{(n-1)!}, \;\; n \geq 1$	$\frac{1}{s^n}$	$\frac{1}{\sqrt{\pi t}}$	$\frac{1}{\sqrt{s}}$
$2\sqrt{t/\pi}$	$s^{-3/2}$	$\frac{2^n t^{n-1/2}}{1\cdot3\cdots(2n-1)\sqrt{\pi}}, \;\; n \geq 1$	$s^{-(n+1/2)}$
$t^{k-1}, \;\; k=1,2,\ldots$	$\frac{(k-1)!}{s^k}$	e^{-at}	$\frac{1}{s+a}$
te^{-at}	$\frac{1}{(s+a)^2}$	$\frac{t^{n-1}e^{-at}}{(n-1)!}, \;\; n \geq 1$	$\frac{1}{(s+a)^n}$
$\frac{e^{-at}-e^{-bt}}{b-a}, \;\; a \neq b$	$\frac{1}{(s+a)(s+b)}$	$\frac{ae^{-at}-be^{-bt}}{a-b}, \;\; a \neq b$	$\frac{s}{(s+a)(s+b)}$
$\ln(t)$	$-\frac{\ln(\mathbf{C}s)}{s}, \;\; \mathbf{C} \approx .57721$	$\frac{e^{-bt}-e^{-at}}{t}$	$\ln\left(\frac{s+a}{s+b}\right)$
$\frac{\ln(t)}{\sqrt{t}}$	$-\sqrt{\pi/s}\ln(4\mathbf{C}s)$	$t^{1/2}e^{-a/4t}$	$\frac{1}{2}\sqrt{\pi/s^3}(1+\sqrt{as})e^{-\sqrt{as}}$
$t^{-1/2}e^{-a/4t}$	$\sqrt{\pi/s}\,e^{-\sqrt{as}}$	$t^{-3/2}e^{-a/4t}$	$2\sqrt{\pi/a}\,e^{-\sqrt{as}}$
$\frac{1}{\sqrt{t}(t+a)}$	$\frac{\pi}{\sqrt{a}}e^{as}(1-\mathrm{erf}\sqrt{as})$	$\frac{1}{\sqrt{t+a}}$	$\sqrt{\frac{\pi}{s}}\,e^{as}(1-\mathrm{erf}\sqrt{as})$
e^{-at^2}	$\frac{1}{2}\sqrt{\frac{\pi}{a}}\,e^{s^2/4a}\left(1-\mathrm{erf}\frac{2}{2\sqrt{a}}\right)$	$\mathrm{erf}\sqrt{at}$	$\frac{\sqrt{a}}{s\sqrt{s+a}}$
$\mathrm{erf}\,at$	$\frac{1}{s}e^{s^2/4a^2}\left(1-\mathrm{erf}\frac{s}{2a}\right)$	$1-\mathrm{erf}\frac{a}{\sqrt{t}}$	$\frac{1}{s}e^{-2a\sqrt{s}}$
$e^{-at}\int_0^{\sqrt{at}}e^{x^2}dx$	$\frac{\sqrt{a\pi}}{2\sqrt{s}(s+a)}$	$\sqrt{at}-e^{-at}\int_0^{\sqrt{at}}e^{x^2}dx$	$\frac{\sqrt{a^3\pi}}{2\sqrt{s^3}(s+a)}$

TABLE C.2. Functions and Their Laplace Transforms

Written in these terms we recognize the transform pairs

$$1 \iff \frac{1}{s}, \quad t - a \iff \frac{e^{-as}}{s}, \quad \delta_0(t - a) \iff e^{-as}.$$

This implies that the $f(t)$ equals the convolution of 1 with $t - a$ minus a step function at $t - a$,

$$f(t) = 1 + t - a - \delta_0(t - a).$$

Even if a transform cannot be inverted using these techniques useful information can still be obtained using elementary techniques. For example, all of the moments of a random variable X can be obtained from the transform by simple differentiation (see (5.60))

$$\left. \frac{d^n \mathcal{L}_X(s)}{ds^n} \right|_{s=0} = \left. \frac{d^n f_X^*(s)}{ds^n} \right|_{s=0} = (-1)^n E[X^n].$$

We close this section by mentioning a second way to invert Laplace transforms. This technique, which lies outside the scope of our text, uses the Cauchy residue theorem to evaluate a closed integral in the complex plane. This technique is used when table lookup techniques fail and is not typically encountered in modeling applications. We refer readers to [78] for a description of its use.

C.1 Generating Functions

Recall from Section 3.3.3 that the definition of a generating function for the sequence of values $f_i, i = 0, 1, \ldots$, is given by

$$F(x) \stackrel{\text{def}}{=} \sum_{i=0}^{\infty} f_i x^i$$

provided the sum converges. There is a relationship between the generating function of a sequence and its Laplace transform, which is given by

$$F(x) = f^*(e^{-x}).$$

This relationship implies that properties of generating functions are similar to those of Laplace transforms and thus in this section we provide an abbreviated discussion of generating functions.

If X is a discrete random variable with density $P[X = i] = f_X(i), i = 0, 1, \ldots$, then the generating function of $f_X(i)$ is called the probability generating function and is given by

$$\mathcal{G}_X(z) \stackrel{\text{def}}{=} \sum_{i=0}^{\infty} f_X(i) z^i = E[z^X].$$

The value of $\mathcal{G}_X(z)$ is bounded for all $z < 1$. Let $h_i, i = 0, 1, \ldots,$ be a sequence of values and let

$$c = \sum_{i=0}^{\infty} h_i.$$

Then if $c < \infty$ the normalized function, h_i/c, is a density function. Letting X be a random variable with this density implies that

$$H(z) = c\mathcal{G}_X(z).$$

In this way, the functions in Tables C.4 and C.5 can be converted into probability generating functions of discrete density functions.

The properties of generating functions given in Table C.3 (and repeated in Table 3.6) can be used to expand the entries in Tables C.4 and C.5 as above with Laplace transforms. Generating functions are almost always inverted by decomposing the generating functions into components that match entries in a table. As with Laplace transforms, a method for inverting generating functions that are expressed as the ratio of two polynomials is to use a partial fraction expansion as illustrated in Section 5.4. In general, inversions that use table lookup require ad hoc techniques that are pertinent to the particular problem at hand.

Similar to transforms, useful information can be obtained from probability generating functions even if it cannot be inverted. Factorial moments of discrete random variable X, for example, can be obtained using simple differentiation

$$E\left[X(X-1)\cdots(X-k+1)\right] = \frac{d^k}{dz^k}\mathcal{G}_X(z).$$

Property		Sequence	Generating Function
	Definition	$f_i, i = 0, 1, \ldots$	$F(x) \overset{\text{def}}{=} \sum_{i=0}^{\infty} f_i x^i$
Addition	Addition	$af_i + bg_i$	$aF(x) + bG(x)$
	Convolution	$\sum_{j=0}^{i} f_j g_{i-j}$	$F(x)G(x)$
Shifting	Positive	$f_{i-n}, i \geq n$	$x^n F(x)$
	Negative	f_{i+n}	$\frac{F(x) - \sum_{j=0}^{n-1} f_j x^j}{x^n}$
Scaling	Geometric	$f_i a^i$	$F(ax)$
	Linear	if_i	$x\frac{d}{dx}F(x)$
	Factorial	$(i)_k f_i$	$x^k \frac{d^k}{dx}F(x)$
	Harmonic	f_{i-1}/i	$\int_0^x F(t)dt$
Summations	Cumulative	$\sum_{j=0}^{i} f_j$	$\frac{F(x)}{1-x}$
	Complete	$\sum_{i=0}^{\infty} f_i$	$F(1)$
	Alternating	$\sum_{i=0}^{\infty} (-1)^i f_i$	$F(-1)$
Sequence Values	Initial	f_0	$F(0)$
	Intermediate	f_n	$\frac{1}{n!} \left. \frac{d^n}{dx}F(x)\right|_{x=0}$
	Final	$\lim_{i \to \infty} f_i$	$\lim_{x \to 1}(1-x)F(x)$

TABLE C.3. Algebraic Properties of Ordinary Generating Functions

f_i	$F(x)$	f_i	$F(x)$
$f_i = \begin{cases} 1, i = 0, \\ 0, i \neq 0, \end{cases}$	1	$f_i = \begin{cases} 1, i = n, \\ 0, i \neq n, \end{cases}$	x^n
$f_i = 1$	$\frac{1}{1-x}$	$f_i = \begin{cases} 1, & i \text{ even}, \\ -1, & i \text{ odd} \end{cases}$	$\frac{1}{1+x}$
$f_i = \begin{cases} 1, & i \text{ even}, \\ 0, & i \text{ odd} \end{cases}$	$\frac{1}{1-x^2}$	$f_i = \begin{cases} 1, i = kn, \\ 0, i \neq kn, \end{cases} k = 0, 1, 2, \ldots$	$\frac{1}{1-x^n}$
$f_i = \begin{cases} \beta^{i/2}, & i \text{ even}, \\ 0, & i \text{ odd} \end{cases}$	$\frac{1}{1-\beta x^2}$	$f_i = (-1)^i \beta^i$	$\frac{1}{1+\beta x}$
$f_i = 1 - \beta^i$	$\frac{(1-\beta)x}{(1-x)(1-\beta x)}$	$f_i = \beta^{2i}$	$\frac{1}{1-\beta^2 x}$
$f_i = \beta^i$	$\frac{1}{1-\beta x}$	$f_i = i\beta^i$	$\frac{\beta x}{(1-\beta x)^2}$
$f_i = \beta^{i+1}$	$\frac{\beta}{1-\beta x}$	$f_i = i\beta^{i-1}$	$\frac{x}{1-\beta x}$
$f_i = i^2 \beta^i$	$\frac{\beta x(1+\beta x)}{(1-\beta x)^3}$	$f_i = i^3 \beta^i$	$\frac{\beta x(1+4\beta x+x^2)}{(1-\beta x)^4}$
$f_i = (i+1)\beta^i$	$\frac{1}{(1-\beta x)^2}$	$f_i = i^2 \beta^i$	$\frac{\beta x(1+\beta x)}{(1-\beta x)^3}$
$f_i = i$	$\frac{x}{(1-x)^2}$	$f_i = i+1$	$\frac{1}{(1-x)^2}$
$f_i = i^2$	$\frac{x(1+x)}{(1-x)^3}$	$f_i = i^3$	$\frac{x(1+4x+x^2)}{(1-x)^4}$

TABLE C.4. Sequences and Their Ordinary Generating Functions

f_i	$F(x)$	f_i	$F(x)$
$f_i = \binom{i}{n},\ i \geq n-1$	$\frac{x^n}{(1-x)^{n+1}}$	$f_i = (-1)^i \binom{i}{n},\ i \geq n-1$	$\frac{(-1)^n\, x^n}{(1+x)^{n+1}}$
$f_i = \binom{n}{i}$	$(1+x)^n$	$f_i = \left\langle {n \atop i} \right\rangle$	$\frac{1}{(1-x)^n}$
$f_i = \binom{i+k}{n},\ k \leq n$	$\frac{x^{n-k}}{(1-x)^{n+1}}$	$f_i = \binom{i}{n}\beta^i,\ i \geq n-1$	$\frac{(\beta x)^n}{(1-\beta x)^{n+1}}$
$f_{2i-1} = \frac{1}{i}\,\frac{1}{2^{2i-1}}\binom{2(i-1)}{i-1}$,	$\frac{1-\sqrt{1-x^2}}{x}$	$f_i = \frac{1}{i},\ i \geq 1$	$\ln\left(\frac{1}{1-x}\right)$
$f_i = \frac{(-1)^{i+1}}{i},\ i \geq 1$	$\ln(1+x)$	$f_i = \frac{\beta^i}{i!}$	$e^{\beta x}$
$f_i = \frac{\beta^{i-1}}{i},\ i \geq 1$	$\frac{1}{\beta}\ln\left(\frac{1}{1-\beta x}\right)$	$f_i = \frac{i+1}{i!}\beta^i$	$(1+\beta x)e^{\beta x}$
$f_i = \frac{(-1)^i}{(2i+1)!}$	$\frac{\sin(\sqrt{x})}{\sqrt{x}}$	$f_i = \frac{(-1)^i}{(2i)!}$	$\cos(\sqrt{x})$
$f_i = \frac{\beta^i}{(2i+1)!}$	$\frac{\sinh(\sqrt{\beta x})}{\sqrt{\beta x}}$	$f_i = \frac{\beta^i}{(2i)!}$	$\cosh(\sqrt{\beta x})$
$f_i = H_i \stackrel{\text{def}}{=} \sum_{j=1}^i 1/i$	$\frac{1}{1-x}\ln\left(\frac{1}{1-x}\right)$	$f_i = (H_{n+i} - H_n)\binom{n+i}{i}$	$\frac{1}{(1-x)^{n+1}}\ln\left(\frac{1}{1-x}\right)$

TABLE C.5. Sequences and Their Ordinary Generating Functions - Continued

D

Limits and Order Relationships

In this appendix we provide a brief review of some analysis that is used in the text. Results here are technical and are required to justify steps in derivations. We start with the familiar notion of limits.

D.1

Limits

Let x_n be a sequence of numbers. Then $\{x_n\}$ has a limit of x, denoted by

$$\lim_{n \to \infty} x_n = x, \tag{D.1}$$

or equivalently denoted by

$$x_n \to x, \quad n \to \infty,$$

if for any value of $\epsilon > 0$ there exists a number N_ϵ such that $|x_n - x| < \epsilon$ for all $n > N_\epsilon$. If $x = \infty$ (resp., $x = -\infty$) then the limit means that for any value k there exists a number N such that for all $n > N$, $x_n > k$ (resp., $x_n < k$). All sequences do not necessarily have limits. For example, the sequence $1, -1, 1, -1, \ldots$, has no limiting value, whereas the sequence $1, 1/2, 1/3, \ldots$ has a limiting value of 0. The sequence

$$x_n = 1 + \frac{(-1)^n}{2^n}, \quad n = 1, 2, \ldots, \tag{D.2}$$

clearly has a limiting value of 1 although it alternates below and above 1 (an easy calculation shows that $x_{2k} > 1$ and $x_{2k+1} < 1$ for $k = 1, 2, \ldots$). Intuitively a sequence x_n has a finite limit x if for *all* large values of n, x_n is arbitrarily close to x.

Often we shall require a less restrictive notion of a limit. For example, consider modifying the sequence (D.2) as follows

$$
x_n = \begin{cases} 1+(-1)^n/2^n, & n \text{ not prime}, \\ \\ C, & n \text{ prime}, \end{cases} \qquad n = 1, 2, \ldots, \qquad \text{(D.3)}
$$

and suppose that $C < 1$. In such a case some form of limit exists since for large values of n the value of x_n is arbitrarily close to 1 except if n is a prime number. We redefine our notion of a limit to include such cases and say that $\{x_n\}$ has a *limit superior* written by

$$
\limsup_{n\to\infty} x_n = x \qquad \text{(D.4)}
$$

if

1. for every $\epsilon > 0$ there exists a number N_ϵ such that $x_n < x + \epsilon$ for *all* $n > N_\epsilon$, and

2. for every $\delta > 0$ there exists a number M_δ such that $x_n > x + \delta$ for *some* $n > M_\delta$.

The words *all* and *some* in the above conditions are critical to the definition. In contrast to a limit that might not exist, the two conditions either determine a finite value of x *uniquely* or $x = \infty$. Intuitively, a sequence x_n has a finite limit superior x if for *all* large values of n, $x_n - x$ has arbitrarily small positive values, and for *some* large values of n, $x - x_n$ has arbitrarily large negative values (close to 0). For the sequence (D.3) it is clear that $\limsup_{n\to\infty} x_n = 1$ whereas $\lim_{n\to\infty} x_n$ does not exist.

We similarly define the *limit inferior*, denoted by

$$
\liminf_{n\to\infty} x_n = x, \qquad \text{(D.5)}
$$

if

1. for every $\epsilon > 0$ there exists a number N_ϵ such that $x_n > x - \epsilon$ for *all* $n > N_\epsilon$, and

2. for every $\delta > 0$ there exists a number M_δ such that $x_n < x + \delta$ for *some* $n > M_\delta$.

For the sequence (D.3) it is clear that if we set $C > 1$ then $\liminf_{n\to\infty} x_n = 1$. For the example considered earlier of the sequence $1, -1, 1, -1, \ldots$, it is now clear that the limit superior is 1 and the limit inferior is -1.

The following relationships can be seen to hold between these different forms of limits. First it is clear that the limit inferior cannot be greater than the superior limit:

$$\liminf_{n\to\infty} x_n \leq \limsup_{n\to\infty} x_n.$$

This property is often used to establish that a limit exists. The typical approach is to establish that

$$\liminf_{n\to\infty} x_n \geq x$$

and

$$\limsup_{n\to\infty} x_n = x.$$

Last, by definition it is clear that the limit inferior can be defined in terms of the limit superior limit by

$$\liminf_{n\to\infty} x_n = -\limsup_{n\to\infty}(-x_n).$$

D.2
Order
Relationships

A function $f(x)$ is said to be $o(x^i)$, written

$$f(x) = o(x^i), \tag{D.6}$$

if $f(x)/x^i \to 0$ as $x \to 0$. Hence $h(x) = 4x^3$ is $o(x^2)$ since $h(x)/x^2 = 4x \to 0$ as $x \to 0$, whereas $h(x)$ is not $o(x^5)$ since $h(x)/x^5 \to \infty$.

A function $f(x)$ is said to be $O(x^i)$, written

$$f(x) = O(x^i), \tag{D.7}$$

if $f(x)/x^i$ reaches a finite limit as $x \to 0$. Using the same h defined above we see that $h(x)$ is $O(x^3)$ since $h(x)/x^3 = 4$ whereas $h(x)$ is not $O(x^5)$.

E

List of Common Summations

Rearrangement of Summations

$$\sum_{i=0}^{n}\sum_{j=0}^{i} x_i x_j = \sum_{j=0}^{n}\sum_{i=j}^{n} x_i x_j$$

$$\sum_{i=0}^{n}\sum_{j=0}^{i} x_i x_j = \frac{1}{2}\left(\left(\sum_{i=0}^{n} x_i\right)^2 + \left(\sum_{i=0}^{n} x_i^2\right)\right)$$

$$\left(\sum_{i=1}^{n} x_i y_i\right)^2 = \left(\sum_{i=1}^{n} x_i^2\right)\left(\sum_{i=1}^{n} y_i^2\right)\sum_{k=1}^{n-1}\sum_{j=k+1}^{n}(x_k y_j - x_j y_k)^2$$

Geometric Summations, $0 \le x < 1$

$$\sum_{i=0}^{\infty} x^i = \frac{1}{1-x}$$

$$\sum_{i=0}^{\infty} i x^i = \frac{x}{(1-x)^2}$$

$$\sum_{i=0}^{n} x^i = \frac{1-x^{n+1}}{1-x}$$

$$\sum_{i=0}^{n} i x^i = \frac{nx^{n+2} - (n+1)x^{n+1} + x}{(1-x)^2}$$

Powers of Integers

$$\sum_{i=1}^{n} i = \frac{n(n+1)}{2}$$

$$\sum_{i=1}^{n} i^2 = \frac{n(n+1)(2n+1)}{6}$$

$$\sum_{i=1}^{n} i^3 = \left(\frac{n(n+1)}{2}\right)^2$$

$$\sum_{i=1}^{n} i^4 = \frac{n(n+1)(2n+1)(3n^2+3n-1)}{30}$$

Binomial Identities, $0 \le k \le n, m$

$$\sum_{j=k-1}^{n-1} \binom{j}{k-1} = \binom{n}{k}$$

$$\sum_{j=0}^{k} \binom{n+j-k-1}{j} = \binom{n}{k}$$

$$\sum_{k=0}^{n} \binom{n}{k}^2 = \binom{2n}{n}$$

$$\sum_{k=0}^{2n} (-1)^k \binom{2n}{k}^2 = (-1)^n \binom{2n}{n}$$

$$\sum_{k=0}^{n} \binom{n}{k} x^k = (1+x)^n$$

$$\sum_{k=0}^{n} (k+1) \binom{n}{k} = 2^{n-1}(n+2)$$

$$\sum_{k=0}^{\lfloor n/2 \rfloor} \binom{n}{2k} = 2^{n-1}$$

$$\sum_{k=0}^{\lfloor n/2 \rfloor} \binom{n}{2k+1} = 2^{n-1}$$

$$\sum_{k=0}^{m} \binom{n+k}{k} = \binom{n+m+1}{m}$$

$$\sum_{k=0}^{j} \binom{n}{k} \binom{m}{j-k} = \binom{n+m}{j}$$

$$\sum_{k=0}^{n} (-1)^k \binom{n}{k} k^n = (-1)^n n!$$

$$\sum_{k=0}^{n} (-1)^k \binom{n}{k} (x+k)^n = (-1)^n n!$$

$$\sum_{k=1}^{n} \frac{(-1)^{k+1}}{k} \binom{n}{k} = \sum_{k=1}^{n} 1/k$$

$$\sum_{k=1}^{n} \frac{(-1)^{k+1}}{k+1} \binom{n}{k} = \frac{n}{n+1}$$

$$\sum_{k=0}^{n} \frac{1}{k+1} \binom{n}{k} = \frac{2^{n+1}-1}{n+1}$$

$$\sum_{k=0}^{n} \frac{x^{k+1}}{k+1} \binom{n}{k} = \frac{(x+1)^{n+1}-1}{n+1}$$

$$\sum_{k=0}^{m} (-1)^k \binom{n}{k} = (-1)^m \binom{n-1}{m}$$

$$\sum_{k=0}^{\infty} \binom{n+k}{n} x^k = \frac{1}{(1-x)^{n+1}}$$

$$\sum_{k=0}^{2n+1} (-1)^k \binom{2n+1}{k}^2 = 0$$

$$\sum_{k=1}^{n} k \binom{n}{k}^2 = \frac{(2n-1)!}{((n-1)!)^2}$$

Factorial Summations

$$\sum_{k=0}^{\infty} \frac{1}{k!} = e$$

$$\sum_{k=0}^{\infty} \frac{(-1)^k}{k!} = \frac{1}{e}$$

$$\sum_{k=0}^{\infty} \frac{k}{(2k+1)!} = \frac{1}{e}$$

$$\sum_{k=0}^{\infty} \frac{k}{(k+1)!} = 1$$

$$\sum_{k=0}^{\infty} \frac{1}{(2k)!} = \frac{1}{2}\left(e + \frac{1}{e}\right)$$

$$\sum_{k=0}^{\infty} \frac{1}{(2k+1)!} = \frac{1}{2}\left(e - \frac{1}{e}\right)$$

$$\sum_{k=0}^{\infty} \frac{x^k}{k!} = e^x$$

$$\sum_{k=0}^{\infty} \frac{(x \ln a)^k}{k!} = a^x$$

$$\sum_{k=0}^{\infty} (-1)^k \frac{x^{2k}}{k!} = e^{-x^2}$$

$$\sum_{k=0}^{\infty} \frac{x^k(k+1)}{k!} = (1+x)e^x$$

$$\sum_{k=1}^{\infty} \frac{k^2}{k!} = 2e$$

$$\sum_{k=1}^{\infty} \frac{k^3}{k!} = 5e$$

$$\sum_{k=1}^{\infty} \frac{k^4}{k!} = 15e$$

$$\sum_{k=1}^{\infty} \frac{k^5}{k!} = 52e$$

Miscellaneous Summations

$$\sum_{k=1}^{n} (2k-1) = n^2$$

$$\sum_{k=1}^{n} (2k-1)^2 = \frac{1}{3}n(4n^2-1)$$

$$\sum_{k=1}^{n} (2k-1)^3 = n^2(2n^2-1)$$

$$\sum_{k=1}^{n} k!k = (n+1)! - 1$$

$$\sum_{k=1}^{\infty} \frac{x^{2^{k-1}}}{1 - x^{2^k}} = \frac{x}{1-x}, \quad x^2 < 1$$

$$\sum_{k=1}^{\infty} \frac{2^{k-1}}{x^{2^{k-1}}} = \frac{1}{x-1}, \quad x^2 > 1$$

$$\sum_{k=0}^{n-1} (a+kr)p^k = \frac{a - (a + (n-1)r)p^n}{1-p} + \frac{rp(1 - p^{n-1})}{(1-p)^2}$$

Bibliography

Reference Books

M. Abramowitz and I.A. Stegun. *Handbook of Mathematical Functions.* Dover Publications, 1965.

A. Erdélyi, editor. *Tables of Integral Transforms.* Volume I, McGraw-Hill, 1954.

I.S. Gradshteyn and I.M. Ryzhik. *Table of Integrals, Series, and Products.* Academic Press, 1980.

F.A. Haight. *Handbook of the Poisson Distribution.* John Wiley and Sons, 1967.

N.L. Johnson and S. Kotz. *Discrete Distributions.* Houghton Mifflin, 1969.

N.L. Johnson and S. Kotz. *Continuous Univariate Distributions - 1 and 2.* Houghton Mifflin, 1970.

W.R. LePage. *Complex Variables and the Laplace Transform for Engineers.* Dover Publications, 1980.

Of Historical Interest

E.T. Bell. *Men of Mathematics. A Touchstone Book,* Simon and Schuster, 1986.

C.B. Boyer and U.C. Merzbach. *A History of Mathematics.* John Wiley and Sons, 1989.

A. de Moivre. *The Doctrine of Chances: Or a Method of Calculating the Probabilities of Events in Play.* Chelsea, 1967.

J.H. Eves. *An Introduction to the History of Mathematics. The Saunder Series,* Saunders College Publishing, 1990.

P.S. Laplace. *A Philosophical Essay on Probabilities*. Dover Publications, New York, 1951.

D.E. Smith. *A Source Book in Mathematics*. McGraw-Hill, New York, 1929.

I. Todhunter. *A History of the Mathematical Theory of Probability: From the Time of Pascal to that of Laplace*. Chelsea, 1949.

Combinatorics

D.I.A. Cohen. *Basic Techniques of Combinatorial Theory*. John Wiley and Sons, 1978.

G.M. Constantine. *Combinatorial Theory and Statistical Design*. John Wiley and Sons, 1987.

R.L. Graham, D.E. Knuth, and O. Patashnik. *Concrete Mathematics*. Addison-Wesley, 1989.

J. Riordan. *An Introduction to Combinatorial Analysis*. Princeton University Press, 1978.

N.Y. Vilenkin. *Combinatorics*. Academic Press, 1971.

Probability

Y.S. Chow and H. Teicher. *Probability Theory-Independence, Interchangeability, Martingales*. Springer-Verlag, 1988.

K.L. Chung. *Elementary Probability Theory with Stochastic Processes*. Springer-Verlag, 1979.

A.B. Clarke and R.L. Disney. *Probability and Random Processes: A First Course with Applications*. John Wiley and Sons, 1985.

W. Feller. *An Introduction to Probability Theory and Its Applications-Volume I*. John Wiley and Sons, 1968.

W. Feller. *An Introduction to Probability Theory and Its Applications-Volume II*. John Wiley and Sons, 1968.

B.V. Gnedenko. *The Theory of Probability*, second edition. Chelsea, 1963.

R.W. Hamming. *The Art of Probability-for Scientists and Engineers*. Addison-Wesley, 1991.

P.G. Hoel, S.C. Port, and C.J. Stone. *Introduction to Probability Theory*. Houghton Mifflin, 1971.

A. Kolmogorov. *Foundations of the Theory of Probability*. Chelsea, 1950.

A. Rényi. *Foundations of Probability*. Holden-Day, 1970.

S.M. Ross. *An Introduction to Probability Models*. Academic Press, 1972.

R. von Mises. *Mathematical Theory of Probability and Statistics*. Academic Press, 1964.

Stochastic Processes

E. Çinclar. *Introduction to Stochastic Processes*. Prentice-Hall, 1975.

D.R. Cox. *Renewal Theory*. Methuen, 1962.

S. Karlin and H.M. Taylor. *A First Course in Stochastic Processes*. Academic Press, 1975.

J.G. Kemeny and J.L. Snell. *Finite Markov Chains*. Springer-Verlag, 1976.

J.G. Kemeny, J.L. Snell, and A.W. Knapp. *Denumerable Markov Chains*. Springer-Verlag, 1976.

J. Medhi. *Stochastic Processes*. John Wiley and Sons, 1982.

S.M. Ross. *Stochastic Processes*. John Wiley and Sons, 1983.

L. Takács. *Combinatorial Methods in the Theory of Stochastic Processes*. Robert E. Krieger, 1977.

E. Wong and B. Hajek. *Stochastic Processes in Engineering Systems*. Springer-Verlag, 1985.

Computational Probability and Algorithms

S.C. Bruell and G. Balbo. *Computational Algorithms for Closed Queueing Networks*. North-Holland, 1980.

L. Lipsky. *Queueing Theory-A Linear Algebra Approach*. Macmillan, 1992.

M.F. Neuts. *Matrix Geometric Solutions in Stochastic Models*. John Hopkins University Press, 1981.

M.F. Neuts. *Structure of Stochastic Matrices of M/G/1 Type and Their Applications*. Marcel-Dekker, 1990.

H.C. Tijms. *Stochastic Modelling and Analysis: A Computational Approach*. John Wiley and Sons, 1986.

Queueing Theory

V. E. Beneš. *General Stochastic Processes in the Theory of Queues*. Addison-Wesley, 1963.

J.W. Cohen. *The Single Server Queue*, 2nd edition. North-Holland, 1969.

R.B. Cooper. *Introduction to Queuing Theory*, 2nd edition. Macmillan, 1972.

D.R. Cox and W.L. Smith. *Queues*. Methuen, 1961.

D. Gross and C.M. Harris. *Fundamentals of Queueing Theory*. John Wiley and Sons, 1985.

L. Kleinrock. *Queueing Systems, Volume I: Theory*. John Wiley and Sons, 1975.

L. Kleinrock and R. Gail. *Solutions Manual for Queueing Systems: Volume I: Theory*. Technology Transfer Institute, Santa Monica, CA, 1982.

T.L. Saaty. *Elements of Queueing Theory*. McGraw-Hill, 1961.

L. Takács. *Introduction to the Theory of Queues*. Oxford University Press, 1962.

Queueing Networks

R.L. Disney and P.C. Kiessler. *Traffic Processes in Queueing Networks: A Markov Renewal Approach.* John Hopkins University Press, 1987.

E. Gelenbe and G. Pujolle. *Introduction to Queueing Networks.* John Wiley and Sons, 1987.

F.P. Kelly. *Reversibility and Stochastic Networks.* John Wiley and Sons, 1979.

J. Walrand. *An Introduction to Queueing Networks.* Prentice-Hall, 1988.

R.W. Wolff. *Stochastic Modeling and the Theory of Queues.* Prentice-Hall, 1989.

Computer Performance Evaluation and Other Applications

A.O. Allen. *Probability, Statistics, and Queuing Theory with Computer Science Applications.* Elsevier, 1978.

E.G. Coffman and P.J. Denning. *Operating System Theory.* Prentice-Hall, 1973.

P.J. Curtois. *Decomposability: Queueing and Computer System Applications.* Academic Press, 1977.

D. Ferrari. *Computer Systems Performance Evaluation.* Prentice-Hall, 1978.

E. Gelenbe and I. Mitrani. *Analysis and Synthesis of Computer Systems.* Academic Press, 1980.

P.G. Harrison and N.M. Patel. *Performance Modelling of Communications Networks and Computer Architecture.* Addison-Wesley, 1992.

D.P. Heyman and M.J. Sobel. *Stochastic Models in Operations Research, Volume I.* McGraw-Hill, 1982.

K. Kant. *Introduction to Computer System Performance Evaluation.* McGraw-Hill, 1992.

L. Kleinrock. *Queueing Systems, Volume II: Applications.* John Wiley and Sons, 1975.

L. Kleinrock and R. Gail. *Solutions Manual for Queueing Systems: Volume II: Computer Applications.* Technology Transfer Institute, Santa Monica, CA, 1986.

H. Kobayashi. *Modeling and Analysis-An Introduction to System Performance Evaluation Methodology.* Addison-Wesley, 1978.

S.S. Lavenberg. *Computer Performance Modeling Handbook.* Academic Press, 1983.

E.D. Lazowska, J. Zahorjan, G.S. Graham, and K.C. Sevcik. *Quantitative System Performance, Computer System Analysis Using Queueing Network Models.* Prentice-Hall, 1984.

M.K. Molloy. *Fundamentals of Performance Modeling.* Macmillan, 1989.

T.G. Robertazzi. *Computer Networks and Systems: Queueing Theory and Performance Evaluation.* Springer-Verlag, 1990.

S.M. Ross. *Applied Probability Models with Optimization Applications.* Holden-Day, 1970.

C.H. Sauer and K.M. Chandy. *Computer Systems Performance Modeling.* Prentice-Hall, 1981.

C.H. Sauer and E.A. MacNair. *Simulation of Computer Communications Systems.* Prentice-Hall, 1983.

K.S. Trivedi. *Probability and Statistics with Reliability, Queuing, and Computer Science Applications.* Prentice-Hall, 1982.

Special Topics

F. Baccelli and P. Brémaud. *Palm Probabilities and Stationary Queues. Lecture Notes in Statistics* **41**, Springer-Verlag, 1987.

P. Billingsley. *Convergence of Probability Measures.* John Wiley and Sons, 1968.

J.A. Bucklew. *Large Deviation Techniques in Decision, Simulation, and Estimation.* John Wiley and Sons, 1990.

P. Erdös and J. Spencer. *Probabilistic Methods in Combinatorics.* Academic Press, 1974.

W. Feller. *An Introduction to Probability Theory and Its Applications-Volume II.* John Wiley and Sons, 1968.

P. Glasserman and D.D. Yao. *Monotone Structure in Discrete-Event Systems.* John Wiley and Sons, 1994.

G. Samorodnitsky and M.S. Taqqu. *Stable Non-Gaussian Random Processes, Stochastic Models with Infinite Variance.* Chapman and Hall, 1994.

D. Stoyan. *Comparison Methods for Queues and Other Stochastic Models.* John Wiley and Sons, 1983.

A. Swartz and A. Weiss. *Large Deviations for Performance Analysis.* Chapman and Hall, 1995.

H. Takagi. *Analysis of Polling Systems.* The MIT Press, 1986.

P. Whittle. *Systems in Stochastic Equilibrium.* John Wiley and Sons, 1986.

References

[1] M. Abramowitz and I.A. Stegun. *Handbook of Mathematical Functions*. Dover Publications, 1965.

[2] A.O. Allen. *Probability, Statistics, and Queuing Theory with Computer Science Applications*. Elsevier, 1978.

[3] F. Baccelli and P. Brémaud. *Palm Probabilities and Stationary Queues. Lecture Notes in Statistics* **41**, Springer-Verlag, 1987.

[4] F. Baskett, M. Chandy, R. Muntz, and J. Palacios. Open, closed, and mixed networks of queues with different classes of customers. *Journal of the A.C.M.*, 22:248–260, 1975.

[5] E.T. Bell. *Men of Mathematics. A Touchstone Book*, Simon and Schuster, 1986.

[6] V. E. Beneš. *General Stochastic Processes in the Theory of Queues*. Addison-Wesley, 1963.

[7] P. Billingsley. *Convergence of Probability Measures*. John Wiley and Sons, 1968.

[8] C.B. Boyer and U. C. Merzbach. *A History of Mathematics*. John Wiley and Sons, 1989.

[9] S.C. Bruell and G. Balbo. *Computational Algorithms for Closed Queueing Networks*. North-Holland, 1980.

[10] J.A. Bucklew. *Large Deviation Techniques in Decision, Simulation, and Estimation*. John Wiley and Sons, 1990.

[11] E. Catalan. Note sur une Équation aus différences finies. *Journal de Mathématiques pures et appliquées*, 3:508–516, 1838.

[12] E. Çinclar. *Introduction to Stochastic Processes*. Prentice-Hall, 1975.

[13] K.M. Chandy, U. Herzog, and L.S. Woo. Parametric analysis of queueing networks. *IBM Journal of Research and Development*, 19:43–49, 1975.

[14] Y.S. Chow and H. Teicher. *Probability Theory - Independence, Interchangeability, Martingales*. Springer Verlag, 1988.

[15] K.L. Chung. *Elementary Probability Theory with Stochastic Processes*. Springer-Verlag, 1979.

[16] A. Bruce Clarke and R.L. Disney. *Probability and Random Processes: A First Course with Applications*. Wiley, 1985.

[17] E.G. Coffman and P.J. Denning. *Operating System Theory*. Prentice-Hall, 1973.

[18] D. I. A. Cohen. *Basic Techniques of Combinatorial Theory*. John Wiley and Sons, 1978.

[19] J.W. Cohen. *The Single Server Queue*. North-Holland, 2nd edition, 1969.

[20] G. M. Constantine. *Combinatorial Theory and Statistical Design*. John Wiley and Sons, 1987.

[21] R.B. Cooper. *Introduction to Queuing Theory*. Macmillan, 2nd edition, 1972.

[22] T.M. Cover and J.A. Thomas. *Elements of Information Theory*. John Wiley and Sons, 1991.

[23] D.R. Cox. *Renewal Theory*. Methuen, 1962.

[24] D.R. Cox and W.L. Smith. *Queues*. Methunen, 1961.

[25] P.J. Curtois. *Decomposability: Queueing and Computer System Applications*. Academic Press, 1977.

[26] P.J. Davis. Leonard Euler's integral: a historical profile of the gamma function. *American Mathematics Monthly*, 66:849–869, 1959.

[27] A. De Moivre. *The Doctrine of Chances: or A Method of Calculating the Probabilities of Events in Play*. Chelsea, 1967.

[28] R. L. Disney and P. C. Kiessler. *Traffic Processes in Queueing Networks: A Markov Renewal Approach*. The John Hopkins Press, Baltimore, MD, 1987.

[29] R.L. Disney and P.C. Kiessler. *Traffic Processes in Queueing Networks*. Johns-Hopkins, 1987.

[30] A. Erdélyi, editor. *Tables of Integral Transforms*. Volume I, McGraw-Hill, 1954.

[31] P. Erdös and J. Spencer. *Probabilistic Methods in Combinatorics*. Academic Press, 1974.

[32] J. H. Eves. *An Introduction to the History of Mathematics*. The *Saunder Series*, Saunders College Publishing, 1990.

[33] W. Feller. A direct proof of Stirling's formula. *The American Mathematical Monthly*, 1223–1225, 1967.

[34] W. Feller. *An Introduction to Probability Theory and its Applications–Volume I*. John Wiley and Sons, 1968.

[35] W. Feller. *An Introduction to Probability Theory and its Applications–Volume II*. John Wiley and Sons, 1968.

[36] D. Ferrari. *Computer Systems Performance Evaluation*. Prentice-Hall, 1978.

[37] E. Gelenbe and I. Mitrani. *Analysis and Synthesis of Computer Systems*. Academic Press, 1980.

[38] E. Gelenbe and G. Pujolle. *Introduction to Queueing Networks*. John Wiley and Sons, 1987.

[39] P. Glasserman and D.D. Yao. *Monotone Structure in Discrete–Event Systems*. Wiley and Sons, 1994.

[40] B.V. Gnedenko. *The Theory of Probability*. Chelsea, second edition, 1963.

[41] R.W. Gosper. Decision procedure for indefinite hypergeometric summation. *Proceedings of the National Academy of Sciences of the United States of America*, 75:40–42, 1978.

[42] I.S. Gradshteyn and I.M. Ryzhik. *Table of Integrals, Series, and Products*. Academic Press, 1980.

[43] R.L. Graham, D.E. Knuth, and O. Patashnik. *Concrete Mathematics*. Addison-Wesley, 1989.

[44] D. Gross and C.M. Harris. *Fundamentals of Queueing Theory*. John Wiley and Sons, 1985.

[45] F.A. Haight. *Handbook of the Poisson Distribution*. John Wiley and Sons, 1967.

[46] R.W. Hamming. *The Art of Probability - for Scientists and Engineers*. Addison-Wesley, 1991.

[47] P.G. Harrison and N.M. Patel. *Performance Modelling of Communications Networks and Computer Architecture*. Addison-Wesley, 1992.

[48] I.N. Herstein. *Topics in Algebra*. Blaisdell Publishing Company, 1964.

[49] D. P. Heyman and M.J. Sobel. *Stochastic Models in Operations Research*. Volume I, McGraw Hill, 1982.

[50] P.G. Hoel, S. C. Port, and C. J. Stone. *Introduction to Probability Theory*. Houghton Mifflin, 1971.

[51] P. Hoffman. The man who loves only numbers. *The Atlantic Monthly*, November:64, 1987.

[52] R.A. Horn and C.R. Johnson. *Matrix Analysis*. Cambridge, 1987.

[53] J.R. Jackson. Jobshop-like queueing systems. *Management Science*, 10:131–142, 1963.

[54] N.L. Johnson and S. Kotz. *Continuous Univariate Distributions - 1 and 2*. Houghton Mifflin, 1970.

[55] N.L. Johnson and S. Kotz. *Discrete Distributions*. Houghton Mifflin, 1969.

[56] K. Kant. *Introduction to Computer System Performance Evaluation*. McGraw Hill, 1992.

[57] S. Karlin and H. M. Taylor. *A First Course in Stochastic Processes*. Academic Press, 2nd edition, 1975.

[58] S. Karlin and H. M. Taylor. *A Second Course in Stochastic Processes*. Academic Press, 1981.

[59] Z. Kechen. Uniform distribution of initial states: the physical basis of probability. *Physical Review A*, 41(4):1893–1900, 1990.

[60] J.B. Keller. The probability of heads. *American Mathematics Monthly*, 93:191–197, 1986.

[61] F. P. Kelly. *Reversibility and Stochastic Networks.* Wiley, 1979.

[62] F.P. Kelly. Networks of queues. *Adv. Appl. Prob.*, 8:416–432, 1976.

[63] J.G. Kemeny, J. L. Snell, and A. W. Knapp. *Denumerable Markov Chains.* Springer-Verlag, 1976.

[64] J.G. Kemeny and J.L. Snell. *Finite Markov Chains.* Springer-Verlag, 1976.

[65] L. Kleinrock. *Queueing Systems, Volume II:Applications.* Wiley, 1975.

[66] L. Kleinrock. *Queueing Systems, Volume I:Theory.* Wiley, 1975.

[67] L. Kleinrock and R. Gail. *Solutions Manual for Queueing Systems: Volume II:Computer Applications.* Technology Transfer Institute, Santa Monica, CA, 1986.

[68] L. Kleinrock and R. Gail. *Solutions Manual for Queueing Systems: Volume I:Theory.* Technology Transfer Institute, Santa Monica, CA, 1982.

[69] D.E. Knuth. *The Art of Computer Programming–Volume 1.* Addison-Wesley, 1975.

[70] H. Kobayashi. *Modeling and Analysis - An Introduction to System Performance Evaluation Methodology.* Addison-Wesley, 1978.

[71] A. Kolmogorov. *Foundations of the Theory of Probability.* Chelsea, 1950.

[72] A. Kolmogorov. Zur theorie der markoffschen ketten. *Mathematische Annalen*, 112:155–160, 1936.

[73] P.S. Laplace. *A Philosophical Essay on Probabilities.* Dover, New York, 1951.

[74] G. Latouche and V. Ramaswami. A logarithmic reduction algorithm for quasi–birth–and–death processes. *Journal of Applied Probability*, 30:650–674, 1993.

[75] S. S. Lavenberg and M. Reiser. Stationary state probabilities at arrival instants for closed queueing networks with multiple types of customers. *Journal of Applied Probability*, 17:1048–1061, 1980.

[76] S.S. Lavenberg. *Computer Performance Modeling Handbook.* Academic Press, 1983.

[77] E.D. Lazowska, J. Zahorjan, G.S. Graham, and K.C. Sevcik. *Quantitative System Performance, Computer System Analysis using Queueing Network Models.* Prentice Hall, 1984.

[78] W. R. LePage. *Complex Variables and the Laplace Transform for Engineers.* Dover Publications, 1980.

[79] L. Lipsky. *Queueing Theory - A Linear Algebra Approach.* Macmillan, 1992.

[80] J.D.C. Little. A simple proof of $l = \lambda w$. *Operations Research,* 9:383–387, 1961.

[81] J. Medhi. *Stochastic processes.* John Wiley and Sons, 1982.

[82] M.K. Molloy. *Fundamentals of Performance Modeling.* Macmillan, 1989.

[83] R. Muntz. Poisson departure processes and queueing networks. *IBM Research Report, RC 4145,* 1972.

[84] E. Nagel and J.R. Newman. *Gödel's Proof.* New York University Press, 1958.

[85] R. Nelson. The mathematics of product form networks. *Computing Surveys,* 1993.

[86] M.F. Neuts. *Matrix Geometric Solutions in Stochastic Models.* John Hopkins University Press, 1981.

[87] M.F. Neuts. The probabilistic significance of the rate matrix in matrix-geometric invariant vectors. *Journal of Applied Probability,* 17:291–96, 1980.

[88] M.F. Neuts. *Structure of Stochastic Matrices of M/G/1 Type and Their Applications.* Marcel-Deckker, 1990.

[89] D.S. Ornstein. Ergodic theory, randomness, and "chaos". *Science,* 243:182–187, January 13 1989.

[90] D.S. Ornstein and B. Weiss. Geodesic flows are Bernoullian. *Israel Journal of Mathematics,* 14:184, 1973.

[91] D.S. Ornstein and B. Weiss. Statistical properties of chaotic systems. *Bulletin of the American Mathematical Monthly,* 24:11–116, January 1991.

[92] C.E. Pearson. *Handbook of Applied Mathematics.* Van Nostrand Reinhold Company, 1974.

[93] R. Preston. The mountains of π. *The New Yorker*, 36, March 2 1992.

[94] M. Reiser and S. S. Lavenberg. Mean value analysis of closed multichain queueing networks. *Journal of the A.C.M.*, 27:313–322, 1980.

[95] A. Rényi. *Foundations of Probability*. Holden-Day, 1970.

[96] J. Riordan. *An Introduction to Combinatorial Analysis*. Princeton University Press, 1978.

[97] T. G. Robertazzi. *Computer Networks and Systems: Queueing Theory and Performance Evaluation*. Springer-Verlag, 1990.

[98] Joseph P. Romano and Andrew F. Siegel. *Counterexamples in probability and statistics. statistics/probability series*, Wadsworth and Brooks/Cole, 1986.

[99] S.M. Ross. *Applied Probability Models with Optimization Applications*. Holden-Day, 1970.

[100] S.M. Ross. *An Introduction to Probability Models*. Academic Press, 1972.

[101] S.M. Ross. *Stochastic Processes*. Wiley, 1983.

[102] T.L. Saaty. *Elements of Queueing Theory*. McGraw-Hill, 1961.

[103] G. Samorodnitsky and M.S. Taqqu. *Stable Non–Gaussian Random Processes, Stochastic Models with Infinite Variance*. Chapman and Hall, 1994.

[104] C. H. Sauer and K. M. Chandy. *Computer Systems Performance Modeling*. Prentice-Hall, 1981.

[105] C.H. Sauer and E.A. MacNair. *Simulation of Computer Communications Systems*. Prentice-Hall, 1983.

[106] M. Schroeder. *Fractals, Chaos, Power Laws - Minutes from an Infinite Paradise*. W. H. Freeman and Company, 1991.

[107] K.C. Sevcik and Mitrani I. The distribution of queueing network states at input and output instants. *Journal of the A.C.M.*, 28:358–371, 1981.

[108] A Shwartz and A. Weiss. *Large Deviations for Performance Analysis*. Chapman and Hall, 1995.

[109] D.E. Smith. *A Source Book in Mathematics*. McGraw-Hill, New York, 1929.

[110] D. Stoyan. *Comparison Methods for Queues and Other Stochastic Models.* John Wiley and Sons, 1983.

[111] L. Takács. *Combinatorial Methods in the Theory of Stochastic Processes.* Robert E. Krieger, 1977.

[112] L. Takács. *Introduction to the Theory of Queues.* Oxford University Press, 1962.

[113] H. Takagi. *Analysis of Polling Systems.* The MIT Press, 1986.

[114] H.C. Tijms. *Stochastic Modelling and Analysis: A Computational Approach.* John Wiley and Sons, 1986.

[115] I. Todhunter. *A History of the Mathematical Theory of Probability: From the Time of Pascal to that of Laplace.* Chelsea, 1949.

[116] S.C. Tornay. *Ockham:Studies and Selections.* Open Court Publishers, La Salle, Il, 1938.

[117] D. Towsley. Queueing network models with state-dependent routing. *Journal of the ACM*, 27:323–337, 1980.

[118] K.S. Trivedi. *Probability and Statistics with Reliability, Queuing, and Computer Science Applications.* Prentice Hall, 1982.

[119] N. Y. Vilenkin. *Combinatorics.* Academic Press, 1971.

[120] R. von Mises. *Mathematical Theory of Probability and Statistics.* Academic Press, 1964.

[121] J. Walrand. *An Introduction to Queueing Networks.* Prentice Hall, 1988.

[122] W. Whitt. A review of $l = \lambda w$ and extensions. *Queueing Syst. Theory Appl.*, 9(3):5–68, 1991.

[123] P. Whittle. Nonlinear migration processes. *Bull. Int. Inst. Statist.*, 42:642–647, 1967.

[124] P. Whittle. *Systems in Stochastic Equilibrium.* John Wiley and Sons, 1986.

[125] R. W. Wolff. *Stochastic Modeling and the Theory of Queues.* Prentice Hall, 1989.

[126] R.W. Wolff. Poisson arrivals see time averages. *Operations Research*, 223–231, 1982.

[127] E. Wong and B. Hajek. *Stochastic Processes in Engineering Systems.* Springer-Verlag, 1985.

Index of Notation

Symbol	Meaning	Page
Standard Symbols		
$\mathcal{A}, \mathcal{B}, \ldots$	Sets are represented in calligraphic type	
$A, B, \ldots$	Random variables are represented in capital letters	
$\mathbf{A}, \mathbf{B}, \ldots$	Matrices are represented in bold capital letters	
$A(i,j), a_{i,j}$	The entries of matrix $\mathbf{A}$	
$\mathbf{A}^{\star}$	Matrix $\mathbf{A}$ with its first column eliminated	
$\mathbf{I}$	The identity matrix - a diagonal matrix of 1s	
$\mathbf{a}$	Vectors are represented in bold lowercase letters	
$a(i), a_i$	The entries of vector $\mathbf{a}$	
$\mathrm{diag}(a_1, a_2, \ldots, a_n)$	A diagonal matrix with entries $a_i, i = 1, 2, \ldots, n$, on the diagonal	
$\mathbf{A}^T$	The transpose of matrix $\mathbf{A}$	
$\mathbf{e}$	A column vector of all 1s	
Ω	Sample space	102
ω	Sample point in sample space Ω	102
$\emptyset$	The null set	

$F(x)$	Generating function for sequence $f_0, f_1, \ldots$, $F(x) \stackrel{\text{def}}{=} \sum_{i=0}^{\infty} f_i x^i$	74
$\mathbb{R}$	Set of real numbers	
$\mathbb{R}^n$	n tuples of real numbers	
$\mathbb{N}$	Set of integers	
$[x, y)$	Interval set $\{z \mid x \leq z < y\}$; other sets include $[x, y]$, $(x, y]$, and (x, y)	

Mathematical Expressions

$n!$	n factorial, $n! \stackrel{\text{def}}{=} n(n-1)(n-2)\cdots 2\cdot 1$	49
$(n)_k$	Factorial coefficients, $(n)_k \stackrel{\text{def}}{=} n(n-1)(n-2)\cdots(n-k+1)$	50
$\langle n\rangle_k$	Factorial-R coefficients, $\langle n\rangle_k \stackrel{\text{def}}{=} n^k$	51
$\dbinom{n}{k}$	Binomial coefficients, $\dbinom{n}{k} \stackrel{\text{def}}{=} \frac{n!}{k!(n-k)!}$	55
$\dbinom{k}{k_1, k_2, \ldots, k_n}$	Multinomial coefficients, $\dbinom{k}{k_1, \ldots, k_n} \stackrel{\text{def}}{=} \frac{k!}{k_1!\cdots k_n!}$	
	where $k = k_1 + \cdots + k_n$	61
$\left\langle \begin{array}{c} n \\ k \end{array} \right\rangle$	Binomial-R coefficients, $\left\langle \begin{array}{c} n \\ k \end{array} \right\rangle \stackrel{\text{def}}{=} \frac{n+k-1!}{k!(n-1)!}$	63
$\lvert x\rvert$	Absolute value of x for scalar x	
$\lvert\mathbf{x}\rvert$	The sum of the elements of vector $\mathbf{x}$	
$\lceil x\rceil$	The ceiling of x, for example, $\lceil 4.3\rceil = 5$ $\lceil 4\rceil = 4$	
$\lfloor x\rfloor$	The floor of x, for example, $\lfloor 4.3\rfloor = 4$ $\lfloor 4\rfloor = 4$	
$\delta_0(t)$	Impulse function, $\delta_0(t) = \begin{cases} \infty, & t = 0, \\ 0, & t \neq 0 \end{cases}$	114
δ_x	Delta function, $\delta_x = 1$ if $x = 0$ and otherwise equals 0	
C_n	Catalan number	79
$\mathbf{C}$	Euler's constant, $\mathbf{C} = \lim_{n\to\infty} \left(\sum_{k=1}^{n} 1/k - \ln n\right) \approx .57721$	173
e	The exponential constant $e = \lim_{n\to\infty} (1 + 1/n)^n \approx 2.7182818284\ldots$	
H_i	The ith harmonic number, $H_i \stackrel{\text{def}}{=} \sum_{k=1}^{i} 1/k$	
$h^{(j)}(x)$	The jth derivative of function h evaluated at x	

Mathematical Operations

$\mathcal{A} + \mathcal{B}$	Union of sets $\mathcal{A}$ and $\mathcal{B}$	
$\{\mathcal{A}_i\}_{i=1}^n$	Union of sets A_i, $i = 1, 2, \ldots, n$	
$\mathcal{A}\mathcal{B}$	Intersection of sets $\mathcal{A}$ and $\mathcal{B}$	
$\mathcal{A}^c$	Complement of set $\mathcal{A}$	
$\circledast$	Convolution operator	74
$(x)^+$	Maxplus operator, $(x)^+ \stackrel{\text{def}}{=} \max\{0, x\}$	
$\stackrel{\text{def}}{=}$	Equal by definition	
$=_{st}$	Stochastic equality	106
$\Longleftrightarrow$	Transform pairs	198
$o(x)$	Order relationship, $h(x)$ is $o(x^i)$ if $h(x)/x^i \to 0$ as $x \to 0$	547
$O(x)$	Order relationship, $h(x)$ is $O(x^i)$ if $h(x)/x^i$ is finite as $x \to 0$	547
$\lim$	Limit	545
$\lim\sup$	Limit superior	546
$\lim\inf$	Limit inferior	546
$\stackrel{P}{\longrightarrow}$	Convergence in probability	211
$\stackrel{a.s.}{\longrightarrow}$	Almost sure convergence or convergence with probability 1	213
$\stackrel{D}{\longrightarrow}$	Convergence in distribution	140

Notations from Probability: Chapters 4 and 5

$P[\mathcal{A}]$	Probability of event $\mathcal{A}$	30
$N[\mathcal{A}]$	Number of ways event $\mathcal{A}$ can occur	48
$F_X(x)$	Probability distribution function of random variable X	111
$\overline{F_X}(x)$	Survivor function of random variable X, $\overline{F_X}(x) = 1 - F_X(x)$	111
$f_X(x)$	Probability density function of random variable X	111
$F_{(X,Y)}(x, y)$	Joint probability distribution function of random variables (X, Y)	112

| erf x | The error function, erf $x \stackrel{\text{def}}{=} \frac{2}{\sqrt{\pi}} \int_0^x e^{-u^2} du$ | 218 |

Notations from Renewal Theory: Chapter 6

γ	Average renewal rate	254
A_t	Age at time t	241
R_t	Residual life at time t	241
A	Limiting age random variable	242
R	Limiting residual life random variable	242
L	Limiting length random variable	243
$N(t)$	Number of renewal epochs in $[0, t]$	244
$M(t)$	Expected number of renewal epochs in $[0, t]$	257
$m(t)$	Renewal density function	255
$N_e(t)$	Number of renewal epochs in $[0, t]$ for equilibrium renewal process	265
$M_e(t)$	Expected number of renewal epochs in $[0, t]$ for equilibrium renewal process	266
$m_e(t)$	Renewal density function for an equilibrium renewal process	266

Notations from Queueing Theory: Chapter 7

ρ	Utilization of server	296
W_n	Waiting time of nth customer	107
T_n	Response time of nth customer	107
S_n	Service time of nth customer	107
A_n	Interarrival time between the nth and $(n-1)$st customer	107
W	Customer waiting time	295
T	Customer response time	295
S	Customer service time	295
R	Residual service time	296
B	Length of busy period	303

Notations from Queueing Network Theory: Chapter 10

Reversed Processes

Sets of States for Partial Balance

Quasi-Reversible Queues

Index